Birkhäuser

Trends in Mathematics

Trends in Mathematics is a series devoted to the publication of volumes arising from conferences and lecture series focusing on a particular topic from any area of mathematics. Its aim is to make current developments available to the community as rapidly as possible without compromise to quality and to archive these for reference.

Proposals for volumes can be submitted using the Online Book Project Submission Form at our website www.birkhauser-science.com.

Material submitted for publication must be screened and prepared as follows:

All contributions should undergo a reviewing process similar to that carried out by journals and be checked for correct use of language which, as a rule, is English. Articles without proofs, or which do not contain any significantly new results, should be rejected. High quality survey papers, however, are welcome.

We expect the organizers to deliver manuscripts in a form that is essentially ready for direct reproduction. Any version of $\TeX$ is acceptable, but the entire collection of files must be in one particular dialect of $\TeX$ and unified according to simple instructions available from Birkhäuser.

Furthermore, in order to guarantee the timely appearance of the proceedings it is essential that the final version of the entire material be submitted no later than one year after the conference.

Fourier Analysis

Pseudo-differential Operators, Time-Frequency Analysis and Partial Differential Equations

Michael Ruzhansky
Ville Turunen
Editors

Birkhäuser

Editors
Michael Ruzhansky
Department of Mathematics
Imperial College London
London, UK

Ville Turunen
Department of Mathematics
& System Analysis
Aalto University
Aalto, Finland

ISBN 978-3-319-02549-0 ISBN 978-3-319-02550-6 (eBook)
DOI 10.1007/978-3-319-02550-6
Springer Cham Heidelberg New York Dordrecht London

Library of Congress Control Number: 2014930298

Printed on acid-free paper

Springer Basel is part of Springer Science+Business Media (www.springer.com)

Contents

Preface

The interactions between the theory of *pseudo-differential operators*, the *time-frequency analysis*, and the theory of *partial differential equations* have contributed to progress in all these areas and are an active field of current research. To facilitate further developments and links between these fields, the international conference "Fourier Analysis and Pseudo-Differential Operators", with applications to partial differential equations, was held at Aalto University near Helsinki, Finland, on 25–29 June 2012.

It was organised as a satellite meeting to the European Congress of Mathematicians that took place in Krakow the following week, and as the 6th meeting in the series "Fourier Analysis and Partial Differential Equations", with previous meetings taking place at University of Osaka (2008), Imperial College London (2008), Nagoya University (2009), University of Göttingen (2010) and, finally, Imperial College London (2011).

The conference attracted around 90 participants presenting recent results of their work, with a total of around 75 sectional and plenary talks. The papers collected in this volume are authored by participants of that meeting. They focus on different aspects of current research in the above-mentioned subjects and are, in particular, centred around the following topics:

- pseudo-differential operators in different settings;
- microlocal analysis and Fourier integral operators;
- pseudo-differential operators and noncommutative harmonic analysis;
- time-frequency analysis and its applications;
- linear and nonlinear evolution equations;
- hyperbolic equations and systems;
- dispersive, smoothing and Strichartz estimates;
- applications: wave models, control theory, stochastic analysis.

On one hand, the volume is aimed at being a rigorous presentation of recent research developments in these areas, as well as at emphasising interactions between them. As such, all the contributions are full research papers presenting new results. This allows experts in the field to describe the recent developments in their subjects, to present new results, and will hopefully lead to further collaborative work in the area. On the other hand, the volume gives an overview on the great variety of ongoing current research in several broad fields and, therefore, allows researchers as well as students grasping new aspects and broadening their understanding of these areas. Therefore, the papers provide a wide scope of ideas and detailed proofs of results.

It is our pleasure to acknowledge the sponsorship of the conference and contributions by the following organisations:

- Aalto University and Aalto University Department of Mathematics and Systems Analysis;
- Science Factories programme by Aalto Science Institute at Aalto University School of Science;
- the Finnish National Graduate School in Mathematics and its Applications;
- ISAAC (International Society for Analysis, its Applications and Computation);
- Magnus Ehrnrooth Foundation of the Finnish Society of Sciences and Letters.

Finally, we would also like to thank other members of the organising committee of the conference, in particular, Jens Wirth and Mitsuru Sugimoto, for their valuable contributions in different ways.

Michael Ruzhansky and Ville Turunen

Fourier Analysis
Trends in Mathematics, 1–7

Optimal Constant for a Smoothing Estimate of Critical Index

Neal Bez and Mitsuru Sugimoto

Abstract. We generalise a result by Hoshiro [3] which considered a critical case of Kato–Yajima's smoothing estimate

$$\left\| |x|^{a-1} |\nabla|^a \exp(-it\Delta) f \right\|_{L^2_{t,x}(\mathbb{R}\times\mathbb{R}^d)} \leq C \|f\|_{L^2(\mathbb{R}^d)}$$

for the Schorödinger propagator $\exp(-it\Delta)$. An expression for the optimal constant is also given.

Mathematics Subject Classification (2010). Primary 35B45; Secondary 35P10, 35B65.

Keywords. Smoothing estimates, optimal constants, extremisers.

1. Introduction

We consider the Cauchy problem for the Schrödinger equation

$$\begin{cases} (i\partial_t - \Delta)\, u(t,x) = 0, \\ \qquad\qquad u(0,x) = f(x) \in L^2(\mathbb{R}^d) \end{cases}$$

which has solution $u(t,x) = \exp(-it\Delta)f(x)$ for $t \in \mathbb{R}$ and $x \in \mathbb{R}^d$. For each fixed time $t \in \mathbb{R}$ we have

$$\|u(t,\cdot)\|_{L^2_x(\mathbb{R}^d)} = \|f\|_{L^2(\mathbb{R}^d)},$$

so that the L^2-norm of the initial data f is preserved. However, it is well known that if we integrate the solution in time, we get improved regularity in space, and as an illustration of such a smoothing effect, Kato–Yajima [4] established the smoothing estimate

$$\left\| |x|^{a-1} |\nabla|^a \exp(-it\Delta) f \right\|_{L^2_{t,x}(\mathbb{R}\times\mathbb{R}^d)} \leq C \|f\|_{L^2(\mathbb{R}^d)} \tag{1.1}$$

for $a \in [0, \frac{1}{2})$ and $d \geq 3$. Watanabe [15] implicitly pointed out that this estimate in the critical case $a = \frac{1}{2}$ is not true. Nevertheless, Hoshiro [3] gave an interesting estimate

$$\left\| |x|^{a-1}|\nabla|^a(1-\Lambda)^{1/4-a/2}\exp(-it\Delta)f \right\|_{L^2_{t,x}(\mathbb{R}\times\mathbb{R}^d)} \leq C\|f\|_{L^2(\mathbb{R}^d)} \tag{1.2}$$

for $a \in [0, \frac{1}{2})$ and $d \geq 3$, where $-\Lambda$ denotes the Laplace–Beltrami operator on the sphere $\mathbb{S}^{d-1}$ and $(1-\Lambda)^s$ the homogeneous extension of the elliptic self-adjoint operator $(1-\Lambda)^s$ on the sphere $\mathbb{S}^{d-1}$. For example, for $s \in \mathbb{N}$, we have

$$(1-\Lambda)^s = \left(1 - \sum_{i<j}\left(x_j\frac{\partial}{\partial x_i} - x_i\frac{\partial}{\partial x_j}\right)^2\right)^s,$$

which is a partial differential in the direction tangent to the sphere and behaves like $|x|^{2s}|\nabla|^{2s}$ in the sense of the orders of differential and decay. Then

$$|x|^{a-1}|\nabla|^a(1-\Lambda)^{1/4-a/2} \sim |x|^{-1/2}|\nabla|^{1/2},$$

hence estimate (1.2) provides a natural substitute for estimate (1.1) in the critical case $a = \frac{1}{2}$. This fact also means that the gain of regularity is exactly the full order $\frac{1}{2}$ if we consider a partial regularity along the sphere.

We remark that Sugimoto [13] improved the range of a for estimates (1.1) and (1.2) and established them for $a \in (1 - \frac{d}{2}, \frac{1}{2})$ and $d \geq 2$. In Sugimoto [14] and Ruzhansky–Sugimoto [6], the meaning of the operator $(1-\Lambda)^s$ was also discussed by replacing the Laplacian $-\Delta$ in the Schrödinger equation by more general elliptic operators, and it was applied to a nonlinear analysis in Ruzhansky–Sugimoto [8]. In Ruzhansky–Sugimoto [7], estimate (1.1) is generalised to higher-order equations with $-\Delta$ in the Schrödinger equation replaced by $|\nabla|^m$.

Recently there has been significant interest in obtaining optimal constants $C = \mathbf{C}_d$ in space-time estimates for dispersive equations. In [1] the authors discussed the optimal constant for estimate (1.1), and obtained that

$$\mathbf{C}_d = \left(\pi 2^{2a-1}\frac{\Gamma(1-2a)\Gamma(\frac{d}{2}-1+a)}{\Gamma(1-a)^2\Gamma(\frac{d}{2}-a)}\right)^{1/2}, \tag{1.3}$$

which extends the result of Simon [11] who treated the case $a = 0$. We remark that nonzero radial functions are its extremisers. This optimal constant is independently given by Ozawa–Rogers [5] and Watanabe [15] by completely different approaches to ours; we proceed via spectral considerations and this has the advantage that we are able to obtain the optimal constant for estimate (1.2), as well. Indeed, our main result of this paper is the following theorem, and we clarify in advance that the notation $\mathbb{N}_0$ means $\{0, 1, 2, \dots\}$.

Theorem 1.1. *Let $a \in (1 - \frac{d}{2}, \frac{1}{2})$. Then we have the estimate*

$$\left\| |x|^{a-1}(1-\Lambda)^{1/4-a/2}|\nabla|^a\exp(-it\Delta)f \right\|_{L^2_{t,x}(\mathbb{R}\times\mathbb{R}^d)} \leq \mathbf{C}_d\|f\|_{L^2(\mathbb{R}^d)},$$

where

$$\mathbf{C}_d = \sup_{k\in\mathbb{N}_0}\left(\pi 2^{2a-1}\frac{\Gamma(1-2a)\Gamma(\frac{d}{2}+k-1+a)}{\Gamma(1-a)^2\Gamma(\frac{d}{2}+k-a)}\right)^{1/2}(1+k(k+d-2))^{1/4-a/2}$$

and this constant cannot be improved.

We remark that generalisations and extensions of this theorem with explicit computation of the optimal constants will appear in our forthcoming paper [2]. For example, we will show that the optimal constant is the same as the one given by (1.3) in the cases $d=2,3$, and nonzero radial functions are extremisers. The situation is quite different for $d\geq 4$.

In the following, after some preparation in Section 2 which includes the explicit definition of the operator $(1-\Lambda)^{1/4-a/2}$, we will give the proof of Theorem 1.1 in Section 3.

2. Spherical harmonics

We will review here some basics of the theory of spherical harmonics.

Let $\mathcal{A}_k$ be the space of solid spherical harmonics of degree k (these are harmonic polynomials on $\mathbb{R}^d$ which are homogeneous of degree k), and let $\mathcal{H}_k$ be the space of spherical harmonics of degree k (these are restriction of functions in $\mathcal{A}_k$ to the sphere $\mathbb{S}^{d-1}$). Then the eigenvalues of the Laplace–Beltrami operator $-\Lambda$ on the sphere $\mathbb{S}^{d-1}$ are

$$\mu_k = k(k+d-2) \tag{2.1}$$

and the corresponding eigenspaces are $\mathcal{H}_k$. The projection H_k from $L^2(\mathbb{S}^{d-1})$ to $\mathcal{H}_k$ can be written

$$H_k f(\omega) = \gamma_{d,k}\int_{\mathbb{S}^{d-1}} C_{d,k}(\omega\cdot\tilde{\omega})f(\tilde{\omega})\,d\tilde{\omega}\,; \qquad \gamma_{d,k} = \frac{d-2+2k}{(d-2)|\mathbb{S}^{d-1}|}\,,$$

where $|\mathbb{S}^{d-1}|$ is the surface area of the sphere and $C_{d,k}$ is the Gegenbauer polynomial of degree k associated with $\nu=(d-2)/2$, which can be defined using the generating function

$$(1-2st+t^2)^{-\nu} = \sum_{k=0}^{\infty} C_{d,k}(s)t^k\,; \qquad \nu = \frac{d-2}{2}\,,$$

where $|s|\leq 1$ and $|t|<1$. We remark that, for fixed $\omega\in\mathbb{S}^{d-1}$ (resp. $\tilde{\omega}\in\mathbb{S}^{d-1}$), $C_{d,k}(\omega\cdot\tilde{\omega})$ is a spherical harmonic of degree k in $\tilde{\omega}\in\mathbb{S}^{d-1}$ (resp. $\omega\in\mathbb{S}^{d-1}$).

Also, we write $\mathfrak{H}_k$ for the space of all linear combinations of functions

$$\xi \mapsto P(\xi)f_0(|\xi|)|\xi|^{-d/2-k+1/2}$$

where $P\in\mathcal{A}_k$ and $f_0\in L^2(0,\infty)$. These spaces allow us to decompose $L^2(\mathbb{R}^d)$ as

$$L^2(\mathbb{R}^d) = \bigoplus_{k=0}^{\infty}\mathfrak{H}_k\,,$$

where this is a complete orthogonal direct sum decomposition in the sense that the closed subspaces $\mathfrak{H}_k$ are mutually orthogonal in $L^2(\mathbb{R}^d)$ for $k \in \mathbb{N}_0$, and every $f \in L^2(\mathbb{R}^d)$ can be written $f = \sum_{k=0}^{\infty} f_k$ for some $f_k \in \mathfrak{H}_k$. We refer the reader to [10] and [12] for further details.

The operator H_k on $L^2(\mathbb{S}^{d-1})$ can be homogeneously extended to $L^2(\mathbb{R}^d)$ in a natural way by

$$H_k f(x) = \gamma_{d,k} \int_{\mathbb{S}^{d-1}} C_{d,k}(\omega \cdot \tilde{\omega}) f(r\tilde{\omega})\, d\tilde{\omega}\,; \quad r = |x|,\, \omega = x/|x|,$$

and we shall use the same notation H_k as long as there is no confusion. In this way, the Laplace–Beltrami operator $-\Lambda$ can be also regarded as an operator on $L^2(\mathbb{R}^d)$ by using the spectral decomposition

$$-\Lambda = \sum_{k=0}^{\infty} \mu_k H_k.$$

It is easy to see that the eigenvalues of this operator are again $\{\mu_k\}_{k=0}^{\infty}$, and H_k is the projection to the eigenspace $\mathfrak{H}_k$ of μ_k for each $k \in \mathbb{N}_0$. For appropriate functions $g(\rho)$ of $\rho \in [0, \infty)$, we can also define $g(-\Lambda)$ as an operator on $L^2(\mathbb{R}^d)$ by

$$g(-\Lambda) = \sum_{k=0}^{\infty} g(\mu_k) H_k.$$

Typical examples are fractional powers

$$(1-\Lambda)^\alpha = \sum_{k=0}^{\infty} (1+\mu_k)^\alpha H_k \quad \text{and} \quad (-\Lambda)^\alpha = \sum_{k=0}^{\infty} (\mu_k)^\alpha H_k$$

corresponding to $g(\rho) = (1+\rho)^\alpha$ and $g(\rho) = \rho^\alpha$, respectively.

3. Spectral results

We start with a preliminary observation before discussing the proof of our main result. Let X, Y be Hilbert spaces and $T : X \to Y$ a bounded operator. A typical case which we have in mind is

$$X = L^2(\mathbb{R}^d_x), \quad Y = L^2(\mathbb{R}_t \times \mathbb{R}^d_x), \quad T = |x|^{a-1} g(-\Lambda) |\nabla|^a \exp(-it\Delta).$$

If $f \neq 0$ is an extremiser, that is, if the operator norm $\|T\|$ of T is given by

$$\|T\| = \|Tf\|_Y / \|f\|_X,$$

then f is an eigenfunction of the operator $T^*T : X \to X$ with the eigenvalue $\|T\|^2$. In fact, by the Cauchy–Schwarz inequality, we have generally

$$\|Tf\|_Y^2 = (Tf, Tf)_Y = (T^*Tf, f)_X \leq \|T^*Tf\|_X \|f\|_X \leq \|T\|^2 \|f\|_X^2.$$

If f is an extremiser, then all the inequalities in the above should be equalities. By the condition for the Cauchy–Schwarz inequality to be an equality, T^*Tf and

f should be parallel to each other, that is, $T^*Tf = \lambda f$ for some $\lambda \in \mathbb{C}$ and again from the above (in)equality, we have $\lambda = \|T\|^2$.

First we consider the case $g \equiv 1$, and introduce the linear operator $S : L^2(\mathbb{R}^d) \to L^2(\mathbb{R}^{d+1})$ defined by

$$Sf(x,t) = |x|^{a-1} \int_{\mathbb{R}^d} \exp(i(x \cdot \xi + t|\xi|^2))|\xi|^a f(\xi)\, d\xi \quad (x \in \mathbb{R}^d,\, t \in \mathbb{R})$$

for, say, Schwartz functions $f : \mathbb{R}^d \to \mathbb{C}$. We remark that

$$|x|^{a-1}|\nabla|^a \exp(-it\Delta) f(x) = (2\pi)^{-d} S\widehat{f}(x,t),$$

where $\widehat{f}$ is the Fourier transform of f given by

$$\widehat{f}(\xi) = \int_{\mathbb{R}^d} f(x) \exp(-ix \cdot \xi)\, dx.$$

On account of the above observation, we will first discuss the eigenvalues and eigenspaces of the operator S^*S on $L^2(\mathbb{R}^d)$.

Theorem 3.1. *Let $a \in (1 - \frac{d}{2}, \frac{1}{2})$. Then the operator S^*S has the spectral decomposition*

$$S^*S = \sum_{k=0}^{\infty} \lambda_k H_k,$$

where

$$\lambda_k = (2\pi)^{d+1} 2^{2(a-1)} \frac{\Gamma(1-2a)\Gamma(\frac{d}{2} + k - 1 + a)}{\Gamma(1-a)^2 \Gamma(\frac{d}{2} + k - a)}; \quad k \in \mathbb{N}_0. \tag{3.1}$$

Furthermore, we have that

$$\lambda_k = O\left((1+k)^{2a-1}\right)$$

as k tends to infinity.

Proof. The first assertion was proved in [1]. The second assertion on the decay rate of the eigenvalues is a straightforward consequence of Stirling's formula

$$\lim_{x \to \infty} \frac{\Gamma(x+1)}{\sqrt{2\pi x}(x/e)^x} = 1$$

and we omit the details. □

Next we consider the operator

$$S_g f(x,t) = |x|^{a-1} g(-\Lambda) \int_{\mathbb{R}^d} \exp(i(x \cdot \xi + t|\xi|^2))|\xi|^a f(\xi)\, d\xi,$$

where $g(-\Lambda)$ is an operation in the x-variable. We claim that

$$S_g = S \cdot g(-\Lambda).$$

To see this, it is sufficient to show that

$$H_k \int_{\mathbb{R}^d} \exp(i(x \cdot \xi + t|\xi|^2))|\xi|^a f(\xi)\, d\xi = \int_{\mathbb{R}^d} \exp(i(x \cdot \xi + t|\xi|^2))|\xi|^a H_k f(\xi)\, d\xi$$

for each $k \in \mathbb{N}_0$. Shifting to the polar coordinates $r = |x|$, $\omega = x/|x|$ and $s = |\xi|$, $\eta = \xi/|\xi|$, it is enough to have

$$\int_{\mathbb{S}^{d-1}} \int_{\mathbb{S}^{d-1}} C_{d,k}(\omega \cdot \tilde{\omega}) \exp(irs\eta \cdot \tilde{\omega}) f(s\eta)\, d\eta d\tilde{\omega} = \int_{\mathbb{S}^{d-1}} \int_{\mathbb{S}^{d-1}} C_{d,k}(\eta \cdot \tilde{\omega}) \exp(irs\omega \cdot \eta) f(s\tilde{\omega})\, d\eta d\tilde{\omega},$$

which itself follows from

$$\int_{\mathbb{S}^{d-1}} C_{d,k}(\omega \cdot \tilde{\omega}) \exp(irs\eta \cdot \tilde{\omega})\, d\tilde{\omega} = \int_{\mathbb{S}^{d-1}} C_{d,k}(\eta \cdot \tilde{\omega}) \exp(irs\omega \cdot \tilde{\omega})\, d\tilde{\omega}.$$

This last identity can be justified by the Funk–Hecke theorem (see, for example, [9])

$$\int_{\mathbb{S}^{d-1}} P(\tilde{\omega}) F(\omega \cdot \tilde{\omega})\, d\tilde{\omega} = P(\omega) \frac{|\mathbb{S}^{d-2}|}{C_{d,k}(1)} \int_{-1}^{1} F(\rho) C_{d,k}(\rho)(1-\rho^2)^{\frac{d-3}{2}}\, d\rho$$

for spherical harmonics P of degree k, if we use it with $F(\rho) = \exp(irs\rho)$ and $P(\tilde{\omega}) = C_{d,k}(\omega \cdot \tilde{\omega})$, $C_{d,k}(\eta \cdot \tilde{\omega})$.

Theorem 3.2. *Let $a \in (1 - \frac{d}{2}, \frac{1}{2})$. Assume that $|g(\rho)| = O\left(\rho^{1/4-a/2}\right)$ as $\rho \to \infty$. Then the operator $S_g^* S_g$ is bounded on $L^2(\mathbb{R}^d)$ and has the spectral decomposition*

$$S_g^* S_g = \sum_{k=0}^{\infty} \lambda_k \, |g(\mu_k)|^2 \, H_k,$$

where λ_k and μ_k are numbers defined by (3.1) *and* (2.1)*. Furthermore, the operator S_g is bounded from $L^2(\mathbb{R}^d)$ to $L^2(\mathbb{R}^{d+1})$ and its operator norm $\|S_g\|$ is*

$$\|S_g\| = \sup_{k \in \mathbb{N}_0} \lambda_k^{1/2} \, |g(\mu_k)|.$$

If the supremum is attained for $k = k_0$, then $\mathfrak{H}_{k_0} \setminus \{0\}$ is the space of extremisers.

Proof. We remark that $S_g^* S_g = \bar{g}(-\Lambda) \cdot S^* S \cdot g(-\Lambda)$ and we have the first assertion by Theorem 3.1. Then we have

$$\|S_g f\|^2_{L^2(\mathbb{R}^{d+1})} = (S_g f, S_g f)_{L^2(\mathbb{R}^{d+1})} = (S_g^* S_g f, f)_{L^2(\mathbb{R}^d)}$$
$$= \sum_{k=0}^{\infty} \sum_{\ell=0}^{\infty} \lambda_k \, |g(\mu_k)|^2 (H_k f, H_\ell f)_{L^2(\mathbb{R}^d)} = \sum_{k=0}^{\infty} \lambda_k \, |g(\mu_k)|^2 \|H_k f\|^2_{L^2(\mathbb{R}^d)},$$

hence $\|S_g\| \le \sup_{k \in \mathbb{N}_0} \lambda_k^{1/2} \, |g(\mu_k)|$ because $\sum_{k=0}^{\infty} \|H_k f\|^2_{L^2(\mathbb{R}^d)} = \|f\|^2_{L^2(\mathbb{R}^d)}$, and $\|S_g\| \ge \sup_{k \in \mathbb{N}_0} \lambda_k^{1/2} \, |g(\mu_k)|$ because $\|S_g f\|^2_{L^2(\mathbb{R}^{d+1})} = \lambda_k \, |g(\mu_k)|^2 \|f\|^2_{L^2(\mathbb{R}^d)}$ with $f \in \mathfrak{H}_k \setminus \{0\}$ for each $k \in \mathbb{N}_0$. □

From Theorem 3.2 we immediately obtain Theorem 1.1 by taking $g(\rho) = (1+\rho)^{1/4-a/2}$, if we note the relation

$$|x|^{a-1} g(-\Lambda) |\nabla|^a \exp(-it\Delta) f(x) = (2\pi)^{-d} S_g \widehat{f}(x,t),$$

and recall the Plancherel theorem $\|\widehat{f}\|_{L^2(\mathbb{R}^d)} = (2\pi)^{d/2} \|f\|_{L^2(\mathbb{R}^d)}$.

References

[1] N. Bez, M. Sugimoto, *Optimal constants and extremisers for some smoothing estimates*, arXiv:1206.5110.

[2] N. Bez, M. Sugimoto, *Optimal forward and reverse estimates of Morawetz and Kato–Yajima type with angular smoothing index*, in preparation.

[3] T. Hoshiro, *On weighted L^2 estimates of solutions to wave equations*, J. Anal. Math. **72** (1997), 127–140.

[4] T. Kato, K. Yajima, *Some examples of smooth operators and the associated smoothing effect*, Rev. Math. Phys. **1** (1989), 481–496.

[5] T. Ozawa, K. Rogers, *Sharp Morawetz estimates*, preprint.

[6] M. Ruzhansky, M. Sugimoto, *A smoothing property of Schrödinger equations in the critical case*, Math. Ann. **335** (2006), 645–673.

[7] M. Ruzhansky, M. Sugimoto, *Smoothing estimates of evolution equations via canonical transforms and comparison*, Proc. London Math. Soc. **105** (2012), 393–423.

[8] M. Ruzhansky, M. Sugimoto, *Structural resolvent estimates and derivative nonlinear Schrödinger equations*, Comm. Math. Phys. **314** (2012), 281–304.

[9] R.T. Seeley, *Spherical harmonics*, Amer. Math. Monthly **73** (1966), 115–121.

[10] N. Shimakura, *Partial differential operators of elliptic type*, Translations of Mathematical Monographs, Volume 99, American Mathematical Society, Providence, RI, 1992.

[11] B. Simon, *Best constants in some operator smoothness estimates*, J. Funct. Anal. **107** (1992), 66–71.

[12] E.M. Stein, G. Weiss, *Introduction to Fourier Analysis on Euclidean Spaces*, Princeton Univ. Press, Princeton, N.J., 1971.

[13] M. Sugimoto, *Global smoothing properties of generalized Schrödinger equations*, J. Anal. Math. **76** (1998), 191–204.

[14] M. Sugimoto, *A smoothing property of Schrödinger equations along the sphere*, J. Anal. Math. **89** (2003), 15–30.

[15] K. Watanabe, *Smooth perturbations of the self-adjoint operator $|\Delta|^{\alpha/2}$*, Tokyo J. Math. **14** (1991), 239–250.

Neal Bez
School of Mathematics
The Watson Building
University of Birmingham
Edgbaston, Birmingham
B15 2TT, England
e-mail: `n.bez@bham.ac.uk`

Mitsuru Sugimoto
Graduate School of Mathematics
Nagoya University
Furocho, Chikusa-ku
Nagoya 464-8602, Japan
e-mail: `sugimoto@math.nagoya-u.ac.jp`

Fourier Analysis
Trends in Mathematics, 9–45

The Interplay Between Time-dependent Speed of Propagation and Dissipation in Wave Models

Tang Bao Ngoc Bui and Michael Reissig

Abstract. The aim of this paper is to understand the influence of the interplay between a time-dependent increasing speed of propagation and a time-dependent coefficient in the dissipation on qualitative properties of solutions to the wave model

$$u_{tt} - a^2(t)\Delta u + b(t)u_t = 0,\ u(0,x) = u_1(x),\ u_t(0,x) = u_2(x). \tag{0.1}$$

Our considerations are focused to energy estimates. The main difficulty is to find a good description of *non-effective* and *effective* dissipations depending on a given speed of propagation. The obtained energy estimates are optimal as special examples will show. At the end we will sketch very briefly how to get *scattering* and *over-damping* results. So, we propose a classification of different damping terms which is motivated by the thesis of J. Wirth for the case $a \equiv 1$.

Mathematics Subject Classification (2010). Primary 35L15; Secondary 35L05, 35B40, 35B45.

Keywords. Wave models, damping term, WKB analysis, asymptotic profile, scattering.

1. Introduction

The papers [12] and [13] are devoted to the study of the Cauchy problem for the wave equation with time-dependent dissipation

$$u_{tt} - \Delta u + b(t)u_t = 0,\ u(0,x) = u_1(x),\ u_t(0,x) = u_2(x). \tag{1.1}$$

A description of the influence of the coefficient $b = b(t)$ on the qualitative behavior of solutions is given due to the following classification:

- *Scattering:* If $b(t)$ has a very weak influence, then there is a relation to the free wave equation. Such relations are described by so-called scattering results.

- *Non-effective:* If $b(t)$ has a weak influence, then the classical energy decays to 0 and corresponding L^p-L^q decay estimates are the classical Strichartz decay estimates with an additional term as a time-dependent coefficient coming from the decay of the energy itself. Such weak dissipations will be called non-effective.
- *Effective:* If $b(t)$ has a stronger influence, then L^p-L^q decay estimates are similar to those ones for the classical damped wave equation but with an additional decay function related to the dissipation itself. Such dissipations will be called effective.
- *Over-damping:* If $b(t)$ has a "very strong influence", then in general we can not expect any decay estimate of the classical wave type energy.

In both cases, *scattering* or *over-damping*, we have in general no energy decay. Roughly speaking, energy decay only appears for dissipations "between" the conditions $b \notin L_1(\mathbb{R}^+)$ and $1/b \notin L_1(\mathbb{R}^+)$ in (1.1). But we have to be more precise. This leads to distinguish between *non-effective* and *effective* dissipation. Correspondingly, we only cite here two results from [11]. Assuming the coefficient function $b = b(t)$ is a positive, smooth and monotone function of t, which satisfies

$$|b^{(k)}(t)| \leq C_k b(t) \left(\frac{1}{1+t}\right)^k$$

for all $k \in \mathbb{N}_0$.

Result 1.1. *Assume* $\limsup_{t\to\infty} tb(t) < 1$. *Then the solution* $u = u(t,x)$ *of* (1.1) *satisfies the* L^p-L^q *decay estimate*

$$\|(\partial_t, \nabla)u(t,\cdot)\|_{L^q} \leq C\frac{1}{\lambda(t)}(1+t)^{-\frac{n-1}{2}\left(\frac{1}{p}-\frac{1}{q}\right)}\left(\|u_1\|_{W_p^{N_p+1}} + \|u_2\|_{W_p^{N_p}}\right) \tag{1.2}$$

for $p \in (1,2]$, q *is the corresponding dual index,* $N_p = n\left(\frac{1}{p}-\frac{1}{q}\right)$ *and* $\lambda(t)$ *is an auxiliary function which is defined by*

$$\lambda(t) := \exp\left(\frac{1}{2}\int_0^t b(\tau)d\tau\right).$$

Result 1.2. *Assume* $tb(t) \to \infty$ *as* $t \to \infty$. *Then the solution* $u = u(t,x)$ *of* (1.1) *satisfies the* L^p-L^q *decay estimate*

$$\|(\partial_t, \nabla)u(t,\cdot)\|_{L^q} \leq C\left(1+\int_0^t \frac{d\tau}{b(\tau)}\right)^{-\frac{n}{2}\left(\frac{1}{p}-\frac{1}{q}\right)-\frac{1}{2}}\left(\|u_1\|_{W_p^{N_p+1}} + \|u_2\|_{W_p^{N_p}}\right) \tag{1.3}$$

for $p \in (1,2]$, q *is the corresponding dual index and* $N_p = n\left(\frac{1}{p}-\frac{1}{q}\right)$.

What about wave models in (1.1) without any dissipation? In a series of papers (see [7], [8], [9] or [5]) the authors have obtained results about decay estimates for solutions to the Cauchy problem

$$u_{tt} - a^2(t)\Delta u = 0,\ u(0,x) = u_1(x),\ u_t(0,x) = u_2(x). \tag{1.4}$$

Therein $a(t)$ is chosen as $a^2(t) = \lambda^2(t)b^2(t)$, where $\lambda(t)$ is a monotonously increasing function and $b(t)$ is an oscillating function. In this paper we shall treat only the case of an increasing propagation speed and we are not interested in special oscillating parts in the coefficient $a(t)$. We recall that some results from [10] are obtained under the following assumptions to the coefficient $a = a(t)$:

(A1) $a(t) > 0$, $a'(t) > 0$, for $t \in [0, \infty)$,

(A2) $a_0 \dfrac{a(t)}{A(t)} \le \dfrac{a'(t)}{a(t)} \le a_1 \dfrac{a(t)}{A(t)}$, $a_0, a_1 > 0$,

(A3) $|a''(t)| \le a_2 a(t) \Big(\dfrac{a(t)}{A(t)} \Big)^2$, $a_2 \ge 0$,

(A4) $t + \dfrac{C}{\sqrt{a(t)}}$ is strictly increasing with a positive constant C and for large t.

Here $A(t) = 1 + \int_0^t a(s)ds$ is a primitive of $a(t)$. So a combination of the goals of [10] with the goals of [12] and [13] seems to be reasonable. For this reason we devote to the wave model

$$u_{tt} - a^2(t)\Delta u + b(t)u_t = 0, \; u(0,x) = u_1(x), \; u_t(0,x) = u_2(x) \tag{1.5}$$

with time-dependent increasing speed of propagation and dissipation. An interesting issue is to introduce *precise descriptions* for *non-effective* and *effective dissipations* in model (1.5). Such a classification we shall propose in Sections 2 and 3. We derive energy estimates in both cases. Some examples show in Section 3.7 that our estimates are optimal. At the end of the paper we sketch very briefly scattering and over-damping results. Most of the results are part of the thesis of Mr. Bui Tang Bao Ngoc [1].

2. Non-effective dissipation

Let us devote to the wave model

$$u_{tt} - a^2(t)\Delta u + b(t)u_t = 0. \tag{2.1}$$

Our question is the following:

> *Under which assumptions to the coefficient $b = b(t)$ for a given time-dependent speed of propagation $a = a(t)$ can we call b a non-effective dissipation?*

Here non-effective means, that on the one hand we have really a dissipation effect (scattering to wave models is excluded), but on the other hand the wave model itself is hyperbolic like from the point of view of decay estimates for the wave type energy. Motivated by the considerations from [12] we assume:

(B1) $b(t) > 0$, $b(t) = \mu(t)\dfrac{a(t)}{A(t)}$, $b \notin L^1(\mathbb{R}_+)$,

(B2) $|\mu'(t)| \le C_\mu \mu(t)\dfrac{a(t)}{A(t)}$,

(B3) $\limsup_{t\to\infty} \mu(t) < 1$.

Instead of the assumption (B3) we assume sometimes

(B3)′ $\liminf_{t\to\infty} \mu(t) > 1$.

Finally, we suppose

(C) $\limsup_{t\to\infty} \mu(t) + \alpha(t) < 2$,

where $\alpha = \alpha(t)$ is defined by

$$\frac{a'(t)}{a(t)} =: \alpha(t)\frac{a(t)}{A(t)}.$$

Theorem 2.1. *Let us consider the Cauchy problem* (2.1) *under the assumptions* (A1) *to* (A3), (B1), (B2), (B3) *or* (B3)′ *and* (C). *Then we have the following estimates for the kinetic and elastic energy:*

$$\|u_t(t,\cdot)\|_{L^2} \le C\frac{\sqrt{a(t)}}{\lambda(t)}\big(\|u_1\|_{H^1} + \|u_2\|_{L^2}\big),$$

$$\|a(t)\nabla u(t,\cdot)\|_{L^2} \le C\frac{\sqrt{a(t)}}{\lambda(t)}\big(\|u_1\|_{H^1} + \|u_2\|_{L^2}\big).$$

Here $\lambda = \lambda(t)$ *is defined by*

$$\lambda(t) := \exp\Big(\frac{1}{2}\int_0^t b(\tau)d\tau\Big). \tag{2.2}$$

Proof. Applying partial Fourier transformation we have $\hat{u}_{tt} + a^2(t)|\xi|^2\hat{u} + b(t)\hat{u}_t = 0$. Later we will derive estimates for the fundamental solution $E = E(t,s,\xi)$ of an equivalent system of first order by different ways in different parts of the extended phase space $(0,\infty)\times\mathbb{R}^n$: in the dissipative zone and the hyperbolic zone. These zones are defined by

- $Z_{\mathrm{hyp}}(N) := \{(t,\xi) : t \ge t_\xi\}$,
- $Z_{\mathrm{diss}}(N) := \{(t,\xi) : 0 \le t \le t_\xi\}$,

where t_ξ satisfies $A(t_\xi)|\xi| = N$.

2.1. Considerations in the dissipative zone

Let us define the micro-energy $U = (N\delta(t)\hat{u}, D_t\hat{u})^T$, where we denote $\delta(t) := \dfrac{a(t)}{A(t)}$. Then the transformed equation can be written in the form of a system of first order (in D_t)

$$D_tU = A(t,\xi)U,\ A(t,\xi) = \begin{pmatrix} -i\dfrac{d_t\delta(t)}{\delta(t)} & N\delta(t) \\ \dfrac{a^2(t)|\xi|^2}{N\delta(t)} & ib(t) \end{pmatrix}.$$

Thus the solution $U = U(t,\xi)$ can be represented as $U(t,\xi) = E(t,s,\xi)U(s,\xi)$, where $E(t,s,\xi)$ is the fundamental solution, that is, the solution to the system

$$D_tE(t,s,\xi) = A(t,\xi)E(t,s,\xi),\ E(s,s,\xi) = I,\ 0 \le s \le t \le t_\xi.$$

In the further calculations we use the following statement:

Lemma 2.2.

1. *The assumption* (B3) *implies for the auxiliary function* $\lambda = \lambda(t)$ *the properties*
$$\int_0^t \frac{a(\tau)}{\lambda^2(\tau)} d\tau \lesssim \frac{A(t)}{\lambda^2(t)},$$
and $\frac{A(t)}{\lambda^2(t)}$ *is monotonously increasing if* t *tends to infinity.*

2. *The assumption* (B3)$'$ *implies* $\frac{a(t)}{\lambda^2(t)} \in L^1(\mathbb{R}^+)$ *with*
$$\int_t^\infty \frac{a(\tau)}{\lambda^2(\tau)} d\tau \lesssim \frac{A(t)}{\lambda^2(t)}.$$
Furthermore, $\frac{A(t)}{\lambda^2(t)}$ *is monotonously decreasing for large* t.

Proof. To prove these statements we only use integration by parts and straightforward calculations. □

Lemma 2.3. *Assume that* $a = a(t)$ *satisfies* (A1) *and* (A2), *and the function* $\mu = \mu(t)$ *satisfies the condition* (B3). *Then there exists a constant* $\delta \in (0,1)$ *such that*
$$a(t)^\delta \int_0^t \frac{a(\tau)^{1-\delta}}{\lambda^2(\tau)} d\tau \lesssim \frac{A(t)}{\lambda^2(t)}. \tag{2.3}$$

Proof. The statement follows directly after integration of the following inequality:
$$\frac{a(t)^{1-\delta}}{\lambda^2(t)} \lesssim \left(\frac{A(t)}{a(t)^\delta \lambda^2(t)}\right)' = \frac{a(t)^{1-\delta}}{\lambda^2(t)} - \delta \frac{A(t)a'(t)}{a(t)^{1+\delta}\lambda^2(t)} - \frac{A(t)\mu(t)\frac{a(t)}{A(t)}}{a(t)^\delta \lambda^2(t)},$$
where we use that the right-hand side can be estimated to below by
$$(1-c)\frac{a(t)^{1-\delta}}{\lambda^2(t)} - \delta\frac{A(t)a'(t)}{a(t)^{1+\delta}\lambda^2(t)} \quad \text{for large } t,$$
and $c < 1$ due to condition (B3). The desired estimate
$$\frac{a(t)^{1-\delta}}{\lambda^2(t)} \le C\Big((1-c)\frac{a(t)^{1-\delta}}{\lambda^2(t)} - \delta\frac{A(t)a'(t)}{a(t)^{1+\delta}\lambda^2(t)}\Big)$$
is true if it exits a constant $C > \frac{1}{1-c}$ such that
$$A(t)a'(t) \le (1-c-C^{-1})\delta^{-1}a^2(t).$$
From that we can choose any δ satisfying $\delta < (\limsup_t A(t)a'(t)/a^2(t))^{-1}$. This supremum is finite by condition (A2). □

Denoting by $E^{(jk)}$ the entries of E we get for $k=1,2$ the system

$$D_t E^{(1k)} = -i\frac{d_t\delta(t)}{\delta(t)}E^{(1k)} + N\delta(t)E^{(2k)},$$
$$D_t E^{(2k)} = \frac{a^2(t)|\xi|^2}{N\delta(t)}E^{(1k)} + ib(t)E^{(2k)},\ E^{(jk)}(s,s,\xi) = \delta_{jk}.$$

Integration yields

$$E^{(1k)}(t,s,\xi) = \frac{\delta(t)}{\delta(s)}E^{(1k)}(s,s,\xi) + iN\delta(t)\int_s^t E^{(2k)}(\tau,s,\xi)d\tau, \tag{2.4}$$
$$E^{(2k)}(t,s,\xi) = \frac{\lambda^2(s)}{\lambda^2(t)}E^{(2k)}(s,s,\xi) + \frac{i|\xi|^2}{N\lambda^2(t)}\int_s^t \frac{a^2(\tau)}{\delta(\tau)}\lambda^2(\tau)E^{(1k)}(\tau,s,\xi)d\tau.$$

We are going to prove the following lemma:

Lemma 2.4. *Let us assume* (A1) *to* (A3) *and* (B3). *Then we have the following estimates for the entries* $E^{(kl)}(t,0,\xi)$ *of the fundamental solution* $E(t,0,\xi)$ *in the dissipative zone:*

$$(|E(t,0,\xi)|) := \begin{pmatrix} |E^{(11)}(t,0,\xi)| & |E^{(12)}(t,0,\xi)| \\ |E^{(21)}(t,0,\xi)| & |E^{(22)}(t,0,\xi)| \end{pmatrix} \lesssim \begin{pmatrix} \dfrac{a(t)}{A(t)} & \dfrac{a(t)^{1-\delta}}{\lambda^2(t)} \\ \dfrac{|\xi|^2 K(t)}{\lambda^2(t)} & \dfrac{a(t)^{1-\delta}}{\lambda^2(t)} \end{pmatrix} \tag{2.5}$$

with $K(t) := \int_0^t a^2(\tau)\lambda^2(\tau)d\tau \le \lambda^2(t)a(t)A(t)$.

Proof. Let us consider

$$\begin{aligned} E^{(21)}(t,0,\xi) &= \frac{i|\xi|^2}{N\lambda^2(t)}\int_0^t \frac{a^2(\tau)}{\delta(\tau)}\lambda^2(\tau)E^{(11)}(\tau,0,\xi)d\tau \\ &= \frac{i|\xi|^2}{N\lambda^2(t)}\Big(\int_0^t \frac{a^2(\tau)}{\delta(0)}\lambda^2(\tau)d\tau + \int_0^t a^2(\tau)\lambda^2(\tau)iN d\tau\int_0^\tau E^{(21)}(\theta,0,\xi)d\theta d\tau\Big) \\ &= \frac{i|\xi|^2}{N\delta(0)\lambda^2(t)}K(t) - \frac{|\xi|^2}{\lambda^2(t)}\int_0^t a^2(\tau)\lambda^2(\tau)\int_0^\tau E^{(21)}(\theta,0,\xi)d\theta d\tau \\ &= \frac{i|\xi|^2}{C_N\lambda^2(t)}K(t) - \frac{|\xi|^2}{\lambda^2(t)}\int_0^t \Big(\int_\theta^t a^2(\tau)\lambda^2(\tau)d\tau\Big)E^{(21)}(\theta,0,\xi)d\theta. \end{aligned}$$

Rewriting the integral equation gives

$$\frac{C_N\lambda^2(t)E^{(21)}(t,0,\xi)}{|\xi|^2K(t)} = i + \int_0^t k_1(t,\theta,\xi)\frac{C_N\lambda^2(\theta)E^{(21)}(\theta,0,\xi)}{|\xi|^2K(\theta)}d\theta \tag{2.6}$$

with kernel

$$k_1(t,\theta,\xi) = -|\xi|^2\frac{K(\theta)}{K(t)\lambda^2(\theta)}\int_\theta^t a^2(\tau)\lambda^2(\tau)d\tau,\ \theta\in[0,t]. \tag{2.7}$$

Now we estimate

$$\int_0^t \sup_{\theta\le\tilde t\le t} |k_1(\tilde t,\theta,\xi)|d\theta \lesssim |\xi|^2 \int_0^{t_\xi} \sup_{\tilde t} \frac{K(\theta)}{\lambda^2(\theta)K(\tilde t)}\big(K(\tilde t)-K(\theta)\big)d\theta$$
$$\le |\xi|^2 \int_0^{t_\xi} \frac{K(\theta)}{\lambda^2(\theta)}d\theta \lesssim |\xi|^2 \int_0^{t_\xi} a(\theta)A(\theta)d\theta = \frac{1}{2}|\xi|^2A^2(t_\xi) \lesssim 1$$

uniformly in $Z_{\text{diss}}(N)$. Therefore, we obtained

$$|E^{(21)}(t,0,\xi)| \lesssim \frac{|\xi|^2K(t)}{\lambda^2(t)}. \tag{2.8}$$

Substituting this estimate into the first integral equation implies

$$|E^{(11)}(t,0,\xi)| \le \frac{\delta(t)}{\delta(0)} + N\delta(t)\int_0^t \frac{|\xi|^2K(\tau)}{\lambda^2(\tau)}d\tau \lesssim \delta(t) + |\xi|^2\delta(t)A^2(t) \lesssim \delta(t) = \frac{a(t)}{A(t)}.$$

Next, we consider

$$E^{(22)}(t,0,\xi) = \frac{\lambda^2(0)}{\lambda^2(t)} + \frac{i|\xi|^2}{N\lambda^2(t)}\int_0^t \frac{a^2(\tau)}{\delta(\tau)}\lambda^2(\tau)E^{(12)}(\tau,0,\xi)d\tau$$
$$= \frac{\lambda^2(0)}{\lambda^2(t)} - \frac{|\xi|^2}{\lambda^2(t)}\int_0^t a^2(\tau)\lambda^2(\tau)\int_0^\tau E^{(22)}(\theta,0,\xi)d\theta d\tau,$$
$$\lambda^2(t)E^{(22)}(t,0,\xi) = 1 - |\xi|^2\int_0^t \Big(\int_\theta^t a^2(\tau)\lambda^2(\tau)d\tau\Big)E^{(22)}(\theta,0,\xi)d\theta,$$

respectively. Our goal is to show that $|E^{(22)}(t,0,\xi)| \lesssim \frac{a(t)^{1-\delta}}{\lambda^2(t)}$. Therefore, we rewrite the integral equation as

$$\frac{\lambda^2(t)E^{(22)}(t,0,\xi)}{a(t)^{1-\delta}} = \frac{1}{a(t)^{1-\delta}} + \int_0^t k_2(t,\theta,\xi)\frac{\lambda^2(\theta)E^{(22)}(\theta,0,\xi)}{a(\theta)^{1-\delta}}d\theta \tag{2.9}$$

with the kernel

$$k_2(t,\theta,\xi) = -|\xi|^2\frac{a(\theta)^{1-\delta}}{a(t)^{1-\delta}\lambda^2(\theta)}\int_\theta^t a^2(\tau)\lambda^2(\tau)d\tau,\ \theta\in[0,t]. \tag{2.10}$$

The following integral over the kernel satisfies the desired estimate. It holds

$$\int_0^t \sup_{\theta\le\tilde t\le t} |k_2(\tilde t,\theta,\xi)|d\theta \lesssim |\xi|^2 \int_0^{t_\xi} \sup_{\tilde t} \frac{(a(\theta))^{1-\delta}}{(a(\tilde t)^{1-\delta})\lambda^2(\theta)}\big(K(\tilde t)-K(\theta)\big)d\theta$$
$$\le |\xi|^2\int_0^{t_\xi} \sup_{\tilde t}\frac{(a(\theta))^{1-\delta}K(\tilde t)}{(a(\tilde t))^{1-\delta}\lambda^2(\theta)}d\theta \le |\xi|^2\lambda^2(t_\xi)A(t_\xi)(a(t_\xi))^\delta\int_0^{t_\xi}\frac{(a(\theta))^{1-\delta}}{\lambda^2(\theta)}d\theta$$
$$\lesssim |\xi|^2\lambda^2(t_\xi)A(t_\xi)\frac{A(t_\xi)}{\lambda^2(t_\xi)} \le |\xi|^2A^2(t_\xi) \lesssim 1.$$

Here we have used Lemma 2.3 and, therefore,

$$|E^{(22)}(t,0,\xi)| \lesssim \frac{a(t)^{1-\delta}}{\lambda^2(t)}. \tag{2.11}$$

After plugging this estimate into the first integral equation and using Lemma 2.3 again we have

$$|E^{(12)}(t,0,\xi)| \lesssim \delta(t)\int_0^t \frac{a(\tau)^{1-\delta}}{\lambda^2(\tau)}d\tau \lesssim \frac{a(t)^{1-\delta}}{A(t)}\frac{A(t)}{\lambda^2(t)} \lesssim \frac{a(t)^{1-\delta}}{\lambda^2(t)}. \tag{2.12}$$

This completes the proof. □

Now let us come back to

$$U(t,\xi) = E(t,0,\xi)U(0,\xi) \text{ for all } 0 \le t \le t_\xi. \tag{2.13}$$

Because of $a(t)|\xi||\hat{u}(t,\xi)| \le N\dfrac{a(t)}{A(t)}|\hat{u}(t,\xi)|$ in $Z_{\text{diss}}(N)$ the following statement can be concluded from (2.13) and Lemma 2.4:

Corollary 2.5. *We have in the dissipative zone $Z_{\text{diss}}(N)$ the following estimates for all $0 \le t \le t_\xi$:*

$$a(t)|\xi||\hat{u}(t,\xi)| \le C_N\frac{a(t)}{A(t)}|\hat{u}(0,\xi)| + C_N\frac{a(t)^{1-\delta}}{\lambda^2(t)}|D_t\hat{u}(0,\xi)|,$$

$$|D_t\hat{u}(t,\xi)| \le C_N\frac{|\xi|^2K(t)}{\lambda^2(t)}|\hat{u}(0,\xi)| + C_N\frac{a(t)^{1-\delta}}{\lambda^2(t)}|D_t\hat{u}(0,\xi)|.$$

Lemma 2.6. *Let us assume* (A1) *to* (A3) *and* (B3)′. *Then we have the following estimates for the entries $E^{(kl)}(t,0,\xi)$ of the fundamental solution $E(t,0,\xi)$:*

$$(|E(t,0,\xi)|) := \begin{pmatrix} |E^{(11)}(t,0,\xi)| & |E^{(12)}(t,0,\xi)| \\ |E^{(21)}(t,0,\xi)| & |E^{(22)}(t,0,\xi)| \end{pmatrix} \lesssim \begin{pmatrix} \dfrac{a(t)}{A(t)} & \dfrac{a(t)}{A(t)} \\ \dfrac{a(t)}{A(t)} & \dfrac{a(t)}{A(t)} \end{pmatrix}. \tag{2.14}$$

Proof. We start by estimating the first column. Plugging the representation for $E^{(21)}(t,s,\xi)$ into the integral equation for $E^{(11)}(t,s,\xi)$ gives

$$\frac{\delta(0)}{\delta(t)}E^{(11)}(t,0,\xi) = 1 - |\xi|^2\int_0^t\int_0^\tau \frac{\lambda^2(\theta)}{\lambda^2(\tau)}a^2(\theta)\frac{\delta(0)}{\delta(\theta)}E^{(11)}(\theta,0,\xi)d\theta d\tau,$$

$$\frac{1}{\delta(t)}|E^{(11)}(t,0,\xi)| \lesssim 1 + |\xi|^2\int_0^t\int_0^\tau \underbrace{\frac{\lambda^2(\theta)}{\lambda^2(\tau)}}_{\le 1}a^2(\theta)\frac{1}{\delta(\theta)}|E^{(11)}(\theta,0,\xi)|d\theta d\tau,$$

$$\frac{1}{\delta(t)}|E^{(11)}(t,0,\xi)| \lesssim \exp\left(|\xi|^2\int_0^t\int_0^\tau a^2(\theta)d\theta d\tau\right) \lesssim \exp\left(|\xi|^2A^2(t)\right) \lesssim 1,$$

$$|E^{(11)}(t,0,\xi)| \lesssim \delta(t) = \frac{a(t)}{A(t)}.$$

Here we have used the definition of dissipative zone and assumption (A2) for $a(t)$. Let us consider $E^{(21)}(t,0,\xi)$. We have

$$|E^{(21)}(t,0,\xi)| \lesssim \frac{|\xi|^2}{\lambda^2(t)} \int_0^t \frac{a^2(\tau)}{\delta(\tau)} \lambda^2(\tau) |E^{(11)}(\tau,0,\xi)| d\tau$$
$$\lesssim |\xi|^2 \int_0^t \frac{a^2(\tau)}{\delta(\tau)} \underbrace{\frac{\lambda^2(\tau)}{\lambda^2(t)}}_{\leq 1} \delta(\tau) d\tau \lesssim |\xi|^2 A(t) a(t) \leq C_N \frac{a(t)}{A(t)}.$$

Now we will estimate the entries of the second column. We get

$$\frac{1}{\delta(t)} E^{(12)}(t,0,\xi) = iN\lambda^2(0) \int_0^t \frac{d\tau}{\lambda^2(\tau)} - |\xi|^2 \int_0^t \int_0^\tau \underbrace{\frac{\lambda^2(\theta)}{\lambda^2(\tau)}}_{\leq 1} \frac{a^2(\theta)}{\delta(\theta)} E^{(12)}(\theta,0,\xi) d\theta d\tau.$$

Because the first integral is uniformly bounded by the second statement from Lemma 2.2 we can obtain by the above reasoning together with assumption (A1) the desired estimate for $E^{(12)}$. For $E^{(22)}$ we have

$$\frac{1}{\delta(t)} |E^{(22)}(t,0,\xi)| \lesssim \frac{A(t)}{a(t)\lambda^2(t)} + \frac{|\xi|^2 A(t)}{a(t)\lambda^2(t)} \int_0^t a^2(\tau)\lambda^2(\tau) d\tau$$
$$\lesssim \frac{A(t)}{a(t)\lambda^2(t)} + \underbrace{\frac{|\xi|^2 A(t)}{a(t)} \int_0^t a^2(\tau) d\tau}_{\leq C_N}.$$

If we notice the second statement of Lemma 2.2 the function $\frac{A(t)}{\lambda^2(t)}$ is decreasing. Taking account of the increasing behavior of $a(t)$ this implies that $\frac{A(t)}{a(t)\lambda^2(t)}$ is uniformly bounded for large t. This completes the proof. □

2.2. Considerations in the hyperbolic zone

Here we use the hyperbolic micro-energy $U = (a(t)|\xi|\hat{u}, D_t\hat{u})^T$. Then U satisfies

$$D_tU = A(t,\xi)U := \begin{pmatrix} \frac{D_t a}{a} & a(t)|\xi| \\ a(t)|\xi| & ib(t) \end{pmatrix} U. \tag{2.15}$$

Let us carry out the first step of diagonalization. For this reason we introduce

$$M = \begin{pmatrix} 1 & -1 \\ 1 & 1 \end{pmatrix}, \quad M^{-1} = \frac{1}{2}\begin{pmatrix} 1 & 1 \\ -1 & 1 \end{pmatrix} \quad \text{and } U^{(0)} := M^{-1}U.$$

So $D_tU^{(0)} = \mathcal{D}(t,\xi)U^{(0)} + \mathcal{R}(t)U^{(0)}$, where

$$\mathcal{D}(t,\xi) = \begin{pmatrix} \tau_1(t,\xi) & 0 \\ 0 & \tau_2(t,\xi) \end{pmatrix} := \begin{pmatrix} a(t)|\xi| & 0 \\ 0 & -a(t)|\xi| \end{pmatrix},$$

$$\mathcal{R}(t) = (R_{kl}(t)) := \frac{1}{2}\begin{pmatrix} \frac{D_ta}{a} + ib(t) & -\frac{D_ta}{a} + ib(t) \\ -\frac{D_ta}{a} + ib(t) & \frac{D_ta}{a} + ib(t) \end{pmatrix}.$$

Let $F_0(t)$ be the diagonal part of $\mathcal{R}(t)$. Now we carry out the second step of diagonalization procedure. Therefore we introduce the matrices

$$N^{(1)} = \begin{pmatrix} 0 & \dfrac{R_{12}}{\tau_1-\tau_2} \\ \dfrac{R_{21}}{\tau_2-\tau_1} & 0 \end{pmatrix} = \begin{pmatrix} 0 & i\dfrac{\delta_1(t)}{4a(t)|\xi|} \\ -i\dfrac{\delta_1(t)}{4a(t)|\xi|} & 0 \end{pmatrix}, \quad N_1 = I + N^{(1)}.$$

Here $\delta_1 := \dfrac{a'}{a} + b$. We have

$$\left(\frac{\frac{a'}{a}(t)}{4a(t)|\xi|}\right)^2 \lesssim \left(\frac{\frac{a}{A}(t)}{4a(t)|\xi|}\right)^2 \lesssim \Big(\frac{1}{A(t)|\xi|}\Big)^2 \leqslant \frac{C}{N^2}.$$

If we use the ansatz $b(t) = \mu(t)\dfrac{a(t)}{A(t)}$ and the assumptions (B3) or (B3)$'$ and (C), i.e., we have $\limsup_{t\to\infty}\mu(t) \lesssim 1$, then

$$\Big(\frac{b(t)}{4a(t)|\xi|}\Big)^2 = \Big(\frac{\mu(t)a(t)}{4a(t)A(t)|\xi|}\Big)^2 \lesssim \Big(\frac{1}{A(t)|\xi|}\Big)^2 \leq \frac{C}{N^2}.$$

Thus we can choose a sufficiently large N such that the determinant of N_1 is $\det N_1 = 1 - \Big(\dfrac{\delta_1(t)}{4a(t)|\xi|}\Big)^2 < 1/2$. Hence, the matrix N_1 is invertible. Setting

$$B^{(1)} = D_tN^{(1)} - (\mathcal{R}-F_0)N^{(1)} = \begin{pmatrix} -\dfrac{\delta_1^2(t)}{8a(t)|\xi|} & \partial_t\dfrac{\delta_1(t)}{4a(t)|\xi|} \\ -\partial_t\dfrac{\delta_1(t)}{4a(t)|\xi|} & \dfrac{\delta_1^2(t)}{8a(t)|\xi|} \end{pmatrix},$$

and $\mathcal{R}_1(t,\xi) = -N_1^{-1}B^{(1)}(t,\xi)$. We can conclude that

$$\big(D_t - \mathcal{D}(t,\xi) - \mathcal{R}(t)\big)N_1(t,\xi)U^{(1)} = N_1(t,\xi)\big(D_t - \mathcal{D}(t,\xi) - F_0(t) - \mathcal{R}_1(t,\xi)\big)U^{(1)}.$$

Now we shall find the solution $U^{(0)}(t,\xi) := N_1(t,\xi)U^{(1)}(t,\xi)$, where $U^{(1)}(t,\xi)$ is the solution to the system

$$\big(D_t - \mathcal{D}(t,\xi) - F_0(t) - \mathcal{R}_1(t,\xi)\big)U^{(1)}(t,\xi) = 0.$$

We can write $U^{(1)}(t,\xi) = E(t,t_\xi,\xi)U^{(1)}(t_\xi,\xi)$, where $E(t,s,\xi)$ is the fundamental solution, that is, the solution of the system

$$D_tE(t,s,\xi) = \big(D_t - \mathcal{D}(t,\xi) - F_0(t) - \mathcal{R}_1(t,\xi)\big)E(t,s,\xi), E(s,s,\xi) = I, t \geq s \geq t_\xi.$$

The solution $E_0 = E_0(t,s,\xi)$ of the "principal diagonal part" of the last system fulfils

$$D_t E_0(t,s,\xi) = (\mathcal{D}(t,\xi) + F_0(t))E_0(t,s,\xi), \quad E_0(s,s,\xi) = I, \ t \geq s \geq t_\xi.$$

Thus

$$E_0(t,s,\xi) = \frac{\sqrt{a(t)}}{\sqrt{a(s)}} \frac{\lambda(s)}{\lambda(t)} \begin{pmatrix} \exp\Big(\int_s^t ia(\tau)|\xi|d\tau\Big) & 0 \\ 0 & \exp\Big(-\int_s^t ia(\tau)|\xi|d\tau\Big) \end{pmatrix}.$$

Let us set

$$\mathcal{R}_2(t,s,\xi) = E_0(t,s,\xi)^{-1}\mathcal{R}_1(t,\xi)E_0(t,s,\xi),$$

$$Q(t,s,\xi) = I + \sum_{k=1}^{\infty} \int_s^t \mathcal{R}_2(t_1,s,\xi) \int_s^{t_1} \mathcal{R}_2(t_2,s,\xi) \dots \int_s^{t_{k-1}} \mathcal{R}_2(t_k,s,\xi)dt_k \dots dt_2 dt_1.$$

Then $Q(t,s,\xi)$ solves the Cauchy problem

$$D_t Q(t,s,\xi) = \mathcal{R}_2(t,s,\xi)Q(t,s,\xi), \quad Q(s,s,\xi) = I, \ t \geq s \geq t_\xi.$$

The fundamental solution $E = E(t,s,\xi)$ is representable in the form $E(t,s,\xi) = E_0(t,s,\xi)Q(t,s,\xi)$. Taking into consideration the above representation for $Q(t,s,\xi)$ we are able to prove the following estimate:

$$|Q(t,s,\xi)| \leq \exp\Big(\int_s^t |\mathcal{R}_1(\tau,\xi)|d\tau\Big) \leq \exp\Big(\frac{1}{|\xi|}\Big(\frac{1}{A(\tau)}\Big)\Big|_s^t\Big) \leq C_N.$$

The backward transformation yields

$$U(t,\xi) = MN_1(t,\xi)E_0(t,s,\xi)Q(t,s,\xi)N_1^{-1}(s,\xi)M^{-1}U(s,\xi),$$

therefore, we may conclude

$$\left|\begin{pmatrix} a(t)|\xi|\hat{u}(t,\xi) \\ D_t\hat{u}(t,\xi) \end{pmatrix}\right| \leq \frac{\sqrt{a(t)}}{\sqrt{a(s)}} \frac{\lambda(s)}{\lambda(t)} \left|\begin{pmatrix} a(s)|\xi|\hat{u}(s,\xi) \\ D_t\hat{u}(s,\xi) \end{pmatrix}\right| \quad \text{for all } t \geq s \geq t_\xi.$$

Corollary 2.7. *We have in the hyperbolic zone* $Z_{\text{hyp}}(N)$ *the estimate*

$$\left|\begin{pmatrix} a(t)|\xi|\hat{u}(t,\xi) \\ D_t\hat{u}(t,\xi) \end{pmatrix}\right| \leq C \frac{\sqrt{a(t)}}{\sqrt{a(t_\xi)}} \frac{\lambda(t_\xi)}{\lambda(t)} \left|\begin{pmatrix} a(t_\xi)|\xi|\hat{u}(t_\xi,\xi) \\ D_t\hat{u}(t_\xi,\xi) \end{pmatrix}\right|$$

for all $t \geq t_\xi$.

2.3. Gluing procedure and L^2-L^2 estimates

Before gluing the estimates in the dissipative zone and the hyperbolic zone we state the following lemma:

Lemma 2.8. *Assume that the functions* $\mu = \mu(t)$ *and* $\alpha = \alpha(t)$ *satisfy the assumption*

$$\limsup_{t\to\infty} \big(\mu(t) + \alpha(t)\big) < 2.$$

Then the following inequality holds:

$$\frac{\lambda(t)\sqrt{a(t)}}{A(t)} \leq C.$$

Proof. We have from the definition of λ and α

$$\frac{\lambda(t)\sqrt{a(t)}}{A(t)} \lesssim \frac{\exp\left(\frac{1}{2}\int_0^t \mu(s)\frac{a(s)}{A(s)}ds\right)\exp\left(\frac{1}{2}\int_0^t \alpha(s)\frac{a(s)}{A(s)}ds\right)}{\exp\left(\int_0^t \frac{a(s)}{A(s)}ds\right)}$$

$$= \exp\left(\frac{1}{2}\int_0^t (\mu(s)+\alpha(s)-2)\frac{a(s)}{A(s)}ds\right).$$

According to the assumption it holds $\mu(t)+\alpha(t)-2 \leq 0$ for $t \geq t_0$ with a suitable t_0. From that we may conclude

$$\frac{\lambda(t)\sqrt{a(t)}}{A(t)} \lesssim \exp\left(\int_0^{t_0} (\mu(s)+\alpha(s)-2)\frac{a(s)}{A(s)}ds\right) \leq C(t_0). \tag{2.16}$$

This completes the proof. □

From the statements of Corollaries 2.5 and 2.7 we derive the statement of our theorem.

Case 1: $\{|\xi| \geq N\}$. Then the statement of Corollary 2.7 implies immediately

$$\left|\begin{pmatrix} a(t)|\xi|\hat{u}(t,\xi) \\ D_t\hat{u}(t,\xi) \end{pmatrix}\right| \leq C\frac{\sqrt{a(t)}}{\lambda(t)}\left|\begin{pmatrix} |\xi|\hat{u}(0,\xi) \\ D_t\hat{u}(0,\xi) \end{pmatrix}\right|$$

for all $t \geq 0$.

Case 2: $\{|\xi| \leq N\}$ *and* $\{t \geq t_\xi\}$. Then the statements of Corollary 2.7 imply immediately

$$a(t)|\xi||\hat{u}(t,\xi)| + |D_t\hat{u}(t,\xi)| \leq C\frac{\sqrt{a(t)}}{\sqrt{a(t_\xi)}}\frac{\lambda(t_\xi)}{\lambda(t)}\Big(a(t_\xi)|\xi||\hat{u}(t_\xi,\xi)| + |D_t\hat{u}(t_\xi,\xi)|\Big)$$

$$\leq C_N\frac{\sqrt{a(t)}}{\lambda(t)}\Big(\frac{\lambda(t_\xi)}{\sqrt{a(t_\xi)}}a(t_\xi)|\xi||\hat{u}(t_\xi,\xi)| + \frac{\lambda(t_\xi)}{\sqrt{a(t_\xi)}}|D_t\hat{u}(t_\xi,\xi)|\Big).$$

From Corollary 2.5 we have for $t = t_\xi$

$$a(t_\xi)|\xi||\hat{u}(t,\xi)| + |D_t\hat{u}(t_\xi,\xi)| \leq C\frac{a(t_\xi)}{A(t_\xi)}|\hat{u}(0,\xi)| + C\frac{a(t_\xi)^{1-\delta}}{\lambda^2(t_\xi)}|D_t\hat{u}(0,\xi)|.$$

Summarizing we get

$$a(t)|\xi||\hat{u}(t,\xi)| + |D_t\hat{u}(t_\xi,\xi)| \leq C\frac{\sqrt{a(t)}}{\lambda(t)}\left(\frac{\sqrt{a(t_\xi)}\lambda(t_\xi)}{A(t_\xi)}|\hat{u}(0,\xi)| + \frac{a(t_\xi)^{\frac{1}{2}-\delta}}{\lambda(t_\xi)}|D_t\hat{u}(0,\xi)|\right)$$

for all admissible (t,ξ). If we choose $\delta \geq \frac{1}{2}$ and apply Lemma 2.8, then we may conclude

$$a(t)|\xi||\hat{u}(t,\xi)|+|D_t\hat{u}(t,\xi)|\leq C_N\frac{\sqrt{a(t)}}{\lambda(t)}\big(|\hat{u}(0,\xi)|+|D_t\hat{u}(0,\xi)|\big)\text{ for all admissible }(t,\xi).$$

Case 3: $\{|\xi| \leq N\}$ and $\{t \leq t_\xi\}$. Then the statements of Corollary 2.5 imply immediately

$$a(t)|\xi||\hat{u}(t,\xi)| + |D_t\hat{u}(t,\xi)| \leq C_N\frac{a(t)}{A(t)}|\hat{u}(0,\xi)| + C_N\frac{a(t)^{1-\delta}}{\lambda^2(t)}|D_t\hat{u}(0,\xi)|.$$

If we choose $\delta \geq \frac{1}{2}$ and apply Lemma 2.8, then we may conclude

$$a(t)|\xi||\hat{u}(t,\xi)|+|D_t\hat{u}(t,\xi)|\leq C_N\frac{\sqrt{a(t)}}{\lambda(t)}\big(|\hat{u}(0,\xi)|+|D_t\hat{u}(0,\xi)|\big)\text{ for all admissible }(t,\xi).$$

This completes the proof to Theorem 2.1. □

Remark 2.9. If we choose in Theorem 2.1 the coefficient $a(t) \equiv 1$, then the obtained estimates coincide with the estimates from Result 1.1 for $p = q = 2$.

Example. Let $\mu \in (0,1)$ or $\mu \in \big(1, 2 - \frac{l}{l+1}\big)$. We choose with $l > 0$

$$a(t) = (1+t)^l,\ A(t) = \frac{1}{l+1}(1+t)^{l+1},\ b(t) = \frac{\mu(l+1)}{1+t}.$$

These coefficients satisfy the assumptions of Theorem 2.1. Taking into consideration $\lambda(t) = (1+t)^{\frac{\mu(l+1)}{2}}$ we may conclude

$$\big\|\big((1+t)^l\nabla u(t,\cdot), u_t(t,\cdot)\big)\big\|_{L^2} \lesssim (1+t)^{\frac{l}{2}-\frac{\mu(l+1)}{2}}\big(\|u_1\|_{H^1} + \|u_2\|_{L^2}\big).$$

Example. Let $\mu \in (0,1)$. We choose

$$a(t) = e^t,\ A(t) = e^t,\ b(t) = \mu.$$

These coefficients satisfy the assumptions of Theorem 2.1. We have $\lambda(t) = e^{\frac{\mu}{2}t}$. So, we may conclude

$$\big\|\big(e^t\nabla u(t,\cdot), u_t(t,\cdot)\big)\big\|_{L^2} \lesssim e^{\frac{t}{2}-\frac{\mu}{2}t}\big(\|u_1\|_{H^1} + \|u_2\|_{L^2}\big).$$

Example. Let $\mu > 0$ and $m \geq 1$. We choose

$$a(t) = (e^{[m]}+t)^l,\ A(t) = \frac{1}{l+1}(e^{[m]}+t)^{l+1},$$

$$\text{and }\mu(t) = \frac{\mu}{(l+1)\log(e^{[m]}+t)\cdots\log^{[m]}(e^{[m]}+t)}.$$

These coefficients satisfy the assumptions of Theorem 2.1.

We have $\lambda(t) = \big(\log^{[m]}(e^{[m]}+t)\big)^{\frac{\mu}{2}}$. So, we may conclude

$$\big\|\big((e^{[m]}+t)^l\nabla u(t,\cdot), u_t(t,\cdot)\big)\big\|_{L^2} \lesssim \frac{(e^{[m]}+t)^{\frac{l}{2}}}{\big(\log^{[m]}(e^{[m]}+t)\big)^{\frac{\mu}{2}}}\big(\|u_1\|_{H^1} + \|u_2\|_{L^2}\big).$$

3. Effective dissipation

We consider the following Cauchy problem

$$u_{tt} - a^2(t)\Delta u + b(t)u_t = 0,\ \ u(0,x) = u_1(x),\ \ u_t(0,x) = u_2(x). \tag{3.1}$$

In the previous section we have concerned with the influence of the dissipation term $b(t)u_t$ for a given speed of propagation $a(t)$ such that the equation (3.1) is from the point of view of long time behavior of solutions and its energies in some sense close to the *wave equation with increasing speed of propagation*. Now our question is the following:
Under which assumptions to the coefficient $b = b(t)$ for a given time-dependent speed of propagation $a = a(t)$ can we call b an effective dissipation?
Here effective means, that on the one hand we have really a dissipation effect (overdamping is excluded), but on the other hand the model is parabolic like from the point of view of decay estimates for the wave type energy.
We will apply a transformation of the damped wave equation from (3.1) to a wave equation with time-dependent speed of propagation and potential. Thus, we define the new function

$$v(t,x) := \exp\Big(\frac{1}{2}\int_0^t b(\tau)d\tau\Big)u(t,x).$$

After some calculations we get

$$v_{tt} - a^2(t)\Delta v - \Big(\frac{1}{4}b^2(t) + \frac{1}{2}b'(t)\Big)v = 0.$$

Applying partial Fourier transformation we have

$$\hat{v}_{tt} + m(t,\xi)\hat{v} = 0, \tag{3.2}$$

here

$$m(t,\xi) = a^2(t)|\xi|^2 - \frac{1}{4}b^2(t) - \frac{1}{2}b'(t). \tag{3.3}$$

To study the interacting between $a(t)$ and $b(t)$ we assume:

(B′1) $b(t) > 0$, $b(t) = \mu(t)\dfrac{a(t)}{A(t)}$, $\dfrac{a^2(t)}{b(t)} \notin L^1(\mathbb{R}^+)$,

(B′2) $\big|d_t^k\mu(t)\big| \le C_k\mu(t)\left(\dfrac{a(t)}{A(t)}\right)^k$ for $k = 1,2$,

(B′3) $\mu(t) \to \infty$ as $t \to \infty$, $\dfrac{\mu(t)}{A(t)}$ is monotonous.

Using assumption (B′1) we can rewrite the formula (3.3) by

$$m(t,\xi) = a^2(t)|\xi|^2 - \frac{1}{4}\mu^2(t)\frac{a^2(t)}{A^2(t)} - \frac{1}{2}\Big(\mu(t)\frac{a(t)}{A(t)}\Big)'.$$

Assumptions (B′2) and (B'3) show that $b'(t)$ is a negligible term in comparison with $b^2(t)$, this means $|b'(t)| = o(b^2(t))$ as $t \to \infty$.

We introduce the auxiliary symbol

$$\langle\xi\rangle_{b(t)} := \sqrt{\Big|a^2(t)|\xi|^2 - \frac{b^2(t)}{4}\Big|} = \sqrt{\Big|a^2(t)|\xi|^2 - \frac{\mu^2(t)}{4}\frac{a^2(t)}{A^2(t)}\Big|}. \tag{3.4}$$

The main result of this section is the following statement:

Theorem 3.1. *Let us assume the conditions* (A1) *to* (A3) *and* (B′1) *to* (B′3). *Then we have the following* L^2-L^2 *estimates:*
For the kinetic energy we have

$$\|u_t(t,\cdot)\|_{L^2} \lesssim a(t)\Big(1+\int_0^t \frac{a^2(\tau)}{b(\tau)}d\tau\Big)^{-\frac{1}{2}}\big(\|u_1\|_{H^1}+\|u_2\|_{L^2}\big).$$

For the "elastic" energy we have

$$\|a(t)\nabla u(t,\cdot)\|_{L^2} \lesssim a(t)\Big(1+\int_0^t \frac{a^2(\tau)}{b(\tau)}d\tau\Big)^{-\frac{1}{2}}\big(\|u_1\|_{H^1}+\|u_2\|_{L^2}\big).$$

3.1. Regions and zones

We define the separating curve $t_\xi = t(|\xi|)$ by

$$\Gamma = \Big\{(t,\xi) : |\xi| = \frac{1}{2}\frac{\mu(t)}{A(t)}\Big\},$$

and introduce the following regions in the extended phase space $(0,\infty)\times\mathbb{R}^n_\xi$:

the hyperbolic region: $\Pi_{\text{hyp}} = \Big\{(t,\xi) : |\xi| > \frac{1}{2}\frac{\mu(t)}{A(t)}\Big\}$,

the elliptic region: $\Pi_{\text{ell}} = \Big\{(t,\xi) : |\xi| < \frac{1}{2}\frac{\mu(t)}{A(t)}\Big\}$.

The auxiliary symbol $\langle\xi\rangle_{b(t)}$ is differentiable in these regions and satisfies

$$\partial_t\langle\xi\rangle_{b(t)} = \pm\frac{a'(t)a(t)|\xi|^2 - \frac{\mu(t)a(t)}{2A(t)}\Big(\frac{\mu(t)a(t)}{2A(t)}\Big)'}{\langle\xi\rangle_{b(t)}}, \quad \partial_{|\xi|}\langle\xi\rangle_{b(t)} = \pm\frac{a^2(t)|\xi|}{\langle\xi\rangle_{b(t)}}, \tag{3.5}$$

where the upper sign is taken in the hyperbolic region.

We will also divide both regions of the extended phase space into zones. For this reason we define

the hyperbolic zone:

$$Z_{\text{hyp}}(N) = \Big\{(t,\xi) : \langle\xi\rangle_{b(t)} \ge N\mu(t)\frac{a(t)}{2A(t)}\Big\}\cap\Pi_{\text{hyp}},$$

the pseudo-differential zone:

$$Z_{\text{pd}}(N,\varepsilon) = \Big\{(t,\xi) : \varepsilon\mu(t)\frac{a(t)}{2A(t)} \le \langle\xi\rangle_{b(t)} \le N\mu(t)\frac{a(t)}{2A(t)}\Big\}\cap\Pi_{\text{hyp}},$$

the dissipative zone:

$$Z_{\text{diss}}(c_0) = \Big\{(t,\xi) : |\xi| \le c_0\frac{1}{A(t)}\Big\}\cap\Pi_{\text{ell}},$$

the elliptic zone:

$$Z_{\text{ell}}(c_0,\varepsilon)=\Big\{(t,\xi):|\xi|\geq c_0\frac{1}{A(t)}\Big\}\cap\Big\{\langle\xi\rangle_{b(t)}\geq\varepsilon\mu(t)\frac{a(t)}{2A(t)}\Big\}\cap\Pi_{\text{ell}},$$

the reduced zone:

$$Z_{\text{red}}(\varepsilon)=\Big\{(t,\xi):\langle\xi\rangle_{b(t)}\leq\varepsilon\mu(t)\frac{a(t)}{2A(t)}\Big\}.$$

Remark 3.2. The dissipative zone can be skipped if we assume the further assumption

$$\text{(S1)}\qquad \frac{a^2(t)}{b(t)A^2(t)}=\frac{a(t)}{\mu(t)A(t)}\in L^1(\mathbb{R}^+).$$

Under this additional assumption we define $Z_{\text{ell}}(\varepsilon):=Z_{\text{ell}}(0,\varepsilon)$.

3.2. The hyperbolic region

3.2.1. Symbols in Π_{hyp}.

Definition 3.3. Let us define the following classes of symbols in the hyperbolic zone:

$$S_l\{m_1,m_2,m_3\}=\Big\{c(t,\xi):|D_\xi^\alpha D_t^k c(t,\xi)|\leq C_{\alpha,k}\langle\xi\rangle_{b(t)}^{m_1-|\alpha|}a(t)^{m_2+|\alpha|}\Big(\frac{a(t)}{A(t)}\Big)^{m_3+k}\quad\text{for all }(t,\xi)\in Z_{\text{hyp}}(N),\ \alpha,\ \text{ and }k\leq l\Big\}.$$

Lemma 3.4. *The family of symbol classes $S_l\{m_1,m_2,m_3\}$ generates a hierarchy having the following properties:*

- $S_l\{m_1,m_2,m_3\}$ *is a vector space,*
- $S_l\{m_1,m_2,m_3\}S_l\{m_1',m_2',m_3'\}\subset S_l\{m_1+m_1',m_2+m_2',m_3+m_3'\}$,
- $D_t^kD_\xi^\alpha S_l\{m_1,m_2,m_3\}\subset S_{l-k}\{m_1-|\alpha|,m_2+|\alpha|,m_3+k\}$,
- $S_0\{-1,0,2\}\subset L_\xi^\infty L_t^1(Z_{\text{hyp}}(N))$.

Proof. We only verify the fourth property. Indeed, if $c=c(t,\xi)\in S_0\{-1,0,2\}$, then

$$\int_{t_\xi}^\infty|c(\tau,\xi)|d\tau\lesssim\int_{t_\xi}^\infty\frac{a^2(\tau)d\tau}{\langle\xi\rangle_{b(\tau)}A^2(\tau)}\sim\int_{t_\xi}^\infty\frac{a(\tau)d\tau}{|\xi|A^2(\tau)}\leq\frac{C}{A(t_\xi)|\xi|}\leq\frac{C}{N\mu(t_\xi)}<\infty$$

due to the definition of the hyperbolic zone and assumption (B′3). Remark, that here we used

$$\langle\xi\rangle_{b(t)}\sim a(t)|\xi|\quad\text{uniformly on }Z_{\text{hyp}}(N)\tag{3.6}$$

to conclude what we wanted to have. □

3.2.2. Consideration in the hyperbolic zone. Now we consider the micro-energy

$$V = (\langle\xi\rangle_{b(t)}\hat{v}, D_t\hat{v})^T. \tag{3.7}$$

Then it holds

$$D_tV = \begin{pmatrix} 0 & \langle\xi\rangle_{b(t)} \\ \langle\xi\rangle_{b(t)} & 0 \end{pmatrix} V + \begin{pmatrix} \frac{D_t\langle\xi\rangle_{b(t)}}{\langle\xi\rangle_{b(t)}} & 0 \\ -\frac{\left(\mu(t)\frac{a(t)}{A(t)}\right)'}{2\langle\xi\rangle_{b(t)}} & 0 \end{pmatrix} V. \tag{3.8}$$

Lemma 3.5. *Let us assume the conditions* (B′1), (B′2) *and* (B′3). *Then the following estimate holds for the fundamental solution* $E_V(t,s,\xi)$ *with* $(s,\xi),t,\xi) \in Z_{\text{hyp}}(N), s \leq t$:

$$|E_V(t,s,\xi)| \lesssim \frac{\sqrt{a(t)}}{\sqrt{a(s)}}.$$

Proof. Let us carry out the first step of diagonalization. For this reason we set

$$M = \begin{pmatrix} 1 & -1 \\ 1 & 1 \end{pmatrix}, \quad M^{-1} = \frac{1}{2}\begin{pmatrix} 1 & 1 \\ -1 & 1 \end{pmatrix}, \quad \text{and } V^{(0)} := M^{-1}V.$$

Hence,

$$D_tV^{(0)} = \mathcal{D}(t,\xi)V^{(0)} + \mathcal{R}(t,\xi)V^{(0)}, \tag{3.9}$$

where

$$\mathcal{D}(t,\xi) = \begin{pmatrix} \langle\xi\rangle_{b(t)} & 0 \\ 0 & -\langle\xi\rangle_{b(t)} \end{pmatrix} \in S_2\{1,0,0\}, \tag{3.10}$$

$$\mathcal{R}(t,\xi) = \begin{pmatrix} \frac{D_t\langle\xi\rangle_{b(t)}}{2\langle\xi\rangle_{b(t)}} - \frac{b'(t)}{4\langle\xi\rangle_{b(t)}} & -\frac{D_t\langle\xi\rangle_{b(t)}}{2\langle\xi\rangle_{b(t)}} + \frac{b'(t)}{4\langle\xi\rangle_{b(t)}} \\ -\frac{D_t\langle\xi\rangle_{b(t)}}{2\langle\xi\rangle_{b(t)}} - \frac{b'(t)}{4\langle\xi\rangle_{b(t)}} & \frac{D_t\langle\xi\rangle_{b(t)}}{2\langle\xi\rangle_{b(t)}} + \frac{b'(t)}{4\langle\xi\rangle_{b(t)}} \end{pmatrix} \in S_1\{0,0,1\}. \tag{3.11}$$

Let $F_0(t,\xi)$ be the diagonal part of $\mathcal{R}(t,\xi)$. Now we carry out the second step of diagonalization procedure. Therefore we introduce the matrices

$$\begin{aligned} N^{(1)} &= \begin{pmatrix} 0 & \frac{\mathcal{R}_{12}}{\tau_1-\tau_2} \\ \frac{\mathcal{R}_{21}}{\tau_2-\tau_1} & 0 \end{pmatrix} \\ &= \begin{pmatrix} 0 & -\frac{D_t\langle\xi\rangle_{b(t)}}{4\langle\xi\rangle^2_{b(t)}} - \frac{b'(t)}{8\langle\xi\rangle^2_{b(t)}} \\ -\frac{D_t\langle\xi\rangle_{b(t)}}{4\langle\xi\rangle^2_{b(t)}} + \frac{b'(t)}{8\langle\xi\rangle^2_{b(t)}} & 0 \end{pmatrix} \in S_1\{-1,0,1\}, \end{aligned}$$

$N_1(t,\xi) = I + N^{(1)}(t,\xi) \in S_1\{0,0,0\}$. For sufficiently large time $t_0 = t_0(\varepsilon)$ the matrix $N_1(t,\xi)$ is invertible with uniformly bounded inverse $N_1^{-1}(t,\xi)$ for $t \geq t_0$ in $Z_{\text{hyp}}(N)$ (see Remark 3.6). Now we can follow the usual procedure to diagonalize. Let

$$\begin{aligned} B^{(1)}(t,\xi) &= D_tN^{(1)}(t,\xi) - \big(\mathcal{R}(t,\xi) - F_0(t,\xi)\big)N^{(1)}(t,\xi) \in S_0\{-1,0,2\}, \\ \mathcal{R}_1(t,\xi) &= -N_1^{-1}(t,\xi)B^{(1)}(t,\xi) \in S_0\{-1,0,2\}. \end{aligned}$$

Then we can conclude

$$\big(D_t-\mathcal{D}(t,\xi)-\mathcal{R}(t,\xi)\big)N_1(t,\xi)V^{(1)}=N_1(t,\xi)\big(D_t-\mathcal{D}(t,\xi)-F_0(t,\xi)-\mathcal{R}_1(t,\xi)\big)V^{(1)}.$$

Now we shall find the solution $V^{(0)}(t,\xi)=:N_1(t,\xi)V^{(1)}(t,\xi)$, where $V^{(1)}(t,\xi)$ is the solution to the system

$$\big(D_t-\mathcal{D}(t,\xi)-F_0(t,\xi)-\mathcal{R}_1(t,\xi)\big)V^{(1)}(t,\xi)=0.$$

We can write $V^{(1)}(t,\xi)=E_V(t,t_\xi,\xi)V^{(1)}(t_\xi,\xi)$. Here $E_V(t,s,\xi)$ is the fundamental solution to the following system:

$$\big(D_t-\mathcal{D}(t,\xi)-F_0(t,\xi)-\mathcal{R}_1(t,\xi)\big)E_V(t,s,\xi)=0,\quad E_V(s,s,\xi)=I,\ t\ge s\ge t_\xi.$$

The solution $E_0=E_0(t,s,\xi)$ of the "principal diagonal part" of this system fulfils

$$D_tE_0(t,s,\xi)=\big(\mathcal{D}(t,\xi)+F_0(t,\xi)\big)E_0(t,s,\xi),\quad E_0(s,s,\xi)=I,\ t\ge s\ge t_\xi.$$

Thus

$$E_0(t,s,\xi)=\exp\left(i\int_s^t\big(\mathcal{D}(\tau,\xi)+F_0(\tau,\xi)\big)d\tau\right),$$

and we can get

$$|E_0(t,s,\xi)|\lesssim\exp\left(\int_s^t\frac{\partial_t\langle\xi\rangle_{b(\tau)}}{2\langle\xi\rangle_{b(\tau)}}d\tau\right)=\frac{\sqrt{\langle\xi\rangle_{b(t)}}}{\sqrt{\langle\xi\rangle_{b(s)}}}\sim\frac{\sqrt{a(t)|\xi|}}{\sqrt{a(s)|\xi|}}\sim\frac{\sqrt{a(t)}}{\sqrt{a(s)}}.$$

Let us set

$$\mathcal{R}_2(t,s,\xi)=E_0^{-1}(t,s,\xi)\mathcal{R}_1(t,\xi)E_0(t,s,\xi),$$

$$Q(t,s,\xi)=I+\sum_{k=1}^{\infty}\int_s^t\mathcal{R}_2(t_1,s,\xi)\int_s^{t_1}\mathcal{R}_2(t_2,s,\xi)\cdots\int_s^{t_{k-1}}\mathcal{R}_2(t_k,s,\xi)dt_k\cdots dt_2dt_1.$$

Then $Q=Q(t,s,\xi)$ solves the Cauchy problem

$$D_tQ(t,s,\xi)=\mathcal{R}_2(t,s,\xi)Q(t,s,\xi),\quad Q(s,s,\xi)=I,\ t\ge s\ge t_\xi.$$

The fundamental solution $E_V(t,s,\xi)$ is representable in the form $E_V(t,s,\xi)=E_0(t,s,\xi)Q(t,s,\xi)$. Furthermore, applying the fourth property from Lemma 3.4 to $\mathcal{R}_1\in S_0\{-1,0,2\}\subset L^\infty_\xi L^1_t(Z_{\text{hyp}})$ we see that

$$|Q(t,s,\xi)|\le\exp\left(\int_s^t|\mathcal{R}_1(\tau,\xi)|d\tau\right)\le C_N.$$

This completes the proof. □

Remark 3.6. The large constant N guarantees the invertibility in the whole hyperbolic zone. The remaining problem consists in the understanding of invertibility in the pseudo-differential zone. For $t\ge t_0(\varepsilon)$ this zone can be skipped after the choice $N=\varepsilon$. The other set $\{t\in(0,t_0(\varepsilon)]:(t,\xi)\in Z_{\text{pd}}(N,\varepsilon)\}$ is compact.

3.3. The elliptic region

3.3.1. Symbols in Π_{ell}. The symbols in the elliptic zone are constructed in a similar manner as in the hyperbolic zone with a little change for the auxiliary symbol $\langle\xi\rangle_{b(t)}$ which can be estimated by

$$\langle\xi\rangle_{b(t)} \sim \frac{b(t)}{2} \sim \mu(t)\frac{a(t)}{2A(t)} \qquad \text{uniformly on } Z_{\mathrm{ell}}(c_0,\varepsilon). \tag{3.12}$$

Definition 3.7. Let us define the following classes of symbols in the elliptic zone:

$$S_l\{m_1,m_2,m_3\} = \Big\{c = c(t,\xi) : |D_\xi^\alpha D_t^k c(t,\xi)| \leq C_{\alpha,k}\langle\xi\rangle_{b(t)}^{m_1-|\alpha|} a(t)^{m_2+|\alpha|}$$
$$\Big(\frac{a(t)}{A(t)}\Big)^{m_3+k} \qquad \text{for all } (t,\xi)\in Z_{\mathrm{ell}}(c_0,\varepsilon),\ \alpha \ \text{ and } k \leq l\Big\}.$$

Lemma 3.8. *The family of symbol classes $S_l\{m_1,m_2,m_3\}$ generates a hierarchy having the following properties:*

- $S_l\{m_1,m_2,m_3\}$ *is a vector space,*
- $S_l\{m_1,m_2,m_3\}S_l\{m_1',m_2',m_3'\} \subset S_l\{m_1+m_1',m_2+m_2',m_3+m_3'\}$,
- $D_t^k D_\xi^\alpha S_l\{m_1,m_2,m_3\} \subset S_{l-k}\{m_1-|\alpha|,m_2+|\alpha|,m_3+k\}$,
- $S_0\{-1,0,2\} \subset L_\xi^\infty L_t^1(Z_{\mathrm{ell}}(c_0,\varepsilon))$.

Proof. We only verify the fourth property. Indeed, if $c\in S_0\{-1,0,2\}$, then

$$\int_{t_{\xi_1}}^{t_{\xi_2}} |c(\tau,\xi)|d\tau \lesssim \int_{t_{\xi_1}}^{t_{\xi_2}} \frac{a^2(\tau)}{\langle\xi\rangle_{b(t)}A^2(\tau)}d\tau \sim \int_{t_{\xi_1}}^{t_{\xi_2}} \frac{a(\tau)}{\mu(\tau)A(\tau)}d\tau$$
$$\lesssim \int_{t_{\xi_1}}^{t_{\xi_2}} \frac{\sqrt{1-\varepsilon^2}a(\tau)}{|\xi|A^2(\tau)}d\tau \lesssim \frac{1}{|\xi|A(t_{\xi_1})} \lesssim C(\varepsilon,c_0),$$

where t_{ξ_1}, t_{ξ_2} denotes the lower, upper boundary of the elliptic zone, respectively. From the definitions of the elliptic zone and dissipative zone we have $\mu(t) \geq \dfrac{2|\xi|A(t)}{\sqrt{1-\varepsilon^2}}$ for all $t\in[t_{\xi_1},t_{\xi_2}]$ and $|\xi|A(t_{\xi_1})\sim 1$. □

3.3.2. Consideration in the elliptic zone. In this region we introduce again the micro-energy

$$V = (\langle\xi\rangle_{b(t)}\hat{v}, D_t\hat{v})^T.$$

Then we can get the system of differential equations

$$D_tV = \left(\begin{pmatrix} 0 & \langle\xi\rangle_{b(t)} \\ -\langle\xi\rangle_{b(t)} & 0 \end{pmatrix} + \begin{pmatrix} \frac{D_t\langle\xi\rangle_{b(t)}}{\langle\xi\rangle_{b(t)}} & 0 \\ -\frac{b'(t)}{2\langle\xi\rangle_{b(t)}} & 0 \end{pmatrix}\right)V. \tag{3.13}$$

In a first step we use the diagonalizer of the first matrix which is defined as follows:

$$M = \begin{pmatrix} i & -i \\ 1 & 1 \end{pmatrix}, \quad M^{-1} = \frac{1}{2}\begin{pmatrix} -i & 1 \\ i & 1 \end{pmatrix}, \quad \text{and } V^{(0)} := M^{-1}V.$$

Hence,

$$D_tV^{(0)} = \mathcal{D}(t,\xi)V^{(0)} + \mathcal{R}(t,\xi)V^{(0)}, \tag{3.14}$$

where

$$\mathcal{D}(t,\xi) = \begin{pmatrix} -i\langle\xi\rangle_{b(t)} & 0 \\ 0 & i\langle\xi\rangle_{b(t)} \end{pmatrix} \in S_2\{1,0,0\}, \tag{3.15}$$

$$\mathcal{R}(t,\xi) = \frac{1}{2}\begin{pmatrix} \frac{D_t\langle\xi\rangle_{b(t)}}{2\langle\xi\rangle_{b(t)}} - i\frac{b'(t)}{4\langle\xi\rangle_{b(t)}} & -\frac{D_t\langle\xi\rangle_{b(t)}}{2\langle\xi\rangle_{b(t)}} + i\frac{b'(t)}{4\langle\xi\rangle_{b(t)}} \\ -\frac{D_t\langle\xi\rangle_{b(t)}}{2\langle\xi\rangle_{b(t)}} - i\frac{b'(t)}{4\langle\xi\rangle_{b(t)}} & \frac{D_t\langle\xi\rangle_{b(t)}}{2\langle\xi\rangle_{b(t)}} + i\frac{b'(t)}{4\langle\xi\rangle_{b(t)}} \end{pmatrix} \in S_1\{0,0,1\}. \tag{3.16}$$

Let $F_0 = F_0(t,\xi)$ be the diagonal part of $\mathcal{R} = \mathcal{R}(t,\xi)$. Now we carry out the second step of diagonalization procedure. Thus, we consider the difference δ of the entries of $\mathcal{D}(t,\xi) + F_0(t,\xi)$. We have

$$i\delta(t,\xi) = 2\langle\xi\rangle_{b(t)} + \frac{b'(t)}{2\langle\xi\rangle_{b(t)}} \sim 2\langle\xi\rangle_{b(t)} + \frac{b^2(t)}{2\langle\xi\rangle_{b(t)}} \sim \langle\xi\rangle_{b(t)} \in S_2(1,0,0) \tag{3.17}$$

for $t \geq t_0$ with a sufficiently large $t_0 = t_0(\varepsilon)$ by using $|b'(t)| = o(b^2(t))$. Now we can follow the usual procedure of diagonalization. Therefore, we introduce the matrices

$$N^{(1)} = \begin{pmatrix} 0 & -\frac{\mathcal{R}_{12}}{\delta} \\ \frac{\mathcal{R}_{21}}{\delta} & 0 \end{pmatrix} \sim \begin{pmatrix} 0 & i\frac{D_t\langle\xi\rangle_{b(t)}}{4\langle\xi\rangle^2_{b(t)}} - \frac{b'(t)}{8\langle\xi\rangle^2_{b(t)}} \\ i\frac{D_t\langle\xi\rangle_{b(t)}}{4\langle\xi\rangle^2_{b(t)}} + \frac{b'(t)}{8\langle\xi\rangle^2_{b(t)}} & 0 \end{pmatrix} \in S_1\{-1,0,1\},$$

$N_1(t,\xi) = I + N^{(1)}(t,\xi) \in S_1\{0,0,0\}$. For a sufficiently large time $t \geq t_0$ the matrix $N_1(t,\xi)$ is invertible with uniformly bounded inverse $N_1^{-1}(t,\xi)$ in $Z_{\rm ell}(c_0,\varepsilon)$. Let

$$B^{(1)}(t,\xi) = D_t N^{(1)}(t,\xi) - \big(\mathcal{R}(t,\xi) - F_0(t,\xi)\big)N^{(1)}(t,\xi) \in S_0\{-1,0,2\},$$
$$\mathcal{R}_1(t,\xi) = -N_1^{-1}(t,\xi)B^{(1)}(t,\xi) \in S_0\{-1,0,2\}.$$

We can conclude that

$$\big(D_t - \mathcal{D}(t,\xi) - \mathcal{R}(t,\xi)\big)N_1(t,\xi)V^{(1)} = N_1(t,\xi)\big(D_t - \mathcal{D}(t,\xi) - F_0(t,\xi) - \mathcal{R}_1(t,\xi)\big)V^{(1)}.$$

Now we shall find the solution $V^{(1)}(t,\xi) := N_1^{-1}(t,\xi)V^{(0)}(t,\xi)$, where $V^{(1)}(t,\xi)$ is the solution to the system

$$\big(D_t - \mathcal{D}(t,\xi) - F_0(t,\xi) - \mathcal{R}_1(t,\xi)\big)V^{(1)}(t,\xi) = 0.$$

We can write $V^{(1)}(t,\xi) = E_{V,1}(t,s,\xi)V^{(1)}(s,\xi)$. Here $E_{V,1}(t,s,\xi)$ is the fundamental solution to the following system:

$$\big(D_t - \mathcal{D}(t,\xi) - F_0(t,\xi) - \mathcal{R}_1(t,\xi)\big)E_{V,1}(t,s,\xi) = 0, \quad E_{V,1}(s,s,\xi) = I,\ t \geq s \geq t_\xi.$$

We transform the system for $E_{V,1}(t,s,\xi)$ to an integral equation for a new matrix-valued function $Q_{\rm ell}(t,s,\xi)$ by considering

$$\exp\left(-i\int_s^t \big(\mathcal{D}(\tau,\xi) + F_0(\tau,\xi)\big)d\tau\right)E_{V,1}(t,s,\xi).$$

Using this ansatz we have after differentiation

$$D_t\Big(\exp\Big(-i\int_s^t (\mathcal{D}(\tau,\xi)+F_0(\tau,\xi))d\tau\Big)E_{V,1}(t,s,\xi)\Big)$$

$$= -(\mathcal{D}(t,\xi)+F_0(t,\xi))\exp\Big(-i\int_s^t (\mathcal{D}(\tau,\xi)+F_0(\tau,\xi))d\tau\Big)E_{V,1}(t,s,\xi)$$

$$+\exp\Big(-i\int_s^t (\mathcal{D}(\tau,\xi)+F_0(\tau,\xi))d\tau\Big)(\mathcal{D}(t,\xi)+F_0(t,\xi)+\mathcal{R}_1(t,\xi))E_{V,1}(t,s,\xi)$$

$$= \exp\Big(-i\int_s^t (\mathcal{D}(\tau,\xi)+F_0(\tau,\xi))d\tau\Big)\mathcal{R}_1(t,\xi)E_{V,1}(t,s,\xi).$$

Consequently,

$$E_{V,1}(t,s,\xi) = \exp\Big(i\int_s^t (\mathcal{D}(\tau,\xi)+F_0(\tau,\xi))d\tau\Big)E_{V,1}(s,s,\xi)$$
$$+\,i\int_s^t \exp\Big(i\int_\theta^t (\mathcal{D}(\tau,\xi)+F_0(\tau,\xi))d\tau\Big)\mathcal{R}_1(\theta,\xi)E_{V,1}(\theta,s,\xi)d\theta.$$

We introduce an unknown weight factor to represent $Q_{\mathrm{ell},1}$ in the following way:

$$Q_{\mathrm{ell},1}(t,s,\xi) = \exp\Big(-\int_s^t w(\tau,\xi)d\tau\Big)E_{V,1}(t,s,\xi).$$

Then we get

$$Q_{\mathrm{ell},1}(t,s,\xi) = \exp\Big(\int_s^t (i\mathcal{D}(\tau,\xi)+iF_0(\tau,\xi)-w(\tau,\xi)I)d\tau\Big)$$
$$+\int_s^t \exp\Big(\int_\theta^t (i\mathcal{D}(\tau,\xi)+iF_0(\tau,\xi)-w(\tau,\xi)I)d\tau\Big)\mathcal{R}_1(\theta,\xi)Q_{\mathrm{ell},1}(\theta,s,\xi)d\theta.$$

The main entries of the diagonal matrix $i\mathcal{D}(t,\xi)+iF_0(t,\xi)$ are given by

$$(\mathrm{I}) = \langle\xi\rangle_{b(t)} + \frac{\partial_t\langle\xi\rangle_{b(t)}}{2\langle\xi\rangle_{b(t)}} + \frac{b'(t)}{4\langle\xi\rangle_{b(t)}},\quad (\mathrm{II}) = -\langle\xi\rangle_{b(t)} + \frac{\partial_t\langle\xi\rangle_{b(t)}}{2\langle\xi\rangle_{b(t)}} - \frac{b'(t)}{4\langle\xi\rangle_{b(t)}}.$$

For the difference (II)−(I) we get

$$(\mathrm{II})-(\mathrm{I}) = -2\langle\xi\rangle_{b(t)} - \frac{b'(t)}{2\langle\xi\rangle_{b(t)}} = -\frac{b^2(t)+b'(t)-4a^2(t)|\xi|^2}{2\langle\xi\rangle_{b(t)}} \le 0$$

in $Z_{\mathrm{ell}}(c_0,\varepsilon)$ for $t \ge t_0$ by using $|b'(t)| = o(b^2(t))$. We are choosing the weight $w(t,\xi) = (I)$. By this choice the matrix

$$H(t,s,\xi) = \exp\Big(\int_s^t (i\mathcal{D}(\tau,\xi)+iF_0(\tau,\xi)-w(\tau,\xi)I)d\tau\Big)$$
$$= \mathrm{diag}\Big(1,\exp\Big(\int_s^t\Big(-2\langle\xi\rangle_{b(\tau)} - \frac{b'(\tau)}{2\langle\xi\rangle_{b(\tau)}}\Big)d\tau\Big)\Big) \to \begin{pmatrix} 1 & 0 \\ 0 & 0 \end{pmatrix}$$

as $t \to \infty$ with a fixed s. Hence, the matrix $H(t,s,\xi)$ is uniformly bounded for (s,ξ), $(t,\xi) \in Z_{\mathrm{ell}}(c_0,\varepsilon)$. Taking account of $\mathcal{R}_1 \in S_0\{-1,0,2\}$ is uniformly integrable over the elliptic zone and that the matrix which can be represented by

$$Q_{\mathrm{ell},1}(t,s,\xi) = H(t,s,\xi) + \sum_{k=1}^{\infty} i^k \int_s^t H(t,t_1,\xi)\mathcal{R}_1(t_1,s,\xi)$$
$$\times \int_s^{t_1} H(t_1,t_2,\xi)\mathcal{R}_1(t_2,s,\xi) \cdots \int_s^{t_{k-1}} H(t_{k-1},t_k,\xi)\mathcal{R}_1(t_k,s,\xi)dt_k \cdots dt_2 dt_1$$

is uniformly bounded in $Z_{\mathrm{ell}}(c_0,\epsilon)$ from the last considerations we may conclude

$$\begin{aligned} E_{V,1}(t,s,\xi) &= \exp\Big(\int_s^t w(\tau,\xi)d\tau\Big) Q_{\mathrm{ell},1}(t,s,\xi) \\ &= \exp\Big(\int_s^t \Big(\langle\xi\rangle_{b(\tau)} + \frac{\partial\langle\xi\rangle_{b(\tau)}}{2\langle\xi\rangle_{b(\tau)}} + \frac{b'(\tau)}{4\langle\xi\rangle_{b(\tau)}}\Big)d\tau\Big) Q_{\mathrm{ell},1}(t,s,\xi) \\ &\sim \exp\Big(\int_s^t \Big(\langle\xi\rangle_{b(\tau)} + \frac{\partial\langle\xi\rangle_{b(\tau)}}{2\langle\xi\rangle_{b(\tau)}} + \frac{b'(\tau)}{2b(\tau)}\Big)d\tau\Big) Q_{\mathrm{ell},1}(t,s,\xi) \\ &\sim \frac{\langle\xi\rangle_{b(t)}}{\langle\xi\rangle_{b(s)}} \exp\Big(\int_s^t \langle\xi\rangle_{b(\tau)}d\tau\Big) Q_{\mathrm{ell},1}(t,s,\xi). \end{aligned}$$

Summarizing the considerations of this section we have proved the following lemma:

Lemma 3.9. *Under the assumptions* (B′1), (B′2) *and* (B′3) *the fundamental solution* $E_V(t,s,\xi)$ *to the operator* $D_t - \mathcal{D}(t,\xi) - F_0(t,\xi) - \mathcal{R}_1(t,\xi)$, *with* $(t,\xi),(s,\xi) \in Z_{\mathrm{ell}}(c_0,\varepsilon) \cap \{t \geq t_0(\varepsilon)\}, s \leq t$ *can be represented as*

$$E_{V,1}(t,s,\xi) \sim \frac{\langle\xi\rangle_{b(t)}}{\langle\xi\rangle_{b(s)}} \exp\Big(\int_s^t \langle\xi\rangle_{b(\tau)}d\tau\Big) Q_{\mathrm{ell},1}(t,s,\xi).$$

3.4. The reduced zone

In this zone we can estimate $\langle\xi\rangle_{b(t)}$ by $\varepsilon\frac{b(t)}{2}$. Thus, we consider the micro-energy

$$V = \Big(\varepsilon\frac{b(t)}{2}\hat{v}, D_t\hat{v}\Big)^T. \tag{3.18}$$

We get the following system of first order

$$D_tV = \begin{pmatrix} \frac{D_tb(t)}{b(t)} & \varepsilon\frac{b(t)}{2} \\ \frac{a^2(t)|\xi|^2 - \frac{1}{4}b^2(t) - \frac{1}{2}b'(t)}{\varepsilon\frac{1}{2}b(t)} & \end{pmatrix} V. \tag{3.19}$$

To estimate the entries of this matrix we will use

- $|b'(t)| = o(b^2(t))$,
- $\langle\xi\rangle_{b(t)} \lesssim \varepsilon\frac{b(t)}{2}$,
- consequently, $\frac{a^2(t)|\xi|^2 - \frac{1}{4}b^2(t) - \frac{1}{2}b'(t)}{\varepsilon\frac{1}{2}b(t)} \lesssim \varepsilon\frac{b(t)}{2} - \frac{b'(t)}{\varepsilon b(t)} \lesssim \varepsilon b(t)$.

Thus, we can estimate the norm of the coefficient matrix by $\varepsilon b(t)$ for sufficiently large t.

Summarizing the following statement holds:

Lemma 3.10. *If we assume* (B′1) *to* (B′3), *then the fundamental solution* $E_V(t,s,\xi)$ *to* (3.19) *can be estimated by*

$$|E_V(t,s,\xi)| \lesssim \exp\Big(\varepsilon\int_s^t b(\tau)d\tau\Big)$$

for $t_0 \le s \le t$ *with sufficiently large* $t_0 = t_0(\varepsilon)$ *and* $(t,\xi),(s,\xi) \in Z_{\mathrm{red}}(\varepsilon)$.

Remark 3.11. We can make the reduced zone as small as we want by the control of the constant ε.

3.4.1. The dissipative zone. By Remark 3.2 the dissipative zone can be skipped if we assume $(S1)$. Now, let us assume, that the assumption $(S1)$ does not hold, i.e., $\dfrac{a(t)}{\mu(t)A(t)} \notin L^1(\mathbb{R}^+)$. Thus, we introduced the dissipative zone to ensure integrability of $S_0\{-1,0,2\}$ over the elliptic zone $Z_{\mathrm{ell}}(c_0,\epsilon)$. In this case we can apply directly Lemma 2.6 (because of assumption $\lim_{t\to\infty}\mu(t)\to\infty$) to estimate the fundamental solution $E(t,s,\xi)$ related to the micro-energy $U=(\frac{a(t)}{A(t)}\hat{u}, D_t\hat{u})^T$ and relate this estimate to the fundamental solution $E_V(t,s,\xi)$ related to $V=(\langle\xi\rangle_{b(t)}\hat{v}, D_t\hat{v})^T$.

3.5. Estimates for the fundamental solution

We want to obtain estimates for the energy of the solution to our original Cauchy problem. For this reason we need to transform back to estimate the fundamental solution $E(t,s,\xi)$ which is related to the micro-energy $(a(t)|\xi|\hat{u}, D_t\hat{u})$. Outside the reduced zone it holds

$$E(t,s,\xi) = T(t,\xi)E_V(t,s,\xi)T^{-1}(s,\xi), \tag{3.20}$$

where the matrix $T(t,\xi)$ is defined in the following way:

$$\begin{pmatrix} h(t,\xi)\hat{u} \\ D_t\hat{u}\end{pmatrix} = \underbrace{\begin{pmatrix} \frac{h(t,\xi)}{\lambda(t)\langle\xi\rangle_{b(t)}} & 0 \\ i\frac{b(t)}{2\lambda(t)\langle\xi\rangle_{b(t)}} & \frac{1}{\lambda(t)}\end{pmatrix}}_{T(t,\xi)} \begin{pmatrix} \langle\xi\rangle_{b(t)}\hat{v} \\ D_t\hat{v}\end{pmatrix} \tag{3.21}$$

with inverse

$$T^{-1}(t,\xi) = \begin{pmatrix} \frac{\lambda(t)\langle\xi\rangle_{b(t)}}{h(t,\xi)} & 0 \\ -i\frac{b(t)\lambda(t)}{2h(t,\xi)} & \lambda(t)\end{pmatrix}. \tag{3.22}$$

Recall that outside of the dissipative zone we have $h(t,\xi) = a(t)|\xi|$ and, especially, in the dissipative zone we use $h(t,\xi) = \dfrac{a(t)}{A(t)}$. Inside the reduced zone we have estimated $\langle\xi\rangle_{b(t)}$ by $\varepsilon\frac{b(t)}{2}$. Therefore, we change the definition of the matrix $T(t,\xi)$ by

$$\begin{pmatrix} \frac{2h(t,\xi)}{\varepsilon\lambda(t)b(t)} & 0 \\ i\frac{1}{\lambda(t)} & \frac{1}{\lambda(t)}\end{pmatrix}, \qquad |T(t,\xi)| \sim \lambda^{-1}(t) \tag{3.23}$$

for all $(t,\xi) \in Z_{\mathrm{red}}(\varepsilon)$.

Remark 3.12. We may conclude that in the hyperbolic and reduced zones the fundamental solution to our original Cauchy problem in the extended phase space can be estimated by

$$(|E(t,s,\xi)|) \lesssim \frac{\lambda(s)}{\lambda(t)}(|E_V(t,s,\xi)|).$$

Some auxiliary estimates. We continue with some auxiliary estimates which are used to obtain energy estimates later on.

Lemma 3.13. *Let us suppose* (B′1) *to* (B′3) *and let* $\lambda(t) = \exp\left(\frac{1}{2}\int_0^t b(\tau)d\tau\right)$. *Then the following holds:*

1. *The definition of* $\langle\xi\rangle_{b(t)}$ *in the elliptic zone implies* $\langle\xi\rangle_{b(t)} - \dfrac{b(t)}{2} \leq -\dfrac{a^2(t)|\xi|^2}{b(t)}$.
2. $\dfrac{\lambda(s)}{\lambda(t)} \exp\left(\int_s^t \langle\xi\rangle_{b(\tau)}d\tau\right) \lesssim \exp\left(-|\xi|^2 \int\limits_s^t \dfrac{a^2(\tau)}{b(\tau)}d\tau\right)$.
3. *With* $A(t_{\xi_1})|\xi| \sim 1$ *(separating line between dissipative and elliptic zone) it holds*
$$\exp\left(-|\xi|^2 \int_0^{t_{\xi_1}} \frac{a^2(\tau)}{b(\tau)}d\tau\right) \sim 1.$$

Proof. The first statement is equivalent to the following inequality

$$\sqrt{\frac{b^2(t)}{4} - a^2(t)|\xi|^2} - \frac{b(t)}{2} \leq -\frac{a^2(t)|\xi|^2}{b(t)}.$$

The second statement follows directly from the first one together with the definition of $\lambda(t)$. The third statement can be directly obtained from the following estimate:

$$\begin{aligned} |\xi|^2 \int_0^{t_{\xi_1}} \frac{a^2(\tau)}{b(\tau)}d\tau &= |\xi|^2 \int_0^{t_{\xi_1}} \frac{a(\tau)A(\tau)}{\mu(\tau)}d\tau \\ &\leq |\xi|^2 \int_0^{t_{\xi_1}} \frac{a(\tau)A(\tau)}{\mu_0}d\tau \lesssim |\xi|^2 A^2(t_{\xi_1}) \lesssim 1. \end{aligned}$$

The proof is complete. □

A refined estimate for the fundamental solution in the elliptic zone. Inside the elliptic zone we have

$$|E_V(t,s,\xi)| \lesssim \frac{b(t)}{b(s)} \exp\left(\int_s^t \langle\xi\rangle_{b(\tau)}d\tau\right).$$

This yields in combination with (3.20) the estimate

$$\begin{aligned} (|E(t,s,\xi)|) &\lesssim \begin{pmatrix} \frac{a(t)|\xi|}{b(t)} & 0 \\ & b(t) \end{pmatrix} \exp\left(\int_s^t \left(\langle\xi\rangle_{b(\tau)} - \frac{b(\tau)}{2}\right)d\tau\right) \begin{pmatrix} \frac{1}{a(s)|\xi|} & 0 \\ \frac{1}{a(s)|\xi|} & \frac{1}{b(s)} \end{pmatrix} \\ &\lesssim \exp\left(-|\xi|^2 \int_s^t \frac{a^2(\tau)}{b(\tau)}d\tau\right) \begin{pmatrix} \frac{a(t)}{a(s)} & \frac{a(t)|\xi|}{b(s)} \\ \frac{b(t)}{a(s)|\xi|} & \frac{b(t)}{b(s)} \end{pmatrix}. \end{aligned} \tag{3.24}$$

Here we have used the first statement from Lemma 3.13. The estimate for the first row seems to be optimal while the estimate for the second row is not optimal, because at least for an increasing coefficient function $b(t)$ for fixed ξ some entries of the matrix from the right-hand side of (3.24) become unbounded for increasing t. This estimate contradicts to our a priori knowledge that the wave type energy itself is decreasing. For this reason we need a refined estimate which will be presented in the following steps. Let us assume that $\Phi_k(t,s,\xi)$, $k=1,2$, are solutions to the equation

$$\Phi_{tt}+a^2(t)|\xi|^2\Phi+b(t)\Phi_t=0$$

with initial values

$$\Phi_k(s,s,\xi)=\delta_{1k}, \quad \partial_t\Phi_k(s,s,\xi)=\delta_{2k}.$$

Then we have

$$\begin{pmatrix} a(t)|\xi|v(t,\xi) \\ D_tv(t,\xi) \end{pmatrix}=\begin{pmatrix} \dfrac{a(t)}{a(s)}\Phi_1(t,s,\xi) & ia(t)|\xi|\Phi_2(t,s,\xi) \\ \dfrac{D_t\Phi_1(t,s,\xi)}{a(s)|\xi|} & iD_t\Phi_2(t,s,\xi) \end{pmatrix}\begin{pmatrix} a(s)|\xi|v(s,\xi) \\ D_tv(s,\xi) \end{pmatrix}.$$

Hence, if we compare with the estimate (3.24), then

$$|\Phi_1(t,s,\xi)|\lesssim \exp\Big(-|\xi|^2\int_s^t\frac{a^2(\tau)}{b(\tau)}d\tau\Big), \tag{3.25}$$

$$|\Phi_2(t,s,\xi)|\lesssim \frac{1}{b(s)}\exp\Big(-|\xi|^2\int_s^t\frac{a^2(\tau)}{b(\tau)}d\tau\Big), \tag{3.26}$$

$$|\partial_t\Phi_1(t,s,\xi)|\lesssim b(t)\exp\Big(-|\xi|^2\int_s^t\frac{a^2(\tau)}{b(\tau)}d\tau\Big), \tag{3.27}$$

$$|\partial_t\Phi_1(t,s,\xi)|\lesssim \frac{b(t)}{b(s)}\exp\Big(-|\xi|^2\int_s^t\frac{a^2(\tau)}{b(\tau)}d\tau\Big). \tag{3.28}$$

Let $\Psi_k(t,s,\xi)=\partial_t\Phi_k(t,s,\xi)$, $k=1,2$. Then we obtain the equations of first order

$$\partial_t\Psi_k+b(t)\Psi_k=-a^2(t)|\xi|^2\Phi_k(t,s,\xi), \qquad \Psi_k(s,s,\xi)=\delta_{k2}.$$

Applying Duhamel's principle we get

$$\begin{aligned}
\Psi_1(t,s,\xi)&=-|\xi|^2\int_s^t a^2(\tau)\frac{\lambda^2(\tau)}{\lambda^2(t)}\Phi_1(\tau,s,\xi)d\tau,\\
|\Psi_1(t,s,\xi)|&\lesssim\frac{|\xi|^2}{\lambda^2(t)}\int_s^t a^2(\tau)\lambda^2(\tau)\exp\Big(-|\xi|^2\int_s^\tau\frac{a^2(\theta)}{b(\theta)}d\theta\Big)d\tau\\
&\lesssim\frac{a^2(t)|\xi|^2}{\lambda^2(t)}\int_s^t \lambda^2(\tau)\exp\Big(-|\xi|^2\int_s^\tau\frac{a^2(\theta)}{b(\theta)}d\theta\Big)d\tau\\
&\lesssim\frac{a^2(t)|\xi|^2}{\lambda^2(t)}\int_s^t \underbrace{b(\tau)\lambda^2(\tau)}_{\partial_\tau\lambda^2(\tau)}\frac{1}{b(\tau)}\exp\Big(-|\xi|^2\int_s^\tau\frac{a^2(\theta)}{b(\theta)}d\theta\Big)d\tau
\end{aligned}$$

$$\lesssim \frac{a^2(t)|\xi|^2}{\lambda^2(t)}\left(\lambda^2(\tau)\frac{1}{b(\tau)}\exp\left(-|\xi|^2\int_s^\tau \frac{a^2(\theta)}{b(\theta)}d\theta\right)\right)\Big|_s^t$$
$$+\frac{a^2(t)|\xi|^2}{\lambda^2(t)}\int_s^t \lambda^2(\tau)\underbrace{\left(\frac{|\xi|^2a^2(\tau)}{b(\tau)}+\frac{b'(\tau)}{b^2(\tau)}\right)}_{\lesssim C(\tau)<1}\exp\left(-|\xi|^2\int_s^\tau \frac{a^2(\theta)}{b(\theta)}d\theta\right)d\tau$$
$$\lesssim \frac{a^2(t)|\xi|^2}{b(t)}\exp\left(-|\xi|^2\int_s^t \frac{a^2(\tau)}{b(\tau)}d\tau\right)-\frac{a^2(t)|\xi|^2}{b(s)}\frac{\lambda^2(s)}{\lambda^2(t)}.$$

Here we have used $a^2(t)|\xi|^2/b^2(t) \le 1/2$ from the definition of the elliptic zone and $\frac{b'(t)}{b^2(t)} = o(1)$. We see that the second summand is subordinate to the first one because

$$\frac{b(s)}{b(t)}\exp\left(-|\xi|^2\int_s^t \frac{a^2(\tau)}{b(\tau)}d\tau\right)\frac{\lambda^2(t)}{\lambda^2(s)}=\exp\Bigg(\int_s^t \underbrace{\left(b(\tau)-\frac{a^2(\tau)|\xi|^2}{b(\tau)}-\frac{b'(\tau)}{b(\tau)}\right)}_{>0,\ \text{if}\,\tau\ge t_0}d\tau\Bigg)$$

for $t_0 \le s \le t$ with t_0 sufficiently large. Similarly, we can represent Ψ_2 in the following way:

$$\Psi_2(t,s,\xi)=\frac{\lambda^2(s)}{\lambda^2(t)}-|\xi|^2\int_s^t a^2(\tau)\frac{\lambda^2(\tau)}{\lambda^2(t)}\Phi_2(\tau,s,\xi)d\tau,$$
$$|\Psi_2(t,s,\xi)|\lesssim \frac{\lambda^2(s)}{\lambda^2(t)}+\frac{|\xi|^2}{\lambda^2(t)}\int_s^t a^2(\tau)\lambda^2(\tau)\frac{1}{b(s)}\exp\left(-|\xi|^2\int_s^\tau \frac{a^2(\theta)}{b(\theta)}d\theta\right)d\tau$$
$$\lesssim \frac{\lambda^2(s)}{\lambda^2(t)}+\frac{a^2(t)|\xi|^2}{b(t)b(s)}\exp\left(-|\xi|^2\int_s^t \frac{a^2(\tau)}{b(\tau)}d\tau\right).$$

Thus, we have proved the following lemma:

Lemma 3.14. *Let $(s,\xi),(t,\xi)\in Z_{\mathrm{ell}}(c_0,\varepsilon)$ with $s\le t$. Then the fundamental solution $E(t,s,\xi)$ can be estimated in the following way:*

$$(|E(t,s,\xi)|)\lesssim \exp\left(-|\xi|^2\int_s^t \frac{a^2(\tau)}{b(\tau)}d\tau\right)\begin{pmatrix}\frac{a(t)}{a(s)} & \frac{a(t)|\xi|}{b(s)}\\ \frac{a^2(t)|\xi|}{a(s)b(t)} & \frac{a^2(t)|\xi|^2}{b(s)b(t)}\end{pmatrix}+\frac{\lambda^2(s)}{\lambda^2(t)}\begin{pmatrix}0&0\\0&1\end{pmatrix}. \tag{3.29}$$

Remark 3.15. Let us choose a fixed s. Then the second summand in (3.29) is dominated by the first one. If we set $s = t_{\xi_2}$, then in the two cases $\Big(\frac{\mu(t)}{A(t)}$ is increasing or decreasing$\Big)$ we can use $a(t_{\xi_2})|\xi|\sim b(t_{\xi_2})$ to get the following estimate:

$$(|E(t,t_{\xi_2},\xi)|)\lesssim \exp\left(-|\xi|^2\int_{t_{\xi_2}}^t \frac{a^2(\tau)}{b(\tau)}d\tau\right)\begin{pmatrix}\frac{a(t)}{a(t_{\xi_2})} & \frac{a(t)}{a(t_{\xi_2})}\\ \frac{a^2(t)|\xi|}{a(t_{\xi_2})b(t)} & \frac{a^2(t)|\xi|}{a(t_{\xi_2})b(t)}\end{pmatrix}. \tag{3.30}$$

3.6. Gluing procedure

Case 1: The function $\dfrac{\mu(t)}{A(t)}$ *is monotonously decreasing.*

In the previous sections we have considered the fundamental solution in different zones. Now we have to glue the estimates from Lemmas 2.6, 3.14, 3.10 and 3.5. We obtain for the part of the hyperbolic zone which contains large frequencies $\{\xi : |\xi| \geq c > 0\}$ the following estimate for the fundamental solution:

$$(|E(t,0,\xi)|) \lesssim \sqrt{a(t)} \exp\Big(-\frac{1}{2}\int_0^t b(\tau)d\tau\Big)\begin{pmatrix} 1 & 1 \\ 1 & 1 \end{pmatrix}$$

to our original problem in the extended phase space, cf. Lemma 3.5 and Remark 3.12. It remains to consider the influence of the dissipative zone, the elliptic zone, the reduced zone and the hyperbolic zone for small frequencies. We denote by $t_{\xi_k}, k = 1,2,3$, the separating lines between the dissipative zone and the elliptic zone ($k = 1$), between the elliptic zone and the reduced zone ($k = 2$) and between the reduced zone and the hyperbolic zone ($k = 3$).

Case 1.1: $t \leq t_{\xi_1}$.

In this case we follow directly Lemma 2.6.

Case 1.2: $t_{\xi_1} \leq t \leq t_{\xi_2}$.

Now we have to glue the estimates from Lemmas 2.6 and 3.10. We have the following corollary:

Corollary 3.16. *The following estimates hold for all* $t \in [t_{\xi_1}, t_{\xi_2}]$:

$$(|E(t,0,\xi)|) \lesssim \exp\Big(-|\xi|^2\int_0^t \frac{a^2(\tau)}{b(\tau)}d\tau\Big)\begin{pmatrix} a(t)|\xi| & a(t)|\xi| \\ \frac{a^2(t)|\xi|^2}{b(t)} & \frac{a^2(t)|\xi|^2}{b(t)} \end{pmatrix}$$
$$+ \exp\Big(-\int_{t_{\xi_1}}^t b(\tau)d\tau\Big)\, a(t_{\xi_1})|\xi| \begin{pmatrix} 0 & 0 \\ 1 & 1 \end{pmatrix}.$$

Proof. The fundamental solution $E(t,0,\xi)$ can be represented as

$$E(t,t_{\xi_1},\xi)E(t_{\xi_1},0,\xi).$$

This yields for all $(t,\xi) \in Z_{\mathrm{ell}}(c_0,\varepsilon)$

$$(|E(t,0,\xi)|) \lesssim (|E(t,t_{\xi_1},\xi)|)(|E(t_{\xi_1},0,\xi)|)$$
$$\lesssim \Bigg(\exp\Big(-|\xi|^2\int_{t_{\xi_1}}^t \frac{a^2(\tau)}{b(\tau)}d\tau\Big)\begin{pmatrix} \frac{a(t)}{a(t_{\xi_1})} & \frac{a(t)|\xi|}{b(t_{\xi_1})} \\ \frac{a^2(t)|\xi|}{a(t_{\xi_1})b(t)} & \frac{a^2(t)|\xi|^2}{b(t_{\xi_1})b(t)} \end{pmatrix}$$
$$+\frac{\lambda^2(t_{\xi_1})}{\lambda^2(t)}\begin{pmatrix} 0 & 0 \\ 0 & 1 \end{pmatrix}\Bigg) \times \frac{a(t_{\xi_1})}{A(t_{\xi_1})}\begin{pmatrix} 1 & 1 \\ 1 & 1 \end{pmatrix}$$

$$\lesssim \exp\Big(-|\xi|^2\int_0^t \frac{a^2(\tau)}{b(\tau)}d\tau\Big)\begin{pmatrix} a(t)|\xi| & a(t)|\xi| \\ \frac{a^2(t)|\xi|^2}{b(t)} & \frac{a^2(t)|\xi|^2}{b(t)} \end{pmatrix}$$
$$+\exp\Big(-\int_{t_{\xi_1}}^t b(\tau)d\tau\Big)a(t_{\xi_1})|\xi|\begin{pmatrix} 0 & 0 \\ 1 & 1 \end{pmatrix}.$$

Here we used $a(t_{\xi_1})|\xi| \lesssim b(t_{\xi_1})$, $|\xi| \sim \dfrac{c_0}{A(t_{\xi_1})}$ together with the third statement from Lemma 3.13 to extend the above integral. This completes the proof. □

Case 1.3: $t_{\xi_2} \le t \le t_{\xi_3}$.

Now we will glue the estimates from Lemma 3.10 and Corollary 3.16.

Corollary 3.17. *The following estimate holds for all* $t \in [t_{\xi_2}, t_{\xi_3}]$:

$$(|E(t,0,\xi)|) \lesssim \exp\Big(-|\xi|^2\int_0^t \frac{a^2(\tau)}{b(\tau)}d\tau\Big)a(t)|\xi|\begin{pmatrix} 1 & 1 \\ 1 & 1 \end{pmatrix}.$$

Proof. From Lemma 3.10 and Remark 3.12 we have the following estimate:

$$(|E(t,t_{\xi_2},\xi)|) \lesssim \frac{\lambda(t_{\xi_2})}{\lambda(t)}\exp\Big(\varepsilon\int_{t_{\xi_2}}^t b(\tau)d\tau\Big)\begin{pmatrix} 1 & 1 \\ 1 & 1 \end{pmatrix}.$$

Taking account of the representation of the fundamental solution $E(t,0,\xi)$ as $E(t,t_{\xi_2},\xi)E(t_{\xi_2},0,\xi)$ gives after application of Corollary 3.16 the following estimate:

$$(|E(t,0,\xi)|) \lesssim (|E(t,t_{\xi_2},\xi)|)(|E(t_{\xi_2},0,\xi)|)$$
$$\lesssim \exp\Big(\Big(\varepsilon-\frac{1}{2}\Big)\int_{t_{\xi_2}}^t b(\tau)d\tau\Big)\begin{pmatrix} 1 & 1 \\ 1 & 1 \end{pmatrix}\Big(\exp\Big(-\int_{t_{\xi_1}}^{t_{\xi_2}} b(\tau)d\tau\Big)a(t_{\xi_1})|\xi|\begin{pmatrix} 0 & 0 \\ 1 & 1 \end{pmatrix}$$
$$+\exp\Big(-|\xi|^2\int_0^{t_{\xi_2}} \frac{a^2(\tau)}{b(\tau)}d\tau\Big)\begin{pmatrix} a(t_{\xi_2})|\xi| & a(t_{\xi_2})|\xi| \\ \frac{a^2(t_{\xi_2})|\xi|^2}{b(t_{\xi_2})} & \frac{a^2(t_{\xi_2})|\xi|^2}{b(t_{\xi_2})} \end{pmatrix}\Big)$$
$$\lesssim \Big(\exp\Big(\Big(\varepsilon-\frac{1}{2}\Big)\int_{t_{\xi_2}}^t b(\tau)d\tau\Big)\exp\Big(-|\xi|^2\int_0^{t_{\xi_2}} \frac{a^2(\tau)}{b(\tau)}d\tau\Big)\Big(a(t_{\xi_2})|\xi|+\frac{a^2(t_{\xi_2})|\xi|^2}{b(t_{\xi_2})}\Big)$$
$$+\exp\Big(\Big(\varepsilon-\frac{1}{2}\Big)\int_{t_{\xi_2}}^t b(\tau)d\tau\Big)\exp\Big(-\int_{t_{\xi_1}}^{t_{\xi_2}} b(\tau)d\tau\Big)a(t_{\xi_1})|\xi|\Big)\begin{pmatrix} 1 & 1 \\ 1 & 1 \end{pmatrix}.$$

From the definition of $Z_{\text{red}}(\varepsilon)$ with a sufficiently small ε we have

$$a^2(t)|\xi|^2 \le \Big(\frac{1}{2}-\varepsilon\Big)b^2(t).$$

For $t \le t_{\xi_2}$ we use

$$a(t)|\xi| \lesssim b(t).$$

Hence, the integral

$$\exp\Big(- \int_{t_{\xi_1}}^{t_{\xi_2}} b(\tau)d\tau\Big)$$

can be estimated by

$$\exp\Big(- |\xi|^2 \int_{t_{\xi_1}}^{t_{\xi_2}} \frac{a^2(\tau)}{b(\tau)} d\tau\Big),$$

and the last integral can be extended up to $t = 0$. Using $t \geq t_{\xi_2}$ and the increasing behavior of a we conclude from the last estimates the desired statement. □

Case 1.4: $t_{\xi_3} \leq t < \infty$.

From Lemma 3.5 and Remark 3.12 we obtain the following statement:

Corollary 3.18. *The following estimate holds for all* $t \in [t_{\xi_3}, \infty)$:

$$(|E(t, t_{\xi_3}, \xi)|) \lesssim \frac{\sqrt{a(t)}}{\sqrt{a(t_{\xi_3})}} \exp\left(-\frac{1}{2} \int_{t_{\xi_3}}^t b(\tau)d\tau\right) \begin{pmatrix} 1 & 1 \\ 1 & 1 \end{pmatrix}.$$

Finally, we have to glue the estimates from Corollaries 3.17 and 3.18.

Corollary 3.19. *The following estimate holds for all* $t \in [t_{\xi_3}, \infty)$:

$$(|E(t, 0, \xi)|) \lesssim \exp\left(-|\xi|^2 \int_0^{t_{\xi_3}} \frac{a^2(\tau)}{b(\tau)} d\tau\right) \exp\left(-\frac{1}{2} \int_{t_{\xi_3}}^t b(\tau)d\tau\right)$$
$$\times \sqrt{a(t)}\sqrt{a(t_{\xi_3})|\xi|} \begin{pmatrix} 1 & 1 \\ 1 & 1 \end{pmatrix}.$$

Case 2: *The function* $\dfrac{\mu(t)}{A(t)}$ *is monotonously increasing.*

The elliptic region lies in this case on the top of the hyperbolic region. For small frequencies the set $\{\xi : |\xi| \leq c_0\}$ lies completely inside the elliptic zone. For this reason we can use the estimates from the elliptic zone and obtain immediately

$$(|E(t, 0, \xi)|) \lesssim \exp\left(-|\xi|^2 \int_0^t \frac{a^2(\tau)}{b(\tau)} d\tau\right) \begin{pmatrix} \frac{a(t)}{a(0)} & \frac{a(t)}{a(0)} \\ \frac{a^2(t)|\xi|}{a(0)b(t)} & \frac{a^2(t)|\xi|}{a(0)b(t)} \end{pmatrix}. \tag{3.31}$$

It remains to consider the influence of the elliptic zone, the reduced zone and the hyperbolic zone for large frequencies. We denote by $t_{\xi_k}, k = 1, 2$, the separating lines between the hyperbolic zone and the reduced zone ($k = 1$) and between the reduced zone and the elliptic zone ($k = 2$).

Case 2.1: $t \leq t_{\xi_1}$.

In this case we follow directly Lemma 3.5 and Remark 3.12 to obtain

$$(|E(t, 0, \xi)|) \lesssim \frac{\sqrt{a(t)}}{\sqrt{a(0)}} \exp\left(-\frac{1}{2} \int_0^t b(\tau)d\tau\right) \begin{pmatrix} 1 & 1 \\ 1 & 1 \end{pmatrix}.$$

Case 2.2: $t_{\xi_1} \le t \le t_{\xi_2}$.

In this case we need to glue the estimates in the hyperbolic zone and in the reduced zone. We have

$$|E(t,0,\xi)| \lesssim \exp\left(\left(\varepsilon - \frac{1}{2}\right)\int_{t_{\xi_1}}^{t} b(\tau)d\tau - \frac{1}{2}\int_0^{t_{\xi_1}} b(\tau)d\tau\right)\frac{\sqrt{a(t_{\xi_1})}}{\sqrt{a(0)}}.$$

Case 2.3: $t_{\xi_2} \le t$.

In this case we need to glue the estimates in the elliptic zone and in the reduced zone. Summarizing yields the following corollary:

Corollary 3.20. *The following estimate holds for all* $t \in [t_{\xi_2}, \infty)$:

$$(|E(t,0,\xi)|) \lesssim \exp\left(-|\xi|^2\int_{t_{\xi_2}}^{t}\frac{a^2(\tau)}{b(\tau)}d\tau + \left(\varepsilon - \frac{1}{2}\right)\int_{t_{\xi_1}}^{t_{\xi_2}} b(\tau)d\tau - \frac{1}{2}\int_0^{t_{\xi_1}} b(\tau)d\tau\right)$$
$$\times \frac{\sqrt{a(t_{\xi_1})}}{\sqrt{a(0)}}\begin{pmatrix} \frac{a(t)}{a(t_{\xi_2})} & \frac{a(t)}{a(t_{\xi_2})} \\ \frac{a^2(t)|\xi|}{a(t_{\xi_2})b(t)} & \frac{a^2(t)|\xi|}{a(t_{\xi_2})b(t)} \end{pmatrix}.$$

3.6.1. L^2-L^2 estimates – end of the proof.

Case 1: *The function* $\dfrac{\mu(t)}{A(t)}$ *is monotonously decreasing.*

In the case $t \in [0, t_{\xi_1}]$ we have from Lemma 2.6 the estimate

$$|E(t,0,\xi)| \lesssim \frac{a(t)}{A(t)} \lesssim \frac{a(t)}{\sqrt{1+\int_0^t \frac{a^2(\tau)}{b(\tau)}d\tau}}.$$

This follows directly from

$$\int_0^t \frac{a^2(\tau)}{b(\tau)}d\tau = \int_0^t \frac{a(\tau)A(\tau)}{\mu(\tau)}d\tau \lesssim \int_0^t a(\tau)A(\tau)d\tau \lesssim A^2(t)$$

for large t.

In the case $t \in [t_{\xi_1}, t_{\xi_2}]$ we will estimate separately each row in the estimate from Corollary 3.16. Let us consider the first row. It holds

$$a(t)|\xi|\exp\left(-|\xi|^2\int_0^t\frac{a^2(\tau)}{b(\tau)}d\tau\right) \lesssim a(t)\left(1+\int_0^t\frac{a^2(\tau)}{b(\tau)}d\tau\right)^{-\frac{1}{2}},$$

therefore, we get the desired decay estimate. Using the monotonicity of a for the second row we can estimate by the first one

$$\frac{a^2(t)}{b(t)}|\xi|^2 = a(t)|\xi|\frac{a(t)|\xi|}{b(t)} \lesssim a(t)|\xi|,$$
$$a(t_{\xi_1})\exp\left(-\int_0^t b(\tau)d\tau\right) \lesssim a(t)\exp\left(-|\xi|^2\int_0^t\frac{a^2(\tau)}{b(\tau)}d\tau\right).$$

In the case $t \in [t_{\xi_2}, t_{\xi_3}]$ by using Corollary 3.17 we can estimate like in the case $t \in [t_{\xi_1}, t_{\xi_2}]$.

To derive the corresponding estimates from Corollary 3.19 we have in the case $t \in [t_{\xi_3}, \infty)$ to estimate the term

$$S(t,|\xi|) := |\xi| \exp\Big(-|\xi|^2 \int_0^{t_{\xi_3}} \frac{a^2(\tau)}{b(\tau)} d\tau\Big) \exp\Big(-\frac{1}{2}\int_{t_{\xi_3}}^t b(\tau)d\tau\Big).$$

Lemma 3.21. *The maximum of the function $S(t,|\xi|)$ is taken at a point $|\xi|$ independent of $t \geq t_{\xi_3}$ and*

$$S(t,|\xi|) \leq \max_{\xi\in\mathbb{R}^n} \Big\{ |\xi| \exp\Big(-|\xi|^2 \int_0^t \frac{a^2(\tau)}{b(\tau)} d\tau\Big)\Big\}.$$

Proof. To estimate the function $S(t,|\xi|)$ it is important that we will prove that the first partial derivative $\partial_{|\xi|}S(t,|\xi|)$ is negative for $|\xi|$ small. This follows from

$$\begin{aligned}
&\partial_{|\xi|}S(t,|\xi|)\\
&= S(t,|\xi|)\Big(\frac{1}{|\xi|} - 2|\xi|\int_0^{t_{\xi_3}} \frac{a^2(\tau)}{b(\tau)}d\tau - \frac{a^2(t_{\xi_3})|\xi|^2}{b(t_{\xi_3})} d_{|\xi|}t_{\xi_3} + \frac{b(t_{\xi_3})}{2} d_{|\xi|}t_{\xi_3}\Big)\\
&< S(t,|\xi|)\Big(\frac{1}{|\xi|} + \Big(\frac{b(t_{\xi_3})}{2} - \frac{a^2(t_{\xi_3})|\xi|^2}{b(t_{\xi_3})}\Big) d_{|\xi|}t_{\xi_3}\Big)\\
&< S(t,|\xi|)\Big(\frac{1}{|\xi|} + \frac{(1-\varepsilon^2)b(t_{\xi_3})}{4} d_{|\xi|}t_{\xi_3}\Big).
\end{aligned}$$

Here we have used

$$\frac{a^2(t_{\xi_3})|\xi|^2}{b(t_{\xi_3})} = \frac{(1+\varepsilon^2)b(t_{\xi_3})}{4}.$$

Hence, a sufficiently small ε guarantees $\frac{(1-\varepsilon^2)b(t_{\xi_3})}{4} > 0$. Taking account of $d_{|\xi|}t_{\xi_3} < 0$, $|d_{|\xi|}t_{\xi_3}| \geq \frac{\mu(t_{\xi_3})}{|\xi| b(t_{\xi_3})}$ and $\mu(t_{\xi_3}) \to \infty$ for $|\xi| \to 0$ we have the desired decreasing behavior of the function $S(t,|\xi|)$ in $|\xi|$. Now let us fix $t > 0$. Then the function $S(t,|\xi|)$ takes its maximum for $|\tilde{\xi}|$ satisfying $t = t_{\tilde{\xi}_3}$, that is, the second integral vanishes in $S(t,|\xi|)$. This completes the proof. □

Corollary 3.19 and Lemma 3.21 yield the following result:

$$|E(t,0,\xi)| \lesssim a(t)\Big(1 + \int_0^t \frac{a^2(\tau)}{b(\tau)}d\tau\Big)^{-\frac{1}{2}} \quad \text{for } t \in [t_{\xi_3}, \infty).$$

In this way all statements are proved.

Case 2: The function $\dfrac{\mu(t)}{A(t)}$ is monotonously increasing.

For small frequencies $\{\xi : |\xi| \leq c_0\}$ we can apply the estimate in (3.31). Here we use that $\dfrac{A(t)}{\mu(t)}$ is monotonously decreasing. For large frequencies $\{\xi : |\xi| \geq c_0\}$

we consider the estimates from Corollary 3.20, that is, we have

$$\exp\left(-|\xi|^2\int_{t_{\xi_2}}^t \frac{a^2(\tau)}{b(\tau)}d\tau + \left(\varepsilon-\frac{1}{2}\right)\int_{t_{\xi_1}}^{t_{\xi_2}} b(\tau)d\tau - \frac{1}{2}\int_0^{t_{\xi_1}} b(\tau)d\tau\right)$$
$$\lesssim \exp\left(-c_0^2\int_0^t \frac{a^2(\tau)}{b(\tau)}d\tau\right).$$

Here we use for ε sufficiently small the inequality

$$\left(\frac{1}{2}-\varepsilon\right)\int_{t_{\xi_1}}^{t_{\xi_2}} b(\tau)d\tau \geq |\xi|^2\int_{t_{\xi_1}}^{t_{\xi_2}} \frac{a^2(\tau)}{b(\tau)}d\tau.$$

Moreover, the following estimate holds for $c_0 < \dfrac{1}{\sqrt{2}}\dfrac{\mu(0)}{A(0)}$:

$$\frac{b(t)}{2} \geq c_0^2\frac{a^2(t)}{b(t)} \text{ iff } \frac{b^2(t)}{2} \geq c_0^2 a^2(t) \text{ iff } \frac{1}{2}\frac{\mu^2(t)}{A^2(t)} \geq c_0^2.$$

We can see that the first row in the estimate from Corollary 3.20 has its maximum for large t inside of $\{\xi : |\xi| \leq c_0\}$. From that the theorem is completely proved.

Remark 3.22. If we choose in Theorem 3.1 the coefficient $a(t) \equiv 1$, then the obtained estimates coincide with the estimates from Result 1.2 for $p = q = 2$.

Examples. We will give some examples for special coefficients.

Example. Let $a(t) = (1+t)^l$, $b(t) = (1+t)^k$, $k \in (-1, 2l+1]$ with $l > 0$. Then we have

$$\left\|\left((1+t)^l\nabla u(t,\cdot), u_t(t,\cdot)\right)\right\|_{L^2} \lesssim (1+t)^{\frac{k-1}{2}}\left(\|u_1\|_{H^1} + \|u_2\|_{L^2}\right).$$

Example. Let $a(t) = e^t$, $b(t) = e^{\beta t}$, $\beta \in (0, 2]$. Then we have

$$\left\|\left(e^t\nabla u(t,\cdot), u_t(t,\cdot)\right)\right\|_{L^2} \lesssim e^{\frac{\beta}{2}t}\left(\|u_1\|_{H^1} + \|u_2\|_{L^2}\right).$$

3.7. Comparison with known results

In [1] some results for scale-invariant models are proved by applying direct calculations and the theory of special functions. Now let us compare these results with the main results Theorem 2.1 for non-effective dissipations from Section 2 and Theorem 3.1 for effective dissipations from Section 3 to see that our estimates are optimal.

3.7.1. Speed of potential order. We start with the Cauchy problem

$$u_{tt} - (1+t)^{2l}\Delta u + \frac{\mu(l+1)}{(1+t)}u_t = 0,\ u(0,x) = u_1(x),\ u_t(0,x) = u_2(x) \qquad (3.32)$$

with $l > 0$. Then the following result can be found in [1]:

Proposition 3.23. *We have the following estimate for the energies of solutions to* (3.32)*:*

$$\|u_t(t,\cdot),(1+t)^l\nabla u(t,\cdot)\|_{L^2} \lesssim (1+t)^{l+(l+1)\max\{\rho-\frac{1}{2},-1\}}\left(\|u_1\|_{H^1}+\|u_2\|_{L^2}\right) \quad (3.33)$$

with $\rho=\frac{1}{2}(1-\mu-\frac{l}{l+1})$.

Case 1: *Non-effective dissipation* $\left(\max\{\rho-\frac{1}{2},-1\}=\rho-\frac{1}{2}\right)$.

With $\mu\neq 1$ we can see that $a(t)=(1+t)^{1+l}$, $b(t)=\dfrac{\mu(l+1)}{(1+t)}$ satisfy all assumptions from Theorem 2.1. Otherwise, from the definition of ρ and the condition $\left(\max\{\rho-\frac{1}{2},-1\}=\rho-\frac{1}{2}\right)$ we obtain $\mu+\frac{l}{l+1}<2$, i.e., this condition satisfies the condition (C): $\limsup_{t\to\infty}\mu(t)+\alpha(t)<2$.

Applying Theorem 2.1 in the case of non-effective dissipation the asymptotic profile for the "kinetic energy" $\|u_t(t,\cdot)\|_{L^2}$ and for the "elastic energy" $\|(1+t)^l\nabla u(t,\cdot)\|_{L^2}$ is determined by

$$\frac{\sqrt{a(t)}}{\lambda(t)}=\frac{(1+t)^{\frac{l}{2}}}{e^{\frac{1}{2}\int_0^t\frac{\mu(l+1)}{1+s}ds}}=(1+t)^{\frac{l}{2}-\frac{\mu(l+1)}{2}}.$$

This profile coincides with the profiles from the estimates in Proposition 3.23.

Case 2: *Effective dissipation* $\left(\max\{\rho-\frac{1}{2},-1\}=-1\right)$.

From the definition of ρ we can see that the above condition implies $\mu+\frac{l}{l+1}\geq 2$. Applying Theorem 3.1 for the case of effective dissipation the asymptotic profile of the "kinetic energy" $\|u_t(t,\cdot)\|_{L^2}$ and for the "elastic energy" $\|(1+t)^l\nabla u(t,\cdot)\|_{L^2}$ is determined by

$$a(t)\left(1+\int_0^t\frac{a^2(\tau)}{b(\tau)}d\tau\right)^{-\frac{1}{2}}=\frac{(1+t)^l}{\sqrt{1+\int_0^t\frac{(1+\tau)^{2l+1}}{\mu(l+1)}d\tau}}\sim\frac{1}{1+t}.$$

Due to assumption (B$'$3) it is not allowed to apply Theorem 3.1 directly to the Cauchy problem (3.32). But, if we formally do it for $\mu\geq 2-\frac{l}{l+1}$, then this profile coincides with the profiles from the estimates of Proposition 3.23. For the case $\mu=0$ some L^p-L^q estimates on the conjugate line are proposed in [6].

3.7.2. Speed of exponential order. Now we consider another model case to compare with the general results Theorem 2.1 for non-effective dissipations and Theorem 3.1 for effective dissipations. We devote the Cauchy problem

$$u_{tt}-e^{2t}\Delta u+\mu u_t=0,\ u(0,x)=u_1(x),\ u_t(0,x)=u_2(x). \quad (3.34)$$

Then the following result can be found in [1]:

Proposition 3.24. *We have the following estimates for the solutions to* (3.34)*:*

$$\|u_t(t,\cdot),(e^t\nabla u(t,\cdot)\|_{L^2}\lesssim e^{t+t\max\{\rho-\frac{1}{2},-1\}}\left(\|u_1\|_{H^1}+\|u_2\|_{L^2}\right) \quad (3.35)$$

with $\rho=-\frac{\mu}{2}$.

Case 1: *Non-effective dissipation* $\left(\max\{\rho - \frac{1}{2}, -1\} = \rho - \frac{1}{2}\right)$.

The assumptions from Theorem 2.1 are satisfied for $\mu \neq 1$. Keep in mind that $\rho - \frac{1}{2} > -1 \Leftrightarrow -\frac{\mu}{2} - \frac{1}{2} > -1$, this condition implies $\mu + 1 < 2$, i.e., it satisfies the condition $(C) : \limsup_{t\to\infty} \mu(t) + \alpha(t) < 2$.

Applying Theorem 2.1 in the case of non-effective dissipations the asymptotic profile for the "kinetic energy" $\|u_t(t,\cdot)\|_{L^2}$ and for the "elastic energy" $\|e^t \nabla u(t,\cdot)\|_{L^2}$ is determined by

$$\frac{\sqrt{a(t)}}{\lambda(t)} = \frac{e^{\frac{t}{2}}}{e^{\frac{1}{2}\int_0^t \mu ds}} = e^{\frac{t}{2} - \frac{\mu t}{2}}.$$

This profile coincides with the profiles from the estimates from Proposition 3.24.

Case 2: *Effective dissipation* $\left(\max\{\rho - \frac{1}{2}, -1\} = -1\right)$.

From the definition of ρ we can see that the above condition implies $\mu + 1 \geq 2$. Hence, this condition does not satisfy the condition (C). Applying Theorem 3.1 in the case of effective dissipations the asymptotic profile of the "kinetic energy" $\|u_t(t,\cdot)\|_{L^2}$ and of the "elastic energy" $\|e^t \nabla u(t,\cdot)\|_{L^2}$ is determined by

$$a(t)\left(1 + \int_0^t \frac{a^2(s)}{b(s)} ds\right)^{-\frac{1}{2}} = \frac{e^t}{\sqrt{1 + \int_0^t \frac{e^{2s}}{\mu} ds}} \sim C.$$

Due to assumption (B$'$3) it is not allowed to apply Theorem 3.1 to the Cauchy problem (3.35). But if we formally do it for $\mu \geq 1$, then this profile coincides with the profiles from the estimates of Proposition 3.24. For the case $\mu = 0$ some L^p-L^q estimates on the conjugate line are proposed in [4].

4. Scattering and over-damping

From the thesis of J. Wirth [11] we expect scattering and over-damping results, too.

4.1. Scattering

We will concern with conditions for $b = b(t)$ that the solutions $u = u(t,x)$ of

$$u_{tt} - a^2(t)\Delta u + b(t)u_t = 0, \; u(0,x) = u_1(x), \; u_t(0,x) = u_2(x) \tag{4.1}$$

behave asymptotically equal to the solution of the corresponding wave equation with strictly increasing speed of propagation

$$v_{tt} - a^2(t)\Delta v = 0, \; v(0,x) = v_1(x), \; v_t(0,x) = v_2(x) \tag{4.2}$$

with some suitable Cauchy data (v_1, v_2).

Here we introduce the energy space $E(\mathbb{R}^n) = \dot{H}^1(\mathbb{R}^n) \times L^2(\mathbb{R}^n)$ and we assume $(u_1, u_2) \in E(\mathbb{R}^n)$, this means, $(|D|u_1, u_2) \in L^2(\mathbb{R}^n)$.

Result 4.1. *We assume that the coefficient $b = b(t)$ satisfies $b \in L^1(\mathbb{R}^+)$. Then there exists the Møller wave operator $W_+ : E \to E$ mapping the Cauchy data $(a(0)u_1, u_2) \in E$ from (4.1) to Cauchy data $(a(0)v_1, v_2)$ from (4.2) by*

$$(a(0)v_1, v_2)^T = W_+(a(0)u_1, u_2)^T$$

such that the asymptotic equivalence of solutions of the problems (4.1) and (4.2) holds in the following way:

$$\frac{1}{\sqrt{a(t)}} \|(a(t)u, D_t u) - (a(t)v, D_t v)\|_E \to 0 \tag{4.3}$$

while $t \to \infty$. Moreover, we have the decay estimate

$$\frac{1}{\sqrt{a(t)}} \|(a(t)u, D_t u) - (a(t)v, D_t v)\|_E \lesssim \|(u_1, u_2)\|_E \int_t^\infty b(\tau) d\tau \tag{4.4}$$

with the convergence rate $\int_t^\infty b(\tau)d\tau$ to 0 as $t \to \infty$.

4.2. Over-damping

We consider now "large" coefficients $b = b(t)$ in the damping term. For this reason we may assume

$$\text{(OD)} \qquad \int_0^\infty \frac{a^2(\tau)}{b(\tau)} d\tau < \infty.$$

Then the formal application of Theorem 3.1 implies among other things

$$\|\nabla u(t, \cdot)\|_{L^2} \leq C(\|u_1\|_{H^1} + \|u_2\|_{L^2}).$$

The following result shows that no more can be expected in this case of so-called over-damping.

Result 4.2. *Assume* (A1) *to* (A3), (B′1) *to* (B′3) *and* (OD). *Then for $(u_1, u_2) \in L^2(\mathbb{R}^n) \times H^{-1}(\mathbb{R}^n)$ the limit*

$$u(\infty, x) = \lim_{t \to \infty} u(t, x)$$

exists in $L^2(\mathbb{R}^n)$ and is different from zero for non-zero data. Furthermore, under the regularity assumption $(u_1, u_2) \in H^2(\mathbb{R}^n) \times H^1(\mathbb{R}^n)$ it holds

$$\|u(t, \cdot) - u(\infty, \cdot)\|_{L^2} = \mathcal{O}\left(\int_t^\infty \frac{a^2(\tau)}{b(\tau)} d\tau \right).$$

5. Concluding remarks

Remark 5.1. There are several papers which are devoted to the Cauchy problem for the following non-linear wave equations

$$u_{tt} - a(t)^2\Delta u = u_t^2 - a(t)^2|\nabla u|^2,\ u(0,x) = u_1(x),\ \ u_t(0,x) = u_2(x). \qquad (5.1)$$

In particular, in two papers [14] and [15] it is explained how the above class of special semi-linear Cauchy problems can be reduced by Nirenberg's transformation to a linear model with constrain condition. The above papers and the paper [2] concern with the problem of global existence (in time) for small data solutions to the semi-linear Cauchy problem

$$u_{tt} - a(t)^2\Delta u = u_t^2 - a(t)^2|\nabla u|^2,\ u(0,x) = u_1(x),\ u_t(0,x) = u_2(x). \qquad (5.2)$$

It would be a challenge to apply this approach to the case of *non-effective* dissipations to the following semi-linear Cauchy problem

$$u_{tt} - a(t)^2\Delta u + b(t)u_t = u_t^2 - a(t)^2|\nabla u|^2,\ u(0,x) = u_1(x),\ u_t(0,x) = u_2(x). \ (5.3)$$

Remark 5.2. Another interesting application to the case of *effective* dissipations is the question for global small data solutions to the following semi-linear model

$$u_{tt} - a(t)^2\Delta u + b(t)u_t = f(u),\ u(0,x) = u_1(x),\ \ u_t(0,x) = u_2(x), \qquad (5.4)$$

where $f(u) \approx |u|^p$. In a recent paper [3] the authors have constructed counter-examples which provide a nonexistence result for weak solutions to (5.4).

Both problems are attacked in forthcoming papers.

Acknowledgment

Both authors are grateful to DAAD that the PhD project of Mr. Bui under supervision of Prof. Reissig is supported at Technical University Bergakademie Freiberg by a scholarship during the period 2010–2014.

References

[1] Tang Bao Ngoc Bui, *Wave models with time-dependent speed and dissipation*, PhD thesis, Technical University Bergakademie Freiberg, in preparation, 2013.

[2] M.R. Ebert, M. Reissig, *The influence of oscillations on global existence for a class of semi-linear wave equations*, Math. Meth. Appl. Sci. **34** (2009) 11, 1289–1307.

[3] D'Abbicco and Lucente, *A modified test function method for damped wave equations*, Preprint, arXiv:1211.0453 [math.AP], 2012.

[4] Anahit Galstian, *L_p-L_q decay estimate for the wave equations with exponentially growing speed of propagation*, Applicable Analysis **82** (2003) 3, 197–214.

[5] F. Hirosawa, J. Wirth, *Generalised energy conservation law for wave equations with variable propagation speed*, J. Math. Appl. **358** (2009), 56–74.

[6] M. Reissig, *On L_p-L_q estimates for solutions of a special weakly hyperbolic equation*, Ed. Li Ta-Tsien, Nonlinear Evolution Equations and Infinite-Dimensional Dynamical Systems, 153–164, World Scientific (1997).

[7] M. Reissig, K. Yagdjian, *One application of Floquet's theory to L_p-L_q estimates for hyperbolic equations with very fast oscillations*, Math. Meth. Appl. Sci. **22** (1999), 937–951.

[8] M. Reissig, K. Yagdjian, *About the influence of oscillations on Strichartz-type decay estimates*, Rendiconti Del Seminario Matematico Torino **58** (2000) 3, 375–388.

[9] M. Reissig, K. Yagdjian, *L_p-L_q decay estimates for hyperbolic equations with oscillations in coefficients*, Chin. Ann. of Math., Series B, **21** (2000) 2, 153–164.

[10] M. Reissig, *Optimality of the asymptotic behavior of the energy for wave models*, Modern Aspects of Partial Differential Equations, Birkhäuser, Vol. **216** (2011), 291–315.

[11] J. Wirth, *Asymptotic properties of solutions to wave equations with time dependent dissipation*, PhD thesis, Technical University Bergakademie Freiberg, 2005.

[12] J. Wirth, *Wave equations with time-dependent dissipation I. Non-effective dissipation*, J. Differential Equations **222** (2006), 487–514.

[13] J. Wirth, *Wave equations with time-dependent dissipation II. Effective dissipation*, J. Differential Equations **232** (2007), 74–103.

[14] K. Yagdjian, *Parametric resonance and nonexistence of global solution to nonlinear wave equations*, J. Math. Anal. Appl. **260** (2001) 1, 251–268.

[15] K. Yagdjian, *Global existence in the Cauchy problem for nonlinear wave equations with variable speed of propogation*, Operator Theory: Advances and Applications, Birkhäuser, Vol. **159** (2005), 301–385.

Tang Bao Ngoc Bui and Michael Reissig
Institute of Applied Analysis
TU Bergakademie Freiberg
D-09596 Freiberg, Germany
e-mail: buitangngockstn@yahoo.com
reissig@math.tu-freiberg.de

Fourier Analysis
Trends in Mathematics, 47–72

A Note on Complete Hyperbolic Operators with log-Zygmund Coefficients

Ferruccio Colombini, Daniele Del Santo,
Francesco Fanelli and Guy Métivier

Abstract. The present paper is the continuation of the recent work [7], and it is devoted to strictly hyperbolic operators with non-regular coefficients. We focus here on the case of complete operators whose second-order coefficients are log-Zygmund continuous in time, and we investigate the H^∞ well-posedness of the associate Cauchy problem.

Mathematics Subject Classification (2010). Primary 35L15; Secondary 35B65, 35S50, 35B45.

Keywords. Hyperbolic operators, non-Lipschitz coefficient, log-Zygmund regularity, energy estimates, well-posedness.

1. Introduction

The present paper is, in a certain sense, the continuation of the recent work [7]: it is devoted to the study of strictly hyperbolic operators with log-Zygmund in time coefficients. In particular, we want to prove the H^∞ well-posedness of the Cauchy problem related to a complete second-order operator,

$$Lu = \partial_t^2 u - \sum_{j,k=1}^{N} \partial_j \left(a_{jk}(t,x)\, \partial_k u\right) + \sum_{j=0}^{N} b_j(t,x)\, \partial_j u + c(t,x)\, u\,, \tag{1.1}$$

(with $\partial_0 = \partial_t$) whose highest-order coefficients satisfy such a regularity assumption (see also relation (2.3) below).

The core of the proof is to establish suitable energy estimates for L.

It is well known (see [14]; see also, e.g., [13, Ch. IX] or [18, Ch. 6] for analogous results) that, if the coefficients a_{jk} are Lipschitz continuous with respect to t and only measurable in x, then the Cauchy problem for L is well posed in $H^1 \times L^2$. If the a_{jk}'s are Lipschitz continuous with respect to t and $\mathcal{C}_b^\infty$ (i.e., $\mathcal{C}^\infty$ and bounded with all their derivatives) with respect to the space variables, one can recover the well-posedness in $H^{s+1} \times H^s$ for all $s \in \mathbb{R}$. Moreover, in the latter case, one gets, for all $s \in \mathbb{R}$ and for a constant C_s depending only on it, the following energy estimate:

$$\begin{aligned} &\sup_{0\leq t\leq T}\left(\|u(t,\cdot)\|_{H^{s+1}} + \|\partial_t u(t,\cdot)\|_{H^s}\right) \\ &\qquad \leq C_s\left(\|u(0,\cdot)\|_{H^{s+1}} + \|\partial_t u(0,\cdot)\|_{H^s} + \int_0^T \|Lu(t,\cdot)\|_{H^s}\,dt\right) \end{aligned} \tag{1.2}$$

for all $u \in \mathcal{C}([0,T];H^{s+1}(\mathbb{R}^N)) \cap \mathcal{C}^1([0,T];H^s(\mathbb{R}^N))$ for which Lu belongs to $L^1([0,T];H^s(\mathbb{R}^N))$. Let us explicitly remark that previous inequality involves no loss of regularity for the function u: estimate (1.2) holds for every u of class $\mathcal{C}^2([0,T];H^\infty(\mathbb{R}^N))$ and the Cauchy problem for L is well posed in H^∞ *with no loss of derivatives.*

If the Lipschitz continuity (in time) hypothesis is not fulfilled, then (1.2) is no more true. Nevertheless, one can still try to recover H^∞ well-posedness, with a *loss of derivatives* in the energy estimate.

The first case to consider is the case of the coefficients a_{jk} depending only on t:

$$Lu \;=\; \partial_t^2 u \;-\; \sum_{j,k=1}^N a_{jk}(t)\,\partial_j\partial_k u\,.$$

In [5], Colombini, De Giorgi and Spagnolo assumed the coefficients to satisfy an integral log-Lipschitz condition:

$$\int_0^{T-\varepsilon} |a_{jk}(t+\varepsilon) \,-\, a_{jk}(t)|\,dt \;\leq\; C\,\varepsilon\,\log\left(1+\frac{1}{\varepsilon}\right), \tag{1.3}$$

for some constant $C>0$ and all $\varepsilon \in\,]0,T]$. They obtained an energy estimate *with a fixed loss of derivatives*: there exists a constant $\delta>0$ such that, for all $s\in\mathbb{R}$, the inequality

$$\begin{aligned} &\sup_{0\leq t\leq T}\left(\|u(t,\cdot)\|_{H^{s+1-\delta}} + \|\partial_t u(t,\cdot)\|_{H^{s-\delta}}\right) \\ &\qquad \leq C_s\left(\|u(0,\cdot)\|_{H^{s+1}} + \|\partial_t u(0,\cdot)\|_{H^s} + \int_0^T \|Lu(t,\cdot)\|_{H^s}\,dt\right) \end{aligned} \tag{1.4}$$

holds true for all $u \in \mathcal{C}^2([0,T];H^\infty(\mathbb{R}^N))$, for some constant C_s depending only on s.

Let us explain briefly the main ideas of the proof, which are actually the basis to establish also the next results we are going to quote. The authors first smoothed out the coefficients using a mollifier kernel $(\rho_\varepsilon)_\varepsilon$. Then, by Fourier transform, they defined an approximated energy $E_\varepsilon(\xi,t)$ in the phase space, where the problem becomes a family of ordinary differential equations. At that point, the key step was to perform a different approximation of the coefficients in different zones of the phase space, which was realized setting $\varepsilon = |\xi|^{-1}$. This particular choice allowed them to obtain the result.

Let us remark that if the coefficients a_{jk} are not Lipschitz continuous then a loss of regularity cannot be avoided. In fact, in paper [3], Cicognani and Colombini proved that, if the regularity of the coefficients a_{jk} is measured by a modulus of continuity, any intermediate modulus of continuity between the Lipschitz and the log-Lipschitz ones entails necessarily a loss of regularity, which, however, can be made arbitrarily small.

Recently Tarama (see paper [19]) was able to prove an analogous result, and the consequent H^∞ well-posedness of the related Cauchy problem, when the coefficients satisfy an integral log-Zygmund condition: there exists a constant $C > 0$ such that, for all j, k and all $\varepsilon \in]0, T/2[$, one has

$$\int_\varepsilon^{T-\varepsilon} |a_{jk}(t+\varepsilon) + a_{jk}(t-\varepsilon) - 2\,a_{jk}(t)|\,dt \leq C\,\varepsilon\,\log\left(1 + \frac{1}{\varepsilon}\right). \tag{1.5}$$

It is obvious that, if the a_{jk}'s satisfy (1.3), then they satisfy also (1.5): so, a more general class of functions is considered. The improvement with respect to paper [5], however, was obtained defining a new energy, which involved (by derivation in time) second derivatives of the approximated coefficients.

In paper [10], Colombini and Lerner considered instead the case when the coefficients a_{jk} depend both on time and space variables. In particular, they assumed an isotropic punctual log-Lipschitz condition, i.e., there exists a constant $C > 0$ such that, for all $\zeta = (\tau, \xi) \in \mathbb{R} \times \mathbb{R}^N$, $\zeta \neq 0$, one has

$$\sup_{z=(t,x)\in\mathbb{R}\times\mathbb{R}^N} |a_{jk}(z+\zeta) - a_{jk}(z)| \leq C\,|\zeta|\,\log\left(1 + \frac{1}{|\zeta|}\right).$$

The analysis was based on a broad use of the Littlewood–Paley decomposition and paradifferential calculus. As a matter of facts, one cannot use the Fourier transform, due to the dependence of a_{jk} on x. Also in this case, a different approximation of the coefficients in different zones of the phase space was required: this was obtained setting $\varepsilon = 2^{-\nu}$, where 2^ν is the "size" of the frequencies in the νth ring of the Littlewood–Paley decomposition (see also Section 3). In the end, they got the following statement: for all $s \in]0, 1/4]$, there exist positive constants β and C_s and a time $T^* \in]0, T]$ such that

$$\begin{aligned} &\sup_{0\leq t\leq T^*}\left(\|u(t,\cdot)\|_{H^{-s+1-\beta t}} + \|\partial_t u(t,\cdot)\|_{H^{-s-\beta t}}\right) \\ &\quad\leq C_s\left(\|u(0,\cdot)\|_{H^{-s+1}} + \|\partial_t u(0,\cdot)\|_{H^{-s}} + \int_0^{T^*} \|Lu(t,\cdot)\|_{H^{-s-\beta t}}\,dt\right) \end{aligned} \tag{1.6}$$

for all $u \in \mathcal{C}^2([0,T]; H^\infty(\mathbb{R}^N))$. Let us point out that the bound on s was due to this reason: the product by a log-Lipschitz function is well defined in H^s if and only if $|s| < 1$. Note also that this fact gives us a bound on the lifespan of the solution: the regularity index $-s + 1 - \beta T^*$ has to be strictly positive, so one can expect only local in time existence of a solution. Moreover in the case the coefficients a_{jk} are $\mathcal{C}_b^\infty$ in space, the authors proved inequality (1.6) for all s: so, they got well-posedness in H^∞ *with a loss of derivatives increasing in time.*

The case of a complete strictly hyperbolic second-order operator,

$$Lu = \sum_{j,k=0}^{N} \partial_{y_j} (a_{jk}\, \partial_{y_k} u) + \sum_{j=0}^{N} \big(b_j\, \partial_{y_j} u + \partial_{y_j}(c_j\, u)\big) + d\, u$$

(here we set $y = (t,x) \in \mathbb{R}_t \times \mathbb{R}_x^N$), was considered by Colombini and Métivier in [11]. They assumed the same isotropic log-Lipschitz condition of [10] on the coefficients of the second-order part of L, while b_j and c_j were supposed to be α-Hölder continuous (for some $\alpha \in]1/2,1[$) and d to be only bounded. The authors headed towards questions such as local-in-space existence and uniqueness, and finite propagation speed. Note that the restriction on α was due to the local-in-space study the authors performed (in particular, in order to have traces on the boundary well-defined): see for instance papers [10] and [9], were no bounds on the Hölder regularity were required.

Trying to mix up a Tarama-like hypothesis (concerning the dependence on the time variable) with the one of Colombini and Lerner (with respect to x) is at this point a natural step. More precisely, we want to consider a pointwise log-Zygmund condition in time and a pointwise log-Lipschitz condition in space, uniformly with respect to the other variable (see conditions (2.3) and (2.4) below). Under this assumption, it is possible to prove an energy estimate analogous to (1.6). The first result in this direction was due to Colombini and Del Santo, in paper [6] (see also [12], where smoothness in space was however required). Nevertheless, they had to restrict themselves to the case of space dimension $N = 1$ (see also [9] for the case of a complete operator). The result in the general instance $N \geq 1$ was only recently proved in paper [7]. Even if the difficulty of the additional space variables seems to be just technical, the approach assumed in [7], i.e., of paradifferential calculus depending on parameters, actually gives a more general and complete point of view of the problem.

The aim of the present note is to extend the result of [7] for homogeneous second-order operators to the case where also lower-order terms come into play. In particular, according with the results of [10], [11] and [9], we will assume the first-order coefficients to be Hölder continuous in the space variable, and the zero-order coefficient to be just bounded. From an energy estimate analogous to (1.6), we will infer, for smooth in space coefficients, the H^∞ well-posedness of the related Cauchy problem with a finite loss of derivatives.

To complete the general picture of the state of the art, let us recall that in the already quoted paper [19], Tarama proved also an energy estimate with no loss of derivatives for coefficients $a_{jk}(t)$ which fulfill a (integral) Zygmund hypothesis. This result has been recently generalized in [8] to the case of a_{jk}'s Zygmund continuous both in t and x (assuming also Hölder first-order coefficients and bounded zero-order coefficient), but only in the space $H^{1/2} \times H^{-1/2}$. If an isotropic Zygmund assumption implies or not loss of regularity for general initial data in $H^s \times H^{s-1}$ is still an open problem.

Our paper is organized in the following way.

In the next section, we will state our main results.

A brief presentation of the tools we need in order to prove them will follow, in Section 3: we refer to [7] for a complete description. These tools mostly come from Fourier analysis. In particular, we will recall some basic facts on Littlewood–Paley decomposition and paradifferential calculus, both in its classical version (see, e.g., [1], [2] and [16]) and in the one depending on a parameter (see [15] and [17]). We will also recall some basic facts on functions with low regularity modulus of continuity.

Section 4 is devoted to the proof of energy estimates for our strictly hyperbolic operator with non regular (both in t and x) coefficients: this is the most involving part. We will just give a sketch of the analysis of the higher-order terms (already treated in [7]), while we will give the most of the details about the lower-order ones.

Finally, in Section 5 we will show how the just proved energy estimates allow to recover the H^∞ well-posedness of the related Cauchy problem, in the case of coefficients which are smooth in the space variables.

2. Main results

Let us consider the operator over $[0, T_0] \times \mathbb{R}^N$ (for some $T_0 > 0$ and $N \geq 1$)

$$Lu = \partial_t^2 u - \sum_{j,k=1}^{N} \partial_j \left(a_{jk}(t,x)\, \partial_k u\right) + \sum_{j=0}^{N} b_j(t,x)\, \partial_j u + c(t,x)\, u \tag{2.1}$$

(with the notation $\partial_0 = \partial_t$) and let us suppose L to be strictly hyperbolic with bounded coefficients, i.e., there exist two positive constants $0 < \lambda_0 \leq \Lambda_0$ such that, for all $(t,x) \in \mathbb{R}_t \times \mathbb{R}^N_x$ and all $\xi \in \mathbb{R}^N$, one has

$$\lambda_0\, |\xi|^2 \leq \sum_{j,k=1}^{N} a_{jk}(t,x)\, \xi_j\, \xi_k \leq \Lambda_0\, |\xi|^2 . \tag{2.2}$$

Moreover, assume that the a_{jk}'s are log-Zygmund continuous in the time variable, uniformly with respect to x: there exists a constant $K_0 > 0$ such that, for all $\tau > 0$, one has

$$\sup_{(t,x)} |a_{jk}(t+\tau,x) + a_{jk}(t-\tau,x) - 2a_{jk}(t,x)| \leq K_0\, \tau\, \log\left(1 + \frac{1}{\tau}\right) . \tag{2.3}$$

The goal of the present paper is to prove the following H^∞ well-posedness result.

Theorem 2.1. *Consider the operator L defined in (2.1), under the strictly hyperbolicity condition (2.2).*
Let us suppose the following conditions on the coefficients of L:

(i) *the a_{jk}'s are* log-*Zygmund continuous in time, i.e.,* (2.3) *holds;*
(ii) *the a_{jk}'s are also of class $\mathcal{C}_b^\infty$ with respect to x;*
(iii) *b_j, for all $0 \leq j \leq N$, and c belong to $L^\infty([0,T_0];\mathcal{C}_b^\infty(\mathbb{R}^N))$.*

Then the Cauchy problem for L,

$$\text{(CP)} \qquad \begin{cases} Lu = f \\ u_{|t=0} = u_0\,, \quad \partial_t u_{|t=0} = u_1\,, \end{cases}$$

is well posed in $H^\infty(\mathbb{R}^N)$, globally in time.

Recall that the notation $f \in \mathcal{C}_b^\infty$ means that f is of class $\mathcal{C}^\infty$, and it is bounded with all its derivatives.

The proof of the previous result is essentially based on finding energy estimates for L: we will establish them when the coefficients satisfy minimal regularity assumptions also with respect to the space variable. This means that we will focus on log-Lipschitz continuous a_{jk}'s and Hölder continuous b_j's, while the 0th-order coefficient c can be taken just bounded (see also [10], [11] and [9] about these hypothesis).

So, assume that there exists a constant $K_1 > 0$ such that

$$\sup_{(t,x)} |a_{jk}(t,x+y) - a_{jk}(t,x)| \leq K_1\,|y|\,\log\left(1 + \frac{1}{|y|}\right) \tag{2.4}$$

for all $y \in \mathbb{R}^N \setminus \{0\}$. Suppose also the following on the lower-order terms:

$$b_j \in L^\infty([0,T_0];\mathcal{C}^\omega(\mathbb{R}^N))\,, \quad 0 \leq j \leq N\,, \quad \text{and} \quad c \in L^\infty([0,T_0]\times\mathbb{R}^N)\,, \tag{2.5}$$

for some real number $0 < \omega \leq 1$ (with $\mathcal{C}_*^\omega := B_{\infty,\infty}^\omega$ if $\omega = 1$).

Theorem 2.2. *Let L be the operator defined in (2.1), whose coefficients fulfill conditions (2.2), (2.3), (2.4) and (2.5).*

Then, for all fixed $\theta \in\,]0,\min\{\omega,1\}[$, there exist a constant $\beta^* > 0$, a time $T > 0$ and a constant $C > 0$ such that the estimate

$$\begin{aligned} &\sup_{0\leq t\leq T}\left(\|u(t,\cdot)\|_{H^{-\theta+1-\beta^* t}} + \|\partial_t u(t,\cdot)\|_{H^{-\theta-\beta^* t}}\right) \\ &\quad \leq C\left(\|u(0,\cdot)\|_{H^{-\theta+1}} + \|\partial_t u(0,\cdot)\|_{H^{-\theta}} + \int_0^T \|Lu(t,\cdot)\|_{H^{-\theta-\beta^* t}}\,dt\right) \end{aligned} \tag{2.6}$$

holds true for all $u \in \mathcal{C}^2([0,T];H^\infty(\mathbb{R}^N))$.

We will explain in Section 5 how one can recover the H^∞ well-posedness for (*CP*) from the energy estimates of Theorem 2.2.

Remark 2.3. It is natural to wonder if a similar result still holds true under an isotropic log-Zygmund condition. In our proof, we will use the fact that the high frequencies of a log-Lipschitz function behave as the ones of a log-Zygmund one (see Propositions 3.10 and 3.11). The low frequencies of a log-Zygmund function, instead, have an additional logarithmic loss with respect to a log-Lipschitz function: because of this loss, our method doesn't work anymore for such a regularity assumption. Therefore, the answer to the previous question is not clear at the present time.

3. Functional toolbox

The present section aims to give an overview of the tools, based on Fourier analysis techniques, we need to prove our results. We refer to paper [7] (in particular Section 3) and the references therein for a more complete and detailed treatment.

Let us start with a definition. We set[1] $\Pi(D) := \log(2+|D|)$, i.e., its symbol is $\pi(\xi) := \log(2+|\xi|)$.

Definition 3.1. For all $\alpha \in \mathbb{R}$, we define the space $H^{s+\alpha\log}$ as $\Pi^{-\alpha}H^s$, i.e.,

$$f \in H^{s+\alpha\log} \quad \Leftrightarrow \quad \Pi^\alpha f \in H^s \quad \Leftrightarrow \quad \pi^\alpha(\xi)\left(1+|\xi|^2\right)^{s/2}\widehat{f}(\xi) \in L^2 .$$

These logarithmic Sobolev spaces were introduced in [11]. Due to the low regularity of the coefficients of L, this kind of spaces will come into play in our computations.

3.1. Littlewood–Paley theory and paradifferential calculus

Let us now define the so-called Littlewood–Paley decomposition, based on a non-homogeneous dyadic partition of unity with respect to the Fourier variable. We refer to [1], [2] and [16] for the details.

So, fix a smooth radial function χ supported in the ball $B(0,2)$, equal to 1 in a neighborhood of $B(0,1)$ and such that $r \mapsto \chi(r\,e)$ is nonincreasing over $\mathbb{R}_+$ for all unitary vectors $e \in \mathbb{R}^N$. Set also $\varphi(\xi) = \chi(\xi) - \chi(2\xi)$.

The dyadic blocks $(\Delta_j)_{j\in\mathbb{Z}}$ are defined by

$$\Delta_j := 0 \ \text{ if } \ j \leq -1, \quad \Delta_0 := \chi(D) \quad \text{and} \quad \Delta_j := \varphi(2^{-j}D) \ \text{ if } \ j \geq 1.$$

We also introduce the following low frequency cut-off operator:

$$S_j u := \chi(2^{-j}D) = \sum_{k\leq j}\Delta_k \quad \text{for} \quad j \geq 0.$$

The following classical properties will be used freely throughout the paper:

- for any $u \in \mathcal{S}'$, the equality $u = \sum_j \Delta_j u$ holds true in $\mathcal{S}'$;
- for all u and v in $\mathcal{S}'$, the sequence $(S_{j-3}u\ \Delta_j v)_{j\in\mathbb{N}}$ is spectrally supported in dyadic annuli.

---

[1] Throughout we agree that $f(D)$ stands for the pseudo-differential operator $u \mapsto \mathcal{F}^{-1}(f\,\mathcal{F}u)$.

Let us also mention a fundamental result, which explains, by the so-called *Bernstein's inequalities*, the way in which derivatives act on spectrally localized functions.

Lemma 3.2. *Let $0 < r < R$. A constant C exists so that, for any nonnegative integer k, any couple (p,q) in $[1,+\infty]^2$ with $p \leq q$ and any function $u \in L^p$, we have, for all $\lambda > 0$,*

(i) $\operatorname{supp}\widehat{u} \subset B(0,\lambda R) \quad \Longrightarrow \quad \|\nabla^k u\|_{L^q} \leq C^{k+1}\,\lambda^{k+N\left(\frac{1}{p}-\frac{1}{q}\right)}\,\|u\|_{L^p}$;
(ii) $\operatorname{supp}\widehat{u} \subset \{\xi \in \mathbb{R}^N \,|\, r\lambda \leq |\xi| \leq R\lambda\}$
$\Longrightarrow \quad C^{-k-1}\,\lambda^k\|u\|_{L^p} \leq \|\nabla^k u\|_{L^p} \leq C^{k+1}\,\lambda^k\|u\|_{L^p}$.

By the Littlewood–Paley decomposition, we can characterize the new Sobolev class introduced in Definition 3.1 (see [16, Prop. 4.1.11]).

Proposition 3.3. *Let s, $\alpha \in \mathbb{R}$. A tempered distribution u belongs to the space $H^{s+\alpha\log}$ if and only if:*

(i) *for all $k \in \mathbb{N}$, $\Delta_k u \in L^2(\mathbb{R}^N)$;*
(ii) *setting $\delta_k := 2^{ks}\,(1+k)^\alpha\,\|\Delta_k u\|_{L^2}$ for all $k \in \mathbb{N}$, then $(\delta_k)_k \in \ell^2(\mathbb{N})$.*

Moreover, $\|u\|_{H^{s+\alpha\log}}$ is equivalent to $\|(\delta_k)_k\|_{\ell^2}$.

Littlewood–Paley decomposition allows also to define operators associated to low regularity symbols, via the so-called paradifferential calculus. For reasons which will appear clear in the sequel, we have to resort not to the classical version, as introduced by Bony in [2], but to the one depending on a parameter $\gamma \geq 1$ (see, e.g., [17] and [15]). Let us recall here the main ideas of the construction.

Hence, fix $\gamma \geq 1$ and take a cut-off function $\psi_\gamma \in \mathcal{C}^\infty(\mathbb{R}^N \times \mathbb{R}^N)$ which verifies the following properties:

- there exist $0 < \varepsilon_1 < \varepsilon_2 < 1$ such that

$$\psi_\gamma(\eta,\xi) = \begin{cases} 1 & \text{for} \quad |\eta| \leq \varepsilon_1\,(\gamma + |\xi|) \\ 0 & \text{for} \quad |\eta| \geq \varepsilon_2\,(\gamma + |\xi|)\,; \end{cases}$$

- for all $(\beta,\alpha) \in \mathbb{N}^N \times \mathbb{N}^N$, there exists a constant $C_{\beta,\alpha}$ such that

$$\left|\partial_\eta^\beta \partial_\xi^\alpha \psi_\gamma(\eta,\xi)\right| \leq C_{\beta,\alpha}\,(\gamma + |\xi|)^{-|\alpha|-|\beta|}\,.$$

Remark 3.4. We remark that ε_1, ε_2 and the different $C_{\beta,\alpha}$'s occurring in the previous estimates must not depend on γ.

For instance, if $\gamma = 1$, one can take

$$\psi_\gamma(\eta,\xi) \equiv \psi_1(\eta,\xi) := \sum_{l=0}^{+\infty} \chi_{l-3}(\eta)\,\varphi_l(\xi)\,,$$

where χ and φ are the localization (in the phase space) functions associated to a Littlewood–Paley decomposition. Similarly, if $\gamma > 1$ it is possible to find a suitable integer $\mu \geq 0$ such that

$$\psi_\gamma(\eta,\xi) := \chi_\mu(\eta)\,\chi_{\mu+2}(\xi) + \sum_{l=\mu+3}^{+\infty} \chi_{l-3}(\eta)\,\varphi_l(\xi) \tag{3.1}$$

is a function with the properties just described.

Define now G^{ψ_γ} as the inverse Fourier transform of ψ_γ with respect to the variable η:

$$G^{\psi_\gamma}(x,\xi) := \left(\mathcal{F}_\eta^{-1}\psi_\gamma\right)(x,\xi).$$

Lemma 3.5. *For all $(\beta,\alpha) \in \mathbb{N}^N \times \mathbb{N}^N$, there exist constants $C_{\beta,\alpha}$, independent of γ, such that:*

$$\left\|\partial_x^\beta \partial_\xi^\alpha G^{\psi_\gamma}(\cdot,\xi)\right\|_{L^1(\mathbb{R}_x^N)} \leq C_{\beta,\alpha}\left(\gamma+|\xi|\right)^{-|\alpha|+|\beta|}, \tag{3.2}$$

$$\left\||\cdot|\,\log\left(2+\frac{1}{|\cdot|}\right)\,\partial_x^\beta \partial_\xi^\alpha G^{\psi_\gamma}(\cdot,\xi)\right\|_{L^1(\mathbb{R}_x^N)} \leq C_{\beta,\alpha}\left(\gamma+|\xi|\right)^{-|\alpha|+|\beta|-1} \cdot \log(1+\gamma+|\xi|). \tag{3.3}$$

Thanks to G^{ψ_γ}, we can smooth out a symbol a in the x variable and then define the paradifferential operator associated to a as the pseudodifferential operator related to this smooth function. We set the classical symbol associated to a to be

$$\sigma_a(x,\xi) := \left(\psi_\gamma(D_x,\xi)\,a\right)(x,\xi) = \left(G^{\psi_\gamma}(\cdot,\xi) *_x a(\cdot,\xi)\right)(x),$$

and then the paradifferential operator associated to a:

$$T_a := \sigma_a(x,D_x),$$

where we have omitted ψ_γ because the definition is independent of it, up to lower-order terms (see [17]).

Remark 3.6. Let us note that if $a = a(x) \in L^\infty$ and if we take the cut-off function ψ_1, then T_a is actually the usual paraproduct operator. If we take ψ_γ as defined in (3.1), instead, we get a paraproduct operator which starts from high enough frequencies, which will be indicated with T_a^μ(see [11, Par. 3.3]).

Let us point out that we can also define a γ-dyadic decomposition. First of all, we set

$$\Lambda(\xi,\gamma) := \left(\gamma^2 + |\xi|^2\right)^{1/2}.$$

Then, taken the usual smooth function χ associated to a Littlewood–Paley decomposition, we define

$$\chi_\nu(\xi,\gamma) := \chi\left(2^{-\nu}\Lambda(\xi,\gamma)\right), \quad S_\nu^\gamma := \chi_\nu(D_x,\gamma), \quad \Delta_\nu^\gamma := S_{\nu+1}^\gamma - S_\nu^\gamma.$$

The usual properties of the support of the localization functions still hold, and for all fixed $\gamma \geq 1$ and all tempered distributions u, we have

$$u = \sum_{\nu=0}^{+\infty} \Delta_\nu^\gamma u \quad \text{in } \mathcal{S}'.$$

Moreover, with natural modifications in definitions, we can introduce the space $H_\gamma^{s+\alpha\log}$ as the set of tempered distributions for which

$$\|u\|^2_{H_\gamma^{s+\alpha\log}} := \int_{\mathbb{R}^N_\xi} \Lambda^{2s}(\xi,\gamma)\,\log^{2\alpha}(1+\gamma+|\xi|)\,|\widehat{u}(\xi)|^2\,d\xi \;<\;+\infty\,.$$

For the details see [17, Appendix B]. What is important to retain is that, once we fix $\gamma \geq 1$ (for example, to obtain positivity of paradifferential operators involved in our computations), all the previous construction is equivalent to the classical one; in particular, the space $H_\gamma^{s+\alpha\log}$ coincides with $H^{s+\alpha\log}$, the respective norms are equivalent and the characterization given by Proposition 3.3 still holds true.

3.2. On log-Lipschitz and log-Zygmund functions

Let us now recall some definitions and properties about the low regularity conditions we assume on the second-order coefficients of L. If not otherwise specified, one can find all the proofs to next results in [7, Section 3].

Definition 3.7. A function $f \in L^\infty(\mathbb{R}^N)$ is said to be log-Lipschitz, and we write $f \in LL(\mathbb{R}^N)$, if the quantity

$$|f|_{LL} := \sup_{x,y\in\mathbb{R}^N,\,|y|<1}\left(\frac{|f(x+y)\;-\;f(x)|}{|y|\,\log\left(1\,+\,\frac{1}{|y|}\right)}\right) \;<\;+\infty\,.$$

We set $\|f\|_{LL} := \|f\|_{L^\infty} + |f|_{LL}$.

Let us define also the space of log-Zygmund functions. We will give the general definition in $\mathbb{R}^N$, even if the one-dimensional case will be the only relevant one for our purposes.

Definition 3.8. A function $g \in L^\infty(\mathbb{R}^N)$ is said to be log-Zygmund, and we write $g \in LZ(\mathbb{R}^N)$, if the quantity

$$|g|_{LZ} := \sup_{x,y\in\mathbb{R}^N,\,|y|<1}\left(\frac{|g(x+y)\,+\,g(x-y)\;-\;2\,g(x)|}{|y|\,\log\left(1\,+\,\frac{1}{|y|}\right)}\right) \;<\;+\infty\,.$$

We set $\|g\|_{LZ} := \|g\|_{L^\infty} + |g|_{LZ}$.

Remark 3.9. Let us immediately point out that, by monotonicity of logarithmic function, we can replace the factor $\log(1+1/|y|)$ in previous definitions with $\log(1+\gamma+1/|y|)$, for all parameters $\gamma \geq 1$. As paradifferential calculus with parameters will play a fundamental role in our computations, it is convenient to perform such a change, and so does also in hypothesis (2.3) and (2.4) of Section 2.

Let us recall also the following characterization of the space LZ.

Proposition 3.10. *The space $LZ(\mathbb{R}^N)$ coincides with the logarithmic Besov space $B^{1-\log}_{\infty,\infty}$, i.e., the space of tempered distributions u such that*

$$\sup_{\nu\geq 0}\left(2^\nu\,(\nu+1)^{-1}\,\|\Delta_\nu u\|_{L^\infty}\right) \;<\;+\infty\,. \tag{3.4}$$

On the one hand, from Definitions 3.7 and 3.8, it is obvious that $LL(\mathbb{R}^N) \hookrightarrow LZ(\mathbb{R}^N)$: Proposition 3.3 of [10] explains this property in terms of dyadic decomposition.

Proposition 3.11. *There exists a constant C such that, for all $a \in LL(\mathbb{R}^N)$ and all integers $k \geq 0$, we have*

$$\|\Delta_k a\|_{L^\infty} \leq C\,(k+1)\,2^{-k}\,\|a\|_{LL}\,. \tag{3.5}$$

Moreover, for all $k \in \mathbb{N}$ we have

$$\|a\,-\,S_k a\|_{L^\infty} \leq C\,(k+1)\,2^{-k}\,\|a\|_{LL} \tag{3.6}$$

$$\|S_k a\|_{C^{0,1}} \leq C\,(k+1)\,\|a\|_{LL}\,. \tag{3.7}$$

Recall that, again from Proposition 3.3 of [10], Property (3.7) is a characterization of the space LL.

On the other hand, we have also $B^1_{\infty,\infty} \hookrightarrow LL$: the Zygmund condition entails a logarithmic loss in the first variation. As the next lemma states, this instance actually occurs also for log-Zygmund functions. For our purposes, it is enough to consider a log-Zygmund function a depending only on the time variable t, but the same reasoning holds also in higher dimensions.

Lemma 3.12. *For all $a \in LZ(\mathbb{R})$, there exists a constant C, depending only on the LZ norm of a, such that, for all $\gamma \geq 1$ and all $0 < |\tau| < 1$ one has*

$$\sup_{t\in\mathbb{R}} |a(t+\tau) - a(t)| \,\leq\, C\,|\tau|\,\log^2\left(1+\gamma+\frac{1}{|\tau|}\right). \tag{3.8}$$

Now, given a log-Zygmund function $a(t)$, we can regularize it by convolution. So, take an even function $\rho \in \mathcal{C}_0^\infty(\mathbb{R}_t)$, $0 \leq \rho \leq 1$, whose support is contained in the interval $[-1,1]$ and such that $\int \rho(t)dt = 1$, and define the mollifier kernel

$$\rho_\varepsilon(t) \,:=\, \frac{1}{\varepsilon}\,\rho\left(\frac{t}{\varepsilon}\right) \qquad\qquad \forall\,\varepsilon \in\,]0,1]\,.$$

We smooth out the function a setting, for all $\varepsilon \in\,]0,1]$,

$$a_\varepsilon(t) \,:=\, (\rho_\varepsilon\,*\,a)\,(t) \,=\, \int_{\mathbb{R}_s} \rho_\varepsilon(t-s)\,a(s)\,ds\,. \tag{3.9}$$

The following proposition holds.

Proposition 3.13. *Let $a \in LZ(\mathbb{R})$. There exist constants C such that, for all $\gamma \geq 1$,*

$$|a_\varepsilon(t) - a(t)| \leq C\,\|a\|_{LZ}\;\varepsilon\,\log\left(1+\gamma+\frac{1}{\varepsilon}\right) \tag{3.10}$$

$$|\partial_t a_\varepsilon(t)| \leq C\,\|a\|_{LZ}\;\log^2\left(1+\gamma+\frac{1}{\varepsilon}\right) \tag{3.11}$$

$$\left|\partial_t^2 a_\varepsilon(t)\right| \leq C\,\|a\|_{LZ}\;\frac{1}{\varepsilon}\;\log\left(1+\gamma+\frac{1}{\varepsilon}\right). \tag{3.12}$$

3.3. Low regularity symbols and calculus

Let us now consider symbols $a(t,x,\xi)$, which are $LZ_t - LL_x$, and investigate properties of the corresponding paradifferential operators, built as explained above.

As we will have to make use of the paradifferential calculus with parameters, we will allow the symbols to depend also on some $\gamma \geq 1$.

We point out that in our calculus the time can be treated as an additional parameter, while ξ represents, as usual, the dual variable to x.

Definition 3.14. A *symbol of order* $m+\delta \log$ (for $m, \delta \in \mathbb{R}$) is a function $a(t,x,\xi,\gamma)$ which is locally bounded on $[0,T_0]\times\mathbb{R}^N\times\mathbb{R}^N\times[1,+\infty[$, of class $\mathcal{C}^\infty$ with respect to ξ and which satisfy the following property: for all $\alpha \in \mathbb{N}^N$, there exists a constant $C_\alpha > 0$ such that

$$\left|\partial_\xi^\alpha a(t,x,\xi,\gamma)\right| \leq C_\alpha \left(\gamma + |\xi|\right)^{m-|\alpha|} \log^\delta \left(1+\gamma+|\xi|\right) \tag{3.13}$$

for all (t,x,ξ,γ).

So, take a symbol $a(t,x,\xi,\gamma)$ of order $m \geq 0$, which is log-Zygmund in t and log-Lipschitz in x, uniformly with respect to the other variables. Now we smooth a out with respect to time, as done in (3.9). Next lemma, whose proof can be found in [7], provides us estimates on classical symbols associated to a_ε and its time derivatives.

Lemma 3.15. *The classical symbols associated to* a_ε *and its time derivatives satisfy:*

$$\begin{aligned}
\left|\partial_\xi^\alpha \sigma_{a_\varepsilon}\right| &\leq C_\alpha \left(\gamma+|\xi|\right)^{m-|\alpha|} \\
\left|\partial_x^\beta \partial_\xi^\alpha \sigma_{a_\varepsilon}\right| &\leq C_{\beta,\alpha} \left(\gamma+|\xi|\right)^{m-|\alpha|+|\beta|-1} \log\left(1+\gamma+|\xi|\right) \\
\left|\partial_\xi^\alpha \sigma_{\partial_t a_\varepsilon}\right| &\leq C_\alpha \left(\gamma+|\xi|\right)^{m-|\alpha|} \log^2\left(1+\gamma+\frac{1}{\varepsilon}\right) \\
\left|\partial_x^\beta \partial_\xi^\alpha \sigma_{\partial_t a_\varepsilon}\right| &\leq C_{\beta,\alpha} \left(\gamma+|\xi|\right)^{m-|\alpha|+|\beta|-1} \log\left(1+\gamma+|\xi|\right) \frac{1}{\varepsilon} \\
\left|\partial_\xi^\alpha \sigma_{\partial_t^2 a_\varepsilon}\right| &\leq C_\alpha \left(\gamma+|\xi|\right)^{m-|\alpha|} \log\left(1+\gamma+\frac{1}{\varepsilon}\right) \frac{1}{\varepsilon} \\
\left|\partial_x^\beta \partial_\xi^\alpha \sigma_{\partial_t^2 a_\varepsilon}\right| &\leq C_{\beta,\alpha} \left(\gamma+|\xi|\right)^{m-|\alpha|+|\beta|-1} \log\left(1+\gamma+|\xi|\right) \frac{1}{\varepsilon^2},
\end{aligned}$$

where all the constants which occur here don't depend on the parameter γ.

Note that first and second inequalities are satisfied also by the symbol a (not smoothed in time).

Now let us quote some basic facts on symbolic calculus, which follow from previous lemma. Before doing this, we recall a definition (see also [17, Def. B.8]).

Definition 3.16. We say that an operator P is of order $m+\delta\log$ if, for every $(s,\alpha)\in\mathbb{R}^2$ and every $\gamma \geq 1$, P maps $H_\gamma^{s+\alpha\log}$ into $H_\gamma^{(s-m)+(\alpha-\delta)\log}$ continuously.

Proposition 3.17.

(i) *Let* a *be a symbol of order* m *which is* L^∞ *in the* x *variable. Then* T_a *is of order* m, *i.e., it maps* $H_\gamma^{s+\alpha\log}$ *into* $H_\gamma^{s-m+\alpha\log}$.

(ii) *Let us take two symbols a, b of order m and m' respectively. Suppose that a, b are LL in the x variable. The composition of the associated operators can be approximated by the symbol associated to the product of these symbols, up to a remainder term:*

$$T_a \circ T_b = T_{ab} + R .$$

R has order $m + m' - 1 - \log$: it maps $H_\gamma^{s+\alpha \log}$ into $H_\gamma^{s-m-m'+1+(\alpha+1)\log}$.

(iii) *Let a be a symbol of order m which is LL in the x variable. The adjoint (over L^2) operator of T_a is, up to a remainder operator, $T_{\overline{a}}$. The remainder operator is of order $m-1+\log$, and it maps $H_\gamma^{s+\alpha \log}$ into $H_\gamma^{s-m+1+(\alpha+1)\log}$.*

Let us end this subsection stating a basic positivity estimate. In this situation, paradifferential calculus with parameters comes into play.

Proposition 3.18. *Let $a(t,x,\xi,\gamma)$ be a symbol of order m, which is LL in the x variable and such that*

$$\mathrm{Re}\, a(t,x,\xi,\gamma) \geq \lambda_0 \, (\gamma + |\xi|)^m .$$

Then, there exists a constant λ_1, depending only on $|a|_{LL}$ and on λ_0 (so not on γ), such that, for γ large enough, one has

$$\mathrm{Re}\, (T_a u, u)_{L^2(\mathbb{R}^N_x)} \geq \lambda_1 \, \|u\|^2_{H_\gamma^{m/2}} .$$

Remark 3.19. If the positive symbol a has low regularity in time and we smooth it out by convolution with respect to this variable, we obtain a family $(a_\varepsilon)_\varepsilon$ of positive symbols, with same constant λ_0. Now, all the paradifferential operators associated to these symbols will be positive operators, uniformly in ε: the constant λ_1 of the previous inequality can be chosen independently of ε.

Then we have the following corollary, which turns out to be fundamental in our energy estimates.

Corollary 3.20. *Let a be a positive symbol of order 1 and suppose that a is LL in the x variable.*

Then there exists $\gamma \geq 1$, depending only on the symbol a, such that

$$\|T_a u\|_{L^2} \sim \|u\|_{H^1_\gamma}$$

for all $u \in H^1_\gamma(\mathbb{R}^N)$.

4. Proof of the energy estimates

Let us now tackle the proof of Theorem 2.2.

The analysis of the higher-order terms, which is the most involving part, was already performed in [7]: so, we will give just a sketch of it.

We will give the most of the details, instead, when treating the lower-order terms. In doing this, we will follow a different approach than the one used in [9], based on the Bony's paraproduct decomposition (as done in the quoted work [7] for the second-order part).

4.1. Approximate and total energy

We first regularize coefficients a_{jk} in the time variable by convolution, as done in (3.9), and define the 0th-order symbol

$$\alpha_\varepsilon(t,x,\xi) := \left(\gamma^2+|\xi|^2\right)^{-1/2}\left(\gamma^2+\sum_{j,k} a_{jk,\varepsilon}(t,x)\,\xi_j\,\xi_k\right)^{1/2}.$$

We take $\varepsilon = 2^{-\nu}$ (see also [10] and [6]), and (for convenience) we will miss out the ε in the notation.

We immediately fix a real number $\gamma \geq 1$, which will depend only on λ_0 and on the $\sup_{j,k}|a_{jk}|_{LL_x}$, such that (see Corollary 3.20)

$$\|T_{\alpha^{-1/2}}\,w\|_{L^2} \geq \frac{\lambda_0}{2}\,\|w\|_{L^2} \quad \text{and} \quad \left\|T_{\alpha^{1/2}(\gamma^2+|\xi|^2)^{1/2}}\,w\right\|_{L^2} \geq \frac{\lambda_0}{2}\,\|w\|_{H^1} \tag{4.1}$$

for all $w \in H^\infty$. Recall that the choice of γ is equivalent to the choice of the parameter μ in (3.1) (see Remark 3.6): hence, from now on we will consider paraproducts starting from this μ, according to Definition (3.1), even if we will omit it in the notations.

Consider in (2.1) a function $u \in \mathcal{C}^2([0,T_0];H^\infty)$. We want to get energy estimates for u. We rewrite the equation using paraproduct operators by the coefficients a_{jk}:

$$\begin{aligned}
\partial_t^2 u &= Lu + \sum_{j,k}\partial_j\left(a_{jk}\,\partial_k u\right) - \sum_{j=0}^{N} b_j(t,x)\,\partial_j u - c(t,x)\,u \\
&= Lu + \sum_{j,k}\partial_j\left(\left(T_{a_{jk}}+\widetilde{A}_{jk}\right)\partial_k u\right) - \sum_{j=0}^{N}\left(T_{b_j}+\widetilde{B}_j\right)\partial_j u - \left(T_c+\widetilde{C}\right)u,
\end{aligned}$$

where $\widetilde{A}_{jk}\partial_k u = \left((a_{jk}-T_{a_{jk}})\partial_k u\right)$ and analogous for $\widetilde{B}_j$ and $\widetilde{C}$. Let us apply the operator Δ_ν: we get

$$\begin{aligned}
\partial_t^2 u_\nu = &\sum_{j,k}\partial_j\left(T_{a_{jk}}\,\partial_k u_\nu\right) - \sum_j T_{b_j}\,\partial_j u_\nu - T_c u_\nu \\
&+ (Lu)_\nu + \sum_{j,k}\left(\partial_j\left(\widetilde{A}_{jk}\partial_k u\right)\right)_\nu - \sum_j\left(\widetilde{B}_j\partial_j u\right)_\nu - \left(\widetilde{C}u\right)_\nu \\
&+ \sum_{j,k}\partial_j\left(\left[\Delta_\nu, T_{a_{jk}}\right]\partial_k u\right) - \sum_j\left[\Delta_\nu, T_{b_j}\right]\partial_j u - \left[\Delta_\nu, T_c\right]u,
\end{aligned} \tag{4.2}$$

where we have set $f_\nu = \Delta_\nu f$ and $[\Delta_\nu, T_g]$ is the commutator operator between Δ_ν and the paramultiplication by g.

Now, we define

$$\begin{aligned}
v_\nu(t,x) &:= T_{\alpha^{-1/2}}\,\partial_t u_\nu - T_{\partial_t(\alpha^{-1/2})}\,u_\nu \\
w_\nu(t,x) &:= T_{\alpha^{1/2}(\gamma^2+|\xi|^2)^{1/2}}\,u_\nu \\
z_\nu(t,x) &:= u_\nu
\end{aligned}$$

and the approximate energy associated to u_ν:

$$e_\nu(t) := \|v_\nu(t)\|^2_{L^2} + \|w_\nu(t)\|^2_{L^2} + \|z_\nu(t)\|^2_{L^2} \,. \tag{4.3}$$

Remark 4.1. Note that $\|w_\nu(t)\|^2_{L^2} \sim \|\nabla u_\nu\|^2_{L^2} \sim 2^{2\nu}\|u_\nu\|^2_{L^2}$, thanks to hypothesis (2.2) and the choice of the frequency μ from which the paraproduct starts, recall also (4.1).

Now, we fix a $\theta \in]0,1[$, as required in hypothesis, and we take a $\beta > 0$ to be chosen later; we can define the total energy associated to the solution u as

$$E(t) := \sum_{\nu\geq 0} e^{-2\beta(\nu+1)t}\, 2^{-2\nu\theta}\, e_\nu(t)\,. \tag{4.4}$$

It is not difficult to prove (see also inequality (4.8) below) that there exist constants C_θ and C'_θ, depending only on the fixed θ, for which one has:

$$(E(0))^{1/2} \leq C_\theta \left(\|\partial_t u(0)\|_{H^{-\theta}} + \|u(0)\|_{H^{-\theta+1}}\right) \tag{4.5}$$

$$(E(t))^{1/2} \geq C'_\theta \left(\|\partial_t u(t)\|_{H^{-\theta-\beta^* t}} + \|u(t)\|_{H^{-\theta+1-\beta^* t}}\right), \tag{4.6}$$

where we have set $\beta^* = \beta\,(\log 2)^{-1}$.

4.2. Time derivative of the approximate energy

Let us find an estimate on the time derivative of the energy. We start analysing each term of (4.3).

4.2.1. z_ν term.

For the third term we have:

$$\frac{d}{dt}\|z_\nu(t)\|^2_{L^2} = 2\,\mathrm{Re}\,(u_\nu\,,\,\partial_t u_\nu)_{L^2}\,. \tag{4.7}$$

Now, we have to control the term $\partial_t u_\nu$: using positivity of operator $T_{\alpha^{-1/2}}$, we get

$$\begin{aligned}\|\partial_t u_\nu\|_{L^2} &\leq C\,\|T_{\alpha^{-1/2}}\partial_t u_\nu\|_{L^2}\\ &\leq C\left(\|v_\nu\|_{L^2} + \left\|T_{\partial_t(\alpha^{-1/2})}u_\nu\right\|_{L^2}\right) \leq C\,(e_\nu)^{1/2}\,.\end{aligned} \tag{4.8}$$

So, we find the estimate:

$$\frac{d}{dt}\|z_\nu(t)\|^2_{L^2} \leq C\,e_\nu(t)\,. \tag{4.9}$$

4.2.2. v_ν term.

Straightforward computations show that

$$\partial_t v_\nu(t,x) = T_{\alpha^{-1/2}}\partial_t^2 u_\nu - T_{\partial_t^2(\alpha^{-1/2})}u_\nu\,.$$

Therefore, keeping in mind relation (4.2), we get:

$$\begin{aligned}\frac{d}{dt}\|v_\nu(t)\|^2_{L^2} = &-2\mathrm{Re}\left(v_\nu, T_{\partial_t^2(\alpha^{-1/2})}u_\nu\right)_{L^2}\\ &+2\sum_{j,k}\mathrm{Re}\,\left(v_\nu, T_{\alpha^{-1/2}}\left(\partial_j\left(T_{a_{jk}}\partial_k u_\nu\right) - T_{b_j}\partial_j u_\nu - T_c u_\nu\right)\right)_{L^2}\\ &+2\mathrm{Re}\left(v_\nu, T_{\alpha^{-1/2}}\left(\widetilde{L}u\right)_\nu\right)_{L^2} + 2\mathrm{Re}\,(v_\nu, T_{\alpha^{-1/2}}\mathcal{X}u)_{L^2}\,,\end{aligned} \tag{4.10}$$

where we have defined

$$\widetilde{L}u := Lu + \sum_{j,k} \partial_j \left(\widetilde{A}_{jk}\partial_k u\right) - \sum_j \widetilde{B}_j\partial_j u - \widetilde{C}u$$

$$\mathcal{X}u := \sum_{j,k} \partial_j \left[\Delta_\nu, T_{a_{jk}}\right] \partial_k u \,+\, \sum_j \left[\Delta_\nu, T_{b_j}\right] \partial_j u \,+\, \left[\Delta_\nu, T_c\right] u\,.$$

Obviously, we have

$$\left|2\,\mathrm{Re}\,\left(v_\nu\,,\,T_{\alpha^{-1/2}}\left(\widetilde{L}u\right)_\nu\right)_{L^2}\right| \leq C\,(e_\nu)^{1/2}\,\left\|\left(\widetilde{L}u\right)_\nu\right\|_{L^2}\,, \tag{4.11}$$

while from Lemma 3.15 and the choice $\varepsilon = 2^{-\nu}$ one immediately recovers

$$\left|2\,\mathrm{Re}\,\left(v_\nu\,,\,T_{\partial_t^2(\alpha^{-1/2})}u_\nu\right)_{L^2}\right| \leq C\,(\nu+1)\,e_\nu\,. \tag{4.12}$$

The other terms (paraproducts, commutators) of (4.10) will be treated later.

4.2.3. w_ν term. We now differentiate w_ν in time. A systematic use of symbolic calculus (see [7, Par. 4.2.3] for the details) leads us to

$$\frac{d}{dt}\,\|w_\nu\|^2_{L^2} \;=\; 2\,\mathrm{Re}\,\left(v_\nu\,,\,T_{\alpha^{-1/2}}T_{\alpha^2(\gamma^2+|\xi|^2)}u_\nu\right)_{L^2} \,+\, \mathcal{R}_\nu\,, \tag{4.13}$$

where $\mathcal{R}_\nu$ is a suitable remainder term which can be bounded by the localized energy, at the cost of a logarithmic loss:

$$|\mathcal{R}_\nu| \;\leq\; C\,(\nu+1)\,e_\nu\,. \tag{4.14}$$

Remark 4.2. The analysis of the remainder term $\mathcal{R}_\nu$ is performed by use of the symbolic calculus, introduced in Section 3. Let us point out that, concerning the regularity in x of the symbols, only the first-order derivatives come into play. This fact will be relevant in Section 5, in particular in Lemma 5.1.

Let us now consider the paraproduct terms. First of all, we pair up the second term of (4.10) with the first one of (4.13). As proved in [7], we have

$$\left|2\mathrm{Re}\left(v_\nu, T_{\alpha^{-1/2}}\Big(\sum\nolimits_{j,k} \partial_j\left(T_{a_{jk}}\partial_k u_\nu\right) + T_{\alpha^2(\gamma^2+|\xi|^2)}u_\nu\Big)\right)_{L^2}\right| \leq C\,(\nu+1)\,e_\nu, \tag{4.15}$$

where the constant C depends on the log-Lipschitz norm (with respect to space) of the coefficients a_{jk} and on the fixed parameters μ and γ.

For the lower-order terms, instead, we just use continuity results for the paraproduct operator (and (4.8) for $j=0$ below):

$$\left|2\,\mathrm{Re}\,\left(v_\nu\,,\,T_{\alpha^{-1/2}}T_c u_\nu\right)_{L^2}\right| \leq C\,e_\nu \tag{4.16}$$

$$\left|\sum\nolimits_{j=0}^{N} 2\,\mathrm{Re}\,\left(v_\nu\,,\,T_{\alpha^{-1/2}}T_{b_j}\partial_j u_\nu\right)_{L^2}\right| \leq C\,e_\nu\,, \tag{4.17}$$

where the constants C depend on $\|c\|_{L^\infty([0,T_0]\times\mathbb{R}^N)}$ and $\sup_j \|b_j\|_{L^\infty([0,T_0];\mathcal{C}^\omega)}$.

To sum up, from inequalities (4.9), (4.11), (4.12), from the analysis of the paradifferential operators in (4.15), (4.16) and (4.17), and from the estimate for the remainder term (4.14), we can conclude that

$$\frac{d}{dt}e_\nu(t) \leq C_1\,(\nu+1)\,e_\nu(t)+C_2\,(e_\nu(t))^{1/2}\left\|\left(\widetilde{L}u\right)_\nu(t)\right\|_{L^2}+\left|2\,\mathrm{Re}\,(v_\nu,T_{\alpha^{-1/2}}\mathcal{X}u)_{L^2}\right|. \tag{4.18}$$

4.3. Commutator terms

Let us now focus on the commutator terms. We will recall also the most of the details concerning the analysis of the higher-order terms, because conditions over β^* and T in Theorem 2.2 will follow from it.

So, first of all let us consider the quantity

$$\left|\sum_{j,k}\mathrm{Re}\,\left(v_\nu\,,\,T_{\alpha^{-1/2}}\partial_j\left[\Delta_\nu,T_{a_{jk}}\right]\partial_k u\right)_{L^2}\right|.$$

We start remarking that

$$[\Delta_\nu,T_{a_{jk}}]w=[\Delta_\nu,S_\mu a_{jk}]S_{\mu+2}w+\sum_{l=\mu+3}^{+\infty}[\Delta_\nu,S_{l-3}a_{jk}]\,\Delta_l w,$$

where μ is fixed, as usual (see Remark 3.6). Consequently,

$$\begin{aligned}&\partial_j\left([\Delta_\nu,T_{a_{jk}}]\,\partial_k u\right)\\&\quad=\partial_j\left([\Delta_\nu,S_\mu a_{jk}]\,\partial_k(S_{\mu+2}u)\right)+\partial_j\left(\sum_{l=\mu+3}^{+\infty}[\Delta_\nu,S_{l-3}a_{jk}]\,\partial_k(\Delta_l u)\right).\end{aligned}$$

The support of the Fourier transform of $[\Delta_\nu,S_\mu a_{jk}]\,\partial_k(S_{\mu+2}u)$ is contained in $\{|\xi|\leq 2^{\mu+4}\}$ and moreover $[\Delta_\nu,S_\mu a_{jk}]\,\partial_k(S_{\mu+2}u)$ is identically 0 if $\nu\geq\mu+5$. Then, from Bernstein's inequality and [4, Th. 35] we have

$$\|\partial_j\left([\Delta_\nu,S_\mu a_{jk}]\,\partial_k(S_{\mu+2}u)\right)\|_{L^2}\leq C_\mu\,\left(\sup_{j,k}\|a_{jk}\|_{LL_x}\right)\,\|S_{\mu+2}u\|_{L^2}\,;$$

hence, putting all these facts together, we find

$$\begin{aligned}&\left|\sum_{\nu=0}^{+\infty}e^{-2\beta(\nu+1)t}2^{-2\nu\theta}\sum_{j,k}2\mathrm{Re}\,(v_\nu,T_{\alpha^{-1/2}}\,\partial_j\left([\Delta_\nu,S_\mu a_{jk}]\partial_k(S_{\mu+2}u)\right))_{L^2}\right|\\&\leq C_\mu\left(\sup_{j,k}\|a_{jk}\|_{LL_x}\right)\sum_{\nu=0}^{\mu+4}e^{-2\beta(\nu+1)t}\,2^{-2\nu\theta}\|v_\nu\|_{L^2}\left(\sum_{h=0}^{\mu+2}\|u_h\|_{L^2}\right)\\&\leq C_\mu\,\left(\sup_{j,k}\|a_{jk}\|_{LL_x}\right)e^{\beta(\mu+5)T}\,2^{(\mu+4)\theta}\\&\quad\times\sum_{\nu=0}^{\mu+4}e^{-\beta(\nu+1)t}\,2^{-\nu\theta}\|v_\nu\|_{L^2}\sum_{h=0}^{\mu+4}e^{-\beta(h+1)t}\,2^{-h\theta}\|u_h\|_{L^2}\\&\leq C_\mu\,\left(\sup_{j,k}\|a_{jk}\|_{LL_x}\right)\,e^{\beta(\mu+5)T}\,2^{(\mu+4)\theta}\sum_{\nu=0}^{\mu+4}e^{-2\beta(\nu+1)t}\,2^{-2\nu\theta}e_\nu(t)\,.\end{aligned} \tag{4.19}$$

Next, let us consider

$$\partial_j\left(\sum_{l=\mu+3}^{+\infty}[\Delta_\nu,S_{l-3}a_{jk}]\,\partial_k(\Delta_l u)\right).$$

Looking at the support of the Fourier transform, it is possible to see that

$$[\Delta_\nu,S_{l-3}a_{jk}]\,\partial_k(\Delta_l u)$$

is identically 0 if $|l-\nu| \geq 3$. Consequently the sum over l is reduced to at most 5 terms: $\partial_j([\Delta_\nu, S_{\nu-5}a_{jk}]\,\partial_k(\Delta_{\nu-2}u)) + \cdots + \partial_j([\Delta_\nu, S_{\nu-1}a_{jk}]\,\partial_k(\Delta_{\nu+2}u))$, each of them having the support of the Fourier transform contained in $\{|\xi| \leq 2^{\nu+1}\}$. Let us consider one of these terms, e.g., $\partial_j([\Delta_\nu, S_{\nu-3}a_{jk}]\,\partial_k(\Delta_\nu u))$, the computation for the other ones being similar. We have, from Bernstein's inequality,

$$\left\|\partial_j\left([\Delta_\nu, S_{\nu-3}a_{jk}]\,\partial_k(\Delta_\nu u)\right)\right\|_{L^2} \leq C\,2^\nu\,\|[\Delta_\nu, S_{\nu-3}a_{jk}]\,\partial_k(\Delta_\nu u)\|_{L^2}\,.$$

On the other hand, using [4, Th. 35] again, we get

$$\|[\Delta_\nu, S_{\nu-3}a_{jk}]\partial_k(\Delta_\nu u)\|_{L^2} \leq C\,\|\nabla S_{\nu-3}a_{jk}\|_{L^\infty}\,\|\Delta_\nu u\|_{L^2}\,,$$

where C does not depend on ν. Consequently, using also (3.7), we deduce

$$\left\|\partial_j\left([\Delta_\nu, S_{\nu-3}a_{jk}]\,\partial_k(\Delta_\nu u)\right)\right\|_{L^2} \leq C\,2^\nu\,(\nu+1)\,\left(\sup_{j,k}\|a_{jk}\|_{LL_x}\right)\,\|\Delta_\nu u\|_{L^2}\,.$$

From this last inequality and similar ones for the other terms, we infer

$$\begin{aligned}&\left|\sum_{j,k}\mathrm{Re}\left(v_\nu, T_{\alpha^{-1/2}}\partial_j\left(\sum_{l=\mu+3}^{+\infty}[\Delta_\nu, S_{l-3}a_{jk}]\partial_k(\Delta_l u)\right)\right)_{L^2}\right| \\ &\qquad \leq C\left(\sup_{j,k}\|a_{jk}\|_{LL_x}\right)(\nu+1)e_\nu(t)\end{aligned}$$

and then, keeping in mind also (4.19),

$$\begin{aligned}&\left|\sum_{\nu=0}^{+\infty} e^{-2\beta(\nu+1)t}\,2^{-2\nu\theta}\sum_{j,k}2\,\mathrm{Re}\left(v_\nu\,,\,T_{\alpha^{-1/2}}\partial_j\left[\Delta_\nu, T_{a_{jk}}\right]\partial_k u\right)_{L^2}\right| \\ &\qquad \leq C\sum_{\nu=0}^{+\infty}(\nu+1)\,e^{-2\beta(\nu+1)t}\,2^{-2\nu\theta}\,e_\nu(t)\,,\end{aligned} \tag{4.20}$$

where C depends on μ, $\sup_{j,k}\|a_{jk}\|_{LL_x}$, on θ and on the product βT.

Let us now deal with the commutators coming from the lower-order parts. As the analysis is analogous to just performed one, we will only sketch the line of the argument. Moreover, let us consider only the b_j's terms, which are actually worse than the commutator $[\Delta_\nu, T_c]$.

First of all, we focus on the case of space derivatives: so, we take $1 \leq j \leq N$. As before, we can write

$$\left[\Delta_\nu\,,\,T_{b_j}\right]\partial_j u = [\Delta_\nu\,,\,S_\mu b_j]\,\partial_j S_{\mu+2}u + \sum_{l=\mu+3}^{+\infty}[\Delta_\nu\,,\,S_{l-3}b_j]\,\partial_j\Delta_l u\,.$$

From Bernstein's inequality and [4, Th. 35], one gets

$$\begin{aligned}\|[\Delta_\nu\,,\,S_\mu b_j]\,\partial_j S_{\mu+2}u\|_{L^2} &\leq C\,\|\nabla S_\mu b_j\|_{L^\infty}\,\|S_{\mu+2}u\|_{L^2} \\ &\leq C_\mu\,\|b_j\|_{C^\omega}\,\|S_{\mu+2}u\|_{L^2}\,;\end{aligned}$$

hence, exactly as done in estimate (4.19), we finally have

$$\begin{aligned}&\left|\sum_{\nu=0}^{+\infty} e^{-2\beta(\nu+1)t}2^{-2\nu\theta}\sum_{j=1}^{N}2\mathrm{Re}\left(v_\nu, T_{\alpha^{-1/2}}\,[\Delta_\nu, S_\mu b_j]\,\partial_j S_{\mu+2}u\right)_{L^2}\right| \\ &\qquad \leq C_\mu\left(\sup_j\|b_j\|_{C^\omega}\right)\,e^{\beta(\mu+5)T}\,2^{(\mu+4)\theta}\,E(t)\,.\end{aligned}$$

Moreover, also in this case, the series over l reduces to a finite sum, due to spectral localization of operators S_l and Δ_l. As before, it is enough to consider only the term for $l = \nu$: [4, Th. 35] tells us, once again, that

$$\|[\Delta_\nu, S_{\nu-3}b_j]\,\partial_j\Delta_\nu u\|_{L^2} \le C\|\nabla S_{\nu-3}b_j\|_{L^\infty}\|\Delta_\nu u\|_{L^2} \le C\,2^\nu\|b_j\|_{C^\omega}\|\Delta_\nu u\|_{L^2},$$

and therefore we have

$$\left|\sum_{\nu=0}^{+\infty} e^{-2\beta(\nu+1)t}2^{-2\nu\theta}\sum_{j=1}^{N} 2\mathrm{Re}\,\left(v_\nu, T_{\alpha^{-1/2}}\sum_{l=\mu+3}^{+\infty}[\Delta_\nu, S_{l-3}b_j]\,\partial_j\Delta_k u\right)_{L^2}\right|$$
$$\le C\left(\sup_j\|b_j\|_{C^\omega}\right)E(t)\,.$$

The term involving the time derivative of u, instead, requires us a little bit harder work because we can no more shift the derivative on the coefficient b_0 (as done before thanks to [4, Th. 35]). It is not difficult to see that the kernel of the operator

$$[\Delta_\nu\,,\,S_\kappa b_0]\,,$$

for all $\kappa \ge 0$, is the function

$$k(x,y) \;=\; 2^{\kappa N}\,\check{\varphi}(2^\kappa\,(x-y))\,\left(S_\kappa b_0(x) \;-\; S_\kappa b_0(y)\right)\,,$$

where we have set $\check{\varphi}$ to be the inverse Fourier transform of the localization function φ (defined in Subsection 3.1). So, by Schur's Lemma the operator norm of the previous commutator over L^2 is controlled by

$$\|[\Delta_\nu\,,\,S_\kappa b_0]\|_{\mathcal{L}(L^2)} \;<\; C\,\|b_0\|_{C^\omega}\,2^{-\omega\kappa}\,.$$

This having been established, we consider now the term $[\Delta_\nu\,,\,S_\mu b_0]\,S_{\mu+2}\partial_t u$. From the previous estimate we get

$$\|[\Delta_\nu\,,\,S_\mu b_0]\,S_{\mu+2}\partial_t u\|_{L^2} \;\le\; C\,\|b_0\|_{C^\omega}\,2^{-\omega\mu}\;\|S_{\mu+2}\partial_t u\|_{L^2}\,,$$

and arguing as for the (4.19) leads us to

$$\left|\sum_{\nu=0}^{+\infty} e^{-2\beta(\nu+1)t}\,2^{-2\nu\theta}2\mathrm{Re}\,\left(v_\nu, T_{\alpha^{-1/2}}\,[\Delta_\nu, S_\mu b_0]\,S_{\mu+2}\partial_t u\right)_{L^2}\right|$$
$$\le C\,\|b_0\|_{C^\omega}\,e^{\beta(\mu+5)T}\,2^{(\mu+4)(\theta-\omega)}\,E(t)\,.$$

The same arguments give an analogous estimate for $[\Delta_\nu, S_{\nu-3}b_0]\,\Delta_\nu\partial_t u$. Therefore the commutator term with first-order coefficients can be bounded by

$$\left|\sum_{\nu=0}^{+\infty} e^{-2\beta(\nu+1)t}2^{-2\nu\theta}\sum_{j=0}^{N} 2\mathrm{Re}\,\left(v_\nu, T_{\alpha^{-1/2}}\left[\Delta_\nu, T_{b_j}\right]\partial_j u\right)_{L^2}\right| \le C\,E(t)\,, \qquad (4.21)$$

where C depends also on $\sup_j\|b_j\|_{C^\omega}$.

As already said, the commutator term $[\Delta_\nu, T_c]$ is actually better, even if c is only L^∞, because there are no derivatives acting on u. The estimate is finally

$$\left|\sum_{\nu=0}^{+\infty} e^{-2\beta(\nu+1)t}\,2^{-2\nu\theta}\,2\,\mathrm{Re}\,\left(v_\nu, T_{\alpha^{-1/2}}\,[\Delta_\nu, T_c]\,u\right)_{L^2}\right| \le C\,\|c\|_{L^\infty}\;E(t)\,. \qquad (4.22)$$

In the end, putting inequalities (4.20), (4.21) and (4.22) all together, we get

$$\left|\sum_{\nu=0}^{+\infty} e^{-2\beta(\nu+1)t}2^{-2\nu\theta}2\mathrm{Re}\,(v_\nu, T_{\alpha^{-1/2}}\mathcal{X}u)_{L^2}\right| \leq C_3\sum_{\nu=0}^{+\infty}(\nu+1)e^{-2\beta(\nu+1)t}2^{-2\nu\theta}e_\nu(t)\,, \tag{4.23}$$

where C_3 depends on μ, on the norms of the coefficients on the respective functional spaces, on θ and on the product βT.

4.4. The analysis of $\widetilde{L}u$

From (4.18) and (4.23) we get

$$\begin{aligned}\frac{d}{dt}E(t) \leq (C_1 + C_3 - 2\beta)\sum_{\nu=0}^{+\infty}(\nu+1)\,e^{-2\beta(\nu+1)t}\,2^{-2\nu\theta}\,e_\nu(t)\\ + C_2\sum_{\nu=0}^{+\infty} e^{-2\beta(\nu+1)t}\,2^{-2\nu\theta}\,(e_\nu(t))^{1/2}\left\|\left(\widetilde{L}u(t)\right)_\nu\right\|_{L^2}.\end{aligned} \tag{4.24}$$

Let us now consider each term of $\widetilde{L}u$ separately. From the Hölder inequality we immediately gather

$$\sum_{\nu=0}^{+\infty} e^{-2\beta(\nu+1)t}2^{-2\nu\theta}\,(e_\nu(t))^{1/2}\,\|(Lu)_\nu\|_{L^2} \leq C\,(E(t))^{1/2}\,\|Lu(t)\|_{H^{-\theta-\beta^*t}}. \tag{4.25}$$

Applying the Hölder inequality for series again implies (see also [7, Sect. 4.4] for the details)

$$\begin{aligned}&\sum_{\nu=0}^{+\infty} e^{-2\beta(\nu+1)t}\,2^{-2\nu\theta}\,(e_\nu(t))^{1/2}\left\|\left(\sum_{j,k}\partial_j\left(\widetilde{A}_{jk}\partial_k u\right)\right)_\nu\right\|_{L^2}\\ &\leq \left(\sum_{\nu=0}^{+\infty}(\nu+1)\,e^{-2\beta(\nu+1)t}\,2^{-2\nu\theta}\,e_\nu(t)\right)^{1/2}\left\|\sum_{j,k}\partial_j\left(\widetilde{A}_{jk}\partial_k u\right)\right\|_{H^{-\theta-\beta^*t-\frac{1}{2}\log}},\end{aligned}$$

and from [11, Prop. 3.4] we have

$$\left\|\sum_{j,k}\partial_j\left(\widetilde{A}_{jk}\partial_k u\right)\right\|_{H^{-s-\frac{1}{2}\log}} \leq C\,\left(\sup_{j,k}\|a_{jk}\|_{LL_x}\right)\|u\|_{H^{1-s+\frac{1}{2}\log}}, \tag{4.26}$$

with C uniformly bounded for s in a compact set of $]0,1[$. Consequently,

$$\begin{aligned}\left\|\sum_{j,k}\partial_j\left(\widetilde{A}_{jk}\partial_k u\right)\right\|_{H^{-\theta-\beta^*t-\frac{1}{2}\log}} &\leq C\,\left(\sup_{j,k}\|a_{jk}\|_{LL_x}\right)\,\|u\|_{H^{1-\theta-\beta^*t+\frac{1}{2}\log}}\\ &\leq C\left(\sum_{\nu=0}^{+\infty}(\nu+1)\,e^{-2\beta(\nu+1)t}\,2^{-2\nu\theta}\,e_\nu(t)\right)^{1/2},\end{aligned}$$

and finally

$$\begin{aligned}&\sum_{\nu=0}^{+\infty} e^{-2\beta(\nu+1)t}\,2^{-2\nu\theta}\,(e_\nu(t))^{1/2}\left\|\left(\sum_{j,k}\partial_j\left(\widetilde{A}_{jk}\partial_k u\right)\right)_\nu\right\|_{L^2}\\ &\leq C_4\sum_{\nu=0}^{+\infty}(\nu+1)e^{-2\beta(\nu+1)t}\,2^{-2\nu\theta}\,e_\nu(t)\,,\end{aligned} \tag{4.27}$$

with C_4 uniformly bounded for $\beta^*t+\theta$ in a compact set of $]0,1[$.

For the lower-order terms, let us state a lemma, which is the analogous of [11, Prop. 3.4].

Lemma 4.3. *Given a function f, we define the remainder operator*

$$\widetilde{F}u := f\,u - T_f\,u\,.$$

(i) *If $f \in \mathcal{C}^\omega(\mathbb{R}^N)$ for some $\omega > 0$ (with $f \in \mathcal{C}^\omega_* := B^\omega_{\infty,\infty}$ if $\omega \in \mathbb{N}$), then $\widetilde{F}$ maps $H^{-s}(\mathbb{R}^N)$ into $H^{\omega-s}(\mathbb{R}^N)$ continuously for all $s \in]0,\omega[$.*

(ii) *If f is only $L^\infty(\mathbb{R}^N)$, then $\widetilde{F}$ maps continuously $H^s(\mathbb{R}^N)$ into $L^2(\mathbb{R}^N)$ for all $s \geq 0$.*

Moreover, the operator norms of $\widetilde{F}$ depend on the norm of f in the respective functional space, and they are uniformly bounded for s varying in a compact subset of $]0,\omega[$ and $[0,+\infty[$ respectively.

Proof. (i) We can write

$$\widetilde{F}u = \sum_{\kappa \geq \mu-3} S_{\kappa+2}u\,\Delta_\kappa f = \sum_\kappa T_\kappa\,.$$

As each term T_κ is spectrally supported in a ball of size 2^κ and $\omega - s > 0$, by [1, Lemma 2.84] it is enough to estimate its L^2 norm. By [1, Prop. 2.79] and dyadic characterization of Hölder spaces (see, e.g., [16]), we have

$$\|S_{\kappa+2}u\,\Delta_\kappa f\|_{L^2} \leq \|S_{\kappa+2}u\|_{L^2}\,\|\Delta_\kappa f\|_{L^\infty} \leq C\,\|f\|_{\mathcal{C}^\omega}\,\|u\|_{H^{-s}}\,2^{(s-\omega)\kappa}\,\lambda_\kappa\,,$$

for a sequence $(\lambda_\kappa)_\kappa \in \ell^2$. Therefore we immediately get

$$\left\|\widetilde{F}u\right\|_{H^{\omega-s}} \leq C\,\|f\|_{\mathcal{C}^\omega}\,\|u\|_{H^{-s}}\,.$$

(ii) Thanks to the continuity properties of the paraproduct operator, we have

$$\|T_f\,u\|_{H^s} \leq C\,\|f\|_{L^\infty}\,\|u\|_{H^s}$$

for any $s \in \mathbb{R}$: in particular, if $s \geq 0$ the L^2 norm can be bounded by the previous quantity. Moreover, if $s \geq 0$, obviously

$$\|f\,u\|_{L^2} \leq \|f\|_{L^\infty}\,\|u\|_{L^2} \leq \|f\|_{L^\infty}\,\|u\|_{H^s}\,.$$

The lemma is completely proved. □

Let us come back to our estimates. As done before, we find

$$\begin{aligned}&\sum_{\nu=0}^{+\infty} e^{-2\beta(\nu+1)t} 2^{-2\nu\theta} \left|\sum_{j=0}^{N} 2\mathrm{Re}\left(v_\nu, T_{\alpha^{-1/2}}\left(\widetilde{B}_j\partial_j u\right)_\nu\right)_{L^2}\right| \\ &\quad \leq \left(\sum_{\nu=0}^{+\infty} e^{-2\beta(\nu+1)t}\,2^{-2\nu\theta}\,2^{-2\nu\omega}\,e_\nu\right)^{1/2} \left\|\sum_{j=0}^{N} \widetilde{B}_j\partial_j u\right\|_{H^{\omega-\theta-\beta^* t}}.\end{aligned}$$

At this point, Lemma 4.3 immediately tells us

$$\sum_{\nu=0}^{+\infty} e^{-2\beta(\nu+1)t}\,2^{-2\nu\theta} \left|\sum_{j=0}^{N} 2\,\mathrm{Re}\left(v_\nu\,,\,T_{\alpha^{-1/2}}\left(\widetilde{B}_j\partial_j u\right)_\nu\right)_{L^2}\right| \leq C_5\,E(t)\,, \tag{4.28}$$

where the constant C_5 depends also on $\sup_j \|b_j\|_{\mathcal{C}^\omega}$.

For $\widetilde{C}u$ we argue as before: by the Hölder inequality we have

$$\sum_{\nu=0}^{+\infty} e^{-2\beta(\nu+1)t}2^{-2\nu\theta}\left|\sum_{j=0}^{N} 2\mathrm{Re}\,\left(v_\nu, T_{\alpha^{-1/2}}\left(\widetilde{C}u\right)_\nu\right)_{L^2}\right| \leq C\,E^{1/2}\left\|\widetilde{C}u\right\|_{H^{-\theta-\beta^* t}},$$

and so, applying Lemma 4.3 again,

$$\sum_{\nu=0}^{+\infty} e^{-2\beta(\nu+1)t}\,2^{-2\nu\theta}\left|\sum_{j=0}^{N} 2\,\mathrm{Re}\,\left(v_\nu\,,\,T_{\alpha^{-1/2}}(\widetilde{C}u)_\nu\right)_{L^2}\right| \leq C_6\,E(t)\,, \qquad (4.29)$$

for C_6 depending on $\|c\|_{L^\infty([0,T_0]\times\mathbb{R}^N)}$.

4.5. Final estimates

Now we take $\beta > 0$ (recall that $\beta^* = \beta(\log 2)^{-1}$) and $T \in]0, T_0]$ such that

$$\beta^*\,T \;=\; \delta \;<\; \min\{1\,,\,\omega\} \;-\; \theta\,; \qquad (4.30)$$

in particular, we have $0 < \theta \leq \theta + \beta^* t \leq \theta + \delta < 1$.

Therefore, putting inequalities (4.25), (4.27), (4.28) and (4.29) into (4.24) we finally find

$$\frac{d}{dt}E(t) \leq (K\,-\,2\beta)\sum_{\nu=0}^{+\infty}(\nu+1)\,e^{-2\beta(\nu+1)t}\,2^{-2\nu\theta}\,e_\nu(t) + C_2\sum_{\nu=0}^{+\infty} e^{-2\beta(\nu+1)t}\,2^{-2\nu\theta}\,\left(e_\nu(t)\right)^{1/2}\,\|(Lu(t))_\nu\|_{L^2}\,,$$

where we have set $K = C_1 + C_4C_2 + C_3 + C_5C_2 + C_6C_2$.

Now we fix β large enough, such that $K - 2\beta \leq 0$: this corresponds to take $T > 0$ small enough in (4.30). Hence, we finally get the estimate

$$\frac{d}{dt}E(t) \;\leq\; C_2\,\left(E(t)\right)^{1/2}\,\|Lu(t)\|_{H^{-\theta-\beta^* t}}\;;$$

by Gronwall's Lemma and relations (4.5) and (4.6) we infer estimate (2.6). □

Remark 4.4. Let us point out that condition (4.30) gives a bound on the lifespan T of a solution to the Cauchy problem for (2.1). It depends on $\theta \in]0, 1[$ and on $\beta^* > 0$, hence on constants $C_1 \ldots C_6$. Going after the guideline of the proof, one can see that, in the end, T depends only on the index θ, on the parameter μ defined by conditions (4.1), on constants λ_0 and Λ_0 defined by (2.2), and on the quantities $\sup_{j,k}|a_{jk}|_{LZ_t}$, $\sup_{j,k}|a_{jk}|_{LL_x}$, $\sup_j\|b_j\|_{L^\infty([0,T_0];\mathcal{C}^\omega)}$ and $\|c\|_{L^\infty([0,T_0]\times\mathbb{R}^N)}$.

5. The well-posedness in H^∞

In this section we want to sketch how recovering the thesis of Theorem 2.1 from Theorem 2.2. We will follow the outline of the proof in [10] for isotropic log-Lipschitz coefficients.

So, let us impose more regularity with respect to x on the coefficients. A preliminary lemma is in order.

Lemma 5.1. *Consider the operator L defined by (2.1), with coefficients a_{jk} fulfilling conditions (2.2) and (2.3).*
Assume that $a_{jk} \in L^\infty([0,T_0];\mathcal{C}^2_b(\mathbb{R}^N))$, while b_j and $c \in L^\infty([0,T_0];\mathcal{C}^1_b(\mathbb{R}^N))$.

Then there exist a $\beta^* > 0$, a time $T > 0$ and a constant $C > 0$ such that estimate (2.6) holds true also for $\theta = 0$.

Proof. First of all, we note that, thanks to the additional regularity on the principal part of L, it makes sense to consider the problem in $H^1 \times L^2$.

Now, it is enough to retake the same steps of the previous proof.

The fundamental fact is that, taking into account the additional regularity of the coefficients of L, the logarithmic loss disappears whenever space derivatives of the coefficients are involved. Nevertheless, this is not the case when the time derivative ∂_t acts on the symbol α: the loss in the energy estimates comes from these terms. □

Taking into account Remark 4.4, note that the lifespan T depends only on the constants λ_0 and Λ_0 (recall (2.2)) and on the norms of the coefficients of L in their respective functional spaces.

Let us now analyse what happens to the space derivatives of u.

Lemma 5.2. *Consider the operator L defined by (2.1), with coefficients a_{jk} fulfilling conditions (2.2) and (2.3).*
Assume that coefficients a_{jk}, b_j and c all belong to the space $L^\infty([0,T_0];\mathcal{C}^\infty_b(\mathbb{R}^N))$.

Then, for all $m \geq 0$, there exists a constant $C_m > 0$ such that, for all, $u \in \mathcal{C}^2([0,T];H^\infty(\mathbb{R}^N))$, one has the following estimate:

$$\sup_{0\leq t\leq T}\left(\|u(t,\cdot)\|_{H^{m+1-\beta^*t}} + \|\partial_t u(t,\cdot)\|_{H^{m-\beta^*t}}\right) \tag{5.1}$$
$$\leq C\left(1+e^{C_mT}\right)\left(\|u(0,\cdot)\|_{H^{m+1}} + \|\partial_t u(0,\cdot)\|_{H^m} + \int_0^T \|Lu(t,\cdot)\|_{H^{m-\beta^*t}}dt\right).$$

Here, the "loss index" β^*, the time T and the constant C are the same of those in Lemma 5.1.

Proof. The proof is analogous to that of [10, Th. 2.3]. Hence, let us just give a sketch of it.

For notation convenience, we define

$$X_m(t) := \int_0^t \left(\|u(\tau)\|_{H^{m+1-\beta^*\tau}} + \|\partial_t u(\tau)\|_{H^{m-\beta^*\tau}}\right)d\tau$$

and the energy of the data

$$\mathcal{U}(t) := \|u(0)\|_{H^{m+1}} + \|\partial_t u(0)\|_{H^m} + \int_0^t \|Lu(\tau)\|_{H^{m-\beta^*\tau}}\,d\tau\,.$$

We set moreover $\Lambda(D_x) := \left(1+|D_x|^2\right)^{1/2}$.

From Lemma 5.1, for any $m \geq 0$ we immediately get

$$X'_m(t) \leq C_0 \left(\mathcal{U}(t) + \int_0^t \left\| \left[L, \Lambda^m(D_x) \right] u(\tau) \right\|_{H^{-\beta^* \tau}} \right) d\tau \, .$$

Now, differentiating equation (2.1) with respect to x, for any multi-index η such that $|\eta| = m$ we find

$$\begin{aligned} L(D_x^\eta u) \, - \, D_x^\eta(Lu) = & \sum_{j,k=1}^{N} \sum_{1 \leq |\varrho| \leq m} \partial_j \left(D_x^\varrho a_{jk} \, \partial_k D_x^{\eta - \varrho} u \right) \\ & - \sum_{j=0}^{N} \sum_{1 \leq |\varrho| \leq m} D_x^\varrho b_j \, \partial_j D_x^{\eta-\varrho} u \, - \, \sum_{1 \leq |\varrho| \leq m} D_x^\varrho c \, D_x^{\eta - \varrho} u \end{aligned}$$

and we apply Lemma 5.1. Second, third and fourth terms on the right-hand side of the previous equality are obviously of lower order. Hence, we finally get

$$X'_m(t) \, \leq \, C_0 \, \mathcal{U}(t) \, + \, C_m \, X_m(t) \, ,$$

for a constant C_m depending only on the L^∞ norms of the coefficients of L and their space derivatives.

Therefore, by Gronwall's inequality we infer

$$X_m(t) \, \leq \, (C_0/C_m) \, e^{C_m t} \, \mathcal{U}(t) \, ,$$

which immediately implies estimate (5.1). □

Now, Theorem 2.1 easily follows from Lemma 5.2.

Let us just point out that the global in time issue can be obtained remarking that the constants which come into play in the previous proof depend only on the operator L (or better, on its coefficients), and not on the initial data.

Acknowledgment

The third author was partially supported by Grant MTM2011-29306-C02-00, MICINN, Spain, ERC Advanced Grant FP7-246775 NUMERIWAVES, ESF Research Networking Programme OPTPDE and Grant PI2010-04 of the Basque Government.

References

[1] H. Bahouri, J.-Y. Chemin and R. Danchin: *"Fourier Analysis and Nonlinear Partial Differential Equations"*, Grundlehren der Mathematischen Wissenschaften (Fundamental Principles of Mathematical Sciences), **343**, Springer, Heidelberg, 2011.

[2] J.-M. Bony: *Calcul symbolique et propagation des singularités pour les équations aux dérivées partielles non linéaires*, Ann. Sci. École Norm. Sup. (4), **14** (1981), 209–246.

[3] M. Cicognani and F. Colombini: *Modulus of continuity of the coefficients and loss of derivatives in the strictly hyperbolic Cauchy problem*, J. Differential Equations, **221** (2006), 143–157.

[4] R. Coifman and Y. Meyer: *Au delà des opérateurs pseudo-différentiels*, Astérisque, **57** (1978), Société Mathématique de France, Paris.

[5] F. Colombini, E. De Giorgi and S. Spagnolo: *Sur les équations hyperboliques avec des coefficients qui ne dépendent que du temps*, Ann. Scuola Norm. Sup. Pisa Cl. Scienze (4), **6** (1979), 511–559.

[6] F. Colombini and D. Del Santo: *A note on hyperbolic operators with log-Zygmund coefficients*, J. Math. Sci. Univ. Tokyo, **16** (2009), 95–111.

[7] F. Colombini, D. Del Santo, F. Fanelli and G. Métivier: *Time-dependent loss of derivatives for hyperbolic operators with non-regular coefficients*, Comm. Partial Differential Equations, **38** (2013), 1791–1817.

[8] F. Colombini, D. Del Santo, F. Fanelli and G. Métivier: *A well-posedness result for hyperbolic operators with Zygmund coefficients*, J. Math. Pures Appl. (9), **100** (2013), 455–475.

[9] F. Colombini and F. Fanelli: *A note on non-homogeneous hyperbolic operators with low regularity coefficients*, Rend. Istit. Mat. Univ. Trieste, **42** (2010), suppl., 1–25.

[10] F. Colombini and N. Lerner: *Hyperbolic operators with non-Lipschitz coefficients*, Duke Math. J., **77** (1995), 657–698.

[11] F. Colombini and G. Métivier, *The Cauchy problem for wave equations with non-Lipschitz coefficients; application to continuation of solutions of some nonlinear wave equations*, Ann. Sci. École Norm. Sup. (4) **41** (2008), 177–220.

[12] D. Del Santo: *The Cauchy problem for a hyperbolic operator with Log-Zygmund coefficients*, Further Progress in Analysis, World Sci. Publ., Hackensack, NJ (2009), 425–433.

[13] L. Hörmander: *"Linear partial differential operators"*, Springer-Verlag, Berlin, 1963.

[14] A.E. Hurd, D.H. Sattinger: *Questions of existence and uniqueness for hyperbolic equations with discontinuous coefficients*, Trans. Amer. Math. Soc., **132** (1968), 159–174.

[15] G. Métivier, *Interactions de deux chocs pour un système de deux lois de conservation, en dimension deux d'espace*, Trans. Amer. Math. Soc., **296** (1986), 431–479.

[16] G. Métivier, *Para-differential calculus and applications to the Cauchy problem for nonlinear systems*, Centro di Ricerca Matematica "Ennio De Giorgi" (CRM) Series, **5**, Edizioni della Normale, Pisa (2008).

[17] G. Métivier and K. Zumbrun, *Large viscous boundary layers for noncharacteristic nonlinear hyperbolic problems*, Mem. Amer. Math. Soc., **175** (2005).

[18] S. Mizohata: *"The Theory of Partial Differential Equations"*, Cambridge University Press, Cambridge, 1973.

[19] S. Tarama: *Energy estimate for wave equations with coefficients in some Besov type class*, Electron J. Differential Equations (2007), Paper No. 85 (electronic).

Ferruccio Colombini
Dipartimento di Matematica
Università di Pisa
Largo Bruno Pontecorvo, 5
I-56127 Pisa, Italy
e-mail: colombini@dm.unipi.it

Daniele Del Santo
Dipartimento di Matematica e Geoscienze
Università degli Studi di Trieste
Via Valerio, 12/1
I-34127 Trieste, Italy
e-mail: delsanto@units.it

Francesco Fanelli
BCAM – Basque Center for Applied Mathematics
Alameda de Mazarredo, 14
E-48009 Bilbao, Basque Country, Spain
e-mail: ffanelli@bcamath.org

Guy Métivier
Institut de Mathématiques de Bordeaux, UMR 5251
Université de Bordeaux 1
351, Cours de la Libération
F-33405 Talence, France
e-mail: guy.metivier@math.u-bordeaux1.fr

Fourier Analysis
Trends in Mathematics, 73–90

A Few Remarks on Time-frequency Analysis of Gevrey, Analytic and Ultra-analytic Functions

Elena Cordero, Fabio Nicola and Luigi Rodino

Abstract. We give a brief survey of recent results concerning almost diagonalization of pseudodifferential operators via Gabor frames. Moreover, we show new connections between symbols with Gevrey, analytic or ultra-analityc regularity and time-frequency analysis of the corresponding pseudodifferential operators.

Mathematics Subject Classification (2010). Primary 35S30; Secondary 47G30.

Keywords. Pseudodifferential operators, Gelfand–Shilov spaces, Gevrey spaces, short-time Fourier transform, Gabor frames.

1. Introduction

The objective of this paper is to report on recent progress concerning time-frequency analysis of pseudodifferential operators with smooth symbols. We discuss in particular connections between, on one side, symbols in the classical Gevrey or (ultra-)analytic spaces, and, on the other, (ultra-)modulation spaces, Gabor frames, which are the framework of time-frequency analysis.

Modulation spaces having moderate weights were introduced by Feichtinger in 1983, and since then have been extensively studied by Feichtinger and many other authors, cf. in particular the textbook [17], to which we address for further references. We recall their definition and main properties in the subsequent Section 2.3, but the main insight to understand their introduction being to consider the decay property of a function with respect to the space variable and the variable of its Fourier transform simultaneously. The time-frequency representation employed for their definition is the so-called short-time Fourier transform (STFT), whose building blocks are the linear operators of translation and modulation (so-called

This work was completed with the support of the Erwin Schrödinger International Institute for Mathematical Physics, Vienna, Austria.

time-frequency shifts) given by

$$T_x f(\cdot) = f(\cdot - x) \quad \text{and} \quad M_\xi f(\cdot) = e^{2\pi i \xi \cdot} f(\cdot), \quad x, \xi \in \mathbb{R}^d.$$

For $z = (x, \xi)$ we shall also write $\pi(z)f = M_\xi T_x f$. Indeed, if g is a non-zero window function in the Schwartz class $\mathcal{S}(\mathbb{R}^d)$, then the short-time Fourier transform (STFT) of a a function/tempered distribution f with respect to the window g is given by

$$V_g f(x, \xi) = \langle f, M_\xi T_x g \rangle = \int_{\mathbb{R}^d} f(t)\, \overline{g(t - x)}\, e^{-2\pi i \xi t}\, dt\,. \tag{1}$$

In short: we say that f is in the modulation space $M^{p,q}_m(\mathbb{R}^d)$, $1 \leq p, q \leq \infty$, where m is a moderate weight function (hence with at most exponential growth at infinity) if $V_g f \in L^{p,q}_m(\mathbb{R}^{2d})$. So it is clear that the bigger the growth of m the faster the decay of $V_g f$ is at infinity.

Starting from signal analysis and quantum mechanics, these spaces have then been proved to be the right spaces for symbols of pseudodifferential operators (see [19, 25, 31, 10, 32] and references therein), in particular localization operators (cf., e.g., [9]), Fourier multipliers ([1, 27]) and recently Fourier integral operators of Schrödinger type [5, 6, 7, 8, 11], providing a good framework for the study of PDE's [12].

If we limit the use of time-frequency shifts on a lattice $\Lambda = \alpha\mathbb{Z}^d \times \beta\mathbb{Z}^d \subset \mathbb{R}^{2d}$ and fix a window function $g \in L^2(\mathbb{R}^d)$, then the sequence $\mathcal{G}(g, \Lambda) = \{g_{m,n} = M_n T_m g,\ (m, n) \in \Lambda\}$ forms a so-called Gabor system. The set $\mathcal{G}(g, \Lambda)$ is a Gabor frame, if there exist constants $A, B > 0$ such that

$$A\|f\|_2^2 \leq \sum_{(m,n)\in\Lambda} |\langle f, g_{m,n}\rangle|^2 \leq B\|f\|_2^2, \qquad \forall f \in L^2(\mathbb{R}^d).$$

For details see the next Section 2, where beside modulation spaces and Gabor frames, we treat classes of symbols and Gelfand–Shilov functions, which represent the natural window functions in our context.

The first paper which uses Gabor frames to approximately diagonalize pseudodifferential operators is [29]. Later, motivated by the study of Sjöstrand [31], Gröchenig proved an almost diagonalization for symbols in the Sjöstrand class, that is the modulation space $M^{\infty,1}$, whose further generalization in [21] is recalled in Theorem 3.1 below. What is remarkable in Theorem 3.1 is that the rate of decay in the almost diagonalization, expressed in terms of Wiener amalgam spaces, characterizes the symbol class, which reveals to be a class of modulation spaces. The discrete representation of a pseudodifferential operator via Gabor matrix is one of the key ingredient for numerical applications, as shown in [12] for the Gabor matrix of Schrödinger propagators.

Motivated by these characterizations and by numerical applications, in the recent work [13] we study diagonalization of pseudodifferential operators with symbols of Gevrey, analytic and ultra-analytic type, and show that Gabor frames surprisingly reveal to be an efficient tool for representing solutions to hyperbolic and parabolic-type differential equations with constant coefficients.

Here we are mainly focused on the theoretical aspects of the almost diagonalization: as a preliminary step we establish the equivalence between Gevrey, analytic and ultra-analytic regularity of symbols, and their membership to modulation spaces with sub-exponential, exponential and super-exponential weights, respectively. For these aspects, our results are strictly related to those of [34]. The main result of the paper, Theorem 3.4, characterizes the corresponding classes of pseudodifferential operators in terms of exponential decay of the almost diagonalization. In the last Section 4, we report on the sparsity result obtained in [13], which motivates this study for numerical applications.

Notations. The Schwartz class is denoted by $\mathcal{S}(\mathbb{R}^d)$, the space of tempered distributions by $\mathcal{S}'(\mathbb{R}^d)$. We use the brackets $\langle f, g\rangle$ to denote the extension to $\mathcal{S}'(\mathbb{R}^d) \times \mathcal{S}(\mathbb{R}^d)$ of the inner product $\langle f, g\rangle = \int f(t)\overline{g(t)}dt$ on $L^2(\mathbb{R}^d)$.

The Euclidean norm of $x \in \mathbb{R}^d$ is given by $|x| = \left(x_1^2 + \cdots + x_d^2\right)^{1/2}$, and $\langle x\rangle = (1+|x|^2)^{1/2}$. We write $xy = x \cdot y$ for the scalar product on $\mathbb{R}^d$, for $x, y \in \mathbb{R}^d$.

For multiindices $\alpha, \beta \in \mathbb{N}^d$, we have $|\alpha| = \alpha_1 + \cdots + \alpha_d$, $\alpha! = \alpha_1! \cdots \alpha_d!$, $x^\alpha = x_1^{\alpha_1} \cdots x_d^{\alpha_d}$, $x \in \mathbb{R}^d$, and, for $\beta \leq \alpha$, i.e., $\beta_j \leq \alpha_j$, $j \in \{1, 2, \ldots, d\}$, $\binom{\alpha}{\beta} = \binom{\alpha_1}{\beta_1} \cdot \cdots \cdot \binom{\alpha_d}{\beta_d}$. The letter C denotes a positive constant, not necessarily the same at every appearance.

The operator of partial differentiation ∂ is given by

$$\partial^\alpha = \partial_x^\alpha = \partial_{x_1}^{\alpha_1} \cdots \partial_{x_d}^{\alpha_d}$$

for all multiindices $\alpha \in \mathbb{N}^d$ and all $x = (x_1, \ldots, x_d) \in \mathbb{R}^d$. If f and g are smooth enough, then the Leibnitz formula holds

$$\partial^\alpha (fg)(x) = \sum_{\beta \leq \alpha} \binom{\alpha}{\beta} \partial^{\alpha-\beta} f(x) \partial^\beta g(x).$$

Observe

$$\sum_{\beta \leq \alpha} \binom{\alpha}{\beta} = 2^{|\alpha|}, \tag{2}$$

and

$$|\alpha|! \leq d^{|\alpha|} \alpha!. \tag{3}$$

The Fourier transform is normalized to be $\hat{f}(\xi) = \mathcal{F}f(\xi) = \int f(t) e^{-2\pi i t\xi} dt$.

For $0 < p \leq \infty$ and a weight m, the space $\ell_m^p(\Lambda)$ is the (quasi-)Banach space of sequences $a = \{a_\lambda\}_{\lambda \in \Lambda}$ on a lattice Λ, such that

$$\|a\|_{\ell_m^p} := \left(\sum_{\lambda \in \Lambda} |a_\lambda|^p m(\lambda)^p\right)^{1/p} < \infty$$

(with obvious changes when $p = \infty$).

We shall use the notation $A \lesssim B$ to express the inequality $A \leq cB$ for a suitable constant $c > 0$, and $A \asymp B$ for the equivalence $c^{-1}B \leq A \leq cB$.

2. Time-frequency analysis and Gelfand–Shilov spaces

2.1. Gelfand–Shilov spaces

The Schwartz class $\mathcal{S}(\mathbb{R}^d)$ does not give enough information about how fast a function $f \in \mathcal{S}(\mathbb{R}^d)$ and its derivatives decay at infinity. This is the main motivation to use subspaces of the Schwartz class, so-called Gelfand–Shilov type spaces, introduced in [16] and now available also in a textbook [28]. Let us recall their definition and some of their properties.

Definition 2.1. Let there be given $s, r \geq 0$. A function $f \in \mathcal{S}(\mathbb{R}^d)$ is in the Gelfand–Shilov type space $S_r^s(\mathbb{R}^d)$ if there exist constants $A, B > 0$ such that

$$|x^\alpha \partial^\beta f(x)| \lesssim A^{|\alpha|} B^{|\beta|} (\alpha!)^r (\beta!)^s, \quad \alpha, \beta \in \mathbb{N}^d. \tag{4}$$

The space $S_r^s(\mathbb{R}^d)$ is nontrivial if and only if $r + s > 1$, or $r + s = 1$ and $r, s > 0$ [16]. So the smallest nontrivial space with $r = s$ is provided by $S_{1/2}^{1/2}(\mathbb{R}^d)$. Every function of the type $P(x)e^{-a|x|^2}$, with $a > 0$ and $P(x)$ polynomial on $\mathbb{R}^d$, is in $S_{1/2}^{1/2}(\mathbb{R}^d)$. Observe that $S_{r_1}^{s_1}(\mathbb{R}^d) \subset S_{r_2}^{s_2}(\mathbb{R}^d)$ for $s_1 \leq s_2$ and $r_1 \leq r_2$. Moreover, if $f \in S_r^s(\mathbb{R}^d)$, for every $\delta, \gamma \in \mathbb{N}^d$, $x^\delta \partial^\gamma f \in S_r^s(\mathbb{R}^d)$. The action of the Fourier transform on $S_r^s(\mathbb{R}^d)$ interchanges the indices s and r, as explained in the following theorem.

Theorem 2.2. *For $f \in \mathcal{S}(\mathbb{R}^d)$ we have $f \in S_r^s(\mathbb{R}^d)$ if and only if $\hat{f} \in S_s^r(\mathbb{R}^d)$.*

Therefore for $s = r$ the spaces $S_s^s(\mathbb{R}^d)$ are invariant under the action of the Fourier transform.

Theorem 2.3. *Assume $s > 0, r > 0, s + r \geq 1$. For $f \in \mathcal{S}(\mathbb{R}^d)$, the following conditions are equivalent:*

a) $f \in S_r^s(\mathbb{R}^d)$.

b) *There exist constants $A, B > 0$, such that*

$$\|x^\alpha f\|_{L^\infty} \lesssim A^{|\alpha|}(\alpha!)^r \quad \textit{and} \quad \|\xi^\beta \hat{f}\|_{L^\infty} \lesssim B^{|\beta|}(\beta!)^s, \quad \alpha, \beta \in \mathbb{N}^d.$$

c) *There exist constants $A, B > 0$, such that*

$$\|x^\alpha f\|_{L^\infty} \lesssim A^{|\alpha|}(\alpha!)^r \quad \textit{and} \quad \|\partial^\beta f\|_{L^\infty} \lesssim B^{|\beta|}(\beta!)^s, \quad \alpha, \beta \in \mathbb{N}^d.$$

d) *There exist constants $h, k > 0$, such that*

$$\|f e^{h|x|^{1/r}}\|_{L^\infty} < \infty \quad \textit{and} \quad \|\hat{f} e^{k|\xi|^{1/s}}\|_{L^\infty} < \infty.$$

A suitable window class for weighted modulation spaces (see Definition 2.8 below) is the Gelfand–Shilov space $\Sigma_1^1(\mathbb{R}^d)$, consisting of functions $f \in \mathcal{S}(\mathbb{R}^d)$ such that for *every* constant $A > 0$ and $B > 0$

$$|x^\alpha \partial^\beta f(x)| \lesssim A^{|\alpha|} B^{|\beta|} \alpha! \beta!, \quad \alpha, \beta \in \mathbb{N}^d. \tag{5}$$

We have $S_s^s(\mathbb{R}^d) \subset \Sigma_1^1(\mathbb{R}^d) \subset S_1^1(\mathbb{R}^d)$ for every $s < 1$. Observe that the characterization of Theorem 2.3 can be adapted to $\Sigma_1^1(\mathbb{R}^d)$ by replacing the words "there exist" by "for every" and taking $r = s = 1$. Similarly, one can define $\Sigma_s^r(\mathbb{R}^d)$ for any $r > 0, s > 0$, with $r + s > 1$.

Let us underline the following property, proved in [28, Proposition 6.1.5] and [13, Proposition 2.4], exhibiting two equivalent ways of expressing the exponential decay of a continuous function f on $\mathbb{R}^d$.

Proposition 2.4. *Consider $r > 0$ and let h be a continuous function on $\mathbb{R}^d$. Then the following conditions are equivalent:*

(i) *There exists a constant $\epsilon > 0$ such that*

$$|h(x)| \lesssim e^{-\epsilon|x|^{\frac{1}{r}}}, \quad x \in \mathbb{R}^d, \tag{6}$$

(ii) *There exists a constant $C > 0$ such that*

$$|x^\alpha h(x)| \lesssim C^{|\alpha|}(\alpha!)^r, \quad x \in \mathbb{R}^d, \alpha \in \mathbb{N}^d. \tag{7}$$

Remark 2.5. To be precise, the relation between the constants ϵ and C is as follows. Assuming (6), then (7) is satisfied with $C = \left(\frac{rd}{\epsilon}\right)^r$. Viceversa, (7) implies (6) for any $\epsilon < r(dC)^{-\frac{1}{r}}$. Also, it follows from the proof that the constant implicit in the notation $\lesssim$ in (6) depends only on the corresponding one in (7) and viceversa.

The strong dual spaces of $S_r^s(\mathbb{R}^d)$ and $\Sigma_1^1(\mathbb{R}^d)$ are the so-called spaces of tempered ultra-distributions, denoted by $(S_r^s)'(\mathbb{R}^d)$ and $(\Sigma_1^1)'(\mathbb{R}^d)$, respectively. Notice that they contain the space of tempered distributions $\mathcal{S}'(\mathbb{R}^d)$. Moreover the spaces $S_r^s(\mathbb{R}^d)$ are nuclear spaces [26]. This provides a kernel theorem for Gelfand Shilov spaces, cf. [35].

Theorem 2.6. *There exists an isomorphism between the space of linear continuous maps T from $S_r^s(\mathbb{R}^d)$ to $(S_r^s)'(\mathbb{R}^d)$ and $(S_r^s)'(\mathbb{R}^{2d})$, which associates to every T a kernel $K_T \in (S_r^s)'(\mathbb{R}^{2d})$ such that*

$$\langle Tu, v\rangle = \langle K_T, v \otimes \bar{u}\rangle, \quad \forall u, v \in S_r^s(\mathbb{R}^d).$$

K_T is called the kernel of T.

2.2. Gabor frames and time-frequency representations

We recall the basic concepts of time-frequency analysis and refer the reader to [17] for the full details. Consider a distribution $f \in \mathcal{S}'(\mathbb{R}^d)$ and a Schwartz function $g \in \mathcal{S}(\mathbb{R}^d) \setminus \{0\}$ (the so-called *window*). The short-time Fourier transform (STFT) of f with respect to g is defined in (1). The short-time Fourier transform is well defined whenever the bracket $\langle \cdot, \cdot \rangle$ makes sense for dual pairs of function or (ultra-)distribution spaces, in particular for $f \in \mathcal{S}'(\mathbb{R}^d)$ and $g \in \mathcal{S}(\mathbb{R}^d)$, $f, g \in L^2(\mathbb{R}^d)$, $f \in (\Sigma_1^1)'(\mathbb{R}^d)$ and $g \in \Sigma_1^1(\mathbb{R}^d)$ or $f \in (S_r^s)'(\mathbb{R}^d)$ and $g \in S_r^s(\mathbb{R}^d)$.

Another time-frequency representation we shall use is the (cross-)Wigner distribution of $f, g \in L^2(\mathbb{R}^d)$, defined as

$$W(f,g)(x,\xi) = \int_{\mathbb{R}^d} f\left(x + \frac{t}{2}\right)\overline{g\left(x - \frac{t}{2}\right)} e^{-2\pi i t\xi}\, dt. \tag{8}$$

If we set $\breve{g}(t) = g(-t)$, then the relation between cross-Wigner distribution and short-time Fourier transform is provided by

$$W(f,g)(x,\xi) = 2^d e^{4\pi i x\xi} V_{\breve{g}} f(2x, 2\xi). \tag{9}$$

For the discrete description of function spaces and operators we use Gabor frames. Let $\Lambda = A\mathbb{Z}^{2d}$ with $A \in GL(2d,\mathbb{R})$ be a lattice of the time-frequency plane. The set of time-frequency shifts $\mathcal{G}(g,\Lambda) = \{\pi(\lambda)g : \lambda \in \Lambda\}$ for a non-zero $g \in L^2(\mathbb{R}^d)$ is called a Gabor system. The set $\mathcal{G}(g,\Lambda)$ is a Gabor frame, if there exist constants $A, B > 0$ such that

$$A\|f\|_2^2 \leq \sum_{\lambda\in\Lambda} |\langle f, \pi(\lambda)g\rangle|^2 \leq B\|f\|_2^2, \qquad \forall f \in L^2(\mathbb{R}^d). \tag{10}$$

If (10) is satisfied, then there exists a dual window $\gamma \in L^2(\mathbb{R}^d)$, such that $\mathcal{G}(\gamma,\Lambda)$ is a frame, and every $f \in L^2(\mathbb{R}^d)$ has the frame expansions

$$f = \sum_{\lambda\in\Lambda} \langle f, \pi(\lambda)g\rangle \pi(\lambda)\gamma = \sum_{\lambda\in\Lambda} \langle f, \pi(\lambda)\gamma\rangle \pi(\lambda)g$$

with unconditional convergence in $L^2(\mathbb{R}^d)$.

Eventually, we list some results about time-frequency analysis of Gelfand–Shilov functions, cf. [14, 22, 33]:

$$f, g \in S_s^s(\mathbb{R}^d),\ s \geq 1/2 \Rightarrow V_g f \in S_s^s(\mathbb{R}^{2d}), \tag{11}$$

If $g \in S_s^s(\mathbb{R}^d)$, $s \geq 1/2$, then

$$f \in S_s^s(\mathbb{R}^d) \Leftrightarrow |V_g(f)(z)| \lesssim e^{-\epsilon|z|^{1/s}} \text{ for some } \epsilon > 0. \tag{12}$$

Since we have not found a precise reference in the literature, we also recall and present the proof of the following properties concerning the Wigner distribution.

Proposition 2.7. *We have:*

$$f, g \in S_s^s(\mathbb{R}^d),\ s \geq 1/2 \Rightarrow W(f,g) \in S_s^s(\mathbb{R}^{2d}), \tag{13}$$

$$f, g \in \Sigma_1^1(\mathbb{R}^d) \Rightarrow W(f,g) \in \Sigma_1^1(\mathbb{R}^{2d}). \tag{14}$$

Proof. We prove (13), formula (14) follows by similar techniques. The assumption $g \in S_s^s(\mathbb{R}^d)$ trivially implies $\breve{g} \in S_s^s(\mathbb{R}^d)$ and (11) proves that the short-time Fourier transform $V_{\breve{g}} g$ is in $S_s^s(\mathbb{R}^{2d})$. To prove that $\Phi = W(g,g) \in S_s^s(\mathbb{R}^{2d})$, we use the equivalence $a) \Leftrightarrow c)$ in Theorem 2.3 and the connection between $V_{\breve{g}} g$ and the Wigner distribution in (9). First, let us prove the first inequality in Theorem 2.3 c). For $z = (x,\xi) \in \mathbb{R}^{2d}$, we have

$$\begin{aligned} |z^\alpha W(g,g)(z)| &= 2^d |z^\alpha V_{\breve{g}} g(2z)| = 2^{d-|\alpha|} |(2z)^\alpha V_{\breve{g}} g(2z)| \\ &\leq 2^{d-|\alpha|} \|z^\alpha V_{\breve{g}} g\|_{L^\infty} \lesssim 2^{d-|\alpha|} A^{|\alpha|} (\alpha!)^s \lesssim \left(\frac{A}{2}\right)^{|\alpha|} (\alpha!)^s. \end{aligned}$$

The second inequality in Theorem 2.3 c) requires more computations but the techniques are the same as before. Using Leibniz formula for the x-derivatives and the ξ-derivatives separately,

$$\begin{aligned}
\partial_\xi^\alpha \partial_x^\beta W(g,g)(x,\xi) &= \partial_\xi^\alpha (\partial_x^\beta (2^d e^{4\pi i x\xi} V_{\check{g}} g(2x, 2\xi)) \\
&= 2^d \partial_\xi^\alpha \left(\sum_{\gamma \le \beta} \binom{\beta}{\gamma} (4\pi i \xi)^{\beta-\gamma} e^{4\pi i x \xi} 2^{|\gamma|} \partial_x^\gamma (V_{\check{g}} g)(2x, 2\xi) \right) \\
&= 2^d \sum_{\gamma \le \beta} \binom{\beta}{\gamma} \sum_{\substack{\delta_1+\delta_2+\delta_3=\alpha \\ \delta_1 \le \beta-\gamma}} \frac{\alpha!}{\delta_1!\delta_2!\delta_3!} \partial_\xi^{\delta_1}((4\pi i \xi)^{\beta-\gamma})(4\pi i x)^{\delta_2} e^{4\pi i x \xi} \\
&\qquad \times \; 2^{|\gamma|+|\delta_3|} \partial_\xi^{\delta_3} \partial_x^\gamma (V_{\check{g}} g)(2x, 2\xi). \\
&= 2^d \sum_{\gamma \le \beta} \binom{\beta}{\gamma} \sum_{\substack{\delta_1+\delta_2+\delta_3=\alpha \\ \delta_1 \le \beta-\gamma}} 2^{|\gamma|+|\delta_3|} (4\pi i)^{|\beta-\gamma|+|\delta_2|} \frac{\alpha!}{\delta_1!\delta_2!\delta_3!} \frac{(\beta-\gamma)!}{(\beta-\gamma-\delta_1)!} \\
&\qquad \times \xi^{\beta-\gamma-\delta_1} x^{\delta_2} e^{4\pi i x \xi} \partial_\xi^{\delta_3} \partial_x^\gamma (V_{\check{g}} g)(2x, 2\xi).
\end{aligned}$$

Now, we have

$$\begin{aligned}
|\xi^{\beta-\gamma-\delta_1} x^{\delta_2} \partial_\xi^{\delta_3} \partial_x^\gamma (V_{\check{g}} g)(2x, 2\xi)| &\lesssim A^{|\alpha|+|\beta|} ((\beta-\gamma-\delta_1+\delta_2+\delta_3+\gamma)!)^s \\
&= A^{|\alpha|+|\beta|} ((\beta+\alpha-2\delta_1)!)^s,
\end{aligned}$$

where we used $\delta_2+\delta_3=\alpha-\delta_1$. Since from (2)

$$\frac{(\beta-\gamma)!}{(\beta-\gamma-\delta_1)!} \le 2^{|\beta-\gamma|} \delta_1! < 2^{|\beta-\gamma|} ((2\delta_1)!)^{1/2} \le 2^{|\beta-\gamma|} ((2\delta_1)!)^s,$$

together with $\sum_{\gamma \le \beta} \binom{\beta}{\gamma} = 2^{|\beta|}$ and $\sum_{\delta_1+\delta_2+\delta_3=\alpha} \frac{\alpha!}{\delta_1!\delta_2!\delta_3!} = 3^{|\alpha|}$, we obtain the desired estimate

$$|\partial_\xi^\alpha \partial_x^\beta W(g,g)(x,\xi)| \lesssim C^{|\alpha+\beta|} ((\alpha+\beta)!)^s. \qquad \square$$

2.3. Modulation spaces

Modulation spaces measure the decay of the STFT on the time-frequency (phase space) plane and were introduced by Feichtinger in the 80s [15], for a weight of sub-exponential growth at infinity. The study of weights of exponential growth at infinity was developed in [14, 33].

Weight functions. In the sequel v will always be a continuous, positive, even, submultiplicative function (submultiplicative weight), i.e., $v(0)=1$, $v(z)=v(-z)$, and $v(z_1+z_2) \le v(z_1)v(z_2)$, for all $z, z_1, z_2 \in \mathbb{R}^{2d}$. Submultiplicativity implies that $v(z)$ is *dominated* by an exponential function, i.e.,

$$\exists C, k > 0 \quad \text{such that} \quad 1 \le v(z) \le C e^{k|z|}, \quad z \in \mathbb{R}^{2d}. \tag{15}$$

For instance, weights of the form

$$v(z) = e^{s|z|^b} (1+|z|)^a \log^r(e+|z|) \tag{16}$$

where $a, r, s \ge 0$, $0 \le b \le 1$, satisfy the above conditions.

We denote by $\mathcal{M}_v(\mathbb{R}^{2d})$ the space of v-moderate weights on $\mathbb{R}^{2d}$; these are positive and measurable functions m satisfying $m(z+\zeta) \le Cv(z)m(\zeta)$ for every $z, \zeta \in \mathbb{R}^{2d}$.

Definition 2.8. Given $g \in \Sigma_1^1(\mathbb{R}^d)$, a weight function $m \in \mathcal{M}_v(\mathbb{R}^{2d})$, and $1 \le p, q \le \infty$, the *modulation space* $M_m^{p,q}(\mathbb{R}^d)$ consists of all tempered ultra-distributions $f \in (\Sigma_1^1)'(\mathbb{R}^d)$ such that $V_g f \in L_m^{p,q}(\mathbb{R}^{2d})$ (weighted mixed-norm spaces). The norm on $M_m^{p,q}(\mathbb{R}^d)$ is

$$\|f\|_{M_m^{p,q}} = \|V_g f\|_{L_m^{p,q}} = \left(\int_{\mathbb{R}^d} \left(\int_{\mathbb{R}^d} |V_g f(x,\xi)|^p m(x,\xi)^p \, dx \right)^{q/p} d\xi \right)^{1/q} \tag{17}$$

(obvious changes if $p = \infty$ or $q = \infty$).

For $f, g \in \Sigma_1^1(\mathbb{R}^d)$ the above integral is convergent and thus $\Sigma_1^1(\mathbb{R}^d) \subset M_m^{p,q}(\mathbb{R}^d)$, $1 \le p, q \le \infty$, cf. [14], with dense inclusion when $p, q < \infty$, cf. [3]. When $p = q$, we simply write $M_m^p(\mathbb{R}^d)$ instead of $M_m^{p,p}(\mathbb{R}^d)$. The spaces $M_m^{p,q}(\mathbb{R}^d)$ are Banach spaces and every nonzero $g \in M_v^1(\mathbb{R}^d)$ yields an equivalent norm in (17) and so $M_m^{p,q}(\mathbb{R}^d)$ is independent on the choice of $g \in M_v^1(\mathbb{R}^d)$.

We observe that these properties of modulation spaces do depend on the fact that the weight functions involved have at most exponential growth at infinity. Indeed, as well known, if we consider super exponential weights at infinity, say, e.g., $m(z) = e^{k|z|^b}$, with $k > 0$ and $1 < b \le 2$, then the related modulation spaces $M_m^{p,q}(\mathbb{R}^d)$ can still be defined by taking Gelfand–Shilov windows, but their definition depends on the choice of the window g, because of the loss of sub-multiplicativity of the weight.

2.4. Ultra-modulation spaces

In this section we present a new definition of modulation spaces, which considers also weights of super-exponential growth at infinity (of a very particular form). For related constructions, we refer to [3, 22] and more recently to [34]. Let us first introduce the weights:

$$w_{s,\epsilon}(z) := e^{\epsilon |z|^{\frac{1}{s}}}, \quad z \in \mathbb{R}^d, \; s \ge 1/2, \; \epsilon > 0, \tag{18}$$

and the weight class $\mathcal{N}(\mathbb{R}^{2d}) = \{m_{s,\epsilon}(x,\xi) = (1 \otimes w_{s,\epsilon})(x,\xi) = w_{s,\epsilon}(\xi), \, x, \xi \in \mathbb{R}^d, \, \epsilon > 0, s \ge 1/2\}$, which we may identify with the set $(0, +\infty) \times [1/2, +\infty)$ of the corresponding ϵ, s parameters. Thus, a weight $m_{s,\epsilon} \in \mathcal{N}(\mathbb{R}^{2d})$ grows faster than exponentially at infinity in the frequency variable whenever $1/2 \le s < 1$. Notice that the bound $s = 1/2$ is admitted. We use the class $\mathcal{N}(\mathbb{R}^{2d})$ as weight class for modulation spaces. We limit our study to weights in the frequency variable for simplicity. On the other hand, observe that a limit in enlarging the definition to the space variables is imposed by Hardy's theorem: if $m(z) \ge Ce^{c|z|^2}$, for $z = (x, \xi) \in \mathbb{R}^{2d}$ and some $c > \pi/2$, then the corresponding modulation spaces are trivial [22].

Definition 2.9. Let $\epsilon > 0$, $s \geq 1/2$ and $m_{s,\epsilon} \in \mathcal{N}(\mathbb{R}^{2d})$. Consider a non-zero *window* function g in $\mathcal{S}^s_s(\mathbb{R}^d)$. For $1 \leq p, q \leq \infty$, we define $M^{p,q}_{m_{s,\epsilon},g}(\mathbb{R}^d)$ the subspace of $f \in (\Sigma^1_1)'(\mathbb{R}^d)$ such that the integrals in (17) are finite (with obvious changes if either $p = \infty$ or $q = \infty$).

Notice that:

(i) If $s > 1/2$, $f, g \in \mathcal{S}^{1/2}_{1/2}(\mathbb{R}^d)$, the integral in (17) is convergent thanks to (11) and Theorem 2.3, item d). Indeed, we can find $h > 0$ such that

$$\|V_g f e^{h|\cdot|^2}\|_{L^\infty} < \infty$$

and

$$\int_{\mathbb{R}^d}\left(\int_{\mathbb{R}^d} |V_g f(x,\xi)|^p m_{s,\epsilon}(x,\xi)^p \, dx\right)^{q/p} d\xi$$

$$\leq C\,\|(V_g f)e^{h|\cdot|^2}\|_{L^\infty}\int_{\mathbb{R}^d}\left(\int_{\mathbb{R}^d} |m_{s,\epsilon}(x,\xi)|^p e^{-hp|(x,\xi)|^2}\, dx\right)^{q/p} d\xi < \infty.$$

(ii) For $m_{s,\epsilon} \in \mathcal{M}_v(\mathbb{R}^{2d})$, hence $s \geq 1$, then $M^{p,q}_{m_{s,\epsilon},g}(\mathbb{R}^d)$ is the subspace of ultra-distribution $(\Sigma^1_1)'(\mathbb{R}^d)$ defined in Definition 2.8. So we come back to the classical modulation spaces. This justify the same notation.

(iii) The definition of $M^{p,q}_{m_{s,\epsilon},g}(\mathbb{R}^d)$ may depend on the choice of the window function g. However, if $s \geq 1$, the definition of $M^{p,q}_{m_{s,\epsilon},g}(\mathbb{R}^d)$ does not depend on g: the class of admissible windows can be enlarged to $M^1_{m_{s,\epsilon}}(\mathbb{R}^d)$ [3, Proposition 1], so we recapture the standard Definition 2.8.

2.5. Time-frequency analysis of Gevrey-analytic and ultra-analytic symbols

Here we present the results obtained in [13, Section 3], where the smoothness and the growth of a function f on $\mathbb{R}^d$ is characterized in terms of the decay of its STFT $V_g f$, for a suitable window g.

Theorem 2.10. *Consider $s > 0$, $m \in \mathcal{M}_v(\mathbb{R}^d)$, $g \in M^1_{v\otimes 1}(\mathbb{R}^d) \setminus \{0\}$ such that there exists $C_g > 0$,*

$$\|\partial^\alpha g\|_{L^1_v(\mathbb{R}^d)} \lesssim C_g^{|\alpha|}(\alpha!)^s, \quad \alpha \in \mathbb{N}^d. \tag{19}$$

For $f \in \mathcal{C}^\infty(\mathbb{R}^d)$ the following conditions are equivalent:

(i) *There exists a constant $C_f > 0$ such that*

$$|\partial^\alpha f(x)| \lesssim m(x) C_f^{|\alpha|}(\alpha!)^s, \quad x \in \mathbb{R}^d,\ \alpha \in \mathbb{N}^d. \tag{20}$$

(ii) *There exists a constant $C_{f,g} > 0$ such that*

$$|\xi^\alpha V_g f(x,\xi)| \lesssim m(x) C_{f,g}^{|\alpha|}(\alpha!)^s, \quad (x,\xi) \in \mathbb{R}^{2d},\ \alpha \in \mathbb{N}^d. \tag{21}$$

(iii) *There exists a constant $\epsilon > 0$ such that*

$$|V_g f(x,\xi)| \lesssim m(x) e^{-\epsilon|\xi|^{\frac{1}{s}}}, \quad (x,\xi) \in \mathbb{R}^{2d},\ \alpha \in \mathbb{N}^d. \tag{22}$$

We say that the function f is Gevrey if $s > 1$, analytic if $s = 1$, and ultra-analytic when $s < 1$.

Remark 2.11. We observe that the assumption $m \in \mathcal{M}_v(\mathbb{R}^d)$ in the previous theorem is essential to obtain the equivalence: indeed in the proof we exploit the v-moderateness of m.

For simplicity, from now on we shall assume $m = v = 1$.

A natural question is whether we may find window functions satisfying (19). To find the answer, we recall the following characterization of Gelfand–Shilov spaces.

Proposition 2.12. *Let $g \in \mathcal{S}(\mathbb{R}^d)$. We have $g \in S_r^s(\mathbb{R}^d)$, with $s, r > 0$, $r + s \geq 1$, if and only if there exist constants $A > 0$, $\epsilon > 0$ such that*

$$|\partial^\alpha g(x)| \lesssim A^{|\alpha|}(\alpha!)^s e^{-\epsilon|x|^{\frac{1}{r}}}, \quad x \in \mathbb{R}^d, \ \alpha \in \mathbb{N}^d.$$

We have $g \in \Sigma_1^1(\mathbb{R}^d)$ if and only if, for every $A > 0$, $\epsilon > 0$,

$$|\partial^\alpha g(x)| \lesssim A^{|\alpha|}\alpha! e^{-\epsilon|x|}, \quad x \in \mathbb{R}^d, \ \alpha \in \mathbb{N}^d.$$

Hence every $g \in S_r^s(\mathbb{R}^d)$ with $s > 0$, $0 < r < 1$, $s + r \geq 1$, satisfies (19) for every submultiplicative weight v (see (15)). The same holds true if $g \in \Sigma_1^1(\mathbb{R}^d)$ and $s \geq 1$.

3. Almost diagonalization for pseudodifferential operators

In this section we first present the almost diagonalization for pseudodifferential operators having (ultra-)analytic symbols obtained in [13, Section 4]. This result can be seen as an extension of the almost diagonalization for pseudodifferential operators obtained in [21], where only the Gevrey-analytic case was discussed.

Secondly, we exhibit a new characterization involving ultra-modulation spaces.

Recall the Weyl form σ^w of a pseudodifferential operator (so-called Weyl operator or Weyl transform) with symbol $\sigma(x, \xi)$ on $\mathbb{R}^{2d}$, formally defined by

$$\sigma^w f(x) = \int_{\mathbb{R}^d} \sigma\left(\frac{x+y}{2}, \eta\right) e^{2\pi i(x-y)\eta} f(y)\, dy d\eta. \tag{23}$$

Using the Kernel Theorem for Gelfand–Shilov spaces (Theorem 2.6), we have a characterization of linear continuous operators $T : S_r^s(\mathbb{R}^d) \to (S_r^s)'(\mathbb{R}^d)$. In particular, any such operator T can be represented as a pseudodifferential operator in the Weyl form, with $\sigma \in (S_s^r)'(\mathbb{R}^{2d})$. Thus we shall exhibit our results for Weyl operators.

It was proved in [18, 21] that Gabor frames allow to discretize a continuous operator from $\mathcal{S}(\mathbb{R}^d)$ to $\mathcal{S}'(\mathbb{R}^d)$ into an infinite matrix that captures the properties of the original operator. In particular, the authors consider symbols in the modulation spaces $M^{\infty,q}_{1\otimes v\circ j^{-1}}$, where v is a continuous and submultiplicative weight function on $\mathbb{R}^{2d}$, satisfying the GRS-condition (Gelfand–Raikov–Shilov)

$$\lim_{n\to\infty} v(nz)^{1/n} = 1, \quad \forall z \in \mathbb{R}^{2d}, \tag{24}$$

and j is the rotation on $\mathbb{R}^{2d}$:

$$j(z_1, z_2) = (z_2, -z_1), \quad (z_1, z_2) \in \mathbb{R}^{2d}.$$

Fix a Banach algebra under convolution of sequences $\ell^q_v(\Lambda)$, where Λ is a lattice on $\mathbb{R}^{2d}$ with relatively compact fundamental domain $\mathcal{Q}$ containing the origin. Then a function $H \in L^\infty_{loc}(\mathbb{R}^{2d})$ belongs to the Wiener amalgam space $W(\ell^q_v)(\mathbb{R}^{2d})$ if the sequence $h(\lambda) := \operatorname{ess\,sup}_{u\in\lambda+\mathcal{Q}} H(u)$ is in $\ell^q_v(\Lambda)$.

The almost diagonalization for pseudodifferential operators in [18, 21] can be extended easily to modulation spaces with exponential weights $v(z) = w_{1,\epsilon} = e^{\epsilon|z|}$, defined in (18). These weights do not fulfill the GRS-condition (24), but this condition is not necessary in the almost diagonalization result below. Indeed, observe that the algebra $\mathcal{A} = \ell^1_v(\Lambda)$ is a Banach algebra under convolution if and only if v is submultiplicative (see [21, Example 3.1]) (hence the GRS-condition is not required). So, the proof of [18, Theorem 3.2] (equivalently [21, Theorem 4.1] with the algebra $\mathcal{A} = \ell^1_v(\Lambda)$) still holds true also for the exponential weights $v = w_{1,\epsilon}$, provided the existence of a Gabor frame for $L^2(\mathbb{R}^d)$ with $g \in M^1_v(\mathbb{R}^d)$, having some dual window $\gamma \in M^1_v(\mathbb{R}^d)$ (observe that in general γ is not the canonical dual window). The so-called Gabor super-frames $\mathcal{G}(g,\Lambda)$, introduced in [20] (and recalled in detail before Theorem 3.6), satisfy this condition because $g \in S^{1/2}_{1/2}(\mathbb{R}^d) \subset M^1_v(\mathbb{R}^d)$ and there exists a dual window $\gamma \in S^{1/2}_{1/2}(\mathbb{R}^d) \subset M^1_v(\mathbb{R}^d)$. Hence we shall state [21, Theorem 4.1] including exponential weights, provided that the symbol space of distributions $\mathcal{S}'(\mathbb{R}^{2d})$ is replaced by the space of ultra-distributions $(\Sigma^1_1)'(\mathbb{R}^{2d})$.

Theorem 3.1 ([18, 21]). *Assume that $\mathcal{G}(g,\Lambda)$ is a frame for $L^2(\mathbb{R}^d)$ with $g \in M^1_v(\mathbb{R}^d)$ if v satisfies (24), whereas $\mathcal{G}(g,\Lambda)$ a Gabor super-frame when v is an exponential weight. Then the following statements are equivalent for $\sigma \in (\Sigma^1_1)'(\mathbb{R}^{2d})$:*

(i) $\sigma \in M^{\infty,q}_{1\otimes v\circ j^{-1}}(\mathbb{R}^{2d})$.

(ii) *There exists a function $H \in W(\ell^q_v)(\mathbb{R}^{2d})$ such that*

$$|\langle \sigma^w \pi(z) g, \pi(w) g\rangle| \lesssim H(w-z), \qquad \forall w, z \in \mathbb{R}^{2d}. \tag{25}$$

(iii) *There exists a sequence $h \in \ell^q_v(\Lambda)$ such that*

$$|\langle \sigma^w \pi(\lambda) g, \pi(\mu) g\rangle| \lesssim h(\mu-\lambda), \qquad \forall \lambda, \mu \in \Lambda. \tag{26}$$

The Gabor kernel of σ^w: $\langle \sigma^w \pi(z) g, \pi(w) g\rangle$ is referred as *continuous* Gabor matrix, whereas $\langle \sigma^w \pi(\lambda) g, \pi(\mu) g\rangle$ is the Gabor matrix of σ^w.

Our concern is in the case of a continuous Gabor matrix dominated by a function $H \in W(\ell^\infty_v)(\mathbb{R}^{2d})$. In this case $\ell^\infty_v(\Lambda)$ is an algebra if and only if $v^{-1} * v^{-1} \leq C v^{-1}$, that is v is subconvolutive (see [21, Example 3.1] and reference therein). Since exponential weights do not satisfy this condition we state the risult in the original setting.

Theorem 3.2. *Assume that $\mathcal{G}(g,\Lambda)$ is a frame for $L^2(\mathbb{R}^d)$ with $g \in M^1_v(\mathbb{R}^d)$ and v satisfying* (24). *Then the following statements are equivalent for $\sigma \in \mathcal{S}'(\mathbb{R}^{2d})$:*

(i) $\sigma \in M^\infty_{1\otimes v\circ j^{-1}}(\mathbb{R}^{2d})$.

(ii) *The continuous Gabor matrix satisfies*

$$|\langle \sigma^w \pi(z)g, \pi(w)g\rangle| \lesssim v(w-z), \qquad \forall w, z \in \mathbb{R}^{2d}. \tag{27}$$

(iii) *The Gabor matrix satisfies*

$$|\langle \sigma^w \pi(\lambda)g, \pi(\mu)g\rangle| \lesssim v(\mu-\lambda), \qquad \forall \lambda, \mu \in \Lambda. \tag{28}$$

Our main aim here is to give a counterpart of Theorem 3.2 for weights of super-exponential growth. First, let us survey the new results of [13, Section 4] which provide the arguments for the characterization of Theorem 3.4 below.

The crucial relation between the action of the Weyl operator σ^w on time-frequency shifts and the short-time Fourier transform of its symbol, contained in [18, Lemma 3.1] can now be extended to Gelfand–Shilov spaces and their dual spaces as follows.

Lemma 3.3. *Consider $s \geq 1/2$, $g \in S^s_s(\mathbb{R}^d)$, $\Phi = W(g,g)$. Then, for $\sigma \in (S^s_s)'(\mathbb{R}^{2d})$,*

$$|\langle \sigma^w \pi(z)g, \pi(w)g\rangle| = \left| V_\Phi \sigma\left(\frac{z+w}{2}, j(w-z)\right)\right| = |V_\Phi\sigma(u,v)| \tag{29}$$

and

$$|V_\Phi\sigma(u,v)| = \left| \langle \sigma^w \pi\left(u - \frac{1}{2}j^{-1}(v)\right) g, \pi\left(u + \frac{1}{2}j^{-1}(v)\right) g\rangle\right|. \tag{30}$$

Proof. Since $\Phi = W(g,g) \in S^s_s(\mathbb{R}^{2d})$ for $g \in S^s_s(\mathbb{R}^d)$ by (13), the duality

$$\langle \sigma, \pi(u,v)\Phi\rangle_{(S^s_s)'\times S^s_s}$$

is well defined so that the short-time Fourier transform $V_\Phi\sigma(u,v)$ makes sense. The rest of the proof is analogous to [18, Lemma 3.1]. □

Given a pseudodifferential operator σ^w, with smooth symbol $\sigma \in \mathcal{C}^\infty(\mathbb{R}^{2d})$, we exhibit the connection between the Gevrey, analytic or ultra-analytic regularity of σ (as considered in Theorem 2.10, the dimension being now $2d$), the (ultra-)modulation space σ belongs to (cf. Definition 2.9), and the decay of the continuous Gabor matrix of σ^w.

Recall the weights $w_{s,\epsilon}$ defined in (18) and here used as functions over $\mathbb{R}^{2d}$.

Theorem 3.4. *Let $s \geq 1/2$ and consider a window function $g \in S^s_s(\mathbb{R}^d)$. Set $\Phi = W(g,g)$. Then the following properties are equivalent for $\sigma \in \mathcal{C}^\infty(\mathbb{R}^{2d})$:*

(i) *There exists $\epsilon > 0$ such that $\sigma \in M^\infty_{1\otimes w_{s,\epsilon},\Phi}(\mathbb{R}^{2d})$.*

(ii) *There exists $C > 0$ such that the symbol σ satisfies*

$$|\partial^\alpha \sigma(z)| \lesssim C^{|\alpha|}(\alpha!)^s, \quad \forall\, z \in \mathbb{R}^{2d}, \forall \alpha \in \mathbb{N}^{2d}. \tag{31}$$

(iii) *There exists* $\epsilon > 0$ *such that*

$$|\langle \sigma^w \pi(z)g, \pi(w)g\rangle| \lesssim w_{s,-\epsilon}(w-z), \qquad \forall\, z, w \in \mathbb{R}^{2d}. \tag{32}$$

Proof. (i) $\Rightarrow$ (iii). Assuming $\sigma \in M^\infty_{1\otimes w_{s,\epsilon},\Phi}(\mathbb{R}^{2d})$, that is

$$\sup_{u,v\in\mathbb{R}^{2d}} |V_\Phi \sigma(u,v)| e^{\epsilon|v|^{\frac{1}{s}}} < \infty$$

and using (30), we obtain the claim:

$$\begin{aligned} |\langle \sigma^w \pi(z)g, \pi(w)g\rangle| &= \left| V_\Phi \sigma(\frac{w+z}{2}, j(w-z)) \right| \\ &\le \sup_{u\in\mathbb{R}^{2d}} |V_\Phi \sigma(u, j(w-z))| \lesssim e^{-\epsilon|w-z|^{\frac{1}{s}}}. \end{aligned}$$

(iii) $\Rightarrow$ (i). Relation (30) and the decay assumption (32) give

$$|V_\Phi \sigma(u,v)| = \left| \langle \sigma^w \pi\left(u - \frac{1}{2} j^{-1}(v)\right) g, \pi\left(u + \frac{1}{2} j^{-1}(v)\right) g\rangle \right| \tag{33}$$

$$\lesssim e^{-\epsilon|j^{-1}(v)|^{\frac{1}{s}}} = e^{-\epsilon|v|^{\frac{1}{s}}}. \tag{34}$$

The equivalences (ii) $\Leftrightarrow$ (iii) are proved in Theorem [13, Theorem 4.2]. For sake of clarity, we give a brief sketch of this result. The techniques are similar to [18, Theorem 3.2].

(ii) $\Rightarrow$ (iii). The window $\Phi = W(g,g) \in S^s_s(\mathbb{R}^{2d})$, for $g \in S^s_s(\mathbb{R}^d)$ by (13), and satisfies the assumptions of Theorem 2.10. Hence, using the equivalence (20) $\Leftrightarrow$ (22), the assumption (31) is equivalent to the following decay estimate of the corresponding short-time Fourier transform

$$|V_\Phi \sigma(u,v)| \lesssim e^{-\epsilon|v|^{\frac{1}{s}}}, \quad u, v \in \mathbb{R}^{2d},$$

for a suitable $\epsilon > 0$, hence

$$|V_\Phi \sigma(\frac{z+w}{2}, j(w-z))| \lesssim e^{-\epsilon|j(w-z)|^{\frac{1}{s}}} = e^{-\epsilon|w-z|^{\frac{1}{s}}}$$

which combined with (29) yields (iii).

(iii) $\Rightarrow$ (ii). We use relation (30) and the decay assumption (32) again, which give (33) and using the equivalence (20) $\Leftrightarrow$ (22) we obtain the claim. □

Remark 3.5.

a) We observe that the constant $\epsilon > 0$, which depends on the choice of $g \in S^s_s(\mathbb{R}^d)$, is the same in (i) and (iii). Whereas the link between $\epsilon > 0$ and the constant $C > 0$ in (ii) is specified in Remark 2.5.
b) If we consider $s \ge 1$, that is the symbol σ is a Gevrey or an analytic symbol, then the equivalence (i) $\Leftrightarrow$ (ii) follows by combining Theorems 3.2 and [13, Theorem 4.2].

Of course the estimate (32) implies the discrete analog (35) below. The vice versa is not obvious and requires the existence of a Gabor frame $\mathcal{G}(g,\Lambda)$ having $g \in S^{1/2}_{1/2}(\mathbb{R}^d)$ and a dual window $\gamma \in S^{1/2}_{1/2}(\mathbb{R}^d)$ as well. We call such a Gabor frame a *Gabor super-frame.* The existence of Gabor super-frames is due to a result obtained by Gröchenig and Lyubarskii in [20]. They find sufficient conditions on the lattice $\Lambda = A\mathbb{Z}^2$, $A \in GL(2,\mathbb{R})$, such that $g = \sum_{k=0}^n c_k H_k$, with H_k Hermite function, forms a Gabor frame $\mathcal{G}(g,\Lambda)$. Besides they prove the existence of dual windows γ that belong to the space $S^{1/2}_{1/2}(\mathbb{R})$ (cf. [20, Lemma 4.4]). This theory transfers to the d-dimensional case simply by taking tensor products $g = g_1 \otimes \cdots \otimes g_d \in S^{1/2}_{1/2}(\mathbb{R}^d)$ of windows as above, which define a Gabor frame on the lattice $\Lambda_1 \times \cdots \times \Lambda_d$ and possess a dual window $\gamma = \gamma_1 \otimes \cdots \otimes \gamma_d$ in the same space $\in S^{1/2}_{1/2}(\mathbb{R}^d)$.

The Gabor super-frames allow the discretization of the kernel in (32).

Theorem 3.6. *Let $\mathcal{G}(g,\Lambda)$ a Gabor super-frame for $L^2(\mathbb{R}^d)$. Consider $s \geq 1/2$ and a symbol $\sigma \in \mathcal{C}^\infty(\mathbb{R}^{2d})$. Then the following properties are equivalent:*

(i) *There exists $\epsilon > 0$ such that the estimate (32) holds.*

(ii) *There exists $\epsilon > 0$ such that*

$$|\langle \sigma^w \pi(\mu) g, \pi(\lambda) g\rangle| \lesssim w_{s,-\epsilon}(\lambda - \mu), \qquad \forall\, \lambda,\mu \in \Lambda. \tag{35}$$

Proof. Let us sketch (ii) $\Rightarrow$ (i). The pattern of [18, Theorem 3.2] can be adapted to this proof by using a Gabor super-frame $\mathcal{G}(g,\Lambda)$, with a dual window $\gamma \in S^{1/2}_{1/2}(\mathbb{R}^d)$ and the following property of the weights (18) ([13, Lemma 3.5]):

$$(w_{s,-\epsilon} * w_{s,-\epsilon})(\lambda) := \sum_{\nu\in\Lambda} w_{s,-\epsilon}(\lambda-\nu) w_{s,-\epsilon}(\nu) \lesssim \begin{cases} w_{s,-\epsilon}(\lambda), & \text{for } s \geq 1 \\ w_{s,\,-\epsilon 2^{-1/s}}(\lambda), & \text{for } \frac{1}{2} \leq s < 1. \end{cases} \tag{36}$$

Let $\mathcal{Q}$ be a symmetric relatively compact fundamental domain of the lattice $\Lambda \subset \mathbb{R}^{2d}$. Given $w, z \in \mathbb{R}^{2d}$, we can write them uniquely as $w = \lambda + u$, $z = \mu + u'$, for $\lambda, \mu \in \Lambda$ and $u, u' \in \mathcal{Q}$. Using the Gabor reproducing formula for the time-frequency shift $\pi(u) g \in S^{1/2}_{1/2}(\mathbb{R}^d)$ we can write

$$\pi(u) g = \sum_{\nu\in\Lambda} \langle \pi(u) g, \pi(\nu)\gamma\rangle \pi(\nu) g.$$

Inserting the prior expansions in the assumption (35),

$$\begin{aligned} &|\langle \sigma^w \pi(\mu + u') g, \pi(\lambda + u) g\rangle| \\ &\quad \leq \sum_{\nu,\nu'\in\Lambda} |\langle \sigma^w \pi(\mu+\nu') g, \pi(\lambda+\nu) g\rangle|\, |\langle \pi(u') g, \pi(\nu')\gamma\rangle|\, |\langle \pi(u) g, \pi(\nu)\gamma\rangle| \\ &\quad \lesssim \sum_{\nu,\nu'\in\Lambda} m\left(\frac{\lambda+\mu+\nu+\nu'}{2}\right) e^{-\epsilon|\lambda+\nu-\mu-\nu'|^{\frac{1}{s}}} |V_\gamma g(\nu'-u')| |V_\gamma g(\nu - u)|. \end{aligned}$$

Since the window functions g, γ are both in $S^{1/2}_{1/2}(\mathbb{R}^d)$, the STFT $V_\gamma g$ is in $S^{1/2}_{1/2}(\mathbb{R}^{2d})$, due to (11). Thus there exists $h > 0$ such that $|V_\gamma g(z)| \lesssim e^{-h|z|^2}$, for every $z \in \mathbb{R}^{2d}$. Inserting this estimate in the previous majorization and using (36) repeatedly,

$$\begin{aligned} |\langle \sigma^w \pi(\mu+u')g, \pi(\lambda+u)g\rangle| &\lesssim \sum_{\nu,\nu'\in\Lambda} e^{-\epsilon|\lambda+\nu-\mu-\nu'|^{\frac{1}{s}}} e^{-h|\nu|^2} e^{-h|\nu'|^2} \\ &\lesssim \sum_{\nu,\nu'\in\Lambda} e^{-b(|\lambda+\nu-\mu-\nu'|^{\frac{1}{s}}+|\nu|^{\frac{1}{s}}+|\nu'|^{\frac{1}{s}})} \\ &= w_{s,b} * w_{s,b} * w_{s,b})(\lambda-\mu) \\ &\lesssim e^{-\tilde{\epsilon}|\lambda-\mu|^{\frac{1}{s}}} \end{aligned}$$

for a suitable $\tilde{\epsilon} > 0$.

If $w, z \in \mathbb{R}^{2d}$ and $w = \lambda + u$, $z = \mu + u'$, $\lambda, \mu \in \Lambda$, $u, u' \in \mathcal{Q}$, then $\lambda - \mu = w - z + u' - u$ and $u' - u \in \mathcal{Q} - \mathcal{Q}$, which is a relatively compact set, thus

$$e^{-\tilde{\epsilon}|\lambda-\mu|^{\frac{1}{s}}} \lesssim \sup_{u\in\mathcal{Q}-\mathcal{Q}} e^{-\tilde{\epsilon}|w-z+u|^{\frac{1}{s}}} \lesssim e^{-\tilde{\epsilon}|w-z|^{\frac{1}{s}}}. \tag{37}$$

This gives the desired implication. □

4. Sparsity of the Gabor matrix

The operators σ^w in Theorem 3.4 enjoy a fundamental sparsity property. Indeed, let $\mathcal{G}(g, \Lambda)$ be a Gabor super-frame for $L^2(\mathbb{R}^d)$. Then, as we saw,

$$|\langle \sigma^w \pi(\mu)g, \pi(\lambda)g\rangle| \leq Ce^{-\epsilon|\lambda-\mu|^{\frac{1}{s}}}, \qquad \forall \lambda, \mu \in \Lambda, \tag{38}$$

with suitable constants $C > 0$, $\epsilon > 0$. This gives at once an exponential-type sparsity, which is proved in [13, Proposition 4.5] (we refer to [2, 23] for the more standard notion of super-polynomial sparsity) and recalled below.

Proposition 4.1. *Let the Gabor matrix $\langle \sigma^w \pi(\mu)g, \pi(\lambda)g\rangle$ satisfy* (38). *Then it is sparse in the following sense. Let a be any column or raw of the matrix, and let $|a|_n$ be the n-largest entry of the sequence a. Then, $|a|_n$ satisfies*

$$|a|_n \leq Ce^{-\epsilon n^{\frac{1}{2ds}}}, \quad n \in \mathbb{N}$$

for some constants $C > 0, \epsilon > 0$.

Proof. By a discrete analog of Proposition 2.4 it suffices to prove that

$$n^\alpha |a|_n \leq C^{\alpha+1}(\alpha!)^{2ds}, \quad \alpha \in \mathbb{N}.$$

On the other hand we have

$$n^{\frac{1}{p}} \cdot |a|_n \leq \|a\|_{\ell^p},$$

for every $0 < p \leq \infty$. Hence by (38) and setting $p = 1/\alpha$ we obtain

$$n^\alpha |a|_n \leq \Big(\sum_{\lambda\in\Lambda} e^{-\epsilon p|\lambda-\mu|^{\frac{1}{s}}}\Big)^{\frac{1}{p}} = \Big(\sum_{\lambda\in\Lambda} e^{-\epsilon p|\lambda|^{\frac{1}{s}}}\Big)^{\frac{1}{p}}.$$

Let $\mathcal{Q}$ be a fundamental domain of the lattice Λ. Then if $x \in \lambda + \mathcal{Q}$, $\lambda \in \Lambda$, we have $|x| \le |\lambda| + C_0$, therefore $|x|^{1/s} \le C_1(|\lambda|^{1/s} + 1)$. Hence

$$\sum_{\lambda\in\Lambda} e^{-\epsilon p|\lambda|^{\frac{1}{s}}} \le C_2 \int_{\mathbb{R}^{2d}} e^{-\epsilon p|x|^{\frac{1}{s}}}\, dx = \int_{\mathbb{S}^{2d-1}} d\sigma \int_0^{+\infty} e^{-\epsilon p\rho^{\frac{1}{s}}} \rho^{2d-1} d\rho$$

$$= \frac{C_3 s}{(\epsilon p)^{2ds}} \int_0^{+\infty} e^{-t} t^{2ds-1} dt = \frac{C_3 s\Gamma(2ds)}{(\epsilon p)^{2ds}} = \frac{C_4}{p^{2ds}}$$

Finally, by Stirling's formula,

$$n^\alpha |a|_n \le \frac{C_4^{1/p}}{p^{\frac{2ds}{p}}} \le C_5^{\alpha+1} (\alpha!)^{2ds}. \qquad \square$$

References

[1] A. Bényi, K. Gröchenig, K.A. Okoudjou and L.G. Rogers. Unimodular Fourier multipliers for modulation spaces. *J. Funct. Anal.*, 246(2):366–384, 2007.

[2] E.J. Candès and L. Demanet. The curvelet representation of wave propagators is optimally sparse. *Comm. Pure Appl. Math.*, 58:1472–1528, 2004.

[3] E. Cordero. Gelfand–Shilov Window Classes for Weighted Modulation Spaces. *Int. Tran. Spec. Funct.*, 18(11):809–817, 2007.

[4] E. Cordero and K. Gröchenig and F. Nicola. Approximation of Fourier integral operators by Gabor multipliers. *J. Fourier Anal. Appl.*, to appear. DOI: 10.1007/s00041-011-9214-1.

[5] E. Cordero and F. Nicola. Strichartz estimates in Wiener amalgam spaces for the Schrödinger equation. *Math. Nachr.*, 281(1): 25–41, 2008.

[6] E. Cordero and F. Nicola. Some new Strichartz estimates for the Schrödinger equation. *J. Differential Equations*, 245(7):1945–1974, 2008.

[7] E. Cordero and F. Nicola. Metaplectic representation on Wiener amalgam spaces and applications to the Schrödinger equation. *J. Funct. Anal.*, 254(2):506–534, 2008.

[8] E. Cordero and F. Nicola. Boundedness of Schrödinger type propagators on modulation spaces. *J. Fourier Anal. Appl.* 16(3):311–339, 2010.

[9] E. Cordero and F. Nicola. Sharp Continuity Results for the Short-Time Fourier Transform and for Localization Operators. *Monatshefte für Mathematik*, 162(3),251–276, 2011.

[10] E. Cordero and F. Nicola. Pseudodifferential Operators on L^p, Wiener Amalgam and Modulation Spaces. *International Mathematics Research Notices*, 2010(10):1860–1893, 2010.

[11] E. Cordero, F. Nicola and L. Rodino. Time-frequency analysis of Fourier integral operators. *Commun. Pure Appl. Anal.*, 9(1):1–21, 2010.

[12] E. Cordero, F. Nicola and L. Rodino. Sparsity of Gabor representation of Schrödinger propagators. *Appl. Comput. Harmon. Anal.*, 26(3):357–370, 2009.

[13] E. Cordero, F. Nicola and L. Rodino. Gabor representations of evolution operators. *Trans. Amer. Math. Soc.*, to appear. Available on arXiv:1209.0945.

[14] E. Cordero, S. Pilipović, L. Rodino and N. Teofanov. Localization operators and exponential weights for modulation spaces. *Mediterranean J. Math.*, 2(4):381–394, 2005.

[15] H.G. Feichtinger, Modulation spaces on locally compact abelian groups, *Technical Report, University Vienna, 1983,* and also in *Wavelets and Their Applications*, M. Krishna, R. Radha, S. Thangavelu, editors, Allied Publishers, 99–140, 2003.

[16] I.M. Gelfand, G.E. Shilov. *Generalized Functions II, III.* Academic Press, 1967.

[17] K. Gröchenig. *Foundations of time-frequency analysis.* Birkhäuser Boston, Inc., Boston, MA, 2001.

[18] K. Gröchenig. Time-Frequency Analysis of Sjöstrand's Class. *Rev. Mat. Iberoamericana*, 22(2):703–724, 2006.

[19] K. Gröchenig and C. Heil. Modulation spaces and pseudodifferential operators. *Integral Equations Operator Theory*, 34:439–457, 1999.

[20] K. Gröchenig and Y. Lyubarskii. Gabor (super)frames with Hermite functions. *Math. Ann.*, 345:267–286, 2009.

[21] K. Gröchenig and Z. Rzeszotnik. Banach algebras of pseudodifferential operators and their almost diagonalization. Ann. Inst. Fourier., 58(7):2279–2314, 2008.

[22] K. Gröchenig and G. Zimmermann. Spaces of test functions via the STFT. *Journal of Function Spaces and Applications*, 2(1): 25–53, 2004.

[23] K. Guo and D. Labate. Sparse shearlet representation of Fourier integral operators *Electron. Res. Announc. Math. Sci.*, 14:7–19, 2007.

[24] L. Hörmander. *The analysis of linear partial differential operators*, Vol. 2, Springer, 1983.

[25] D. Labate. Pseudodifferential operators on modulation spaces. *J. Math. Anal. Appl.*, 262:242–255, 2001.

[26] B.S. Mitjagin. Nuclearity and other properties of spaces of type *S*. *Amer. Math. Soc. Transl.*, 93(2):45–59, 1970.

[27] A. Miyachi, F. Nicola, S. Rivetti, A. Tabacco and N. Tomita. Estimates for unimodular Fourier multipliers on modulation spaces. *Proc. Amer. Math. Soc.*, 137:3869–3883, 2009.

[28] F. Nicola and L. Rodino. *Global Pseudo-Differential Calculus on Euclidean Spaces.* Pseudo-Differential Operators. Theory and Applications, 4, Birkhäuser Verlag, Basel, 2010.

[29] R. Rochberg and K. Tachizawa. Pseudodifferential operators, Gabor frames, and local trigonometric bases. In *Gabor Analysis and Algorithms,* Appl. Numer. Harmon. Anal., Birkhäuser Boston, Boston, MA, 171–192, 1998.

[30] J. Sjöstrand. An algebra of pseudodifferential operators. *Math. Res. Lett.*, 1(2):185–192, 1994.

[31] J. Sjöstrand. Wiener type algebras of pseudodifferential operators. In *Séminaire sur les Équations aux Dérivées Partielles,* 1994–1995, pages Exp. No. IV, 21. École Polytech., Palaiseau, 1995.

[32] T. Strohmer. Pseudodifferential operators and Banach algebras in mobile communications. *Appl. Comput. Harmon. Anal.*, 20(2):237–249, 2006.

[33] N. Teofanov, Ultradistributions and time-frequency analysis, in *Pseudo-differential Operators and Related Topics, Operator Theory: Advances and Applications*, P. Boggiatto, L. Rodino, J. Toft, M.W. Wong, editors, Birkhäuser, 164:173–191, 2006.

[34] J. Toft. The Bargmann transform on modulation and Gelfand–Shilov spaces with applications to Toeplitz and pseudo-differential operators. *J. Pseudo-Differ. Oper. Appl.*, 3:145–227, 2012.

[35] F. Treves. *Topological Vector Spaces, Distributions and Kernels.* Academic Press, New York, 1967.

Elena Cordero and Luigi Rodino
Dipartimento di Matematica
Università di Torino
via Carlo Alberto 10
I-10123 Torino, Italy
e-mail: elena.cordero@unito.it
luigi.rodino@unito.it

Fabio Nicola
Dipartimento di Scienze Matematiche
Politecnico di Torino
corso Duca degli Abruzzi 24
I-10129 Torino, Italy
e-mail: fabio.nicola@polito.it

Fourier Analysis
Trends in Mathematics, 91–106

Global Wave-front Sets of Intersection and Union Type

Sandro Coriasco, Karoline Johansson and Joachim Toft

Abstract. We show that a temperate distribution belongs to an ordered intersection or union of admissible Banach or Fréchet spaces if and only if the corresponding global wave-front set of union or intersection type is empty. We also discuss the situation where intersections and unions of sequences of spaces with two indices are involved. A main situation where the present theory applies is given by sequences of weighted, general modulation spaces.

Mathematics Subject Classification (2010). 35A18,35S30,42B05,35H10.

Keywords. Wave-front, Fourier, Banach space, modulation space, micro-local, pseudo-differential.

1. Introduction

Wave-front sets of global type are a standard tool to investigate the regularity, local and at infinity, of solutions $f \in \mathscr{S}' = \mathscr{S}'(\mathbf{R}^d)$ to equations

$$Tf - g, \tag{1.1}$$

for a (pseudo)differential operator T and some given distribution $g \in \mathscr{S}'$.

In particular, it is often necessary to deal with situations where T is not globally hypoelliptic, and the involved distributions f and g belong to appropriate Banach or Fréchet spaces of temperate distributions $\mathcal{B}$ and $\mathcal{C}$, respectively.

In this paper we complete the analysis carried on in [4] (see also [6]), by giving the detailed proofs of some results involving global wave-front sets associated with sequences of appropriate function spaces, compatible with the class of pseudo-differential operators we use here, namely, the SG-operators. The SG-calculus of pseudo-differential operators was introduced in the 70s, independently by C. Parenti [17] and H.O. Cordes, see, e.g., [2] (for a different approach, see R. Melrose [16]). This calculus allows to treat global problems associated, e.g., with any linear partial differential operator with constant coefficients, the Klein–Gordon's equation, the Schrödinger equations for different atoms, and classes of

Dirac-type operators on $\mathbf{R}^d$. The definition and some basic facts about the SG-operators can be found in Section 2.

The general assumptions on $\mathcal{B}$ and $\mathcal{C}$ in (1.1) are described in detail in Section 1 of [4]. For example, $\mathcal{B}$ and $\mathcal{C}$ can be modulation spaces, a family of Banach spaces of functions and tempered distributions, introduced by H.G. Feichtinger, see, e.g., [8], and developed further and generalized by H.G. Feichtinger and K.H. Gröchenig [7], see also [9, 11]. For the convenience of the reader, we included the essential definitions in Section 2, following the approach in [9]. We remark that the family of modulation spaces is broad, in the sense that it contains the Sobolev spaces H^2_s and the Sobolev–Kato spaces $H^2_{s,t}$, see Remark 2.4 below.

In Section 3 we recall the definition of the global wave-front set $\mathrm{WF}_{\mathcal{B}}(f)$ of the distribution f, with respect to the Banach or Fréchet space $\mathcal{B}$, given in [4]. This object, loosely speaking, gives informations about the local regularity (smoothness) and the behavior at infinity (decay and oscillation properties within certain cones) of f. It is a remarkable fact that, whenever the space $\mathcal{B}$ is SG-admissible (see Section 2 below for the precise definition), then

$$\mathrm{WF}_{\mathcal{B}}(f) = \emptyset \quad \Longleftrightarrow \quad f \in \mathcal{B}. \tag{1.2}$$

This fact, together with the mapping properties which hold for these global wave-front sets under the action of SG-pseudo-differential operators (namely, the so-called microlocality and microellipticity properties), can be used to obtain rather precise relations between the regularity properties of f and g in (1.1).

Note that, if $\mathcal{B}$ equals $\mathscr{S}$ or $H^2_{s,t}$, then $\mathrm{WF}_{\mathcal{B}}(f)$ agrees with the wave-front sets of f with respect to $\mathscr{S}$ and $H^2_{s,t}$, respectively, given in [5] and [16]. Consequently, we recover also all the properties that hold for wave-front sets of Sobolev type introduced by Hörmander [14], and classical wave-front sets with respect to smoothness (cf. Sections 8.1 and 8.2 in [13]), as well as for wave-front sets of Banach function types in [3] (cf. also [18, 19]), and wave-front sets with respect to $\mathscr{S}$ and $H^2_{s_1,s_2}$ in [5, 16].

In order to get even more detailed information on the links between the regularity properties of f and g in (1.1), in [4] we introduced global wave-front sets with respect to *sequences* of SG-admissible spaces, whose definitions we recall in Section 4. Microlocality and microellipticity under the action of SG-pseudo-differential operators hold also for this refined type of wave-front set. Here we deal with the relations between the union/intersection type wave-front sets and the unions/intersections of the wave-front sets of the spaces belonging to the involved sequences. The main focus is the extension of (1.2) to this more general case, which in Section 4 is proved to hold as well. In other words, the property of a distributions to belong to an intersection, union, union of intersections, or intersection of unions of SG-admissible spaces is still equivalent to the emptiness of the corresponding type of wave-front set, under rather mild assumptions on the involved sequences of spaces.

An example where our theory applies is the case where $\mathcal{B} = \mathcal{C}$ equals $\mathscr{S}(\mathbf{R}^d)$, $Q_0(\mathbf{R}^d)$ or $Q(\mathbf{R}^d)$, where

$$Q_0(\mathbf{R}^d) \equiv \{f \in C^\infty(\mathbf{R}^d) : |\partial^\alpha f(x)| \lesssim \langle x\rangle^N \text{for some } N \text{ and every } \alpha \in \mathbf{Z}^d\}$$

and

$$Q(\mathbf{R}^d) \equiv \{f \in C^\infty(\mathbf{R}^d) : |\partial^\alpha f(x)| \lesssim \langle x\rangle^{N_\alpha}, \text{ where } N_\alpha \text{ depends on } \alpha \in \mathbf{Z}^d\},$$

where $A(x) \lesssim B(x)$ means that there exists $C > 0$, independent of x, such that, for any value of x on which A and B are defined, $A(x) \le C \cdot B(x)$.

The proofs of the results mentioned in Sections 2 and 3 can be found in [4].

2. Preliminaries

We start by recalling some basic definition and concept which will be needed throughout the paper. The material in this section comes mainly from [4].

2.1. Weight functions

Let ω and v be positive measurable functions on $\mathbf{R}^d$. Then ω is called v-moderate if

$$\omega(x+y) \lesssim \omega(x)v(y). \tag{2.1}$$

If v in (2.1) can be chosen as a polynomial, then ω is called a function or weight of *polynomial type*. We let $\mathscr{P}(\mathbf{R}^d)$ be the set of all polynomial type functions on $\mathbf{R}^d$. If $\omega(x,\xi) \in \mathscr{P}(\mathbf{R}^{2d})$ is constant with respect to the x-variable or the ξ-variable, then we sometimes write $\omega(\xi)$, respectively $\omega(x)$, instead of $\omega(x,\xi)$. In this case we consider ω as an element in $\mathscr{P}(\mathbf{R}^{2d})$ or in $\mathscr{P}(\mathbf{R}^d)$ depending on the situation. We say that v is submultiplicative if (2.1) holds for $\omega = v$. For convenience we assume that all submultiplicative weights are even, and we always let v and v_j stand for submultiplicative weights, if nothing else is stated.

Without loss of generality we may assume that every $\omega \in \mathscr{P}(\mathbf{R}^d)$ is smooth and satisfies the ellipticity condition $\partial^\alpha \omega/\omega \in L^\infty$. In fact, by Lemma 1.2 in [20] it follows that for each $\omega \in \mathscr{P}(\mathbf{R}^d)$, there is a smooth and elliptic $\omega_0 \in \mathscr{P}(\mathbf{R}^d)$ which is equivalent to ω in the sense

$$\omega \asymp \omega_0, \tag{2.2}$$

where $A \asymp B$ means $A \lesssim B \lesssim A$.

The weights involved in the sequel have to satisfy additional conditions. More precisely let $r, \rho \ge 0$. Then $\mathscr{P}_{r,\rho}(\mathbf{R}^{2d})$ is the set of all $\omega(x,\xi)$ in $\mathscr{P}(\mathbf{R}^{2d}) \bigcap C^\infty(\mathbf{R}^{2d})$ such that

$$\langle x\rangle^{r|\alpha|}\langle \xi\rangle^{\rho|\beta|}\frac{\partial_x^\alpha \partial_\xi^\beta \omega(x,\xi)}{\omega(x,\xi)} \in L^\infty(\mathbf{R}^{2d}),$$

for every multi-indices α and β. Note that $\mathscr{P}_{r,\rho}$ is different here compared to [3], and that there are elements in $\mathscr{P}(\mathbf{R}^{2d})$ which have no equivalent elements in $\mathscr{P}_{r,\rho}(\mathbf{R}^{2d})$. On the other hand, if $s, t \in \mathbf{R}$ and $r, \rho \in [0,1]$, then $\mathscr{P}_{r,\rho}(\mathbf{R}^{2d})$ contains all weights of the form $\omega(x,\xi) = \langle x\rangle^t\langle\xi\rangle^s$, which are one of the most common type of weights in the applications.

2.2. Modulation spaces

Let $\phi \in \mathscr{S}(\mathbf{R}^d)$. Then the *short-time Fourier transform* of $f \in \mathscr{S}(\mathbf{R}^d)$ with respect to (the window function) ϕ is defined by

$$V_\phi f(x,\xi) = (2\pi)^{-d/2}\int_{\mathbf{R}^d} f(y)\overline{\phi(y-x)}e^{-i\langle y,\xi\rangle}\,dy. \tag{2.3}$$

More generallly, the short-time Fourier transform of $f \in \mathscr{S}'(\mathbf{R}^d)$ with respect to $\phi \in \mathscr{S}'(\mathbf{R}^d)$ is defined by

$$V_\phi f = \mathscr{F}_2 F, \quad \text{where} \quad F(x,y) = (f\otimes\overline{\phi})(y,y-x), \tag{2.3$'$}$$

where $\mathscr{F}_2 F$ is the partial Fourier transform of $F(x,y) \in \mathscr{S}'(\mathbf{R}^{2d})$ with respect to the y-variable. We refer to [10, 11] for more facts about the short-time Fourier transform. We now recall the notion of translation invariant BF-space on $\mathbf{R}^d$.

Definition 2.1. Let $\mathscr{B}$ be a Banach space which is continuously embedded in $L^1_{\mathrm{loc}}(\mathbf{R}^d)$, and let $v \in \mathscr{P}(\mathbf{R}^d)$ be submultiplicative. Then $\mathscr{B}$ is called a *translation invariant BF-space on* $\mathbf{R}^d$ (with respect to v), if there is a constant C such that the following conditions are fulfilled:

1. $\mathscr{S}(\mathbf{R}^d) \subseteq \mathscr{B} \subseteq \mathscr{S}'(\mathbf{R}^d)$ (continuous embeddings);
2. if $x \in \mathbf{R}^d$ and $f \in \mathscr{B}$, then $f(\cdot - x) \in \mathscr{B}$, and
$$\|f(\cdot - x)\|_{\mathscr{B}} \le Cv(x)\|f\|_{\mathscr{B}}; \tag{2.4}$$
3. if $f,g \in L^1_{\mathrm{loc}}(\mathbf{R}^d)$ satisfy $g \in \mathscr{B}$ and $|f| \le |g|$ almost everywhere, then $f \in \mathscr{B}$ and
$$\|f\|_{\mathscr{B}} \le C\|g\|_{\mathscr{B}}.$$

The following definition of modulation spaces is due to Feichtinger [9].

Definition 2.2. Let $\mathscr{B}$ be a translation invariant BF-space on $\mathbf{R}^{2d}$ with respect to $v \in \mathscr{P}(\mathbf{R}^{2d})$, $\phi \in \mathscr{S}(\mathbf{R}^d)\backslash 0$ and let $\omega \in \mathscr{P}(\mathbf{R}^{2d})$ be such that ω is v-moderate. The *modulation space* $M(\omega,\mathscr{B})$ consists of all $f \in \mathscr{S}'(\mathbf{R}^d)$ such that $V_\phi f\cdot\omega \in \mathscr{B}$. We note that $M(\omega,\mathscr{B})$ is a Banach space with the norm

$$\|f\|_{M(\omega,\mathscr{B})} \equiv \|(V_\phi f)\omega\|_{\mathscr{B}} \tag{2.5}$$

(cf. [7]).

Remark 2.3. Assume that $p,q \in [1,\infty]$, and let $L^{p,q}_1(\mathbf{R}^{2d})$ and $L^{p,q}_2(\mathbf{R}^{2d})$ be the sets of all $F \in L^1_{\mathrm{loc}}(\mathbf{R}^{2d})$ such that

$$\|F\|_{L^{p,q}_1} \equiv \left(\int\left(\int |F(x,\xi)|^p\,dx\right)^{q/p} d\xi\right)^{1/q} < \infty$$

and

$$\|F\|_{L^{p,q}_2} \equiv \left(\int\left(\int |F(x,\xi)|^q\,d\xi\right)^{p/q} dx\right)^{1/p} < \infty.$$

Then $M(\omega, L_1^{p,q}(\mathbf{R}^{2d}))$ is equal to the *classical* modulation space $M_{(\omega)}^{p,q}(\mathbf{R}^d)$, and $M(\omega, L_2^{p,q}(\mathbf{R}^{2d}))$ is equal to the space $W_{(\omega)}^{p,q}(\mathbf{R}^d)$, related to Wiener-amalgam spaces (cf. [7, 8, 9, 11]). We will set $M_{(\omega)}^p = M_{(\omega)}^{p,p} = W_{(\omega)}^{p,p}$. Furthermore, if $\omega = 1$, then we write $M^{p,q}$, M^p and $W^{p,q}$ instead of $M_{(\omega)}^{p,q}$, $M_{(\omega)}^p$ and $W_{(\omega)}^{p,q}$ respectively.

In what follows we let σ_s and $\sigma_{s,t}$ be the weights

$$\sigma_s(x,\xi) = \langle x,\xi\rangle^s \quad \text{and} \quad \sigma_{s,t}(x,\xi) = \langle x\rangle^t\langle \xi\rangle^s, \quad x,\xi \in \mathbf{R}^d. \tag{2.6}$$

Remark 2.4. Several important spaces agree with certain modulation spaces. In fact, let $s,t \in \mathbf{R}$. Then $M_{(\sigma_{s,t})}^2(\mathbf{R}^d)$ is equal to the weighted Sobolev space (or Sobolev–Kato space) $H_{s,t}^2(\mathbf{R}^d)$ in [5, 16], the set of all $f \in \mathscr{S}'(\mathbf{R}^d)$ such that $\langle x\rangle^t\langle D\rangle^s f \in L^2$. In particular, $M_{(\sigma_{s,0})}^2$ and $M_{(\sigma_{0,t})}^2$ are equal to H_s^2 and L_t^2, respectively. Furthermore, $M_{(\sigma_s)}^2(\mathbf{R}^d)$ is equal to the Shubin–Sobolev space of order s. (Cf., e.g., [15]).

2.3. Pseudo-differential operators and SG-symbol classes

Next we recall some facts in Chapter XVIII in [14] concerning pseudo-differential operators. Let $a \in \mathscr{S}(\mathbf{R}^{2d})$, and $t \in \mathbf{R}$ be fixed. Then the pseudo-differential operator $\operatorname{Op}_t(a)$ is the linear and continuous operator on $\mathscr{S}(\mathbf{R}^d)$ defined by the formula

$$(\operatorname{Op}_t(a)f)(x) = (2\pi)^{-d}\iint a((1-t)x+ty,\xi)f(y)e^{i\langle x-y,\xi\rangle}\,dyd\xi. \tag{2.7}$$

For general $a \in \mathscr{S}'(\mathbf{R}^{2d})$, the pseudo-differential operator $\operatorname{Op}_t(a)$ is defined as the continuous operator from $\mathscr{S}(\mathbf{R}^d)$ to $\mathscr{S}'(\mathbf{R}^d)$ with distribution kernel

$$K_{t,a}(x,y) = (2\pi)^{-d/2}(\mathscr{F}_2^{-1}a)((1-t)x+ty,x-y). \tag{2.8}$$

If $t=0$, then $\operatorname{Op}_t(a)$ is the Kohn–Nirenberg representation $\operatorname{Op}(a) = a(x,D)$, and if $t = 1/2$, then $\operatorname{Op}_t(a)$ is the Weyl quantization.

We now recall the definition of the generalized SG-symbol classes. Let m, μ, r, $\rho \in \mathbf{R}$ be fixed. Then $\mathrm{SG}_{r,\rho}^{m,\mu}(\mathbf{R}^{2d})$ is the set of all $a \in C^\infty(\mathbf{R}^{2d})$ such that

$$|D_x^\alpha D_\xi^\beta a(x,\xi)| \lesssim \langle x\rangle^{m-r|\alpha|}\langle\xi\rangle^{\mu-\rho|\beta|},$$

for all multi-indices α and β. Usually we assume that $r,\rho \ge 0$ and $\rho + r > 0$.

More generally, assume that $\omega \in \mathscr{P}_{r,\rho}(\mathbf{R}^{2d})$. Then $\mathrm{SG}_{r,\rho}^{(\omega)}(\mathbf{R}^{2d})$ consists of all $a \in C^\infty(\mathbf{R}^{2d})$ such that

$$|D_x^\alpha D_\xi^\beta a(x,\xi)| \lesssim \omega(x,\xi)\langle x\rangle^{-r|\alpha|}\langle\xi\rangle^{-\rho|\beta|}, \qquad x,\xi\in\mathbf{R}^d, \tag{2.9}$$

for all multi-indices α and β. We note that

$$\mathrm{SG}_{r,\rho}^{(\omega)}(\mathbf{R}^{2d}) = S(\omega, g_{r,\rho}), \tag{2.10}$$

when $g = g_{r,\rho}$ is the Riemannian metric on $\mathbf{R}^{2d}$, defined by the formula

$$(g_{r,\rho})_{(y,\eta)}(x,\xi) = \langle y\rangle^{-2r}|x|^2 + \langle\eta\rangle^{-2\rho}|\xi|^2 \tag{2.11}$$

(cf. Sect. 18.4–18.6 in [14]). Furthermore, $\mathrm{SG}_{r,\rho}^{(\omega)} = \mathrm{SG}_{r,\rho}^{m,\mu}$ when $\omega(x,\xi) = \langle x\rangle^m\langle\xi\rangle^\mu$.

It is a well-known fact that SG-operators give rise to linear continuous mappings from $\mathscr{S}(\mathbf{R}^d)$ to itself, extendable as linear continuous mappings from $\mathscr{S}'(\mathbf{R}^d)$ to itself. They also act continuously between modulation spaces. Indeed, see [4], if $a \in \mathrm{SG}_{r,\rho}^{(\omega_0)}(\mathbf{R}^{2d})$, then $\mathrm{Op}_t(a)$ is continuous from $M(\omega, \mathscr{B})$ to $M(\omega/\omega_0, \mathscr{B})$. Moreover, there exist $a \in \mathrm{SG}_{r,\rho}^{(\omega_0)}(\mathbf{R}^{2d})$ and $b \in \mathrm{SG}_{r,\rho}^{(1/\omega_0)}(\mathbf{R}^{2d})$ such that for every choice of $\omega \in \mathscr{P}(\mathbf{R}^{2d})$ and every translation invariant BF-space $\mathscr{B}$ on $\mathbf{R}^{2d}$, the mappings

$$\mathrm{Op}_t(a) : \mathscr{S}(\mathbf{R}^d) \to \mathscr{S}(\mathbf{R}^d), \quad \mathrm{Op}_t(a) : \mathscr{S}'(\mathbf{R}^d) \to \mathscr{S}'(\mathbf{R}^d)$$

$$\text{and} \quad \mathrm{Op}_t(a) : M(\omega, \mathscr{B}) \to M(\omega/\omega_0, \mathscr{B}).$$

are continuous bijections with inverses $\mathrm{Op}_t(b)$.

Definition 2.5. Let $r, \rho \in [0,1]$, $t \in \mathbf{R}$, $\mathcal{B}$ be a topological vector space of distributions on $\mathbf{R}^d$ such that

$$\mathscr{S}(\mathbf{R}^d) \subseteq \mathcal{B} \subseteq \mathscr{S}'(\mathbf{R}^d)$$

with continuous embeddings. Then $\mathcal{B}$ is called SG-*admissible* (*with respect to* r, ρ *and* d) when $\mathrm{Op}_t(a)$ maps $\mathcal{B}$ continuously into itself, for every $a \in \mathrm{SG}_{r,\rho}^{0,0}$.

If $\mathcal{B}$ and $\mathcal{C}$ are SG-admissible with respect to r, ρ and d, and $\omega_0 \in \mathscr{P}_{r,\rho}(\mathbf{R}^{2d})$, then the pair $(\mathcal{B}, \mathcal{C})$ is called SG-*ordered* (*with respect to* ω_0), when the mappings

$$\mathrm{Op}_t(a) : \mathcal{B} \to \mathcal{C} \quad \text{and} \quad \mathrm{Op}_t(b) : \mathcal{C} \to \mathcal{B}$$

are continuous for every $a \in \mathrm{SG}_{r,\rho}^{(\omega_0)}(\mathbf{R}^{2d})$ and $b \in \mathrm{SG}_{r,\rho}^{(1/\omega_0)}(\mathbf{R}^{2d})$.

Remark 2.6. Let t, r, ρ, ω and ω_0 be as in Definition 2.5, and let $\mathcal{B}$ be SG-admissible with respect to r, ρ and d. Then there is a unique SG-admissible $\mathcal{C}$ such that $(\mathcal{B}, \mathcal{C})$ is an SG-ordered pair with respect to ω_0. In fact, let a be as above. Then $\mathcal{C}$ is the image of $\mathcal{B}$ under $\mathrm{Op}_t(a)$.

In particular, $\mathscr{S}(\mathbf{R}^d)$, $\mathscr{S}'(\mathbf{R}^d)$ and $M(\omega, \mathscr{B})$ are SG-admissible, and

$$(\mathscr{S}(\mathbf{R}^d), \mathscr{S}(\mathbf{R}^d)), \quad (\mathscr{S}'(\mathbf{R}^d), \mathscr{S}'(\mathbf{R}^d)) \quad \text{and} \quad (M(\omega, \mathscr{B}), M(\omega/\omega_0, \mathscr{B}))$$

are SG-ordered with respect to ω_0.

If $a \in \mathrm{SG}_{r,\rho}^{(\omega_0)}(\mathbf{R}^{2d})$, then

$$|a(x,\xi)| \lesssim \omega_0(x,\xi).$$

On the other hand, a is invertible, in the sense that $\frac{1}{a}$ is a symbol in $\mathrm{SG}_{r,\rho}^{(1/\omega_0)}(\mathbf{R}^{2d})$, if and only if

$$\omega_0(x,\xi) \lesssim |a(x,\xi)|. \tag{2.12}$$

A slightly relaxed condition appears when (2.12) holds for all points (x, ξ), outside a compact set $K \subseteq \mathbf{R}^{2d}$. In this case we say that a is SG-*elliptic* (with respect to ω_0).

Definition 2.7. Let $r, \rho \geq 0$, $\omega_0 \in \mathscr{P}_{r,\rho}(\mathbf{R}^{2d})$ and let $a \in \mathrm{SG}_{r,\rho}^{(\omega_0)}(\mathbf{R}^{2d})$.

1. a is called *locally* or *type*-1 *invertible* with respect to ω_0 at the point $(x_0, \xi_0) \in \mathbf{R}^d \times (\mathbf{R}^d \backslash 0)$, if there exist a neighbourhood X of x_0, an open conical neigh-

bourhood Γ of ξ_0 and a positive constant R such that (2.12) holds for $x \in X$, $\xi \in \Gamma$ and $|\xi| \geq R$.

2. a is called *Fourier-locally* or *type-2 invertible* with respect to ω_0 at the point $(x_0, \xi_0) \in (\mathbf{R}^d \setminus 0) \times \mathbf{R}^d$, if there exist an open conical neighbourhood Γ of x_0, a neighbourhood X of ξ_0 and a positive constant R such that (2.12) holds for $x \in \Gamma$, $|x| \geq R$ and $\xi \in X$.
3. a is called *oscillating* or *type-3 invertible* with respect to ω_0 at the point $(x_0, \xi_0) \in (\mathbf{R}^d \setminus 0) \times (\mathbf{R}^d \setminus 0)$, if there exist open conical neighbourhoods Γ_1 of x_0 and Γ_2 of ξ_0, and a positive constant R such that (2.12) holds for $x \in \Gamma_1$, $|x| \geq R$, $\xi \in \Gamma_2$ and $|\xi| \geq R$.

If $m \in \{1, 2, 3\}$ and a is *not* type-m invertible with respect to ω_0 at (x_0, ξ_0), then (x_0, ξ_0) is called *type-m characteristic* for a with respect to ω_0. The set of type-m characteristic points for a with respect to ω_0 is denoted by $\mathrm{Char}^m_{(\omega_0)}(a)$.

The *(global) set of characteristic points* (the characteristic set), for a symbol $a \in \mathrm{SG}^{(\omega_0)}_{r,\rho}(\mathbf{R}^{2d})$ with respect to ω_0, is

$$\mathrm{Char}(a) = \mathrm{Char}_{(\omega_0)}(a) = \mathrm{Char}^1_{(\omega_0)}(a) \bigcup \mathrm{Char}^2_{(\omega_0)}(a) \bigcup \mathrm{Char}^3_{(\omega_0)}(a).$$

Remark 2.8. In the case $\omega_0 = 1$ we exclude the phrase "with respect to ω_0" in Definition 2.7. For example, $a \in \mathrm{SG}^{0,0}_{r,\rho}(\mathbf{R}^{2d})$ is *type-1 invertible* at $(x_0, \xi_0) \in \mathbf{R}^d \times (\mathbf{R}^d \setminus 0)$ if $(x_0, \xi_0) \notin \mathrm{Char}^1_{(\omega_0)}(a)$ with $\omega_0 = 1$. This means that there exist a neighbourhood X of x_0, an open conical neighbourhood Γ of ξ_0 and $R > 0$ such that (2.12) holds for $\omega_0 = 1$, $x \in X$ and $\xi \in \Gamma$ satisfies $|\xi| \geq R$.

In the next definition we introduce different classes of cutoff functions (see also Definition 1.9 in [3]).

Definition 2.9. Let $X \subseteq \mathbf{R}^d$ be open, $\Gamma \subseteq \mathbf{R}^d \setminus 0$ be an open cone, $x_0 \in X$ and let $\xi_0 \in \Gamma$.

1. A smooth function φ on $\mathbf{R}^d$ is called a *cutoff (function)* with respect to x_0 and X, if $0 \leq \varphi \leq 1$, $\varphi \in C_0^\infty(X)$ and $\varphi = 1$ in an open neighbourhood of x_0. The set of cutoffs with respect to x_0 and X is denoted by $\mathscr{C}_{x_0}(X)$ or $\mathscr{C}_{x_0}$.
2. A smooth function ψ on $\mathbf{R}^d$ is called a *directional cutoff (function)* with respect to ξ_0 and Γ, if there is a constant $R > 0$ and open conical neighbourhood $\Gamma_1 \subseteq \Gamma$ of ξ_0 such that the following is true:
   - $0 \leq \psi \leq 1$ and $\operatorname{supp} \psi \subseteq \Gamma$;
   - $\psi(t\xi) = \psi(\xi)$ when $t \geq 1$ and $|\xi| \geq R$;
   - $\psi(\xi) = 1$ when $\xi \in \Gamma_1$ and $|\xi| \geq R$.

   The set of directional cutoffs with respect to ξ_0 and Γ is denoted by $\mathscr{C}^{\mathrm{dir}}_{\xi_0}(\Gamma)$ or $\mathscr{C}^{\mathrm{dir}}_{\xi_0}$.

Remark 2.10. Let $X \subseteq \mathbf{R}^d$ be open and $\Gamma, \Gamma_1, \Gamma_2 \subseteq \mathbf{R}^d \setminus 0$ be open cones. Then the following is true.

1. If $x_0 \in X$, $\xi_0 \in \Gamma$, $\varphi \in \mathscr{C}_{x_0}(X)$ and $\psi \in \mathscr{C}^{\mathrm{dir}}_{\xi_0}(\Gamma)$, then $c_1 = \varphi \otimes \psi$ belongs to $\mathrm{SG}^{0,0}_{1,1}(\mathbf{R}^{2d})$, and is type-1 invertible at (x_0, ξ_0);

2. if $x_0 \in \Gamma$, $\xi_0 \in X$, $\psi \in \mathscr{C}^{\mathrm{dir}}_{x_0}(\Gamma)$ and $\varphi \in \mathscr{C}_{\xi_0}(X)$, then $c_2 = \psi \otimes \varphi$ belongs to $\mathrm{SG}^{0,0}_{1,1}(\mathbf{R}^{2d})$, and is type-2 invertible at (x_0, ξ_0);
3. if $x_0 \in \Gamma_1$, $\xi_0 \in \Gamma_2$, $\psi_1 \in \mathscr{C}^{\mathrm{dir}}_{x_0}(\Gamma_1)$ and $\psi_2 \in \mathscr{C}^{\mathrm{dir}}_{\xi_0}(\Gamma_2)$, then $c_3 = \psi_1 \otimes \psi_2$ belongs to $\mathrm{SG}^{0,0}_{1,1}(\mathbf{R}^{2d})$, and is type-3 invertible at (x_0, ξ_0).

The next proposition shows that $\mathrm{Op}_t(a)$ for $t \in \mathbf{R}$ satisfies convenient invertibility properties of the form

$$\mathrm{Op}_t(a)\,\mathrm{Op}_t(b) = \mathrm{Op}_t(c) + \mathrm{Op}_t(h), \tag{2.13}$$

outside the set of characteristic points for a symbol a. Here $\mathrm{Op}_t(b)$, $\mathrm{Op}_t(c)$ and $\mathrm{Op}_t(h)$ have the roles of "local inverse", "local identity" and smoothing operators respectively. From these statements it also follows that our set of characteristic points in Definition 2.7 are related to those in [5, 14].

We let $\mathbb{I}_m$ and Ω_m, $m = 1, 2, 3$, be the sets

$$\mathbb{I}_1 \equiv [0,1] \times (0,1], \qquad \mathbb{I}_2 \equiv (0,1] \times [0,1], \qquad \mathbb{I}_3 \equiv (0,1] \times (0,1] = \mathbb{I}_1 \cap \mathbb{I}_2,$$

and

$$\begin{aligned} \Omega_1 &= \mathbf{R}^d \times (\mathbf{R}^d \setminus 0), \qquad \Omega_2 = (\mathbf{R}^d \setminus 0) \times \mathbf{R}^d, \\ \Omega_3 &= (\mathbf{R}^d \setminus 0) \times (\mathbf{R}^d \setminus 0), \end{aligned} \tag{2.14}$$

which will be useful in the formulation of our results.

Proposition 2.11. *Let $m \in \{1,2,3\}$, $(r,\rho) \in \mathbb{I}_m$, $\omega_0 \in \mathscr{P}_{r,\rho}(\mathbf{R}^{2d})$ and let $a \in \mathrm{SG}^{(\omega_0)}_{r,\rho}(\mathbf{R}^{2d})$. Also let Ω_m be as in* (2.14)*, $(x_0,\xi_0) \in \Omega_m$, and let (r_0,ρ_0) be equal to $(r,0)$, $(0,\rho)$ and (r,ρ) when m is equal to* 1*,* 2 *and* 3*, respectively. Then the following conditions are equivalent:*

1. $(x_0,\xi_0) \notin \mathrm{Char}^m_{(\omega_0)}(a)$;
2. *there is an element $c \in \mathrm{SG}^{0,0}_{r,\rho}$ which is type-m invertible at (x_0,ξ_0), and an element $b \in \mathrm{SG}^{(1/\omega_0)}_{r,\rho}$ such that $ab = c$;*
3. (2.13) *holds for some $c \in \mathrm{SG}^{0,0}_{r,\rho}$ which is type-m invertible at (x_0,ξ_0), and some elements $h \in \mathrm{SG}^{-r_0,-\rho_0}_{r,\rho}$ and $b \in \mathrm{SG}^{(1/\omega_0)}_{r,\rho}$;*
4. (2.13) *holds for some $c_m \in \mathrm{SG}^{0,0}_{r,\rho}$ in Remark* 2.10 *which is type-m invertible at (x_0,ξ_0), and some elements h and $b \in \mathrm{SG}^{(1/\omega_0)}_{r,\rho}$, where $h \in \mathscr{S}$ when $m \in \{1,3\}$ and $h \in \mathrm{SG}^{-\infty,0}$ when $m = 2$.*

   Furthermore, if $t = 0$, then the supports of b and h can be chosen to be contained in $X \times \mathbf{R}^d$ when $m = 1$, in $\Gamma \times \mathbf{R}^d$ when $m = 2$, and in $\Gamma_1 \times \mathbf{R}^d$ when $m = 3$.

3. Global wave-front sets and SG-pseudo-differential operators

In this section we recall the definition given in [4] of global wave-front sets for temperate distributions with respect to Banach or Fréchet spaces and state some of their properties. We first introduce the complements of the wave-front sets.

More precisely, let Ω_m, $m \in \{1,2,3\}$, be given by (2.14), $\mathcal{B}$ be a Banach or Fréchet space such that $\mathscr{S}(\mathbf{R}^d) \subseteq \mathcal{B} \subseteq \mathscr{S}'(\mathbf{R}^d)$, and let $f \in \mathscr{S}'(\mathbf{R}^d)$. Then the point $(x_0, \xi_0) \in \Omega_m$ is called *type-m* regular for f with respect to $\mathcal{B}$, if

$$\mathrm{Op}(c_m)f \in \mathcal{B}, \tag{3.1}$$

for some c_m as in Remark 2.10. The set of all type-m regular points for f with respect to $\mathcal{B}$ is denoted by $\Theta^m_{\mathcal{B}}(f)$.

Definition 3.1. Let $m \in \{1,2,3\}$, Ω_m be as in (2.14), and let $\mathcal{B}$ be a Banach or Fréchet space such that $\mathscr{S}(\mathbf{R}^d) \subseteq \mathcal{B} \subset \mathscr{S}'(\mathbf{R}^d)$.

1. The *type-m wave-front set* of $f \in \mathscr{S}'(\mathbf{R}^d)$ with respect to $\mathcal{B}$ is the complement of $\Theta^m_{\mathcal{B}}(f)$ in Ω_m, and is denoted by $\mathrm{WF}^m_{\mathcal{B}}(f)$.
2. The *global wave-front set* $\mathrm{WF}_{\mathcal{B}}(f) \subseteq (\mathbf{R}^d \times \mathbf{R}^d)\backslash 0$ is the set

$$\mathrm{WF}_{\mathcal{B}}(f) \equiv \mathrm{WF}^1_{\mathcal{B}}(f) \bigcup \mathrm{WF}^2_{\mathcal{B}}(f) \bigcup \mathrm{WF}^3_{\mathcal{B}}(f).$$

The sets $\mathrm{WF}^1_{\mathcal{B}}(f)$, $\mathrm{WF}^2_{\mathcal{B}}(f)$ and $\mathrm{WF}^3_{\mathcal{B}}(f)$ in Definition 3.1, are also called the *local*, *Fourier-local* and *oscillating* wave-front set of f with respect to $\mathcal{B}$.

From now on we assume that $\mathcal{B}$ in Definition 3.1 is SG-admissible, and recall that Sobolev–Kato spaces and, more generally, modulation spaces, and $\mathscr{S}(\mathbf{R}^d)$ are SG-admissible. (Cf. Definition 2.5, and Remarks 2.4 and 2.6.)

The next result describes the relation between "regularity with respect to $\mathcal{B}$" of temperate distributions and global wave-front sets, which is the aspect of the theory we are focused on in this paper.

Theorem 3.2. *Let $\mathcal{B}$ be SG-admissible, and let $f \in \mathscr{S}'(\mathbf{R}^d)$. Then*

$$f \in \mathcal{B} \quad \Longleftrightarrow \quad \mathrm{WF}_{\mathcal{B}}(f) = \emptyset.$$

For the sake of completeness, we recall that microlocality and microellipticity hold for our global wave-front sets and pseudo-differential operators in $\mathrm{Op}(\mathrm{SG}^{(\omega_0)}_{r,\rho})$, see [4]. This implies that operators which are elliptic with respect to $\omega_0 \in \mathscr{P}_{\rho,\delta}(\mathbf{R}^{2d})$ when $0 < r, \rho \le 1$ preserve the global wave-front set of temperate distributions. We recall that a and $\mathrm{Op}(a)$ are called SG-*elliptic* with respect to $\mathrm{SG}^{(\omega_0)}_{r,\rho}(\mathbf{R}^{2d})$ or ω_0, if there is a compact set $K \subset \mathbf{R}^{2d}$ such that (2.12) holds when $(x,\xi) \notin K$. By (2.9) it follows that

$$|D^\alpha_x D^\beta_\xi a(x,\xi)| \lesssim |a(x,\xi)| \langle x\rangle^{-r|\alpha|} \langle \xi\rangle^{-\rho|\beta|}, \qquad (x,\xi) \in \mathbf{R}^{2d} \setminus K,$$

for every multi-index α, when a is SG-elliptic (see, e.g., [14, 1]). The following result is an immediate corollary of microlocality and microellipticity for operators in $\mathrm{Op}(\mathrm{SG}^{(\omega_0)}_{r,\rho})$:

Theorem 3.3. *Let $m \in \{1,2,3\}$, $(r,\rho) \in \mathbb{I}_m$, $t \in \mathbf{R}$, $\omega_0 \in \mathscr{P}_{r,\rho}(\mathbf{R}^{2d})$, $a \in \mathrm{SG}^{(\omega_0)}_{r,\rho}(\mathbf{R}^{2d})$ be* SG*-elliptic with respect to ω_0 and let $f \in \mathscr{S}'(\mathbf{R}^d)$. Moreover, let $(\mathcal{B},\mathcal{C})$ be a* SG*-ordered pair with respect to ω_0. Then*

$$\mathrm{WF}^m_{\mathcal{C}}(\mathrm{Op}_t(a)f) = \mathrm{WF}^m_{\mathcal{B}}(f).$$

4. Wave-front sets with respect to sequences of spaces

In this section we recall the definition of wave-front sets based on sequences of admissible spaces, see [4], and prove the result corresponding to Theorem 3.2 in this more general situation. In the first part we consider sequences of spaces which are parameterized with one index. Thereafter we discuss further extensions where we consider sequences of spaces which are parameterized with two indices. Here we also recall wave-front sets which are related to "classical wave-front sets", in the sense that they are wave-front sets with respect to classical spaces of smooth functions. In particular, a refinement of the wave-front set of Schwartz-type treated in [5] can also be obtained as a wave-front set based on sequences of admissible spaces, see [4]. An example is discussed at the end of the section.

4.1. Wave-front sets with respect to sequences with one index parameter

Again we start by introducing the complements of the wave-front sets. More precisely, let J be an index set of integers, Ω_m, $m \in \{1,2,3\}$, be given by (2.14), $(\mathcal{B}_j) = (\mathcal{B}_j)_{j\in J}$, be a sequence of Banach or Fréchet spaces such that $\mathscr{S}(\mathbf{R}^d) \subseteq \mathcal{B}_j \subseteq \mathscr{S}'(\mathbf{R}^d)$, for every j, and let $f \in \mathscr{S}'(\mathbf{R}^d)$. Then the point $(x_0,\xi_0) \in \Omega_m$ is called *type-*$(m,\cup)$ regular (*type-*$(m,\cap)$ regular) for f with respect to $(\mathcal{B}_j)$, if

$$\operatorname{Op}(c_m)f \in \bigcap\nolimits_j \mathcal{B}_j \qquad \Big(\operatorname{Op}(c_m)f \in \bigcup\nolimits_j \mathcal{B}_j\Big), \tag{4.1}$$

for some c_m as in Remark 2.10 The set of all type-$m,\cup$ regular points (type-$m,\cap$ regular points) for f with respect to $(\mathcal{B}_j)$, is denoted by $\Theta^{m,\cup}_{(\mathcal{B}_j)}(f)$ $(\Theta^{m,\cap}_{(\mathcal{B}_j)}(f))$.

It is also desirable that right-hand sides of (4.1) should be a vector space, which is guaranteed by imposing that $(\mathcal{B}_j)$ should be *ordered*, i.e., $\mathcal{B}_j$ should be increasing or decreasing with respect to $j \in J$.

Definition 4.1. Let J be an index set of integers, $m \in \{1,2,3\}$, Ω_m be as in (2.14), and let $(\mathcal{B}_j)_{j\in J}$ be a sequence of Banach or Fréchet space such that $\mathscr{S}(\mathbf{R}^d) \subseteq \mathcal{B}_j \subset \mathscr{S}'(\mathbf{R}^d)$, for every j.

1. the *type-*$(m,\cup)$ *wave-front set* (*type-*$(m,\cap)$ *wave-front set*) of $f \in \mathscr{S}'(\mathbf{R}^d)$ with respect to $(\mathcal{B}_j)$ is the complement of $\Theta^{m,\cup}_{(\mathcal{B}_j)}(f)$ $(\Theta^{m,\cap}_{(\mathcal{B}_j)}(f))$ in Ω_m, and is denoted by $\mathrm{WF}^{m,\cup}_{(\mathcal{B}_j)}(f)$ $(\mathrm{WF}^{m,\cap}_{(\mathcal{B}_j)}(f))$;
2. the *global wave-front sets* $\mathrm{WF}^{\cup}_{(\mathcal{B}_j)}(f) \subseteq (\mathbf{R}^d \times \mathbf{R}^d)\backslash 0$ and $\mathrm{WF}^{\cap}_{(\mathcal{B}_j)}(f) \subseteq (\mathbf{R}^d \times \mathbf{R}^d)\backslash 0$, of $\cup$ and $\cap$ types, respectively, are the sets

$$\mathrm{WF}^{\cup}_{(\mathcal{B}_j)}(f) \equiv \mathrm{WF}^{1,\cup}_{(\mathcal{B}_j)}(f) \bigcup \mathrm{WF}^{2,\cup}_{(\mathcal{B}_j)}(f) \bigcup \mathrm{WF}^{3,\cup}_{(\mathcal{B}_j)}(f),$$

$$\mathrm{WF}^{\cap}_{(\mathcal{B}_j)}(f) \equiv \mathrm{WF}^{1,\cap}_{(\mathcal{B}_j)}(f) \bigcup \mathrm{WF}^{2,\cap}_{(\mathcal{B}_j)}(f) \bigcup \mathrm{WF}^{3,\cap}_{(\mathcal{B}_j)}(f).$$

Example. We can consider wave-front sets with respect to sequences of the form

$$(\mathcal{B}_j) \equiv (\mathcal{B}_j)_{j\in J}, \quad \text{with} \quad \mathcal{B}_j = M(\omega_j, \mathscr{B}_j), \tag{4.2}$$

where $\omega_j \in \mathscr{P}(\mathbf{R}^{2d})$, $\mathscr{B}_j$ is a translation invariant BF-space on $\mathbf{R}^d$, and j belongs to some index set J.

Remark 4.2. Let $p_j, q_j \in [1,\infty]$, $\mathscr{B}_j = L_1^{p_j,q_j}(\mathbf{R}^{2d})$, $\omega_j(x,\xi) = \langle x,\xi\rangle^{-j}$ and let $\mathcal{B}_j$ be as in (4.2) for $j \in J = \mathbf{N}_0$. Then it follows that $\mathrm{WF}^{m,\cup}_{(\mathcal{B}_j)}(f)$, $m = 1,2,3$, in Definition 4.1 are equal to the wave-front sets $\mathrm{WF}^{\psi}(f)$, $\mathrm{WF}^{e}(f)$ and $\mathrm{WF}^{\psi e}(f)$ in [5], respectively. In particular, it follows that $\mathrm{WF}^{\cup}_{(\mathcal{B}_j)}(f)$ is equal to the global wave-front set $\mathrm{WF}_{\mathscr{S}}(f)$, which in [5] is denoted by $\mathrm{WF}_{\mathcal{S}}(f)$.

Remark 4.3. Evidently, if $\mathcal{B}_j = \mathcal{B}$ for every $j \in J$, then

$$\mathrm{WF}^{m,\cup}_{(\mathcal{B}_j)}(f) = \mathrm{WF}^{m,\cap}_{(\mathcal{B}_j)}(f) = \mathrm{WF}^{m}_{\mathcal{B}}(f), \quad m = 1,2,3.$$

Proposition 4.4. *Let $m \in \{1,2,3\}$, $\mathcal{B}_j$ be the same as in Definition* 4.1, *and let $f \in \mathscr{S}'(\mathbf{R}^d)$. Then*

$$\bigcup \mathrm{WF}^{m}_{\mathcal{B}_j}(f) \subseteq \mathrm{WF}^{m,\cup}_{(\mathcal{B}_j)}(f), \quad \bigcup \mathrm{WF}_{\mathcal{B}_j}(f) \subseteq \mathrm{WF}^{\cup}_{(\mathcal{B}_j)}(f)$$

and

$$\bigcap \mathrm{WF}^{m}_{\mathcal{B}_j}(f) = \mathrm{WF}^{m,\cap}_{(\mathcal{B}_j)}(f), \quad \bigcap \mathrm{WF}_{\mathcal{B}_j}(f) = \mathrm{WF}^{\cap}_{(\mathcal{B}_j)}(f).$$

Proof. It suffices to prove the first and third relation. Let Ω_m be as in (2.14), and let $X_0 = (x_0,\xi_0) \in \Omega_m$.

First let $X_0 \in \Theta^{m,\cup}_{\mathcal{B}_j}(f)$. Then $\mathrm{Op}(c_m)f \in \cap \mathcal{B}_j$, for some c_m as in Remark 2.10, giving that $\mathrm{Op}(c_m)f \in \mathcal{B}_j$ for every j. This implies that $X_0 \in \Theta^m_{\mathcal{B}_j}(f)$ for every j, i.e., $X_0 \in \cap \Theta^m_{\mathcal{B}_j}(f)$, and the first relation is proved.

The third relation follows from the relations

$$X_0 \in \bigcup \Theta^m_{\mathcal{B}_j}(f) \iff X_0 \in \Theta^m_{\mathcal{B}_j}(f) \text{ for some } j \iff$$
$$\mathrm{Op}(c_m)f \in \mathcal{B}_j \text{ for some } c_m \text{ as in Remark 2.10, and some } j \iff$$
$$\mathrm{Op}(c_m)f \in \cup \mathcal{B}_j \text{ for some } c_m \text{ as in Remark 2.10} \iff X_0 \in \Theta^{m,\cap}_{(\mathcal{B}_j)}(f).$$

The proof is complete. □

We can now prove the result corresponding to Theorem 3.2 for wave-front sets associated with one-parameter sequences of admissible spaces.

Theorem 4.5. *Let $\mathcal{B}_j$ be* SG*-admissible for every j, and let $f \in \mathscr{S}'(\mathbf{R}^d)$. Then*

$$f \in \bigcap \mathcal{B}_j \iff \mathrm{WF}^{\cup}_{(\mathcal{B}_j)}(f) = \emptyset \iff \bigcup \mathrm{WF}_{\mathcal{B}_j}(f) = \emptyset,$$

and if in addition $(\mathcal{B}_j)$ is ordered, then

$$f \in \bigcup \mathcal{B}_j \iff \mathrm{WF}^{\cap}_{(\mathcal{B}_j)}(f) = \emptyset \iff \bigcap \mathrm{WF}_{\mathcal{B}_j}(f) = \emptyset.$$

Proof. Let (1)–(6) be the statements in the theorem. Then (2) ⇒ (3) and (5) ⇔ (6), by Proposition 4.4.

Next we prove (1) ⇒ (2). Let $f \in \cap \mathcal{B}_j$. Then $\mathrm{Op}(c_m)f \in \cap \mathcal{B}_j$ for every c_m as in Remark 2.10, and then $\mathrm{WF}^{m,\cup}_{(\mathcal{B}_j)}(f) = \emptyset$, and the implication follows.

Next we prove (3) ⇒ (1). We have

$$\cup \mathrm{WF}^m_{\mathcal{B}_j}(f) = \emptyset \quad \Longrightarrow \quad \mathrm{WF}^m_{\mathcal{B}_j}(f) = \emptyset \text{ for all } j \quad \Longrightarrow$$
$$f \in \mathcal{B}_j \text{ for all } j \quad \Longrightarrow \quad f \in \cap \mathcal{B}_j,$$

and we have proved that (1) ⇔ (2) ⇔ (3).

We have

$$f \in \cup \mathcal{B}_j \quad \Longrightarrow \quad f \in \mathcal{B}_j \text{ for some } j = j_0 \quad \Longrightarrow$$
$$\mathrm{Op}(c_m) f \in \mathcal{B}_{j_0} \text{ for every } c_m \text{ as in Remark 2.10} \quad \Longrightarrow$$
$$\mathrm{WF}^m_{\mathcal{B}_{j_0}}(f) = \emptyset \quad \Longrightarrow \quad \cap \mathrm{WF}^m_{\mathcal{B}_j}(f) = \emptyset,$$

which shows that (4) ⇒ (6).

Finally, if $\cap \mathrm{WF}^m_{\mathcal{B}_j}(f) = \emptyset$, then $\cap_{k=1}^N \mathrm{WF}^m_{\mathcal{B}_{j_k}}(f) = \emptyset$ for some $j_1, \dots, j_N$, by compactness. Let $\mathcal{B}' = \cup_1^N \mathcal{B}_{j_k}$. Then $\mathrm{WF}^m_{\mathcal{B}'}(f) = \emptyset$, since $(\mathcal{B}_j)$ are ordered. This implies that $f \in \cup_1^N \mathcal{B}_{j_k}$, which in turn implies that $f \in \cup \mathcal{B}_j$. This gives that (6) ⇒ (4), and the proof is complete. □

4.2. Wave-front sets with respect to sequences of spaces with two indices parameters

Next we shall consider wave-front sets with respect to sequences of spaces, parameterized with two indices, and start by introducing the complements of the wave-front sets. More precisely, let J be an index set of integers, Ω_m, $m \in \{1,2,3\}$, be given by (2.14), $(\mathcal{B}_{j,k}) = (\mathcal{B}_{j,k})_{j,k \in J}$, be a sequence of Banach or Fréchet spaces such that $\mathscr{S}(\mathbf{R}^d) \subseteq \mathcal{B}_{j,k} \subseteq \mathscr{S}'(\mathbf{R}^d)$, for every j,k, and let $f \in \mathscr{S}'(\mathbf{R}^d)$. Then the point $(x_0, \xi_0) \in \Omega_m$ is called *type-*$(m, \cup\cap)$ regular (*type-*$(m, \cap\cup)$ regular) for f with respect to $(\mathcal{B}_{j,k})$, if

$$\mathrm{Op}(c_m) f \in \bigcap\nolimits_j \Big(\bigcup\nolimits_k \mathcal{B}_{j,k} \Big) \qquad \Big(\mathrm{Op}(c_m) f \in \bigcup\nolimits_j \Big(\bigcap\nolimits_k \mathcal{B}_{j,k} \Big) \Big), \tag{4.3}$$

for some c_m as in Remark 2.10. The set of all type-$m, \cup\cap$ regular points (type-$m, \cap\cup$ regular points) for f with respect to $(\mathcal{B}_{j,k})$, is denoted by $\Theta^{m,\cup\cap}_{(\mathcal{B}_{j,k})}(f)$ $(\Theta^{m,\cap\cup}_{(\mathcal{B}_{j,k})}(f))$.

Also in here it is desirable that right-hand sides of (4.3) should be a vector space. For this reason, the sequence $(\mathcal{B}_{j,k})$ is called *ordered with respect to* j, if $\mathcal{B}_{j,k}$ increases with j for every k fixed, or decreases with j for every k fixed. The definition of ordered sequences with respect to k is defined in analogous way.

Definition 4.6. Let J be an index set, $m \in \{1,2,3\}$, Ω_m be as in (2.14), and let $(\mathcal{B}_{j,k})_{j,k \in J}$ be a sequence of Banach or Fréchet space such that $\mathscr{S}(\mathbf{R}^d) \subseteq \mathcal{B}_{j,k} \subset \mathscr{S}'(\mathbf{R}^d)$, for every j.

1. the *type-*$(m, \cup\cap)$ *wave-front set* (*type-*$(m, \cap\cup)$ *wave-front set*) of $f \in \mathscr{S}'(\mathbf{R}^d)$ with respect to $(\mathcal{B}_{j,k})$ is the complement of $\Theta^{m,\cup\cap}_{(\mathcal{B}_{j,k})}(f)$ $(\Theta^{m,\cap\cup}_{(\mathcal{B}_{j,k})}(f))$ in Ω_m, and is denoted by $\mathrm{WF}^{m,\cup\cap}_{(\mathcal{B}_{j,k})}(f)$ $(\mathrm{WF}^{m,\cap\cup}_{(\mathcal{B}_{j,k})}(f))$;

2. the *global wave-front sets* $\mathrm{WF}^{\cup\cap}_{(\mathcal{B}_{j,k})}(f) \subseteq (\mathbf{R}^d \times \mathbf{R}^d)\backslash 0$ and $\mathrm{WF}^{\cap\cup}_{(\mathcal{B}_{j,k})}(f) \subseteq (\mathbf{R}^d \times \mathbf{R}^d)\backslash 0$, of $\cup\cap$ and $\cap\cup$ types, respectively, are the sets

$$\mathrm{WF}^{\cup\cap}_{(\mathcal{B}_{j,k})}(f) \equiv \mathrm{WF}^{1,\cup\cap}_{(\mathcal{B}_{j,k})}(f) \bigcup \mathrm{WF}^{2,\cup\cap}_{(\mathcal{B}_{j,k})}(f) \bigcup \mathrm{WF}^{3,\cup\cap}_{(\mathcal{B}_{j,k})}(f),$$
$$\mathrm{WF}^{\cap\cup}_{(\mathcal{B}_{j,k})}(f) \equiv \mathrm{WF}^{1,\cap\cup}_{(\mathcal{B}_{j,k})}(f) \bigcup \mathrm{WF}^{2,\cap\cup}_{(\mathcal{B}_{j,k})}(f) \bigcup \mathrm{WF}^{3,\cap\cup}_{(\mathcal{B}_{j,k})}(f).$$

Remark 4.7. In analogy with Remark 4.3 we notice that if $\mathcal{B}_{j,k} = \mathcal{B}_j$ is independent of $k \in J$, then

$$\mathrm{WF}^{m,\cup\cap}_{(\mathcal{B}_{j,k})}(f) = \mathrm{WF}^{m,\cup}_{(\mathcal{B}_j)}(f), \quad \mathrm{WF}^{m,\cap\cup}_{(\mathcal{B}_{j,k})}(f) = \mathrm{WF}^{m,\cap}_{(\mathcal{B}_j)}(f), \quad m = 1,2,3.$$

Hence, the families of wave-front sets in Definition 4.6 contain the wave-front sets in Definition 4.1.

Remark 4.8. We observe that if $m \in \{1,2,3\}$, $\mathcal{B}_{j,k}$ is SG-admissible for every j,k, Ω_m is given by (2.14) and $f \in \mathscr{S}'(\mathbf{R}^d)$, then $\mathrm{WF}^{m,\cup\cap}_{(\mathcal{B}_{j,k})}(f)$ and $\mathrm{WF}^{m,\cap\cup}_{(\mathcal{B}_{j,k})}(f)$ are closed subsets of Ω_m.

From now on we assume that the involved sequence spaces, $(\mathcal{B}_{j,k})$, are ordered with respect to k when wave-front sets of the form $\mathrm{WF}^{m,\cup\cap}_{(\mathcal{B}_{j,k})}(f)$ are involved, and ordered with respect to j when wave-front sets of the form $\mathrm{WF}^{m,\cap\cup}_{(\mathcal{B}_{j,k})}(f)$ are involved.

Proposition 4.9. *Let $m \in \{1,2,3\}$, $\mathcal{B}_{j,k}$ be the same as in Definition 4.6, and let $f \in \mathscr{S}'(\mathbf{R}^d)$. Then*

$$\bigcup_j \Big(\bigcap_k \mathrm{WF}^m_{\mathcal{B}_{j,k}}(f) \Big) \subseteq \mathrm{WF}^{m,\cup\cap}_{(\mathcal{B}_{j,k})}(f), \quad \bigcup_j \Big(\bigcap_k \mathrm{WF}_{\mathcal{B}_{j,k}}(f) \Big) \subseteq \mathrm{WF}^{\cup\cap}_{(\mathcal{B}_{j,k})}(f),$$

and

$$\bigcap_j \Big(\bigcup_k \mathrm{WF}^m_{\mathcal{B}_{j,k}}(f) \Big) \subseteq \mathrm{WF}^{m,\cap\cup}_{(\mathcal{B}_{j,k})}(f), \quad \bigcap_j \Big(\bigcup_k \mathrm{WF}_{\mathcal{B}_{j,k}}(f) \Big) \subseteq \mathrm{WF}^{\cap\cup}_{(\mathcal{B}_{j,k})}(f).$$

Proof. It suffices to prove the first and third inclusion. Let $X_0 \in \Theta^{m,\cup\cap}_{(\mathcal{B}_{j,k})}(f)$. Then $\mathrm{Op}(c_m)f \in \cap_j(\cup_k \mathcal{B}_{j,k})$ for some c_m in Remark 2.10, which implies that $\mathrm{Op}(c_m)f \in \cup_k \mathcal{B}_{j,k}$ for every j, i.e., for every j, there is a $k = k(j)$ such that $\mathrm{Op}(c_m)f \in \mathcal{B}_{j,k(j)}$.

This means that $X_0 \in \Theta^m_{j,k(j)}$ for every j, giving that $X_0 \in \cup_k \Theta^m_{\mathcal{B}_{j,k}}(f)$ for every j. Hence $X_0 \in \cap_j(\cup_k \Theta^m_{\mathcal{B}_{j,k}}(f))$, and the first inclusion follows.

Next assume that $X \in \Theta^{m,\cap\cup}_{(\mathcal{B}_{j,k})}(f)$. Then $\mathrm{Op}(c_m)f \in \cup_j(\cap_k \mathcal{B}_{j,k})$ for some c_m as in Remark 2.10. Hence $\mathrm{Op}(c_m)f \in \cap_k \mathcal{B}_{j,k}$, for some $j = j_0$. This implies that $\mathrm{Op}(c_m)f \in \mathcal{B}_{j_0,k}$ for every k, giving that $X_0 \in \Theta^m_{\mathcal{B}_{j_0,k}}(f)$ for every k. Hence $X_0 \in \cap_k \Theta^m_{\mathcal{B}_{j_0,k}}(f) \subseteq \cup_j(\cap_k \Theta^m_{\mathcal{B}_{j,k}}(f))$, and the third inclusion follows. The proof is complete. □

Theorem 4.10. *Let $\mathcal{B}_{j,k}$ be* SG*-admissible for every j and k, and let $f \in \mathscr{S}'(\mathbf{R}^d)$. Then*

$$f \in \bigcap_j \Big(\bigcup_k \mathcal{B}_{j,k} \Big) \quad \Longleftrightarrow \quad \mathrm{WF}^{\cup\cap}_{(\mathcal{B}_{j,k})}(f) = \emptyset \quad \Longleftrightarrow \quad \bigcup_j \Big(\bigcap_k \mathrm{WF}_{\mathcal{B}_{j,k}}(f) \Big) = \emptyset,$$

provided $(\mathcal{B}_{j,k})$ *is ordered with respect to* k*, and if instead* $(\mathcal{B}_{j,k})$ *is ordered with respect to* j*, then*

$$f \in \bigcup_j \Big(\bigcap_k \mathcal{B}_{j,k} \Big) \iff \mathrm{WF}^{\cap\cup}_{(\mathcal{B}_{j,k})}(f) = \emptyset \implies \bigcap_j \Big(\bigcup_k \mathrm{WF}_{\mathcal{B}_{j,k}}(f) \Big) = \emptyset.$$

Remark 4.11. Some steps of the proof are dependent of the parametrix constructions in the framework of Proposition 2.11. We note that if $\mathcal{B}_{jk}$ are as in Proposition 4.9 and $\Theta^m_{\mathcal{B}_{j_0,k}}(f) = \Omega_m$, then for every $X \in \Omega_m$, there are elements $c_m = c_{m,X}$ as in Remark 2.10 such that $\mathrm{Op}(c_{m,X})f \in \cap_j(\cup_k \mathcal{B}_{j,k})$. By compactness, there are $c_{m,X_1}, \dots, c_{m,X_N}$ such that if $a = c_1 + \cdots + c_N$, then $a \geq 1$ outside a compact set in Ω_m. Furthermore, $\mathrm{Op}(a)f \in \cap_j(\cup_k \mathcal{B}_{j,k})$.

Let b be as in Proposition 2.11 (4). Then $\mathrm{Op}(b)$ maps $\cap_j(\cup_k \mathcal{B}_{j,k})$ into itself. Hence Proposition 2.11 gives

$$f = \mathrm{Op}(b)\,\mathrm{Op}(a) f \mod \mathscr{S} \subseteq \cap_j(\cup_k \mathcal{B}_{j,k}).$$

Proof of Theorem 4.10. Let (1)–(6) be the statements in the theorem. It is clear that (2) $\Rightarrow$ (3) and (5) $\Rightarrow$ (6), in view of Proposition 4.9.

If $f \in \cap_j(\cup_k \mathcal{B}_{j,k})$, then $\mathrm{Op}(c_m)f \in \cap_j(\cup_k \mathcal{B}_{j,k})$, for every c_m as in Remark 2.10. Consequently, $\mathrm{WF}^{\cup\cap}_{(\mathcal{B}_{j,k})}(f) = \emptyset$, and we have proved that (1) $\Rightarrow$ (2). By Remark 4.11, it follows that (2) $\Rightarrow$ (1), and the equivalence between (1) and (2) follows.

Next assume that (3) holds, i.e., $\cup_j(\cap_k \mathrm{WF}_{\mathcal{B}_{j,k}}(f)) = \emptyset$. Then $\cap_k \mathrm{WF}_{\mathcal{B}_{j,k}}(f) = \emptyset$ for every j. Hence $f \in \cup_k \mathcal{B}_{j,k}$ for every j, by Theorem 4.5. This implies that $f \in \cap_j(\cup_k \mathcal{B}_{j,k})$, and the equivalences between (1)–(3) follow.

Next we prove (4) $\Rightarrow$ (5). Therefore, let $f \in \cup_j(\cap_k \mathcal{B}_{j,k})$. Then $\mathrm{Op}(c_m)f \in \cup_j(\cap_k \mathcal{B}_{j,k})$ for every c_m in Remark 2.10. Hence $\Theta^{m,\cap\cup}_{(\mathcal{B}_{j,k})}(f) = \Omega_m$, and therefore $\mathrm{WF}^{\cap\cup}_{(\mathcal{B}_{j,k})}(f) = \emptyset$. This proves the inclusion.

Finally we prove that (5) $\Rightarrow$ (4). Assume that $\mathrm{WF}^{\cap\cup}_{(\mathcal{B}_{j,k})}(f) = \emptyset$. Then for every $X \in \Omega_m$, there is a $c_m = c_{m,X}$ as in Remark 2.10 such that $\mathrm{Op}(c_m)f \in \cup_j(\cap_k \mathcal{B}_{j,k})$, giving that $\mathrm{Op}(c_m) \in \cap_k \mathcal{B}_{j,k}$ for some $j = j_0$. This implies that $\mathrm{Op}(c_m)f \in \mathcal{B}_{j_0,k}$, for every k. By similar arguments as in Remark 4.11, it follows that $f \in \mathcal{B}_{j_0,k}$ for every k, i.e., $f \in \cap_k \mathcal{B}_{j_0,k} \subseteq \cup_j(\cap_k \mathcal{B}_{j,k})$, and the assertion follows. The proof is complete. □

4.3. An example

In a similar way as in Remark 4.2, we may construct wave-front sets with respect to the spaces $Q_0(\mathbf{R}^d)$ and $Q(\mathbf{R}^d)$ (see [4] for the definition of these spaces). In fact, let

$$p_{j,k}, q_{j,k} \in [1,\infty], \quad \mathscr{B}_{j,k} = L_1^{p_{j,k},q_{j,k}}(\mathbf{R}^{2d}), \quad \omega_{j,k}(x,\xi) = \langle x\rangle^{-j}\langle\xi\rangle^k,$$
$$\mathcal{B}_{j,k} = M(\omega_{j,k}, \mathscr{B}_{j,k}), \quad \mathcal{C}_{j,k} = \mathcal{B}_{k,j} \quad \text{when} \quad j,k \in J = \mathbf{N}_0.$$

By similar arguments as in [12, Remark 2.18] it follows that

$$Q_0(\mathbf{R}^d) = \bigcup_j \Big(\bigcap_k \mathcal{B}_{j,k} \Big), \quad Q(\mathbf{R}^d) = \bigcap_j \Big(\bigcup_k \mathcal{C}_{j,k} \Big).$$

Now we define the components of the wave-front sets with respect to Q_0 and Q as

$$\mathrm{WF}^m_{Q_0}(f) = \mathrm{WF}^{m,\cap\cup}_{(\mathcal{B}_{j,k})}(f), \quad \mathrm{WF}^m_{Q}(f) = \mathrm{WF}^{m,\cup\cap}_{(\mathcal{C}_{j,k})}(f), \quad m = 1,2,3,$$

when $f \in \mathscr{S}'(\mathbf{R}^d)$. By Theorem 4.10 it follows that (1.2) holds when $\mathcal{B} = Q_0(\mathbf{R}^d)$ or $\mathcal{B} = Q(\mathbf{R}^d)$. We also note that

$$\mathrm{WF}^1_{\mathscr{S}}(f) = \mathrm{WF}^1_{Q_0}(f) = \mathrm{WF}^1_{Q}(f) = \mathrm{WF}^1_{C^\infty}(f)$$

agrees with the classical wave-front set of f (see [4] and Section 8.1 in [14]).

The next result deals with the regularity of the solutions f to $Tf = g$, in terms if the regularity of the datum g, under appropriate hypotheses on the operator T. It is a straightforward consequence of the theory developed above and of the propagation results proved in [4].

Theorem 4.12. *Let $T_1 = 1 - \Delta$ be the harmonic oscillator and $T_2 = \partial_t - a(t, x, D)$ be a generalized heat operator, with the symbol $a(t,.,..) \in \mathrm{SG}^{0,0}_{r,\rho}(\mathbf{R}^{2d})$ chosen in such a way that T_2 is elliptic with respect to the weight $\omega_0 = 1$. Then, both T_1 and T_2 map continuously $\mathscr{S}(\mathbf{R}^d)$ to itself, Q to itself, and Q_0 to itself.*

Acknowledgment

We wish to thank Professor H.G. Feichtinger and Professor L. Rodino for useful discussions and suggestions.

References

[1] P. Boggiatto, E. Buzano, L. Rodino *Global Hypoellipticity and Spectral Theory*, Mathematical Research, 92, Akademie Verlag, Berlin, 1996.

[2] H.O. Cordes *The Technique of Pseudodifferential Operators*, Cambridge Univ. Press, 1995.

[3] S. Coriasco, K. Johansson, J. Toft *Local wave-front sets of Banach and Fréchet types, and pseudo-differential operators*, Monatsh. Math., **169**, 3-4 (2013), 285–316.

[4] S. Coriasco, K. Johansson, J. Toft *Global wave-front sets of Banach, Fréchet and Modulation space types, and pseudo-differential operators*, J. Differential Equations **254**, 8 (2013), 3228–3258.

[5] S. Coriasco, L. Maniccia *Wave front set at infinity and hyperbolic linear operators with multiple characteristics*, Ann. Global Anal. Geom., **24** (2003), 375–400.

[6] S. Coriasco, R. Schulz *The global wave front set of tempered oscillatory integrals with inhomogeneous phase functions*, J. Fourier Anal. Appl., **19**, 5 (2013), 1093–1121.

[7] H.G. Feichtinger and K.H. Gröchenig *Banach spaces related to integrable group representations and their atomic decompositions, I*, J. Funct. Anal., **86** (1989), 307–340.

[8] H.G. Feichtinger *Modulation spaces on locally compact abelian groups. Technical report*, University of Vienna, Vienna, 1983; also in: M. Krishna, R. Radha, S. Thangavelu (eds.) Wavelets and their applications, Allied Publishers Private Limited, NewDehli Mumbai Kolkata Chennai Hagpur Ahmedabad Bangalore Hyderbad Lucknow, 2003, pp. 99–140.

[9] H.G. Feichtinger *Modulation spaces: Looking back and ahead*, Sampl. Theory Signal Image Process. **5** (2006), 109–140.

[10] G.B. Folland *Harmonic analysis in phase space*, Princeton U. P., Princeton, 1989.

[11] K.H. Gröchenig *Foundations of Time-Frequency Analysis*, Birkhäuser, Boston, 2001.

[12] A. Holst, J. Toft, P. Wahlberg *Weyl product algebras and modulation spaces*, J. Funct. Anal., **251** (2007), 463–491.

[13] L. Hörmander *Fourier Integral Operators I.* Acta Math. **127** (1971), 79–183.

[14] L. Hörmander *The Analysis of Linear Partial Differential Operators*, vol I–III, Springer-Verlag, Berlin – Heidelberg – New York – Tokyo, 1983, 1985.

[15] F. Luef, Z. Rahbani *On pseudodifferential operators with symbols in generalized Shubin classes and an application to Landau–Weyl operators*, Banach J. Math. Anal. **5** (2011), 59–72.

[16] R. Melrose Spectral and scattering theory for the Laplacian on asymptotically Euclidean spaces, In *Spectral and scattering theory* (*Sanda*, 1992), number 161 in Lecture Notes in Pure and Appl. Math., pages 85–130. Dekker, New York, 1994.

[17] C. Parenti *Operatori pseudodifferenziali in* $\mathbf{R}^n$ *e applicazioni*, Ann. Mat. Pura Appl., **93** (1972), 359–389.

[18] S. Pilipović, N. Teofanov, J. Toft *Micro-local analysis in Fourier Lebesgue spaces. Part I*, J. Fourier Anal. Appl. **17** (2011), 374–407.

[19] S. Pilipović, N. Teofanov, J. Toft *Micro-local analysis in Fourier Lebesgue and modulation spaces. Part II*, J. Pseudo-Differ. Oper. Appl., **1** (2010), 341–376.

[20] J. Toft *Continuity properties for modulation spaces with applications to pseudo-differential calculus, II*, Ann. Global Anal. Geom., **26** (2004), 73–106.

Sandro Coriasco
Dipartimento di Matematica "G. Peano"
Università degli Studi di Torino, Torino, Italy
e-mail: `sandro.coriasco@unito.it`

Karoline Johansson and Joachim Toft
Department of Mathematics
Linnæus University
Växjö, Sweden
e-mail: `karoline.johansson@lnu.se`
`joachim.toft@lnu.se`

Fourier Analysis
Trends in Mathematics, 107–132

A Pseudo-differential Calculus on Graded Nilpotent Lie Groups

Véronique Fischer and Michael Ruzhansky

Abstract. In this paper, we present first results of our investigation regarding symbolic pseudo-differential calculi on nilpotent Lie groups. On any graded Lie group, we define classes of symbols using difference operators. The operators are obtained from these symbols via the natural quantization given by the representation theory. They form an algebra of operators which shares many properties with the usual Hörmander calculus.

Mathematics Subject Classification (2010). Primary 35S05; Secondary 43A80.

Keywords. Pseudo-differential operators, nilpotent Lie groups.

1. Introduction

In the last five decades pseudo-differential operators have become a standard tool in the study of partial differential equations. It is natural to try to define analogues of the Euclidean pseudo-differential calculus in other settings. On one hand, while it is always possible to obtain a local calculus on any (connected) manifold, the question becomes much harder for global calculi. If, in addition, one requires a notion of symbol, the quasi inherent context is the one of Lie groups of type I where a Plancherel–Fourier analysis is available. On the other hand, from the viewpoint of what can actually be done at the level of operators, the investigation should start in the context of Lie groups with polynomial volume-growth where analysis of integral operators is quite well understood. Therefore the natural setting to start developing global pseudo-differential calculi is nilpotent or compact Lie groups, together with their semi-direct products.

The first author acknowledges the support of the London Mathematical Society via the Grace Chisholm Fellowship. It was during this fellowship held at King's College London in 2011 that the work was initiated. The second author was supported in part by the EPSRC Leadership Fellowship EP/G007233/1 and by the EPSRC Grant EP/K039407/1.

The genesis of this idea began quite some time ago; if a starting line had to be drawn, it would be in the seventies with the work of Elias Stein and his collaborators Folland, Rotschild, etc. (see, e.g., [9, 18]), and continued, in the next decade, with the work of Beals and Greiner amongst many others. Their motivation came from the study of differential operators on CR or contact manifolds, modelling locally the operators on homogeneous left-invariant convolution operators on nilpotent groups (cf [17]). In 'practice' and from this motivation, only nilpotent Lie groups endowed with some compatible structure of dilations, i.e., homogeneous groups and, more particularly, graded Lie groups, are considered. The latter is also the setting of our present investigation.

Since the seventies, several global calculi of operators on homogeneous Lie groups have appeared. However they were often calculi of left-invariant operators with the following notable exceptions to the authors' knowledge. Beside Dynin's construction of certain operators on the Heisenberg group in [6, 11], a non-invariant pseudo-differential calculus on any homogeneous group was developed in [3] but this is not symbolic since the operator classes are defined via properties of the kernel. In the revised version of [22], Taylor describes several (non-invariant) operator calculi and, in a different direction, he also explains a way to develop symbolic calculi: using the representations of the group, he defines a general *quantization* and the natural *symbols* on any unimodular type I group (by quantization, we mean a procedure which associates an operator to a symbol). He illustrates this on the Heisenberg group and obtains there several important applications for, e.g., the study of hypoellipticity. He uses the fact that, because of the properties of the Schrödinger representations of the Heisenberg group, a symbol is a family of operators in the Euclidean space, themselves given by symbols via the Weyl quantization. Recently, the attempt at defining suitable classes of Shubin type for these Weyl-symbols led to another version of the calculus on the Heisenberg group in [1][1].

Recently as well, using the global quantization procedure noted in [22], the second author and Turunen developed a global symbolic calculus on any compact Lie group in [19]. They successfully defined symbol classes so that the quantization procedure makes sense and the resulting operators form an algebra of operators with properties 'close enough' to the one enjoyed by the Euclidean Hörmander calculus (in fact, in a later work with Wirth [21], they showed that the calculus in [19] leads to the usual Hörmander operator classes on $\mathbb{R}^n$ extended to compact connected manifolds). Their approach is valid for any compact Lie group whereas the calculus of [1] is very specific to the Heisenberg group. The crucial and new ingredient in the definition of symbol classes in [19] was defining *difference operators* in order to replace the Euclidean derivatives in the Fourier variables. These difference operators allow expressing the pseudo-differential behaviour directly on the group.

[1]and in its revised version

In our present investigations, we build upon this notion to study operators in the nilpotent setting. However it is not possible to extend readily the results of the compact case developed in [19] to the nilpotent context. Some technical difficulties appear because, for example, the dual of G is no longer discrete and the unitary irreducible representations are almost all infinite dimensional. More problematically there is no Laplace–Beltrami operator and one expects to replace it by a sub-Laplacian on stratified Lie groups or, more generally, by a positive Rockland operator $\mathcal{R}$ on graded Lie groups; such operators are not central. Hence new technical ideas are needed to develop a pseudo-differential calculi on graded Lie groups using the natural quantization together with the notion of difference operators from [19].

The results that we have obtained in our investigation of this question so far were presented in the talk given by the first author at the conference *Fourier analysis and pseudo-differential operators*, Aalto University, 25–30 June, 2012. They are the following (here $1 \geq \rho \geq \delta \geq 0$):

(R1) The symbol classes form an algebra of operators $\cup_{m\in\mathbb{R}} S^m_{\rho,\delta}$ stable by taking the adjoint.

(R2) Let $\rho \neq 0$. The operators obtained by quantization from $\cup_{m\in\mathbb{R}} S^m_{\rho,\delta}$ form an algebra of operators $\cup_{m\in\mathbb{R}} \Psi^m_{\rho,\delta}$ stable by taking the adjoint.

(R3) The set of operators $\cup_{m\in\mathbb{R}} \Psi^m_{\rho,\delta}$ contains the left-invariant calculus.

(R4) The kernels are of Calderon–Zygmund type on homogeneous Lie groups; in particular our operators of order 0 are more singular than their Euclidean counterparts.

(R5) If $\rho \in [0,1)$, then the operators in $\Psi^0_{\rho,\rho}$ are continuous on $L^2(G)$.

(R6) $(\mathrm{Id}+\mathcal{R})^{\frac{m}{\nu}} \in \Psi^m_{1,0}$, where $\mathcal{R}$ is a positive Rockland operator of degree ν, see Section 2.3.

(R7) Positive operators of the calculus satisfy sharp Gårding inequalities.

As a consequence from Results (R2), (R5) and (R6), if $\rho \neq 0$, any pseudo-differential operator is continuous on the Sobolev spaces with the loss of derivatives being controlled by the order. All those properties justify, from our viewpoint, the choice of vocabulary of pseudo-differential calculi.

Since the conference in Aalto University of June 2012, these results, together with their complete proof and other progress made by the authors on the subject have started to be collected in the monograph [8], see also [7].

In this paper, due to the lack of space, we will state and prove the following parts of Results (1–7). Result (R1) is proved in Subsection 4.1. The proof of Result (R2) is given in Subsection 4.4 but relies on Result (R4) which is stated in Subsection 4.2 and only partially proved. In Subsection 3.5, Result (R3) is stated and proved while (R6) is stated in greater generality but not proved in this paper. The precise statements and proofs of Results (R5) and (R7) can be found in [8].

The paper is organised as follows. In Section 2, we explain the precise setting of our investigation for the group, the Sobolev spaces involved here and the group Fourier transform. In Section 3, we precise our definition of quantization and symbol classes. In Section 4, we give the properties of the symbols and of the corresponding kernels and operators stated above.

Convention: All along the paper, C denotes a constant which may vary from line to line. We denote by $\lceil r \rceil$ the smallest integer which is strictly greater than the real r.

2. Preliminaries

In this section, we set some notation and recall some known properties regarding the groups under investigation, the Taylor expansion in this context and representation theory.

2.1. The group G

Here we recall briefly the definition of graded nilpotent Lie groups and their natural homogeneous structure. A complete description of the notions of graded and homogeneous nilpotent Lie group may be found in [12, Chapter 1].

We will be concerned with graded Lie groups G which means that G is a connected and simply connected Lie group whose Lie algebra $\mathfrak{g}$ admits an $\mathbb{N}$-gradation $\mathfrak{g} = \oplus_{\ell=1}^{\infty} \mathfrak{g}_\ell$ where the $\mathfrak{g}_\ell$, $\ell = 1, 2, \ldots$ are vector subspaces of $\mathfrak{g}$, almost all equal to $\{0\}$ and satisfying $[\mathfrak{g}_\ell, \mathfrak{g}_{\ell'}] \subset \mathfrak{g}_{\ell+\ell'}$ for any $\ell, \ell' \in \mathbb{N}$. This implies that the group G is nilpotent. Examples of such groups are the Heisenberg group and more generally any stratified groups (which by definition correspond to the case $\mathfrak{g}_1$ generating the full Lie algebra $\mathfrak{g}$).

Let $\{X_1, \ldots, X_{n_1}\}$ be a basis of $\mathfrak{g}_1$ (this basis is possibly reduced to $\{0\}$), let $\{X_{n_1+1}, \ldots, X_{n_1+n_2}\}$ a basis of $\mathfrak{g}_2$ and so on, so that we obtain a basis $X_1, \ldots, X_n$ of $\mathfrak{g}$ adapted to the gradation. Via the exponential mapping $\exp_G : \mathfrak{g} \to G$, we identify the points $(x_1, \ldots, x_n) \in \mathbb{R}^n$ with the points $x = \exp_G(x_1 X_1 + \cdots + x_n X_n)$ in G. Consequently we allow ourselves to denote by $C(G)$, $\mathcal{D}(G)$ and $\mathcal{S}(G)$ etc... the spaces of continuous functions, of smooth and compactly supported functions or of Schwartz functions on G identified with $\mathbb{R}^n$. This basis also leads to a corresponding Lebesgue measure on $\mathfrak{g}$ and the Haar measure dx on the group G.

The coordinate function $x = (x_1, \ldots, x_n) \in G \mapsto x_j \in \mathbb{R}$ is denoted by x_j. More generally we define for every multi-index $\alpha \in \mathbb{N}_0^n$, $x^\alpha := x_1^{\alpha_1} x_2^{\alpha_2} \cdots x_n^{\alpha_n}$, as a function on G. Similarly we set $X^\alpha = X_1^{\alpha_1} X_2^{\alpha_2} \cdots X_n^{\alpha_n}$ in the universal enveloping Lie algebra $\mathfrak{U}(\mathfrak{g})$ of $\mathfrak{g}$.

For any $r > 0$, we define the linear mapping $D_r : \mathfrak{g} \to \mathfrak{g}$ by $D_r X = r^\ell X$ for every $X \in \mathfrak{g}_\ell$, $\ell \in \mathbb{N}$. Then the Lie algebra $\mathfrak{g}$ is endowed with the family of dilations $\{D_r, r > 0\}$ and becomes a homogeneous Lie algebra in the sense of [12]. The weights of the dilations are the integers $\upsilon_1, \ldots, \upsilon_n$ given by $D_r X_j = r^{\upsilon_j} X_j$, $j = 1, \ldots, n$. The associated group dilations are defined by

$$r \cdot x := (r^{\upsilon_1} x_1, r^{\upsilon_2} x_2, \ldots, r^{\upsilon_n} x_n), \quad x = (x_1, \ldots, x_n) \in G, \ r > 0.$$

In a canonical way this leads to the notions of homogeneity for functions and operators. For instance the degree of homogeneity of x^α and X^α, viewed respectively as a function and a differential operator on G, is $[\alpha] = \sum_j \upsilon_j \alpha_j$. Indeed, let us recall that a vector of $\mathfrak{g}$ defines a left-invariant vector field on G and more generally that the universal enveloping Lie algebra of $\mathfrak{g}$ is isomorphic with the left-invariant differential operators; we keep the same notation for the vectors and the corresponding operators.

Recall that a *homogeneous norm* on G is a continuous function $|\cdot| : G \to [0, +\infty)$ homogeneous of degree 1 on G which vanishes only at 0. Any homogeneous norm satisfies a triangular inequality up to a constant. Any two homogeneous norms are equivalent. For example

$$|x|_{\nu_o} := \left(\sum_{j=1}^n x_j^{2\frac{\nu_o}{\upsilon_j}}\right)^{\frac{1}{2\nu_o}} \tag{2.1}$$

with ν_o a common multiple to the weights $\upsilon_1, \dots, \upsilon_n$.

Various aspects of analysis on G can be developed in a comparable way with the Euclidean setting [4], sometimes replacing the topological dimension

$$n := \dim G = \sum_{\ell=1}^{\infty} \dim \mathfrak{g}_\ell,$$

of the group G by its homogeneous dimension

$$Q := \sum_{\ell=1}^{\infty} \ell \dim \mathfrak{g}_\ell = \upsilon_1 + \upsilon_2 + \dots + \upsilon_n.$$

2.1.1. Taylor expansions on G. In the setting of graded Lie groups one can obtain the left or right mean value theorem and left or right Taylor expansions adapted to the homogeneous structure [12, Theorem 1.42]. Let us give the statement for left invariance. We will need the following definition: the (left) Taylor polynomial of homogeneous degree M of a function $f \in C^{M+1}(G)$ at a point $x \in G$ is by definition the polynomial $P^{(f)}_{x,M}$ satisfying

$$X^\alpha P^{(f)}_{x,M}(0) = \begin{cases} X^\alpha f(x) & \text{whenever} \quad \alpha \in \mathbb{N}^n \text{ with } [\alpha] \leq M, \\ 0 & \text{if } [\alpha] > M. \end{cases}$$

We also define the remainder to be

$$R^{(f)}_{x,M}(z) := f(xz) - P^{(f)}_{x,M}(z).$$

Proposition 2.1 (Mean value and Taylor expansion [12]). *Let us fix a homogeneous norm $|\cdot|$ on G.*

1. (*Mean value property.*) *There exist positive group constants C_0 and b such that for any function $f \in C^1(G)$, we have*

$$|f(xy) - f(x)| \leq C_0 \sum_{j=1}^n |y|^j \sup_{|z| \leq b|y|} |X_j f(xz)|.$$

2. (*Taylor expansion.*) *For each $M \in \mathbb{N}_0$ there exist positive group constants C_M such that for any function $f \in C^{M+1}(G)$, we have*

$$\forall y \in G \qquad |R^{(f)}_{x,M}(y)| \leq C_M \sum_{\substack{[\alpha]>M \\ |\alpha|\leq \lceil M \rceil}} |y|^{[\alpha]} \sup_{|z|\leq b^{M+1}|y|} |X^\alpha f(xz)|,$$

where $\lceil M \rfloor := \max\{|\beta| : \beta \in \mathbb{N}_0^n$ *with* $[\beta] \leq M\}$.

The control can be improved in the stratified case (again see [12]) but we present here the more general case of the graded groups.

Remark 2.2. Proposition 2.1 extends easily to functions which are vector valued in a Banach space, replacing the modulus by operator norms.

The Taylor polynomials can be described in the following way. Let $(q_\alpha)_{\alpha\in\mathbb{N}^n}$ be the basis of polynomials obtained from by the duality $\langle X, p\rangle := Xp(0)$ where $X \in \mathfrak{U}(\mathfrak{g})$ and p is a polynomial. This means that the q_α's are the polynomials satisfying

$$\forall \alpha, \beta \in \mathbb{N}_0^n \qquad X^\beta q_\alpha(0) = \begin{cases} 0 & \text{if } \alpha \neq \beta \\ 1 & \text{if } \alpha = \beta \end{cases}. \tag{2.2}$$

We can then write the Taylor polynomial as

$$P^{(f)}_{x,M} = \sum_{[\alpha]\leq M} X^\alpha f(x) q_\alpha.$$

We will need the following properties of the polynomials (q_α) defined via (2.2):

Lemma 2.3.
- *Each polynomial q_α is homogeneous of degree $[\alpha]$. Moreover, $(q_\alpha)_{[\alpha]=d}$ is a basis of the space of homogeneous polynomials of degree d.*
- *For any $\alpha_1, \alpha_2 \in \mathbb{N}_0^n$, the polynomial $q_{\alpha_1}q_{\alpha_2}$ can be written as a linear combination of q_α with $[\alpha] = [\alpha_1] + [\alpha_2]$.*
- *For any $\alpha \in \mathbb{N}_0^n$ and $x, y \in G$,*

$$q_\alpha(xy) = \sum_{[\alpha_1]+[\alpha_2]=[\alpha]} c_{\alpha_1,\alpha_2} q_{\alpha_1}(x) q_{\alpha_2}(y),$$

where the coefficients c_{α_1,α_2} are real and, moreover,

$$c_{\alpha_1,0} = \begin{cases} 1 & \textit{if } \alpha_1 = \alpha \\ 0 & \textit{otherwise} \end{cases} \quad \textit{and} \quad c_{0,\alpha_2} = \begin{cases} 1 & \textit{if } \alpha_2 = \alpha \\ 0 & \textit{otherwise} \end{cases}.$$

Proof. Clearly $(q_\alpha)_{[\alpha]=d}$ is the dual basis of $(X^\beta)_{[\beta]=d}$. The first point follows. The second point is a direct consequence of the first together with the homogeneity of $q_{\alpha_1}q_{\alpha_2}$. The third point is a consequence of the homogeneity in x and in y and of the Baker–Campbell–Hausdorff formula. □

2.2. The unitary dual and the group Fourier transform

We denote by $\widehat{G}$ the unitary dual of the group G, that is, the set of (strongly continuous) unitary irreducible representations modulo unitary equivalence. We will often identify a unitary irreducible representation π of G and its equivalence class; we denote the representation Hilbert space by $\mathcal{H}_\pi$ and the subspace of smooth vectors by $\mathcal{H}_\pi^\infty$.

The group Fourier transform of a function $f \in L^1(G)$ at $\pi \in \widehat{G}$ is the bounded operator $\widehat{f}(\pi)$ (sometimes this will be also denoted by $\pi(f)$ for longer expressions) on $\mathcal{H}_\pi$ given by

$$(\widehat{f}(\pi)v_1, v_2)_{\mathcal{H}_\pi} := \int_G f(x)(\pi(x)^* v_1, v_2)_{\mathcal{H}_\pi} dx, \quad v_1, v_2 \in \mathcal{H}_\pi.$$

One can readily see the equality $\widehat{f_1 * f_2}(\pi) = \widehat{f_2}(\pi)\widehat{f_1}(\pi)$.

The group Fourier transform of a vector $X \in \mathfrak{g}$ at $\pi \in \widehat{G}$ is the operator $\pi(X)$ on $\mathcal{H}_\pi^\infty$ given by

$$(\pi(X)v_1, v_2)_{\mathcal{H}_\pi} := \partial_{s=0}(\pi(e^{sX})v_1, v_2)_{\mathcal{H}_\pi}, \quad v_1, v_2 \in \mathcal{H}_\pi^\infty.$$

Setting $\pi(X^\alpha) = \pi(X)^\alpha$, this yields the definition of the group Fourier transform of any element of $\mathfrak{U}(\mathfrak{g})$. With this notation we have for any $\alpha \in \mathbb{N}_0^n$,

$$\widehat{X^\alpha f}(\pi) = \pi(X)^\alpha \widehat{f}(\pi) = \pi(X^\alpha)\widehat{f}(\pi) \text{ and } \widehat{\tilde{X}^\alpha f}(\pi) = \widehat{f}(\pi)\pi(X)^\alpha = \widehat{f}(\pi)\pi(X^\alpha),$$

with the convention for $\alpha = 0$ that $\pi(X^0) = \Pi(I) = I = \pi(X)^0$ and that $\tilde{X}$ denotes the right-invariant vector field corresponding to $X \in \mathfrak{g}$.

The properties above help in the systematic computations of certain expressions; for example we see $\pi(\{Xf_1\}*f_2) = \pi(f_2)\pi(X)\pi(f_1) = \pi(\tilde{X}f_2)\pi(f_1)$ and this is coherent with the direct and more tedious computation $\{Xf_1\}*f_2 = f_1*\{\tilde{X}f_2\}$.

2.3. A positive Rockland operator $\mathcal{R}$

We choose $\mathcal{R}$ a positive (left) Rockland operator of homogeneous degree ν. Let us recall that being a Rockland operator means that $\mathcal{R}$ is a differential operator on G which is left-invariant and homogeneous of degree ν and such that for every non-trivial irreducible representation π of G, the operator $\pi(\mathcal{R})$ is injective on smooth vectors (see Section 2.2 for the definition of $\pi(\mathcal{R})$); being positive means

$$\forall f \in \mathcal{S}(G) \qquad (\mathcal{R}f, f)_{L^2(G)} \geq 0.$$

Here as usual

$$(f_1, f_2)_{L^2(G)} = \int_G f_1(x)\overline{f_2(x)}dx.$$

In the stratified case, we choose $\mathcal{R} = -\mathcal{L}$ where $\mathcal{L}$ is the sub-Laplacian $\sum_{i=1}^{n_1} X_i^2$ (and so $\nu = 2$). In the graded case, if ν_o is any common multiple of the weights $\upsilon_1, \ldots, \upsilon_n$, then

$$\sum_{j=1}^n (-1)^{\frac{\nu_o}{\upsilon_j}} c_j X_j^{2\frac{\nu_o}{\upsilon_j}} \quad \text{and} \quad \sum_{j=1}^n c_j X_j^{4\frac{\nu_o}{\upsilon_j}} \quad (\text{with } c_j > 0), \tag{2.3}$$

are positive Rockland operators of degree $2\nu_o$ and $4\nu_o$ respectively.

By the celebrated result of Helffer and Nourrigat [14], any Rockland operator is hypoelliptic and satisfies subelliptic estimates. Furthermore, by [12, Ch. 4.B], any positive Rockland operator $\mathcal{R}$, as a differential operator defined on $\mathcal{D}(G)$, admits an essentially self-adjoint extension on $L^2(G)$ for which we keep the same notation $\mathcal{R}$. Let E denote its spectral measure. For any measurable function ϕ on $[0,\infty)$, we define the operator

$$\phi(\mathcal{R}) := \int_0^\infty \phi(\lambda) dE_\lambda,$$

which is invariant under left-translation. If it maps continuously $\mathcal{S}(G) \to \mathcal{S}'(G)$ (for example if ϕ is bounded), by the Schwartz kernel theorem, it is a convolution operator with kernel $\phi(\mathcal{R})\delta_o \in \mathcal{S}'(G)$, that is,

$$\phi(\mathcal{R})f = f \ * \ \phi(\mathcal{R})\delta_o, \quad f \in \mathcal{S}(G).$$

Recall that the group convolution is defined via

$$f_1 * f_2(g) = \int_G f_1(g') f_2(g'^{-1} g) dg', \quad f_1, f_2 \in \mathcal{S}(G).$$

The above hypotheses ensure the following Marcinkiewicz-type properties proved by A. Hulanicki [15].

Proposition 2.4 (Hulanicki). *For any $\alpha, \beta \in \mathbb{N}_0^n$, there exists $k = k_{\alpha,\beta} \in \mathbb{N}_0$ and $C = C_{\alpha,\beta} > 0$ such that for any $\phi \in C^\infty([0,\infty))$ we have*

$$\|x^\alpha X^\beta \phi(\mathcal{R})\delta_o\|_{L^1(G)} \leq C \sup_{\lambda \geq 0,\, k_1 \leq k} (1+\lambda)^k \partial_\lambda^{k_1} |\phi(\lambda)| \,.$$

By this we mean that if the supremum in the right-hand side is finite, then the distribution $x^\alpha X^\beta \phi(\mathcal{R})\delta_o$ coincides with an integrable function and the inequality holds.

Remark 2.5. Consequently if $\phi \in \mathcal{S}(\mathbb{R})$ is Schwartz, then the kernel $\phi(\mathcal{R})\delta_o$ is also Schwartz on G, i.e., $\phi(\mathcal{R})\delta_o \in \mathcal{S}(G)$.

The proof of Proposition 2.4 relies on using the *$\mathcal{R}$-heat kernel* h_t, defined as the kernel of $\exp(-t\mathcal{R})$ for each $t > 0$. In [12], it is proved that the function $h = h_1$ is Schwartz and that

$$h_t(x) = t^{-\frac{Q}{\nu}} h(t^{-\frac{1}{\nu}} x).$$

As in the Euclidean or stratified cases (see [10]), we can define *Bessel potential kernels* associated with a positive Rockland operator $\mathcal{R}$ of degree ν via the integral

$$\mathcal{B}_a(x) = \frac{1}{\Gamma(\frac{a}{\nu})} \int_0^\infty t^{\frac{a}{\nu}-1} e^{-t} h_t(x) dt.$$

Indeed for $a \in \mathbb{C}$ with $\text{Re}\, a > 0$, this integral converges absolutely for $x \neq 0$ and defines the Bessel potential kernel $\mathcal{B}_a \in C^\infty(G \backslash \{0\})$ which satisfies:

$$\|\mathcal{B}_a\|_{L^1(G)} \leq \frac{\Gamma(\text{Re}\, \frac{a}{\nu})}{|\Gamma(\frac{a}{\nu})|} \|h\|_{L^1} < \infty, \quad a \in \mathbb{C},\ \text{Re}\, a > 0.$$

Using the properties of semigroup of $e^{-t\mathcal{R}}$, one obtains that $\|\mathcal{B}_a\|_{L^2(G)}$ is square integrable if Re $a > Q/2$. The Bessel potential is the convolution kernel of the $L^2(G)$-bounded left-invariant operator $(\text{Id} + \mathcal{R})^{-a/\nu}$ and of the $L^2(G)$-bounded right-invariant operator $(\text{Id} + \tilde{\mathcal{R}})^{-a/\nu}$, so that we have

$$(\text{Id} + \mathcal{R})^{-a/\nu} f = f * \mathcal{B}_a, \quad (\text{Id} + \tilde{\mathcal{R}})^{-a/\nu} f = \mathcal{B}_a * f, \quad f \in L^2(G).$$

2.4. Sobolev spaces

For $a \geq 0$ and $\mathcal{R}$ a positive Rockland operator, we define the $\mathcal{R}$-Sobolev spaces as the domain of $(\text{Id} + \mathcal{R})^{\frac{a}{\nu}}$, that is,

$$L^2_a(G) = \{f \in L^2(G),\ (\text{Id} + \mathcal{R})^{\frac{a}{\nu}} f \in L^2(G)\}.$$

For $a < 0$, $L^2_a(G)$ is the completion of $L^2(G)$ for the norm $f \mapsto \|(\text{Id}+\mathcal{R})^{\frac{a}{\nu}} f\|_{L^2(G)}$. It is easy to see that for any $a \in \mathbb{R}$, the Sobolev space $L^2_a(G)$ is a Hilbert space for the norm

$$\|f\|_{L^2_a(G)} := \|(\text{Id} + \mathcal{R})^{\frac{a}{\nu}} f\|_{L^2(G)}.$$

Adapting the stratified case [10] (see [8]), one obtains:

Proposition 2.6 (Sobolev spaces). *Let $\mathcal{R}$ be a positive Rockland operator of homogeneous degree $\nu_{\mathcal{R}}$.*

1. *If $a \leq b$, then $\mathcal{S}(G) \subset L^2_b(G) \subset L^2_a(G) \subset \mathcal{S}'(G)$ and an equivalent norm for $L^2_b(G)$ is $f \mapsto \|f\|_{L^2_a(G)} + \|\mathcal{R}^{\frac{b-a}{\nu}} f\|_{L^2_a(G)}$.*
2. *If $a \in \nu_{\mathcal{R}}\mathbb{N}_0$, then an equivalent norm is given by $f \mapsto \sum_{[\alpha]\leq a} \|X^\alpha f\|_{L^2(G)}$.*
3. *The dual space of $L^2_a(G)$ is isomorphic to $L^2_{-a}(G)$ via the bilinear form $(f_1, f_2) \mapsto \int_G f_1 f_2 dg$.*
4. *We have the usual property of interpolation for Sobolev spaces: let T be a linear mapping from $L^2_{a_0}(G) + L^2_{a_1}(G)$ to locally integrable functions on G; we assume that T maps $L^2_{a_0}(G)$ and $L^2_{a_1}(G)$ boundedly into $L^2_{b_0}(G)$ and $L^2_{b_1}(G)$, respectively. Then T extends uniquely to a bounded mapping from $L^2_{a_t}(G)$ to $L^2_{b_t}(G)$ with $(a_t, b_t) = t(a_0, b_0) + (1 - t)(a_1, b_1)$.*

Consequently, the Sobolev spaces do not depend on the choice of operators $\mathcal{R}$ as in the statement above. Such operators always exist (see (2.3)) and we fix one of them until the end of the paper.

From the interpolation property of Sobolev spaces (cf. Proposition 2.6), we have:

Lemma 2.7. *Let $\kappa \in \mathcal{S}'(G)$ and $a \in \mathbb{R}$. Let $\{\gamma_n, n \in \mathbb{Z}\}$ be a sequence of real numbers which tends to $\pm\infty$ as $n \to \pm\infty$. Assume that for any $n \in \mathbb{Z}$, the operator T_κ extends continuously to a bounded operator $L^2_{\gamma_n}(G) \to L^2_{a+\gamma_n}(G)$. Then the operator T_κ extends continuously to a bounded operator $L^2_\gamma(G) \to L^2_{a+\gamma}(G)$ for any $\gamma \in \mathbb{R}$.*

As in the Euclidean and stratified cases [10], we can prove the following Sobolev inequalities:

Lemma 2.8 (Sobolev inequality). *If $a > Q/2$ then any function $f \in L^2_a(G)$ admits a continuous bounded representative which satisfies*

$$\|f\|_{L^\infty(G)} \leq C_a \|f\|_{L^2_a(G)},$$

with $C_a = \|\mathcal{B}_a\|_{L^2(G)}$ independent of f.

Sketch of the proof. It suffices to write

$$f = (\mathrm{Id} + \mathcal{R})^{-\frac{a}{\nu}} (\mathrm{Id} + \mathcal{R})^{\frac{a}{\nu}} f = \{(\mathrm{Id} + \mathcal{R})^{\frac{a}{\nu}} f\} * \mathcal{B}_a. \qquad \square$$

2.5. The Plancherel Theorem and the von Neumann algebras $\mathscr{L}_L(L^2(G))$, $\mathcal{K}(G)$ and $L^\infty(\widehat{G})$

About representation theory and the Plancherel theorem, we refer the reader to Dixmier's standard textbook [5], especially §18.8.

Recall that a bounded operator A on a Hilbert space $\mathcal{H}$ is in the Hilbert–Schmidt class whenever $\|A\|_{HS} = \sqrt{\mathrm{Tr}\,(A^*A)}$ is finite. If $f \in L^2(G) \cap L^1(G)$ then $\widehat{f}(\pi)$ is a Hilbert–Schmidt operator, and the *Plancherel formula* holds,

$$\int_G |f(g)|^2 dg = \int_{\widehat{G}} \|\widehat{f}(\pi)\|^2_{HS} d\mu(\pi),$$

where μ is the Plancherel measure on $\widehat{G}$. The group Fourier transform extends unitarily to $L^2(G)$ and a square integrable function $f \in L^2(G)$ gives rise to a μ-square-integrable field of Hilbert–Schmidt operators $\{\widehat{f}(\pi)\}$. Conversely, a μ-square-integrable field of Hilbert–Schmidt operators $\{\sigma_\pi\}$ defines a square integrable function f with

$$(f, f_1)_{L^2(G)} = \int_{\widehat{G}} \mathrm{Tr}\left(\sigma_\pi\, \widehat{f_1}(\pi)^*\right) d\mu(\pi), \quad f_1 \in L^2(G).$$

Let $\mathscr{L}(L^2(G))$ denote the set of bounded linear operators $L^2(G) \to L^2(G)$, and let $\mathscr{L}_L(L^2(G))$ be the subset formed by the operators in $\mathscr{L}(L^2(G))$ which commute with the left regular representation $L(g) : f \in L^2(G) \mapsto f(g^{-1}\cdot)$, $g \in G$. Endowed with the operator norm and composition of operators, $\mathscr{L}_L(L^2(G))$ is a von Neumann algebra.

If $T \in \mathscr{L}_L(L^2(G))$, then there exists a μ-measurable field of uniformly bounded operators $\{\sigma_\pi^{(T)}\}$ such that for any $f \in L^2(G)$ the Hilbert–Schmidt operators $\widehat{Tf}(\pi)$ and $\sigma_\pi^{(T)} \widehat{f}(\pi)$ are equal μ-almost everywhere; the field $\{\sigma_\pi^{(T)}\}$ is unique up to a μ-negligible set.

Let $L^\infty(\widehat{G})$ denote the space of μ-measurable fields of uniformly bounded operators on $\widehat{G}$, modulo equivalence with respect to the Plancherel formula μ. As is usual, we will identify such fields with their classes in $L^\infty(\widehat{G})$. We have that if $T \in \mathscr{L}_L(L^2(G))$, then there exists a unique $\sigma \in L^\infty(\widehat{G})$ as above. Note that by the Schwartz kernel theorem, the operator T is of convolution type with kernel $\kappa \in \mathcal{D}'(G)$,

$$Tf = f * \kappa, \quad f \in \mathcal{D}(G).$$

Conversely given a field $\{\sigma_\pi\} \in L^\infty(\widehat{G})$ there exists a unique bounded linear operator $T \in \mathscr{L}_L(L^2(G))$ satisfying $\widehat{Tf}(\pi) = \sigma_\pi \widehat{f}(\pi)$ μ-almost everywhere for any $f \in L^2(G)$.

If $\kappa \in \mathcal{D}'(G)$ is such that the corresponding convolution operator $f \in \mathcal{D}(G) \mapsto f * \kappa$ extends to a bounded operator $T \in \mathscr{L}(L^2(G))$ then $T \in \mathscr{L}_L(L^2(G))$ and we abuse the notation by setting $\sigma_\pi^{(T)} := \pi(\kappa) \equiv \widehat{\kappa}(\pi)$. We denote by $\mathcal{K}(G)$ the set of such distributions κ. It is a von Neumann algebra isomorphic to $\mathscr{L}_L(L^2(G))$ when equipped with the $*$-product $\kappa \mapsto \kappa^*$ where $\kappa^*(x) = \bar{\kappa}(x^{-1})$, and the operator norm

$$\|\kappa\|_* := \|f \mapsto f * \kappa\|_{\mathscr{L}(L^2(G))}.$$

Note that when we equip $L^\infty(\widehat{G})$ with the operation $\sigma \mapsto \sigma^*$ and the norm

$$\|\sigma\|_* = \sup_{\pi \in \widehat{G}} \|\widehat{\sigma}_\pi\|_{\text{op}},$$

where $\|\cdot\|_{\text{op}}$ denotes the operator norm and the supremum is in fact the essential supremum with respect to the Plancherel measure μ, $L^\infty(\widehat{G})$ becomes a von Neumann algebra isomorphic with $\mathscr{L}_L(L^2(G))$ and $\mathcal{K}(G)$. More precisely the group Fourier transform defined on $\mathcal{K}(G)$ gives the isomorphism between $\mathcal{K}(G)$ and $L^\infty(\widehat{G})$.

Throughout this paper, if $\kappa \in \mathcal{D}'(G)$, T_κ denotes the convolution operator

$$T_\kappa : \mathcal{D}(G) \ni f \mapsto f * \kappa,$$

and we keep the same notation for any of its continuous extensions $L^2_b(G) \to L^2_a(G)$ when they exist. With norms possibly infinite, $\|\kappa\|_*$ is equal to the operator norm of $T_\kappa : L^2(G) \to L^2(G)$ by the Plancherel theorem, and is less than $\|\kappa\|_{L^1(G)}$.

For any $a, b \in \mathbb{R}$, it is easy to see that T_κ admits a continuous extension $L^2_b(G) \to L^2_a(G)$ if and only if $(I + \tilde{\mathcal{R}})^{-\frac{b}{\nu}}(\text{Id} + \mathcal{R})^{\frac{a}{\nu}}\kappa \in \mathcal{K}(G)$, with equality between the $L^2_b(G) \to L^2_a(G)$-operator norm and the $\mathcal{K}(G)$-norm. In this case we may abuse the notation and write

$$\pi(\text{Id} + \mathcal{R})^{\frac{a}{\nu}}\pi(\kappa)\pi(\text{Id} + \mathcal{R})^{-\frac{b}{\nu}} \quad \text{instead of} \quad \pi\left((\text{Id} + \mathcal{R})^{\frac{a}{\nu}}(I + \tilde{\mathcal{R}})^{-\frac{b}{\nu}}\kappa\right).$$

3. Quantization and symbols classes

As recalled in Introduction, there exists a natural quantization which is valid on any Lie group of type I. We will present it in this section after defining symbols for which this quantization makes sense and produces operators $\mathcal{D}(G) \to \mathcal{D}'(G)$ with G graded Lie groups. Moreover, the resulting operators admit integral representations with right convolution kernels and these kernels play a major role in every subsequent proof. We will also define symbol classes and give some examples of symbols.

3.1. The symbols and their kernels

A *symbol* is a family of operators $\sigma = \{\sigma(x,\pi),\ x \in G,\ \pi \in \widehat{G}\}$ satisfying:

1. for each $x \in G$, $\{\sigma(x,\pi),\ \pi \in \widehat{G}\}$ is a μ-measurable field of operators $\mathcal{H}_\pi^\infty \to \mathcal{H}_\pi$,
2. there exist $\gamma_1, \gamma_2 \in \mathbb{R}$ such that for any $x \in G$,
$$\{\pi(\mathrm{Id}+\mathcal{R})^{\gamma_1}\sigma(x,\pi)\pi(\mathrm{Id}+\mathcal{R})^{\gamma_2},\ \pi \in \widehat{G}\} \in L^\infty(\widehat{G}), \tag{3.1}$$
3. for any $\pi \in \widehat{G}$ and any $u, v \in \mathcal{H}_\pi$, the scalar function $x \mapsto (\sigma(x,\pi)u, v)_{\mathcal{H}_\pi}$ is smooth over G.

Consequently at each $x \in G$ and $\pi \in \widehat{G}$, the operator $\sigma(x,\pi)$ is densely defined on $\mathcal{H}_\pi$; it is also the case for $X_x^\beta \sigma(x,\pi)$ for any $\beta \in \mathbb{N}_0^n$.

The second condition implies that for each $x \in G$, the μ-measurable field (3.1) correspond to a distribution $\kappa_{x,\gamma_1,\gamma_2} \in \mathcal{K}(G)$ which depends smoothly on x; hence σ corresponds to a distribution
$$\kappa(x,\cdot) = \kappa_x := (\mathrm{Id}+\mathcal{R})^{-\gamma_1}(I+\tilde{\mathcal{R}})^{-\gamma_2}\kappa_{x,\gamma_1,\gamma_2},$$
which we call its *kernel*. By injectivity of π on $\mathcal{K}(G)$, $\pi(X_x^\beta \kappa_x) = X_x^\beta \sigma(x,\pi)$.

Examples of symbols are the symbols within the classes $S^m_{\rho,\delta}$ defined later on. More concrete examples of symbols which do not depend on $x \in G$ are $\pi(X)^\alpha$, $\alpha \in \mathbb{N}_0^n$ or the multipliers in $\pi(\mathcal{R})$, that is, $\phi(\pi(\mathcal{R}))$ with $\phi \in L^\infty(\mathbb{R})$ (for example). Indeed for any $\pi \in \widehat{G}$ the operator $\pi(\mathcal{R})$ is essentially self-adjoint and we denote by E_π its spectral projection, hence giving a meaning to $\phi(\pi(\mathcal{R}))$. The relation between the spectral projections E and E_π of $\mathcal{R}$ and $\pi(\mathcal{R})$ is
$$\pi(\phi(\mathcal{R})f) = \phi(\pi(\mathcal{R}))\pi(f), \quad \phi \in L^\infty(\mathbb{R}),\ f \in L^2(G).$$
It is known [16] that the spectrum of $\pi(\mathcal{R})$ consists of discrete eigenvalues in $(0,\infty)$. This may add a further justification to using the word *quantization*.

3.2. The quantization mapping $\sigma \mapsto \mathrm{Op}(\sigma)$

Our quantization is analogous to the usual Kohn–Nirenberg quantization in the Euclidean setting, and has already been noticed by Taylor [22], used indirectly on the Heisenberg group [22, 1] and explicitly on compact Lie groups [19]. It associates an operator $T = \mathrm{Op}(\sigma)$ to a symbol σ in the following way (with the same notation as in Subsection 3.1). For any $f \in \mathcal{D}(G)$ and $x \in G$,
$$\int_{\widehat{G}} \mathrm{Tr}\left|\sigma(x,\pi)\widehat{f}(\pi)\right| d\mu(\pi) \leq \sup_{\pi \in \widehat{G}} \|\pi(\mathrm{Id}+\mathcal{R})^{\gamma_1}\sigma(x,\pi)\ \pi(\mathrm{Id}+\mathcal{R})^{\gamma_2}\|_{\mathrm{op}}$$
$$\int_{\widehat{G}} \mathrm{Tr}\left|\pi\left((\mathrm{Id}+\mathcal{R})^{-\gamma_1}(I+\tilde{\mathcal{R}})^{-\gamma_2}f\right)\right| d\mu(\pi),$$
is finite and we can set
$$Tf(x) := \int_{\widehat{G}} \mathrm{Tr}\left(\pi(x)\sigma(x,\pi)\widehat{f}(\pi)\right) d\mu(\pi). \tag{3.2}$$

We have obtained a continuous linear operator $T : \mathcal{D}(G) \to \mathcal{D}'(G)$. By the Schwartz kernel theorem, $T = \mathrm{Op}(\sigma)$ has an integral kernel in the distributional sense. However since σ is a symbol, we obtain directly, still in the distributional sense, the following integral representation in terms of the kernel κ defined in Subsection 3.1,

$$Tf(x) = f * \kappa_x(x) = \int_G f(y)\kappa(x, y^{-1}x)dy.$$

For example, the symbol σ given by the identity operator on each space $\mathcal{H}_\pi$ is associated with the identity operator on G; its kernel is the Dirac measure at 0 denoted by δ_0 (independent of the point $x \in G$). More generally, for any $\alpha \in \mathbb{N}_0^n$, the symbol $\pi(X)^\alpha$ is associated with the operator X^α with kernel $(-1)^{|\alpha|}X^\alpha\delta_0$ defined in the sense of distributions via

$$\int_G f(g)(-1)^{|\alpha|}X^\alpha\delta_0(g)dg = \int_G X^\alpha f(g)\delta_0(g)dg = X^\alpha f(0).$$

It is easy to see that the quantization mapping $\sigma \mapsto T = \mathrm{Op}(\sigma)$ is 1-1 and linear.

Before defining symbol classes, we need to define difference operators.

3.3. Difference operators

Difference operators were defined on compact Lie groups in [19], as acting on Fourier coefficients. Its adaptation to our setting leads us to define difference operators on $L^\infty(\widehat{G})$ viewed as fields. More precisely for any $q \in C^\infty(G)$, we set

$$\Delta_q \widehat{f}(\pi) := \widehat{qf}(\pi) = \pi(qf).$$

This defines an operator Δ_q with domain Dom $(\Delta_q) := \mathcal{F}_G\{f \in \mathcal{K}(G),\ qf \in \mathcal{K}(G)\}$, and more generally $\pi(\mathrm{Id}+\mathcal{R})^{-\gamma_1}\pi(I+\tilde{\mathcal{R}})^{-\gamma_2}\mathrm{Dom}\ (\Delta_q)$ for any $\gamma_1, \gamma_2 \in \mathbb{R}$. Note that in general, it is not possible to define an operator Δ_q on each $\mathcal{H}_\pi$; this can be seen quite easily by considering the multiplication by the central variable on the Heisenberg group for example.

The *difference operators* are

$$\Delta^\alpha := \Delta_{\tilde{q}_\alpha}, \quad \alpha \in \mathbb{N}_0^n,$$

where $\tilde{q}_\alpha(x) = q_\alpha(x^{-1})$ and the q_α's were defined via (2.2).

Lemma 2.3 implies that $\Delta^{\alpha_1}\Delta^{\alpha_2}$ is a linear combination of Δ^α with $[\alpha] = [\alpha_1] + [\alpha_2]$. Furthermore, we have

$$\begin{aligned}
\tilde{q}_\alpha(x)\ f_2 * f_1(x) &= \int_G q_\alpha(x^{-1}y\ y^{-1})\ f_2(y)\ f_1(y^{-1}x)\ dy \\
&= \sum_{[\alpha_1]+[\alpha_2]=[\alpha]} c_{\alpha_1,\alpha_2} \int_G q_{\alpha_2}(y^{-1})f_2(y)\ q_{\alpha_1}(x^{-1}y)f_1(y^{-1}x)\ dy \\
&= \sum_{[\alpha_1]+[\alpha_2]=[\alpha]} c_{\alpha_1,\alpha_2}\ (\tilde{q}_{\alpha_2}f_2) * (\tilde{q}_{\alpha_1}f_1),
\end{aligned}$$

and we get the *Leibniz formula*:

$$\Delta^\alpha\left(\widehat{f}_1(\pi)\widehat{f}_2(\pi)\right) = \sum_{[\alpha_1]+[\alpha_2]=[\alpha]} c_{\alpha_1,\alpha_2}\, \Delta^{\alpha_1}\widehat{f}_1(\pi)\, \Delta^{\alpha_2}\widehat{f}_2(\pi). \tag{3.3}$$

The idea of difference operators appear naturally when considering operators on the torus $\mathbb{T}^n$. In this case one recovers forward and backward difference operators on the lattice $\mathbb{Z}^n$. Difference operators were systematically defined and studied on compact Lie groups in [19]. On the Heisenberg group, expressions of a related nature were used to describe the Schwartz space in [13] and with a hypothesis of unitary invariance in [2].

3.4. The symbol classes $S^m_{\rho,\delta}$

Definition 3.1. Let $m,\rho,\delta\in\mathbb{R}$ with $1\geq\rho\geq\delta\geq 0$ and $\delta\neq 1$. A symbol σ is a *symbol of order m and of type (ρ,δ)* whenever, for each $\alpha,\beta\in\mathbb{N}_0^n$ and $\gamma\in\mathbb{R}$, the field

$$\{\pi(\mathrm{Id}+\mathcal{R})^{\frac{\rho[\alpha]-m-\delta[\beta]+\gamma}{\nu}} X_x^\beta \Delta^\alpha\sigma(x,\pi)\pi(\mathrm{Id}+\mathcal{R})^{-\frac{\gamma}{\nu}},\ \pi\in\widehat{G}\},$$

is in $L^\infty(\widehat{G})$ uniformly in $x\in G$; this means that we have

$$\sup_{\pi\in\widehat{G},\, x\in G} \|\pi(\mathrm{Id}+\mathcal{R})^{\frac{\rho[\alpha]-m-\delta[\beta]+\gamma}{\nu}} X_x^\beta \Delta^\alpha\sigma(x,\pi)\pi(\mathrm{Id}+\mathcal{R})^{-\frac{\gamma}{\nu}}\|_{\mathrm{op}} = C_{\alpha,\beta,\gamma}<\infty.$$

(The supremum over π is in fact the essential supremum with respect to the Plancherel measure μ.)

The *symbol class* $S^m_{\rho,\delta}$ is the set of symbol of order m and of type (ρ,δ). We also define $S^{-\infty}_{\rho,\delta}=\cap_{m\in\mathbb{R}}S^m_{\rho,\delta}$ the class of smoothing symbols.

Let us make some comments on this definition:

1. In the abelian case, that is, $\mathbb{R}^n$ endowed with the addition law and $\mathcal{R}=-\mathcal{L}$, $\mathcal{L}$ being the Laplace operator, $S^m_{\rho,\delta}$ boils down to the usual Hörmander class. In the case of compact Lie groups with $\mathcal{R}=-\mathcal{L}$, $\mathcal{L}$ being the Laplace–Beltrami operator, a similar definition leads to the one considered in [19] since the operator $\pi(\mathrm{Id}+\mathcal{R})$ is scalar. However here, in the case of non-abelian graded groups, the operator $\mathcal{R}$ can not have a scalar Fourier transform.
2. The presence of the parameter γ is required to prove that the space of symbols $\cup_{m\in\mathbb{R}}S^m_{\rho,\delta}$ form an algebra of operators later on.
3. The conditions on α and β are of countable nature and it is also the case for γ. Indeed, by Lemma 2.7, it suffices to prove the property above for one sequence $\{\gamma_n,\ n\in\mathbb{Z}\}$ with $\gamma_n \underset{n\to\pm\infty}{\longrightarrow} \pm\infty$.
4. A symbol class $S^m_{\rho,\delta}$ is a vector space. And we have the inclusions
$$m_1\leq m_2,\quad \delta_1\leq\delta_2,\quad \rho_1\geq\rho_2 \quad\Longrightarrow\quad S^{m_1}_{\rho_1,\delta_1}\subset S^{m_2}_{\rho_2,\delta_2}.$$
5. If $\rho\neq 0$, we will show in Subsections 4.2 and 4.4 that we obtain an algebra of operators with smooth kernels κ_x away from the origin.

If σ is a symbol and $a, b, c \in [0, \infty)$, we set

$$\|\sigma(x,\pi)\|_{S^m_{\rho,\delta},a,b,c} := \sup_{\substack{|\gamma|\leq c \\ [\alpha]\leq a,\, [\beta]\leq b}} \|\pi(\mathrm{Id}+\mathcal{R})^{\frac{\rho[\alpha]-m-\delta[\beta]+\gamma}{\nu}} X_x^\beta \Delta^\alpha \sigma(x,\pi)\pi(\mathrm{Id}+\mathcal{R})^{-\frac{\gamma}{\nu}}\|_{\mathrm{op}},$$

and

$$\|\sigma\|_{S^m_{\rho,\delta},a,b,c} := \sup_{x\in G,\, \pi\in\widehat{G}} \|\sigma(x,\pi)\|_{S^m_{\rho,\delta},a,b,c}.$$

It is a routine exercise to check that for any $m \in \mathbb{R}$, $\rho, \delta \geq 0$, the functions $\|\cdot\|_{S^m_{\rho,\delta},a,b,c}$, $a, b, c \in [0, \infty)$, are semi-norms over the vector space $S^m_{\rho,\delta}$. Furthermore, with Comment 3 above, taking a, b, c as non-negative integers, they endow $S^m_{\rho,\delta}$ of a structure of Fréchet space. The class of smoothing symbols is then equipped with the topology of projective limit.

The pseudo-differential operators of order $m \in \mathbb{R} \cup \{-\infty\}$ and type (ρ, δ) are obtained by quantization from the symbols of the same order and type, that is,

$$\Psi^m_{\rho,\delta} := \mathrm{Op}(S^m_{\rho,\delta}),$$

with the quantization given by (3.2). They inherit a structure of topological vector space from the classes of symbols,

$$\|\mathrm{Op}(\sigma)\|_{\Psi^m_{\rho,\delta},a,b,c} := \|\sigma\|_{S^m_{\rho,\delta},a,b,c}.$$

It is not difficult to see from the computations in Subsection 3.2 that any operator $\mathrm{Op}(\sigma)$ is a continuous operator $\mathcal{D}(G) \to \mathcal{D}'(G)$; in fact, we can show that T is continuous $\mathcal{S}(G) \to \mathcal{S}(G)$ but the complete proof which uses Theorem 4.4 and Proposition 3.4 can be found in [8].

The type $(1, 0)$ is thought of as the basic class of symbols and the types (ρ, δ) as their generalisations, the limitation on the parameters (ρ, δ) coming from reasons similar to the ones in the Euclidean settings. For type $(1, 0)$, we set $S^m := S^m_{1,0}$, $\Psi^m := \Psi^m_{1,0}$ and,

$$\|\sigma(x,\pi)\|_{S^m_{1,0},a,b,c} = \|\sigma(x,\pi)\|_{a,b,c}, \quad \|\sigma\|_{S^m_{1,0},a,b,c} = \|\sigma\|_{a,b,c}, \text{ etc}\ldots$$

Before proving that $\cup_{m\in\mathbb{R}} S^m_{\rho,\delta}$ and $\cup_{m\in\mathbb{R}} \Psi^m_{\rho,\delta}$ are stable by composition, let us give some examples.

3.5. First examples

As it should be, $\cup_{m\in\mathbb{R}} \Psi^m$ contains the calculus of left invariant differential operators. More precisely the following lemma implies that $\sum_{[\beta]\leq m} c_\beta X^\beta \in \Psi^m$. The coefficients c_α here are constant and it is easy to relax this condition with each function c_α being smooth and bounded as well as all its derivatives.

Lemma 3.2. *For any $\beta_o \in \mathbb{N}_0^n$, the operator $X^{\beta_o} = \mathrm{Op}(\pi(X)^{\beta_o})$ is in $\Psi^{[\beta_o]}$.*

Proof. For any $\alpha \in \mathbb{N}_0^n$, in the sense of distributions,

$$\int_G f(g)(\tilde{q}_\alpha(-1)^{|\beta_o|} X^{\beta_o}\delta_0)(g)dg = \int_G X^{\beta_o}\{\tilde{q}_\alpha(g)f(g)\}\delta_0(g)dg,$$

is always zero if $[\alpha] < [\beta_o]$ or $[\alpha] = [\beta_o]$ with $\alpha \neq \beta_o$. If $[\alpha] > [\beta_o]$ or $[\alpha] = [\beta_o]$ with $\alpha = \beta_o$, then it is equal to $X^{\beta_o-\alpha} f$ up to some constant $c_{\alpha,\beta_o} \in \mathbb{R}$. Moreover, in the latter case, we get

$$\|f*(\tilde{q}_\alpha(-1)^{|\beta_o|}X^{\beta_0}\delta_0)\|_{L^2_{[\alpha]-[\beta_o]+\gamma}} = |c_{\alpha,\beta_0}|\|X^{\beta_0-\alpha}f\|_{L^2_{[\alpha]-[\beta_o]+\gamma}(G)} \leq C_{\alpha,\beta_o}\|f\|_{L^2_\gamma(G)}.$$

This shows $\|\pi(\text{Id}+\mathcal{R})^{\frac{[\alpha]-[\beta_o]+\gamma}{\nu}}\Delta^\alpha\pi(X)^{\beta_o}\pi(\text{Id}+\mathcal{R})^{-\frac{\gamma}{\nu}}\|_{\text{op}} \leq C_{\alpha,\beta_o}$. □

An example of smoothing operator is given by convolution with a Schwartz function:

Lemma 3.3. *If $\kappa \in \mathcal{S}(G)$ then $T_\kappa \in \Psi^{-\infty}$. Furthermore, the mapping $\mathcal{S}(G) \ni \kappa \mapsto T_\kappa \in \Psi^{-\infty}$ is continuous.*

Proof. For any $\kappa \in \mathcal{S}(G)$ and $a \geq 0$, we have $(\text{Id}+\mathcal{R})^a\kappa \in L^1(G)$. Indeed, it is true if $a \in \mathbb{N}_0$; if $a \notin \mathbb{N}_0$, then writing

$$(\text{Id}+\mathcal{R})^a\kappa = \left\{(\text{Id}+\mathcal{R})^{\lceil a\rceil}\kappa\right\} * \mathcal{B}_{a-\lceil a\rceil},$$

we get

$$\|(\text{Id}+\mathcal{R})^a\kappa\|_{L^1(G)} \leq \|(\text{Id}+\mathcal{R})^{\lceil a\rceil}\kappa\|_{L^1(G)}\|\mathcal{B}_{a-\lceil a\rceil}\|_{L^1(G)}.$$

We have also the same property for $\tilde{\mathcal{R}}$ by adapting the proof above.

Let $m \in \mathbb{R}$. For any $\gamma \in \mathbb{R}$ and $\alpha \in \mathbb{N}_0^n$ such that γ and $[\alpha]-m+\gamma$ are of the same sign, we have

$$\begin{aligned}&\sup_{\pi\in\widehat{G}} \|\pi(\text{Id}+\mathcal{R})^{\frac{[\alpha]-m+\gamma}{\nu}}\Delta^\alpha\pi(\kappa)\pi(\text{Id}+\mathcal{R})^{-\frac{\gamma}{\nu}}\|_{\text{op}}\\ &\leq \begin{cases} \|\mathcal{B}_\gamma\|_{L^1(G)}\|(\text{Id}+\mathcal{R})^{\frac{[\alpha]-m+\gamma}{\nu}}\tilde{q}_\alpha\kappa\|_{L^1(G)} & \text{if } \gamma, [\alpha]-m+\gamma \geq 0,\\ \|(\text{Id}+\tilde{\mathcal{R}})^{-\frac{\gamma}{\nu}}\tilde{q}_\alpha\kappa\|_{L^1(G)}\|\mathcal{B}_{-([\alpha]-m+\gamma)}\|_{L^1(G)} & \text{if } \gamma, [\alpha]-m+\gamma \leq 0.\end{cases}\end{aligned}$$

It is now clear that $T_\kappa \in \Psi^m$ and that any semi-norm $\|T\|_{\Psi^m,a,b,c}$ is controlled by some Schwartz semi-norm of κ. □

By Lemma 3.3 and Remark 2.5, if $\phi \in \mathcal{S}(\mathbb{R})$ then $\phi(\mathcal{R}) \in \Psi^{-\infty}$. This last consequence could also be obtained via the next example.

The $\mathcal{R}$-multipliers in the following class of functions yields operators in the calculus. We consider the space $\mathcal{M}_m$ of smooth functions ϕ on $[0,\infty)$ satisfying for every $k \in \mathbb{N}_0$:

$$\|\phi\|_{\mathcal{M}_m,k} := \sup_{\lambda\geq 0,\, k_1\leq k} \left|(1+\lambda)^{-m+k_1}\partial_\lambda^{k_1}\phi(\lambda)\right| < \infty.$$

An important example is $\phi(\lambda) = (1+\lambda)^m$, $m \in \mathbb{R}$.

Proposition 3.4. *Let $m \in \mathbb{R}$ and $\phi \in \mathcal{M}_{\frac{m}{\nu}}$. Then $\phi(\mathcal{R})$ is in Ψ^m and its symbol satisfies*

$$\forall a,b,c \in \mathbb{N} \qquad \exists k \in \mathbb{N},\ C>0 : \qquad \|\phi(\pi(\mathcal{R}))\|_{a,b,c} \leq C\|\phi\|_{\mathcal{M}_{\frac{m}{\nu}},k},$$

with k and C independent of ϕ.

The proof of Proposition 3.4 can be found in [8]. It is based on Proposition 2.4 and the Cotlar–Stein Lemma.

4. Some properties of symbols, kernels and operators

In this section, we give more explicitly the properties (R1), (R2) and (R4) given in the introduction.

4.1. First properties of the symbols

The following properties of the symbol $\sigma \in S^m_{\rho,\delta}$ of an operator with kernel κ_x are not difficult to obtain.

1. If $\beta_o \in \mathbb{N}_0^n$ then the symbol $X_x^{\beta_o}\sigma(x,\pi)$ is in $S^{m+\delta[\beta_o]}_{\rho,\delta}$ with kernel $X_x^{\beta_o}\kappa_x$ and,
$$\|X_x^{\beta_o}\sigma(x,\pi)\|_{S^m_{\rho,\delta},a,b,c} \le C_{b,\beta_o}\|\sigma(x,\pi)\|_{S^m_{\rho,\delta},a,b+[\beta_o],c}.$$
2. If $\alpha_o \in \mathbb{N}_0^n$ then the symbol $\Delta^{\alpha_o}\sigma(x,\pi)$ is in $S^{m-\rho[\alpha_o]}_{\rho,\delta}$ with kernel $\tilde{q}_{\alpha_o}\kappa_x$ and,
$$\|\Delta^{\alpha_o}\sigma(x,\pi)\|_{S^m_{\rho,\delta},a,b,c} \le C_{b,\beta_o}\|\sigma(x,\pi)\|_{S^m_{\rho,\delta},a+[\alpha_o],b,c}.$$
3. The symbol $\sigma(x,\pi)^*$ is in $S^m_{\rho,\delta}$ with kernel $\kappa_x^* : y \mapsto \bar{\kappa}_x(y^{-1})$ and,
$$\|\sigma(x,\pi)^*\|_{S^m_{\rho,\delta},a,b,c} = \sup_{\substack{|\gamma|\le c\\ [\alpha]\le a,\ [\beta]\le b}} \|\pi(\mathrm{Id}+\mathcal{R})^{-\frac{\gamma}{\nu}} X_x^\beta\Delta^\alpha\sigma(x,\pi)\pi(\mathrm{Id}+\mathcal{R})^{\frac{\rho[\alpha]-m-\delta[\beta]+\gamma}{\nu}}\|_{\mathrm{op}}\,.$$
4. Let $\sigma_1 \in S^{m_1}_{\rho,\delta}$ and $\sigma_2 \in S^{m_2}_{\rho,\delta}$ with respective kernels κ_{1x} and κ_{2x}. Then $\sigma(x,\pi) := \sigma_1(x,\pi)\sigma_2(x,\pi)$ defines the symbol σ in $S^m_{\rho,\delta}$, $m = m_1+m_2$, with kernel $\kappa_{2x} * \kappa_{1x}$; furthermore
$$\|\sigma(x,\pi)\|_{S^m_{\rho,\delta},a,b,c} \le C\|\sigma_1(x,\pi)\|_{S^{m_1}_{\rho,\delta},a,b,c+\rho a+|m_2|+\delta b}\|\sigma_2(x,\pi)\|_{S^{m_2}_{\rho,\delta},a,b,c}.$$
where the constant $C = C_{a,b,c} > 0$ does not depend on σ.
Indeed from the Leibniz rule for Δ^α and X^β, the operator
$$\pi(\mathrm{Id}+\mathcal{R})^{\frac{[\alpha]-m+\gamma}{\nu}} X_x^\beta\Delta^\alpha\sigma(x,\pi)\pi(\mathrm{Id}+\mathcal{R})^{-\frac{\gamma}{\nu}},$$
is a linear combination over $\beta_1,\beta_2,\alpha_1,\alpha_2 \in \mathbb{N}_0^n$ satisfying $[\beta_1]+[\beta_2]=[\beta]$, $[\alpha_1]+[\alpha_2]=[\alpha]$, of terms
$$\pi(\mathrm{Id}+\mathcal{R})^{\frac{\rho[\alpha]-m-\delta[\beta]+\gamma}{\nu}} X_x^{\beta_1}\Delta^{\alpha_1}\sigma_1(x,\pi)X_x^{\beta_2}\Delta^{\alpha_2}\sigma_2(x,\pi)\pi(\mathrm{Id}+\mathcal{R})^{-\frac{\gamma}{\nu}},$$
whose operator norm is less than
$$\|\pi(\mathrm{Id}+\mathcal{R})^{\frac{\rho[\alpha]-m-\delta[\beta]+\gamma}{\nu}} X_x^{\beta_1}\Delta^{\alpha_1}\sigma_1(x,\pi)\pi(\mathrm{Id}+\mathcal{R})^{-\frac{\rho[\alpha_2]-m_2-\delta[\beta_2]+\gamma}{\nu}}\|_{\mathrm{op}}$$
$$\|\pi(\mathrm{Id}+\mathcal{R})^{\frac{\rho[\alpha_2]-m_2-\delta[\beta_2]+\gamma}{\nu}} X_x^{\beta_2}\Delta^{\alpha_2}\sigma_2(x,\pi)\pi(\mathrm{Id}+\mathcal{R})^{-\frac{\gamma}{\nu}}\|_{\mathrm{op}}.$$
Consequently, the collection of symbols $\cup_{m\in\mathbb{R}}S^m_{\rho,\delta}$ forms an algebra.
5. Using the previous point and the left calculus (see Lemma 3.2), if $\sigma \in S^m_{\rho,\delta}$ with kernel κ_x, then $\pi(X)^\beta\sigma\,\pi(X)^{\tilde{\beta}}$ is in $S^{m+[\beta]+[\tilde{\beta}]}$ with kernel $X_y^\beta\tilde{X}_y^{\tilde{\beta}}\kappa_x(y)$.

4.2. First properties of the kernels

As expected from pseudo-differential calculi on manifolds such as homogeneous Lie groups, the kernels of the operators of order 0 are of Calderon–Zygmund type in the sense of Coifman–Weiss [4, Ch. III]. This claim is a consequence of the following proposition together with the properties of the symbols.

Proposition 4.1. *Assume $\rho \in (0,1]$ and let us fix a homogeneous norm $|\cdot|$ on G. Let $\sigma \in S^m_{\rho,\delta}$ and κ_x the associated kernel.*

Then for each $x \in G$, the distribution κ_x coincides with a smooth function in $G\backslash\{0\}$. Furthermore, $(x,y) \mapsto \kappa(x,y)$ is a smooth function on $G \times (G\backslash\{0\})$, and we have:

1. *There exist $C > 0$ and $a,b,c \in \mathbb{N}_0$ such that for any $y \in G$ with $|y| < 1$,*
$$|\kappa_x(y)| \leq C \sup_{\pi\in\widehat{G}} \|\sigma(x,\pi)\|_{S^m_{\rho,\delta},a,b,c} \begin{cases} |y|^{-\frac{Q+m}{\rho}} & \text{if } Q+m>0 \\ 1 & \text{if } Q+m<0 \\ \ln|y| & \text{if } Q+m=0 \end{cases}.$$
2. *For any $M \in \mathbb{N}_0$, there exist $C > 0$ and $a,b,c \in \mathbb{N}_0$ such that for any $y \in G$ with $|y| \geq 1$,*
$$|\kappa_x(y)| \leq C \sup_{\pi\in\widehat{G}} \|\sigma(x,\pi)\|_{S^m_{\rho,\delta},a,b,c}|y|^{-M}.$$

For example our operators of order 0 have singularities of the type $|y|^{-Q}$ with Q homogeneous dimension strictly greater than the topological dimension. Hence the calculus developed here can not coincide with the Hörmander calculus on $\mathbb{R}^n$ (abelian). This contrasts with the compact case: it was shown in [21] that the calculus developed in [19] on compact Lie groups leads to the usual Hörmander operator classes on $\mathbb{R}^n$ extended to compact connected manifolds.

Before discussing the proof of Proposition 4.1, let us prove the following couple of easy lemmata.

Lemma 4.2. *If $\sigma \in S^m_{\rho,\delta}$ and $a \in \mathbb{R}$ with $m+a < -Q/2$, then the distribution $(\text{Id}+\mathcal{R})^{\frac{a}{\nu}}\kappa_x$ coincides with a square integrable function for each fixed x; furthermore, there exists a constant $C = C_{a,m} > 0$, independent of σ, such that we have*
$$\forall x \in G \qquad \|(\text{Id}+\mathcal{R})^{\frac{a}{\nu}}\kappa_x\|_{L^2(G)} \leq C \sup_{\pi\in\widehat{G}} \|\sigma(x,\pi)\|_{S^m_{\rho,\delta},0,0,0}.$$

Proof of Lemma 4.2. By the Plancherel formula and properties of Hilbert–Schmidt operators,
$$\begin{aligned}\|(\text{Id}+\mathcal{R})^{\frac{a}{\nu}}\kappa_x\|^2_{L^2(G)} &= \int_{\widehat{G}} \|(\text{Id}+\mathcal{R})^{\frac{a}{\nu}}\sigma(x,\pi)\|^2_{HS} d\mu(\pi) \\ &\leq \sup_{\pi\in\widehat{G}} \|\pi(\text{Id}+\mathcal{R})^{\frac{-m}{\nu}}\sigma(x,\pi)\|^2_{\text{op}} \int_{\widehat{G}} \|\pi(\text{Id}+\mathcal{R})^{\frac{m+a}{\nu}}\|^2_{HS} d\mu(\pi) \\ &\leq \sup_{\pi\in\widehat{G}} \|\sigma(x,\pi)\|^2_{S^m_{\rho,\delta},0,0,0} \; \|\mathcal{B}_{-(m+a)}\|^2_{L^2(G)},\end{aligned}$$
which is finite by the properties of Bessel potentials (see Subsection 2.3). □

Lemma 4.3. *If $\sigma \in S^m_{\rho,\delta}$ with $m < -Q$, then the associated kernel $\kappa_x(y)$ coincides with a continuous function in y for each $x \in G$. Furthermore, there exists a constant $C = C_{a,m} > 0$, independent of σ, such that we have*

$$\forall x, y \in G \qquad |\kappa_x(y)| \leq C \sup_{\pi \in \widehat{G}} \|\sigma(x,\pi)\|_{S^m_{\rho,\delta},0,0,0},$$

Proof. By Lemma 2.8,

$$\|\kappa_x\|_{L^\infty(G)} \leq C_a \|(\mathrm{Id} + \mathcal{R})^{\frac{a}{\nu}} \kappa_x\|_{L^2(G)} \quad \text{where} \quad a = \frac{Q+m}{2} < -\frac{Q}{2},$$

and we conclude by Lemma 4.2. □

Consequently, for any $\sigma \in S^m_{\rho,\delta}$ with kernel κ_x, we can apply Lemma 4.3 to the symbol

$$X_x^{\beta_o} \Delta^{\alpha_o} \pi(X)^\beta \sigma \pi(X)^{\tilde{\beta}} \in S^{m+[\beta]+[\tilde{\beta}]+\delta[\beta_o]-\rho[\alpha_o]}_{\rho,\delta}.$$

The kernel is given by $\tilde{X}^{\tilde{\beta}} X^\beta X_x^{\beta_o} \tilde{q}_{\alpha_o} \kappa_x$ (see Subsection 4.1) and, if $m + [\beta] + [\tilde{\beta}] + \delta[\beta_o] - \rho[\alpha_o] < -Q$, it coincides with a continuous and bounded function,

$$\forall x, y \in G \quad |\tilde{X}_y^{\tilde{\beta}} X_y^\beta X_x^{\beta_o} \tilde{q}_{\alpha_o} \kappa_x(y)| \leq C \sup_{\pi \in \widehat{G}} \|\sigma(x,\pi)\|_{S^m_{\rho,\delta},a,b,c}$$

with $a = [\alpha_o]$, $b = [\beta_o]$, $c = \rho[\alpha_o] + \delta[\beta_o] + [\tilde{\beta}]$.

Hence if $\rho > 0$ then the kernel of a symbol in $S^m_{\rho,\delta}$ is smooth on $G \times (G \backslash \{0\})$.

Let $|\cdot|_{\nu_o}$ be the homogeneous norm defined in (2.1) for ν_o the smallest common multiple of the υ_j. Clearly, for any $p \in \mathbb{N}$, $|\cdot|_{\nu_o}^{2\nu_o p}$ is a homogeneous polynomial of degree $p2\nu_o$ and thus can be written as a linear combination of q_α with $[\alpha] = 2\nu_o p$ (see Lemma 2.3). Thus

$$\begin{aligned}
|y|_{\nu_o}^{2\nu_o p} |\kappa_x(y)| &= |y^{-1}|_{\nu_o}^{2\nu_o p} |\kappa_x(y)| \\
&\leq C \sup_{[\alpha]=2\nu_o p} \sup_{z \in G} |q_\alpha(z)|_{\nu_o}^{2\nu_o} |\kappa_x(g)| \\
&\leq C \sup_{\pi \in \widehat{G}} \|\sigma(x,\pi)\|_{S^m_{\rho,\delta},a,b,c}
\end{aligned} \tag{4.1}$$

as long as $m - \rho 2\nu_o p > Q$. Here $C = C_{m,p}$ is independent of σ.

The estimate (4.1) proves the second point in Proposition 4.1 and is a weaker version of the first. We will not show this first point because of its argument's length. The complete proof can be found in [8].

4.3. A pseudo-differential operator as a limit of 'nice' operators

The definition of symbols presented above leads to kernels κ_x in the distributional sense and it is often needed to assume that the kernels are 'nice' functions. In this subsection we explain how we proceed to do so.

We fix a non-negative function $\chi_o \in \mathcal{D}(\mathbb{R})$ supported in $[1/4, 4]$ such that $\chi_o \equiv 1$ on $[1/2, 2]$. For any $\epsilon > 0$, we write $\chi_\epsilon(x) = \chi_o(\epsilon |x|_{\nu_o}^{2\nu_o})$ where $|\cdot|_{\nu_o}$ is the homogeneous norm defined in (2.1) for ν_o the smallest common multiple of the υ_j. Cleary $\chi_\epsilon \in \mathcal{D}(G)$.

We denote by $|\pi|$ a 'norm' on $\widehat{G}$, for example the distance between the co-adjoint orbits of π and 1.

By definition, the $\mathcal{H}_\pi$'s, $\pi \in \widehat{G}$, form a field of Hilbert spaces for the Plancherel measure. So we can choose a generating sequence of vectors on $\mathcal{H}_\pi$ depending measurably on π. We denote by proj_ϵ the orthogonal projection on the $\lceil \epsilon^{-1} \rceil$th first vectors of this sequence.

Let $\sigma \in S^m_{\rho,\delta}$. We consider for any $\epsilon \in (0,1)$, the operator

$$\sigma_\epsilon(x,\pi) := \chi_\epsilon(x) 1_{|\pi|\leq\epsilon} \sigma(x,\pi) \text{proj}_\epsilon.$$

Clearly $\sigma_\epsilon \in S^m_{\rho,\delta}$ and for any $a,b,c \in \mathbb{N}_0$ there exists $C = C_{m,a,b,c} > 0$ such that,

$$\|\sigma_\epsilon\|_{S^m_{\rho,\delta},a,b,c} \leq C \|\sigma\|_{S^m_{\rho,\delta},a,b,c}.$$

The corresponding kernel is

$$\kappa_\epsilon(x,y) = \chi_\epsilon(x) \int_{|\pi|\leq\epsilon} \text{Tr}\left(\sigma(x,\pi)\text{proj}_\epsilon\right) d\mu(\pi),$$

which is smooth in x and y and compactly supported in x. From Proposition 4.1, $\kappa_{\epsilon,x}$ decays rapidly at infinity in y uniformly in x. Furthermore, point-wise for $x \in G$ and $y \in G\backslash\{0\}$, or in the sense of $\mathcal{S}'(G)$-distribution for each $x \in G$, we have the convergence $\kappa_{\epsilon,x}(y) \underset{\epsilon\to 0}{\longrightarrow} \kappa_x(y)$.

Let $T_\epsilon = \text{Op}(\sigma_\epsilon)$ be the corresponding operators. For any $f \in \mathcal{S}(G)$, $T_\epsilon f \in \mathcal{D}(G)$ and

$$Tf(x) = \lim_{\epsilon\to 0} T_\epsilon f(x) \quad \text{where} \quad T = \text{Op}(\sigma).$$

In the proofs of the rest of the paper, we will assume that the kernels of the operators are sufficiently regular and compactly supported in y so that all the performed operations, e.g., composition of operators, convolution of kernels, group Fourier transform of kernels etc... make sense. This can be made rigorous via the procedure described above since we will always obtain controls in $S^m_{\rho,\delta}$-semi-norms.

4.4. Composition

We want to prove

Theorem 4.4. *Let $1 \geq \rho \geq \delta \geq 0$ with $\rho \neq 0$ and $\delta \neq 1$. If $T_1 \in \Psi^{m_1}_{\rho,\delta}$ and $T_2 \in \Psi^{m_2}_{\rho,\delta}$ then $T_1 T_2 \in \Psi^m_{\rho,\delta}$ with $m = m_1 + m_2$.*

Let us start with some formal considerations. Denoting σ_j and κ_j the symbol and kernel of T_j for $j = 1,2$, it is not difficult to compute the following expression for the composition $T = T_1 T_2$,

$$Tf(x) = \int_G \int_G f(z) \kappa_2(y, z^{-1}y) \kappa_1(x, y^{-1}x) dy dz.$$

Thus the kernel of T is

$$\kappa_x(w) = \int_G \kappa_2(xz^{-1}, wz^{-1}) \kappa_1(x,z) dz.$$

Using the Taylor expansion for κ_2 in its first variable, we have

$$\kappa_2(xz^{-1},\cdot) \approx \sum_\alpha \tilde{q}_\alpha(z) X_x^\alpha \kappa_{2x}(\cdot) \quad \text{thus} \quad \kappa_x(w) \approx \sum_\alpha X_x^\alpha \kappa_{2x} * \tilde{q}_\alpha \kappa_1(w).$$

Denoting σ the group Fourier transform of κ, we have

$$\sigma(x,\pi) := \pi(\kappa_x) \approx \sum_\alpha \Delta^\alpha \sigma_1(x,\pi)\ X_x^\alpha \sigma_2(x,\pi).$$

From Subsection 4.1, we know that

$$\sum_{[\alpha]\leq M} \Delta^\alpha \sigma_1(x,\pi)\ X_x^\alpha \sigma_2(x,\pi) \in S_{\rho,\delta}^{m-(\rho-\delta)M}.$$

Hence the main problem is to control the remainder coming from the use of the Taylor expansion; this is the object of the following lemma.

Lemma 4.5. *We keep the notation defined just above and set*

$$\tau_M(x,\pi) := \sigma(x,\pi) - \sum_{[\alpha]\leq M} \Delta^\alpha \sigma_1(x,\pi)\ X_x^\alpha \sigma_2(x,\pi).$$

Let $\beta,\tilde{\beta},\beta_o,\alpha_o \in \mathbb{N}_0^n$. Then there exists $M_o \in \mathbb{N}_0$ such that for any integer $M > M_o$, there exist $C>0$ and computable integers a_1,b_1,c_1,a_2,b_2,c_2 (independent of σ_1 and σ_2) such that we have

$$\|\pi(X)^\beta \left\{X_x^{\beta_o}\Delta^{\alpha_o}\tau_M\right\} \pi(X)^{\tilde{\beta}}\|_{\text{op}} \leq C\|\sigma_1\|_{S_{\rho,\delta}^{m_1},a_1,b_1,c_1}\|\sigma_2\|_{S_{\rho,\delta}^{m_2},a_2,b_2,c_2}.$$

Proof of Lemma 4.5. Proceeding as in the proof of (3.3), we have:

$$\tilde{q}_{\alpha_o}(y)\left(\kappa_x - \sum_{[\alpha]\leq M} (\tilde{q}_\alpha \kappa_{1,x}) * (X_x^\alpha \kappa_{2,x})\right)$$
$$= \overline{\overline{\sum_{[\alpha_{o1}]+[\alpha_{o2}]=[\alpha_o]}}} \int_G \tilde{q}_{\alpha_{o1}}(z)\kappa_{1,x}(z)\ R_{0,M}^{(\tilde{q}_{\alpha_{o2}}\kappa_{2,x\cdot})(yz^{-1})}(z^{-1})\,dz.$$

where the sign $\overline{\overline{\sum}}$ means linear combination, here over $[\alpha_{o1}]+[\alpha_{o2}]=[\alpha_o]$,

Consequently, $\pi(X)^{\beta_1}X_x^{\beta_o}\Delta^{\alpha_o}\tau_M(x,\pi)\pi(X)^{\beta_2}$ is the group Fourier transform of the function of y given by

$$\tilde{X}_y^{\beta_2}X_y^{\beta_1}X_x^{\beta_o} \overline{\overline{\sum_{[\alpha_{o1}]+[\alpha_{o2}]=[\alpha_o]}}} \int_G \tilde{q}_{\alpha_{o1}}(z)\kappa_{1,x}(z)\ R_{0,M}^{(\tilde{q}_{\alpha_{o2}}\kappa_{2,x\cdot})(yz^{-1})}(z^{-1})\,dz$$
$$= \overline{\overline{\sum_{\substack{[\alpha_{o1}]+[\alpha_{o2}]=[\alpha_o]\\ [\beta_{o1}]+[\beta_{o2}]=[\beta_o]}}}} \int_G \tilde{q}_{\alpha_{o1}}(z)X_x^{\beta_{o2}}\kappa_{1,x}(z)\times \tag{4.2}$$
$$\times R_{0,M}^{\tilde{X}_y^{\beta_2}X_y^{\beta_1}(\tilde{q}_{\alpha_{o2}}X_x^{\beta_{o1}}\kappa_{2,x\cdot})(yz^{-1})}(z^{-1})\,dz.$$

After some manipulations involving integration by parts and Leibniz rules, we obtain that $\pi(X)^{\beta_1} X_x^{\beta_o} \Delta^{\alpha_o} \tau_M(x,\pi)\pi(X)^{\beta_2}$ is a linear combination over $[\alpha_{o1}]+[\alpha_{o2}]=[\alpha_o]$, $[\beta_{o1}]+[\beta_{o2}]=[\beta_o]$ and $[\beta_{11}]+[\beta_{12}]=[\beta_1]$ of

$$\int_G X_{z_2=z}^{\beta_{12}} \tilde{q}_{\alpha_{o1}}(z_2) X_x^{\beta_{o2}} \kappa_{1,x}(z_2)\pi(z)^* R_{x'=0,M-[\beta_{11}]}^{((X_x^{\beta_{o1}} X_{x'}^{\beta_{11}} \Delta^{\alpha_{o2}} \sigma_2(xx',\pi)\pi(X)^{\beta_2})}(z^{-1})dz, \tag{4.3}$$

where we have extended the notation of the Taylor remainder to accept vector-valued functions. The adapted statement of Taylor's estimates given in Proposition 2.1 remains valid.

We consider each integral (4.3) and insert $\pi(\text{Id}+\mathcal{R})^{\frac{e_o}{\nu}}$ and its inverse with the exponent e_o to be determined in terms of $\beta_{1j}, \beta_{oj}, \alpha_{o,j}$, $j=1,2$. We decompose $-\frac{e_o}{\nu}=e_1+e_2$ with $e_1\in\mathbb{N}_0$. Therefore, each integral in (4.3) is equal to

$$\int_G X_{z_2=z}^{\beta_{12}} \tilde{q}_{\alpha_{o1}}(z_2) X_x^{\beta_{o2}} \kappa_{1,x}(z_2)\pi(z)^* \pi(\text{Id}+\mathcal{R})^{e_1}\ \pi(\text{Id}+\mathcal{R})^{e_2}$$
$$\left\{ R_{x'=0,M-[\beta_{11}]}^{((\pi(\text{Id}+\mathcal{R})^{\frac{e_o}{\nu}} X_x^{\beta_{o1}} X_{x'}^{\beta_{11}} \Delta^{\alpha_{o2}} \sigma_2(xx',\pi)\pi(X)^{\beta_2})} \right\}(z^{-1})dz.$$

We can always write

$$\pi(z)^*\pi(\text{Id}+\mathcal{R})^{e_1} = \left(\sum_{[\beta']=e_1\nu}^{--} X_z^{\beta'} \pi(z) \right)^*,$$

and, therefore, after integrating by parts, (4.3) is equal to

$$\sum_{[\beta_1']+[\beta_2']=[\beta']=e_1\nu}^{--} \int_G X_{z_2=z}^{\beta_1'} X_{z_2=z}^{\beta_{12}} \tilde{q}_{\alpha_{o1}}(z_2) X_x^{\beta_{o2}} \kappa_{1,x}(z_2)\pi(z)^* \pi(\text{Id}+\mathcal{R})^{e_2}$$
$$X_{z_1=z}^{\beta_2'} \left\{ R_{x'=0,M-[\beta_{11}]}^{((\pi(\text{Id}+\mathcal{R})^{\frac{e_o}{\nu}} X_x^{\beta_{o1}} X_{x'}^{\beta_{11}} \Delta^{\alpha_{o2}} \sigma_2(xx',\pi)\pi(X)^{\beta_2})} \right\}(z_1^{-1})dz.$$

We fix a pseudo-norm $|\cdot|$ on G. By Taylor's estimates and easy manipulations, we obtain:

$$\|\{X_{z_1=z}^{\beta_2'} \left\{ R_{x'=0,M-[\beta_{11}]}^{((\pi(\text{Id}+\mathcal{R})^{\frac{e_o}{\nu}} X_x^{\beta_{o1}} X_{x'}^{\beta_{11}} \Delta^{\alpha_{o2}} \sigma_2(xx',\pi)\pi(X)^{\beta_2})} \right\}(z_1^{-1})\|$$
$$\leq C \sup_{x_1\in G} \|\sigma_2(x,\pi)\|_{S^{m_2}_{\rho,\delta},[\alpha_{o2}],b,[\beta_2]} \sum_{\substack{[\gamma]>M-[\beta_{11}]-[\beta_2'] \\ |\gamma|<\lceil M-[\beta_{11}]-[\beta_2']\rceil+1}} \sum_{\substack{[\beta_{o1}']\geq[\beta_{o1}] \\ |\beta_{o1}'|\leq|\beta_{o1}|}} |z|^{[\gamma]+[\beta_{o1}']-[\beta_{o1}]},$$

for $b:=\max_{|\beta_{o1}'|\leq|\beta_o|}[\beta_{o1}']+\max_{|\gamma|<M+1}[\gamma]+e_1\nu+[\beta_1]$. The last inequality is valid if we have chosen e_o, e_1, e_2 satisfying for all β_{o1}' and β_2' as in the sums above, the condition

$$e_o \leq \rho[\alpha_{o2}] - m_2 - \delta([\beta_{o1}'] + [\gamma] + [\beta_2'] + [\beta_{11}]) - [\beta_2],$$

which is implied, in particular, by

$$e_o \leq \rho[\alpha_{o2}] - m_2 - \delta([\beta_{o1}] + M) - [\beta_2].$$

It is now time to choose:

- $e_o = e_1 = e_2 = 0$ if $\rho[\alpha_{o2}] - m_2 - \delta([\beta_{o1}] + M) - [\beta_2] \geq 0$,
- but if $\rho[\alpha_{o2}] - m_2 - \delta([\beta_{o1}] + M) - [\beta_2] < 0$,

$$e_o := \rho[\alpha_{o2}] - m_2 - \delta([\beta_{o1}] + M) - [\beta_2], \quad e_1 = \left\lceil -\frac{e_o}{\nu} \right\rceil \in \mathbb{N}, \quad e_2 = -\frac{e_o}{\nu} - e_1 < 0.$$

We can now go back to (4.3) and obtain that its operator norm is

$$\leq C \sup_{x\in G} \|\sigma_2(x,\pi)\|_{S^{m}_{\rho,\delta},[\alpha_{o2}],b,[\beta_2]} \sum_{[\beta_1']+[\beta_2']=e_1\nu} \int_G |X^{\beta_{12}}(\tilde{q}_{\alpha_{o1}} X_x^{\beta_{o2}} \kappa_{1,x})(z)|$$
$$\sum_{\substack{[\gamma]>M-[\beta_{11}]-[\beta_2'] \\ |\gamma|<\lceil M-[\beta_{11}]-[\beta_2']\rceil+1}} \sum_{\substack{[\beta_{o1}']\geq[\beta_{o1}] \\ |\beta_{o1}'|\leq|\beta_{o1}|}} |z|^{[\gamma]+[\beta_{o1}']-[\beta_{o1}]} dz.$$

Since $X^{\beta_{12}}(\tilde{q}_{\alpha_{o1}} X_x^{\beta_{o2}} \kappa_{1,x})$ behaves like a Schwartz function away from the origin and like $\leq C|z|^{-p_o}$ with p_o depending on $m_1, \beta_{12}, \alpha_{o1}, \beta_{o2}$ near the origin (see Proposition 4.1 and Section 4.1), we see that all these integrals converge when $-p_o + [\gamma] + [\beta_{o1}'] - [\beta_{o1}] > -Q$ for all indices as above, and for this to be true it suffices that $-p_o + M - [\beta_{11}] + e_o - \nu > -Q$. We can always find M such that this is satisfied and in this case the operator norm of (4.3) is

$$\leq C \sup_{x\in G} \|\sigma_2(x,\pi)\|_{S^{m_2}_{\rho,\delta},[\alpha_{o2}],b,[\beta_2]} \|\sigma_1\|_{S^{m_1}{\rho,\delta,a_1,b_1,c_1}}$$

for some (computable) integers a_1, b_1, c_1.

We choose M the smallest integer which is a linear combination over $\mathbb{N}_0$ of the weights $v_1, \ldots, v_n$ such that all the operator norm of (4.3) over $[\alpha_{o1}]+[\alpha_{o2}] = [\alpha_o]$, $[\beta_{o1}] + [\beta_{o2}] = [\beta_o]$, $[\beta_{11}] + [\beta_{12}] = [\beta_1]$ are finite as above. This proves that the operator norm of $\pi(X)^{\beta_1} X_x^{\beta_o} \Delta^{\alpha_o} \tau_M(x,\pi)\pi(X)^{\beta_2}$ is estimated as stated. □

Hence Lemma 4.5 yields the following more precise version of Theorem 4.4.

Corollary 4.6. *Under the hypotheses of Theorem 4.4, writing* $T_1 = \text{Op}(\sigma_1)$ *and* $T_2 = \text{Op}(\sigma_2)$, *there exists a unique symbol* $\sigma \in S^m_{\rho,\delta}$ *such that* $T_1T_2 = \text{Op}(\sigma)$. *Furthermore*

$$\sigma - \sum_{[\alpha]\leq M} \Delta^\alpha \sigma_1 \, X_x^\alpha \sigma_2 \in S^{m-(\rho-\delta)M}_{\rho,\delta},$$

and the mapping

$$\begin{cases} S^{m_1}_{\rho,\delta} \times S^{m_1}_{\rho,\delta} & \longrightarrow & S^{m-(\rho-\delta)M}_{\rho,\delta} \\ (\sigma_1,\sigma_2) & \longmapsto & \sigma - \sum_{[\alpha]\leq M} \Delta^\alpha \sigma_1 \, X_x^\alpha \sigma_2 \end{cases},$$

is continuous.

With similar methods, we can prove that $\Psi^m_{\rho,\delta}$ is stable by taking the formal adjoint of an operator, that is, if $T \in \Psi^m_{\rho,\delta}$ then T^* defined via $(Tf_1, f_2)_{L^2} = (f_1, T^* f_2)_{L^2}$ is also in $\Psi^m_{\rho,\delta}$:

Theorem 4.7. *Let $1 \geq \rho \geq \delta \geq 0$ with $\rho \neq 0$ and $\delta \neq 1$. If $T \in \Psi^m_{\rho,\delta}$, then its formal adjoint T^* is also in $\Psi^m_{\rho,\delta}$. More precisely, writing $T = \text{Op}(\sigma)$ there exists a unique symbol $\sigma^{(*)} \in S^m_{\rho,\delta}$ such that $T^* = \text{Op}(\sigma^{(*)})$. Furthermore*

$$\sigma^{(*)} - \sum_{[\alpha]\leq M} \Delta^\alpha X_x^\alpha \sigma^* \in S^{m-(\rho-\delta)M}_{\rho,\delta},$$

and the mapping

$$\begin{cases} S^m_{\rho,\delta} & \longrightarrow & S^{m-(\rho-\delta)M}_{\rho,\delta} \\ \sigma & \longmapsto & \sigma^{(*)} - \sum_{[\alpha]\leq M} \Delta^\alpha X_x^\alpha \sigma^* \end{cases},$$

is continuous.

Indeed let us perform formal considerations analogous to the ones for the composition. Let $T = \text{Op}(\sigma) \in \Psi^m_{\rho,\delta}$ with kernel κ_x. It is not difficult to compute that the kernel of T^* is $\kappa_x^{(*)}$ given by

$$\kappa_x^{(*)}(y) = \kappa^*_{xy^{-1}}(y) = \bar{\kappa}_{xy^{-1}}(y^{-1}).$$

Using the Taylor expansion for κ_x^* in x, we obtain

$$\kappa_x^{(*)}(y) \approx \sum_\alpha \tilde{q}_\alpha(y) X_x^\alpha \kappa_x^*(y).$$

Denoting $\sigma^{(*)}$ the group Fourier transform of $\kappa^{(*)}$, we have

$$\sigma^{(*)}(x,\pi) := \pi(\kappa_x^{(*)}) \approx \sum_\alpha \Delta^\alpha X_x^\alpha \sigma(x,\pi)^*.$$

From Subsection 4.1, we know that

$$\sum_{[\alpha]\leq M} \Delta^\alpha X_x^\alpha \sigma(x,\pi)^* \in S^{m-(\rho-\delta)M}_{\rho,\delta}.$$

Hence the main problem is as above to control the remainder coming from the use of the Taylor expansion. The proof proceeds in a similar way and is left to the reader.

Finally, we note that the proof of Theorem 4.4 can be adapted to provide the treatment of the remainder also for composition of operators on compact Lie groups in [19, Theorem 10.7.8].

References

[1] Bahouri, H., Fermanian-Kammerer, C. and Gallagher, I., *Phase-space analysis and pseudodifferential calculus on the Heisenberg group*, Astérisque, **342** (2012). See also revised version of March 2013 of Arxiv:0904.4746.

[2] Benson, Ch., Jenkins, J. and Ratcliff, G., The spherical transform of a Schwartz function on the Heisenberg group, *J. Funct. Anal.*, **154** (1998), 379–423.

[3] Christ, M., Geller, D., Głowacki, P. and Polin, L., Pseudodifferential operators on groups with dilations. *Duke Math. J.*, **68** (1992), 31–65.

[4] Coifman, R. and Weiss, G., *Analyse harmonique non-commutative sur certains espaces homogènes*, Lecture Notes in Mathematics **242**, Springer-Verlag, Berlin, 1971.

[5] Dixmier, J., C^*-algebras, Translated from the French by Francis Jellett, North-Holland Mathematical Library, Vol. 15, 1977.

[6] Dynin, A.S., An algebra of pseudodifferential operators on the Heisenberg groups. Symbolic calculus, *Dokl. Akad. Nauk SSSR*, **227** (1976), 792–795.

[7] V. Fischer and M. Ruzhansky, Lower bounds for operators on graded Lie groups, C. R. Math. Acad. Sci. Paris, **351**, 2013, 1-2, pp. 13–18.

[8] Fischer, V. and Ruzhansky, M., *Quantization on nilpotent Lie groups*, in preparation.

[9] Folland, G.B. and Stein, E.M., Estimates for the $\bar{\partial}_b$ complex and analysis on the Heisenberg group, Comm. Pure Appl. Math., 27, 1974, 429–522.

[10] Folland, G.B., Subelliptic estimates and function spaces on nilpotent Lie groups, *Ark. Mat.*, **13** (1975), 161–207.

[11] Folland, G.B., Meta-Heisenberg groups, Fourier analysis (Orono, ME, 1992), Lecture Notes in Pure and Appl. Math., **157**, 121–147, Dekker, New York, 1994.

[12] Folland, G.B. and Stein, E., *Hardy spaces on homogeneous groups*, Mathematical Notes **28**, Princeton University Press, 1982.

[13] Geller, D., Fourier analysis on the Heisenberg group. I. Schwartz space, *J. Funct. Anal.*, **36**, (1980), 205–254.

[14] Helffer, B. and Nourrigat, J., Caractérisation des opérateurs hypoelliptiques homogènes invariants à gauche sur un groupe de Lie nilpotent gradué, *Comm. Partial Differential Equations*, **4** (1979), 899–958.

[15] Hulanicki, A., A functional calculus for Rockland operators on nilpotent Lie groups, *Studia Math.*, **78** (1984), 253–266.

[16] Hulanicki, A., Jenkins, J.W. and Ludwig, J., Minimum eigenvalues for positive Rockland operators, *Proc. Amer. Math. Soc.*, **94** (1985), 718–720.

[17] Ponge, R., *Heisenberg calculus and spectral theory of hypoelliptic operators on Heisenberg manifolds*, Mem. Amer. Math. Soc., **194** (906), 2008.

[18] Rothschild, L.P. and Stein, E.M., Hypoelliptic differential operators and nilpotent groups, Acta Math., 137, 1976, 3-4, 247–320.

[19] Ruzhansky, M. and Turunen, V., *Pseudo-differential operators and symmetries: Background analysis and advanced topics*, Pseudo-Differential Operators: Theory and Applications **2**, Birkhäuser Verlag, 2010.

[20] Ruzhansky, M. and Turunen, V., Global quantization of pseudo-differential operators on compact Lie groups, SU(2) and 3-sphere, *Int. Math. Res. Notices IMRN* (2012), 58 pages, doi: 10.1093/imrn/rns122.

[21] Ruzhansky, M., Turunen, V., and Wirth, J., Hörmander class of pseudo-differential operators on compact Lie groups and global hypoellipticity. ArXiv:1004.4396.

[22] Taylor, M.E., *Noncommutative microlocal analysis. I.* Mem. Amer. Math. Soc. **52** (1984). Revised version accessible at http://math.unc.edu/Faculty/met/ncmlms.pdf

[23] Taylor, M.E., *Noncommutative harmonic analysis*, American Mathematical Society, 1986.

Véronique Fischer
Università degli studi di Padova
Dip. ICEA
Via Trieste 63
I-35 121 Padova, Italy
e-mail: fischer@dmsa.unipd.it

Michael Ruzhansky
Imperial College London
Department of Mathematics
180 Queen's Gate
London, SW7 2AZ, UK
e-mail: m.ruzhansky@imperial.ac.uk

Fourier Analysis
Trends in Mathematics, 133–148

L^p Microlocal Properties for Vector Weighted Pseudodifferential Operators with Smooth Symbols

Gianluca Garello and Alessandro Morando

Abstract. The authors introduce a class of pseudodifferential operators, whose symbols satisfy completely inhomogeneous estimates at infinity for the derivatives.

Continuity properties in suitable weighted Sobolev spaces of L^p type are given and L^p microlocal properties studied.

Mathematics Subject Classification (2010). Primary 35S05; Secondary 35A17.

Keywords. Pseudodifferential operators, weight vector, weighted Sobolev spaces, microlocal properties.

1. Introduction

A vector weighted pseudodifferential operator is characterized by a smooth symbol which in general satisfies the estimates:

$$|\partial_\xi^\alpha \partial_x^\beta a(x,\xi)| \leq c_{\alpha,\beta} m(\xi)\Lambda(\xi)^{-\alpha}. \tag{1.1}$$

Here $m(\xi)$ is a suitable positive continuous weight function, which indicates the "order" of the symbol, and $\Lambda(\xi) = (\lambda_1(\xi), \dots, \lambda_n(\xi))$ is a weight vector that estimates the decay at infinity of the derivatives; see the next Definition 2.6.

The paper must be considered in the general framework given by the symbol classes $S^\lambda(\phi,\varphi)$ and $S(m,g)$, introduced respectively by R. Beals [1] and L. Hörmander [10], [11]. Particularly we follow here the approach of Rodino [13], where a generalization of the Hörmander smooth wave front set is given, and Garello [4], where the extension to the inhomogeneous microlocal analysis for weighted Sobolev singularities of L^2 type is performed.

In a previous work [7] we studied continuity and microlocal properties of quasi homogeneous L^p type, for pseudodifferential operators of zero order, whose symbol satisfies the decay estimates at infinity:

$$|\partial_\xi^\alpha \partial_x^\beta a(x,\xi)| \le c_{\alpha,\beta} \prod_{j=1}^{n} \langle \xi \rangle_M^{-\frac{\alpha_j}{m_j}}, \tag{1.2}$$

where $M = (m_1, \dots, m_n) \in \mathbb{N}^n$, $\min\limits_{1\le j\le n} m_j = 1$ and $\langle \xi \rangle_M = \sqrt{1 + \sum_{j=1}^n \xi_j^{2m_j}}$.

Considering now the classical Hörmander symbol classes $S^m_{\rho,\delta}$, see [11, Ch. 18], it is a matter of fact that the quasi-homogeneous symbols, characterized by (1.2), satisfy in natural way the condition

$$\xi^\gamma \partial_\xi^\gamma a(x,\xi) \in S^0_{\rho,0}, \quad \gamma \in \{0,1\}^n, \tag{1.3}$$

for suitable $0 < \rho < 1$. The corresponding pseudodifferential operators are L^p bounded, see Taylor [14, Ch. XI, Sect. 4]. Note that the estimate (1.3) follows essentially from the fact that, $\langle \xi \rangle_M^{\frac{1}{m_j}} \ge |\xi_j|$, for any $j = 1, \dots, n$.

Assuming that a similar condition is satisfied by the components of the weight vector in (1.1) (see Definition 2.1), in Section 2 we obtain a family of weight vectors which define, via (2.17), a class of L^p bounded pseudodifferential operators.

In the study of the microlocal properties of these operators, the main problem arises from the lack of any homogeneity of the weight vector $\Lambda(\xi)$; this does not allow us to use in a suitable way conic neighborhoods in $\mathbb{R}^n_\xi$, as done in the classical definition of the Hörmander wave front set, see [11] and the quasi-homogeneous generalization given in [7]. Following now the approach in [13], [4] and [8], suitable neighborhoods of sets $X \subset \mathbb{R}^n_\xi$ are introduced in Section 4; they allow us to derive in Section 5 useful microlocal properties.

Finally, in Section 6 the microlocal results are expressed in terms of m-filter of Sobolev singularities, following the approach in [4].

2. Vector weighted symbol classes

Definition 2.1. A vector-valued function $\Lambda(\xi) = (\lambda_1(\xi), \dots, \lambda_n(\xi))$, $\xi \in \mathbb{R}^n$, with positive continuous components $\lambda_j(\xi)$ for $j = 1, \dots, n$, is a weight vector if there exist positive constants C, c such that for any $j = 1, \dots, n$:

$$c\langle \xi \rangle^c \le \lambda_j(\xi) \le C\langle \xi \rangle^C; \tag{2.1}$$

$$\lambda_j(\xi) \ge c|\xi_j|; \tag{2.2}$$

$$c \le \frac{\lambda_j(\eta)}{\lambda_j(\xi)} \le C \quad \text{when} \quad \sum_{k=1}^{n} |\xi_k - \eta_k| \lambda_k(\eta)^{-1} \le c. \tag{2.3}$$

As usual we denote, for $\xi \in \mathbb{R}^n$: $\langle \xi \rangle = \sqrt{1 + |\xi|^2}$.

Definition 2.2. A positive real continuous function $m(\xi)$ is an admissible weight, associated to the weight vector $\Lambda(\xi)$, if for some positive constants N, C, c

$$m(\eta) \leq C\, m(\xi)\,(1+|\eta-\xi|)^N; \tag{2.4}$$

$$\frac{1}{C} \leq \frac{m(\eta)}{m(\xi)} \leq C \quad \text{when} \quad \sum_{k=1}^{n} |\xi_k - \eta_k| \lambda_k(\eta)^{-1} \leq c. \tag{2.5}$$

We say that a vector-valued function $\Lambda(\xi) = (\lambda_1(\xi), \ldots, \lambda_n(\xi))$ is *slowly varying* if condition (2.3) is satisfied. Analogously, a function $m = m(\xi)$ satisfying condition (2.5) is said to be *slowly varying* with respect to the weight vector Λ, while m is *temperate* when condition (2.4) holds true.

Considering respectively $\xi = 0$ and $\eta = 0$ in (2.4) it follows that $\frac{1}{C}\langle\xi\rangle^{-N} \leq m(\xi) \leq C\langle\xi\rangle^N$.

We say that two weights $m(\xi)$, $\tilde{m}(\xi)$ are equivalent, and write $m(\xi) \asymp \tilde{m}(\xi)$, if $c \leq \dfrac{m(\xi)}{\tilde{m}(\xi)} \leq C$, for some positive constants c, C. Again $\tilde{\Lambda}(\xi) = (\tilde{\lambda}_1(\xi), \ldots, \tilde{\lambda}_n(\xi))$ is equivalent to $\Lambda(\xi)$, if $\tilde{\lambda}_j(\xi) \asymp \lambda_j(\xi)$, for any $j = 1, \ldots, n$. It is trivial that $\tilde{m}(\xi)$ and $\tilde{\Lambda}(\xi)$ are respectively admissible weight and weight vector.

Moreover set $m(\xi) \approx m(\eta)$ if $c \leq \dfrac{m(\eta)}{m(\xi)} \leq C$, for some positive constants c, C.

Example

1. Consider $\langle\xi\rangle_M = \left(1 + \sum_{j=1}^{n} \xi_j^{2m_j}\right)^{1/2}$ *quasi-homogeneous polynomial*, where $M = (m_1, \ldots, m_n) \in \mathbb{N}^n$, and $\min_{1\leq j\leq n} m_j = 1$. Then $\Lambda_M(\xi) = (\langle\xi\rangle_M^{1/m_1}, \ldots, \langle\xi\rangle_M^{1/m_n})$ is a weight vector.
2. For any positive continuous function $\lambda(\xi)$ satisfying (2.1) and the slowly varying condition

$$\lambda(\eta) \approx \lambda(\xi), \text{ when} \sum_{j=1}^{n} |\eta_j - \xi_j| \left(\lambda(\eta)^{\frac{1}{\mu}} + |\eta_j|\right)^{-1} \leq c, \text{ for some } c, \mu > 0, \tag{2.6}$$

   the vector $\Lambda(\xi) := \left(\lambda(\xi)^{\frac{1}{\mu}} + |\xi_1|, \ldots, \lambda(\xi)^{\frac{1}{\mu}} + |\xi_n|\right)$ is a weight vector, see [8, Proposition 1] for the proof. In such frame emphasis is given to the *multi-quasi-homogeneous polynomials* $\lambda_{\mathcal{P}}(\xi) = \left(\sum_{\alpha\in V(\mathcal{P})} \xi^{2\alpha}\right)^{1/2}$, where $V(\mathcal{P})$ is the set of the vertices of a *complete Newton polyedron* $\mathcal{P}$ as introduced in [9], see also [2]; in this case, the value μ in the definition of $\Lambda(\xi)$ and in condition (2.6) is the *formal order* of $\mathcal{P}$.
3. Any positive constant function on $\mathbb{R}^n$ is an admissible weight associated to every weight vector $\Lambda(\xi)$.
4. For any $s \in \mathbb{R}$, the functions $\langle\xi\rangle_M^s$, $\lambda(\xi)^s$ are admissible weights for the weight vectors respectively defined in 1. and 2.

Remark 2.3. Consider a function $\lambda(\xi)$ satisfying the slowly varying condition (2.6). Since $|\xi-\eta|^\mu \le c\lambda(\eta)$ implies $\lambda(\eta) \le C\lambda(\xi) \le C\lambda(\xi)\,(1+|\xi-\eta|)^\mu$, using moreover (2.1), we obtain that $\lambda(\xi)$ satisfies the temperance condition (2.4) with constant $N = \mu$.

Proposition 2.4. *For $\Lambda(\xi) = (\lambda_1(\xi), \dots, \lambda_n(\xi))$ weight vector, the function:*

$$\pi(\xi) = \min_{1\le j\le n} \lambda_j(\xi), \quad \xi \in \mathbb{R}^n \tag{2.7}$$

is an admissible weight associated to $\Lambda(\xi)$ and it moreover satisfies (2.6).

Proof. In view of (2.2) and (2.7), the assumption $\sum_{k=1}^n |\xi_k - \eta_k|\,(\pi(\eta)+|\eta_k|)^{-1} \le c$ directly gives $\sum_{k=1}^n |\xi_k - \eta_k|\lambda_k(\eta)^{-1} \le \tilde{c}$, where $\tilde{c} > 0$ depends increasingly on c. Then for suitably small c, we obtain from the slowly varying condition (2.3) and some $C > 0$: $\frac{1}{C}\lambda_j(\xi) \le \lambda_j(\eta) \le C\lambda_j(\xi)$, for any $j = 1, \dots, n$. It then follows: $\frac{1}{C}\pi(\xi) = \frac{1}{C}\min_j \lambda_j(\xi) \le \pi(\eta) = \min_j \lambda_j(\eta) \le C\min_j \lambda_j(\xi) = C\pi(\xi)$. Thus $\pi(\xi)$ satisfies (2.6) and in the same way we can prove that it fulfils (2.5). Then by means of the previous remark we conclude the proof. □

Lemma 2.5. *If m, m' are admissible weights associated to the weight vector $\Lambda(\xi)$, then the same property is fulfilled by mm' and $1/m$.*

Proof. $m(\eta) \le Cm(\xi)\,(1+|\xi-\eta|)^N \iff 1/m(\xi) \le C1/m(\eta)\,(1+|\xi-\eta|)^N$; then interchanging ξ and η we immediately obtain that $1/m$ is temperate. The remaining part of the proof is then trivial. □

Definition 2.6. For Ω open subset of $\mathbb{R}^n$, $\Lambda(\xi)$ weight vector and $m(\xi)$ admissible weight, the symbol class $S_{m,\Lambda}(\Omega)$ is given by all the smooth functions $a(x,\xi) \in C^\infty(\Omega \times \mathbb{R}^n)$, such that, for any compact subset $K \subset \Omega$ and $\alpha, \beta \in \mathbb{Z}^n_+$, there exists $c_{\alpha,\beta,K} > 0$ such that:

$$\sup_{x\in K} |\partial_\xi^\alpha \partial_x^\beta a(x,\xi)| \le c_{\alpha,\beta,K}\, m(\xi)\Lambda(\xi)^{-\alpha}, \quad \xi \in \mathbb{R}^n \tag{2.8}$$

where, with standard vectorial notation, $\Lambda(\xi)^\gamma = \prod_{k=1}^n \lambda_k(\xi)^{\gamma_k}$.

$S_{m,\Lambda}(\Omega)$ turns out to be a Fréchet space, with respect to the family of natural semi-norms defined as the best constants $c_{\alpha,\beta,K}$ involved in the estimates (2.8).

Remark 2.7. Let $\Lambda(\xi)$ be the weight vector according to Definition 2.1 then:

1. considering now the constants C, c in (2.1) and N in (2.4), the following relation with the usual Hörmander [11] symbol classes $S^m_{\rho,\delta}(\Omega)$, $0 \le \delta < \rho \le 1$, is trivial:

$$S_{m,\Lambda}(\Omega) \subset S^N_{c,0}(\Omega)\,. \tag{2.9}$$

2. If m_1, m_2 are admissible weights such that $m_1 \leq Cm_2$, then $S_{m_1,\Lambda}(\Omega) \subset S_{m_2,\Lambda}(\Omega)$, with continuous imbedding. In particular the identity $S_{m_1,\Lambda}(\Omega) = S_{m_2,\Lambda}(\Omega)$ holds true, as long as $m_1 \asymp m_2$.
   When the admissible weight m is an arbitrary positive constant function, the symbol class $S_{m,\Lambda}(\Omega)$ will be just denoted by $S_\Lambda(\Omega)$ and $a(x,\xi) \in S_\Lambda(\Omega)$ will be called a *zero order symbol.*
3. Since for any $k \in \mathbb{Z}_+$ the admissible weight $\pi(\xi)^{-k}$ is less than $C^k\langle\xi\rangle^{-ck}$, then for m admissible weight we have

$$\bigcap_{k\in\mathbb{Z}+} S_{m\pi^{-k},\Lambda}(\Omega) \subset \bigcap_{N\in\mathbb{Z}_+} S^{-N}_{1,0}(\Omega) =: S^{-\infty}(\Omega).$$

   On the other hand $a(x,\xi) \in S^{-\infty}(\Omega)$ means that

$$\sup_{x\in K} |\partial_x^\beta \partial_\xi^\alpha a(x,\xi)| \leq c_{\alpha,\beta}\langle\xi\rangle^{\mu-|\alpha|}, \qquad \text{for any } \mu \in \mathbb{R},\ K \subset\subset \Omega. \tag{2.10}$$

   Recall now that, for suitable $N, C > 0$, $m(\xi) \geq \frac{1}{C}\langle\xi\rangle^{-N}$, $\pi(\xi) \leq C\langle\xi\rangle$ and $\lambda_j(\xi) \leq C\langle\xi\rangle^C$. Then setting, for any fixed $\alpha \in \mathbb{Z}^n_+$ and arbitrary $k \in \mathbb{Z}_+$, $\mu = -N - k - (C-1)|\alpha|$ in (2.10), we obtain $|\partial_x^\beta \partial_\xi^\alpha a(x,\xi)| \leq c_{\alpha,\beta} m(\xi)\pi(\xi)^{-k}\Lambda(\xi)^{-\alpha}$, for suitable $c_{\alpha,\beta}$, that is $a(x,\xi) \in S_{m\pi^{-k},\Lambda}(\Omega)$ for any $k \in \mathbb{Z}_+$. Then

$$\bigcap_{k\in\mathbb{Z}_+} S_{m\pi^{-k},\Lambda}(\Omega) \equiv S^{-\infty}(\Omega). \tag{2.11}$$

4. Using (2.8), (2.4), (2.1) and (2.2), it immediately follows that for any $\alpha, \gamma \in \mathbb{Z}^n_+$, $K \subset\subset \Omega$,

$$\sup_{x\in K} |\xi^\gamma \partial_\xi^{\alpha+\gamma} a(x,\xi)| \leq M_{\alpha,\gamma,K}\langle\xi\rangle^{N-c|\alpha|}$$

   with some positive constant $M_{\alpha,\gamma,K}$. Then $S_{m,\Lambda}(\Omega) \subset M^N_{c,0}(\Omega)$. Here $M^r_{\rho,0}(\Omega)$, $0 < \rho \leq 1$, are the symbol classes defined in [14] given by all the symbols $a(x,\xi) \in S^r_{\rho,0}(\Omega)$ such that for any $\gamma \in \{0,1\}^n$, $\xi^\gamma \partial_\xi^\gamma a(x,\xi) \in S^r_{\rho,0}(\Omega)$.

By means of the arguments in [12] Proposition 1.1.6 and [1] jointly with Remark 2.7 we obtain the following asymptotic expansion.

Proposition 2.8. *Given a weight vector $\Lambda(\xi) = (\lambda_1(\xi), \ldots, \lambda_n(\xi))$, let $\pi = \pi(\xi)$ be the admissible weight defined by (2.7). For any sequence of symbols $\{a_k\}_{k\in\mathbb{Z}_+}$, $a_k(x,\xi) \in S_{m\pi^{-k},\Lambda}(\Omega)$, there exists $a(x,\xi) \in S_{m,\Lambda}(\Omega)$ such that for every integer $N \geq 1$:*

$$a(x,\xi) - \sum_{k<N} a_k(x,\xi) \in S_{m\pi^{-N},\Lambda}(\Omega). \tag{2.12}$$

Moreover $a(x,\xi)$ is uniquely defined modulo symbols in $S^{-\infty}(\Omega)$.

We write

$$a(x,\xi) \sim \sum_{k=0}^{\infty} a_k(x,\xi), \tag{2.13}$$

if for every $N \geq 1$ (2.12) holds.

Proposition 2.9. *Any admissible weight $m(\xi)$ admits an equivalent smooth admissible weight $\tilde{m}(\xi) \in S_{m,\Lambda}(\mathbb{R}^n) = S_{\tilde{m},\Lambda}(\mathbb{R}^n)$.*

Proof. For fixed $\varepsilon > 0$, in the set of smooth compactly supported functions $C_0^\infty(\mathbb{R}^n)$, consider a non-negative $\varphi(\zeta)$ such that $|\zeta_j| \le \varepsilon$ for every $j = 1, \dots, n$ in $\operatorname{supp}\varphi(\zeta)$ and $\varphi(\zeta) = 1$ when $|\zeta_j| \le \frac{\varepsilon}{2}$, $j = 1, \dots, n$. Taking now the weight vector $\Lambda(\xi) = (\lambda_1(\xi), \dots, \lambda_n(\xi))$, we set:

$$\Phi(\xi, \eta) := \varphi\left(\frac{\xi_1 - \eta_1}{\lambda_1(\eta)}, \dots, \frac{\xi_n - \eta_n}{\lambda_n(\eta)}\right).$$

Notice now that in the support of $\Phi(\xi, \eta)$ one has $|\xi_j - \eta_j| \le \varepsilon\lambda_j(\eta)$, for any $j = 1, \dots, n$, and $\Phi(\xi, \eta)$ is identically equal to 1 when $|\xi_j - \eta_j| \le \frac{\varepsilon}{2}\lambda_j(\eta)$. Then, assuming $\varepsilon < \frac{c}{2n}$ and ξ, η in $\operatorname{supp}\Phi(\xi, \eta)$, (2.3) assures that, for some $C > 0$, $H = \varepsilon C$:

$$c \le \frac{\lambda_j(\eta)}{\lambda_j(\xi)} \le C \quad \text{and} \quad |\xi_j - \eta_j| \le H\lambda_j(\xi), \quad j = 1, \dots, n.$$

The same is true when $\Phi(\xi, \eta) = 1$ by changing H with $\tilde{H} = \dfrac{\varepsilon}{2}C$. Then

$$\int \Phi(\xi, \eta)\, d\eta \le \|\varphi\|_\infty \int \chi_{B(\xi)}(\xi - \eta)\, d\eta = (2H)^n \|\varphi\|_\infty \prod_{j=1}^n \lambda_j(\xi);$$

$$\int \Phi(\xi, \eta)\, d\eta \ge \int \chi_{\tilde{B}(\xi)}(\xi - \eta)\, d\eta = (2\tilde{H})^n \prod_{j=1}^n \lambda_j(\xi).$$

Here $\chi_{B(\xi)}$ and $\chi_{\tilde{B}(\xi)}$ are the characteristic functions respectively of the cube $B(\xi) = \prod_{j=1}^n [-H\lambda_j(\xi), H\lambda_j(\xi)]$ and $\tilde{B}(\xi) = \prod_{j=1}^n \left[-\tilde{H}\lambda_j(\xi), \tilde{H}\lambda_j(\xi)\right]$. It then follows that $\int \Phi(\xi, \eta)\, d\eta \asymp \prod\limits_{j=1}^n \lambda_j(\xi)$. Set now:

$$\tilde{m}(\xi) = \int m(\eta)\Phi(\xi, \eta)\Pi_{j=1}^n \lambda_j(\eta)^{-1}\, d\eta. \tag{2.14}$$

Since for $\varepsilon < \frac{c}{2n}$ and any $j = 1, \dots, n$, $|\xi_j - \eta_j| \le \varepsilon\lambda_j(\eta)$ in $\operatorname{supp}\Phi(\xi, \eta)$, it follows from (2.3) and (2.5), $m(\eta) \approx m(\xi)$ and $\lambda_j(\eta) \approx \lambda_j(\xi)$, for any $j = 1, \dots, n$, then $\tilde{m}(\xi) \asymp m(\xi)$. Moreover $\tilde{m}(\xi)$ is obviously smooth and for any $\alpha \in \mathbb{Z}_+^n$:

$$\partial^\alpha \tilde{m}(\xi) = \int m(\eta)\partial_\zeta^\alpha \varphi\left(\frac{\xi_1 - \eta_1}{\lambda_1(\eta)}, \dots, \frac{\xi_n - \eta_n}{\lambda_n(\eta)}\right) \prod_{j=1}^n \lambda_j(\eta)^{-\alpha_j - 1}\, d\eta. \tag{2.15}$$

Since $\operatorname{supp}\partial_\zeta^\alpha\varphi \subset \operatorname{supp}\varphi$, we obtain, for some positive constant M_α:

$$|\partial^\alpha \tilde{m}(\xi)| \le M_\alpha \tilde{m}(\xi)\Lambda(\xi)^{-\alpha}, \tag{2.16}$$

which concludes the proof. □

Thanks to the relations with the Hörmander symbol classes (2.9), we can define for $a(x,\xi) \in S_{m,\Lambda}(\Omega)$ the pseudodifferential operator

$$a(x,D)u := (2\pi)^{-n} \int e^{ix\cdot\xi} a(x,\xi)\hat{u}(\xi)\, d\xi\,, \quad u \in C_0^\infty(\Omega). \tag{2.17}$$

Op $S_{m,\Lambda}(\Omega)$ denotes the class of all the pseudodifferential operators with symbol in $S_{m,\Lambda}(\Omega)$.

Any symbol $a(x,\xi) \in S_{m,\Lambda}(\Omega)$ defines by means of (2.17) a bounded linear operator: $a(x,D) : C_0^\infty(\Omega) \mapsto C^\infty(\Omega)$, which extends to a linear operator from $\mathcal{E}'(\Omega)$ to $\mathcal{D}'(\Omega)$. Let now $\widetilde{\mathrm{Op}}S_{m,\Lambda}(\Omega)$ be the class of *properly supported* pseudodifferential operators, that is the operators which map $C_0^\infty(\Omega)$ to $\mathcal{E}'(\Omega)$ and the same happens for their transposed.

For any $a(x,\xi) \in S_{m,\Lambda}(\Omega)$, there exists $a'(x,\xi) \in S_{m,\Lambda}(\Omega)$ such that $a'(x,D)$ is properly supported and $a'(x,\xi) \sim a(x,\xi)$, that is $a'(x,\xi) - a(x,\xi) \in S^{-\infty}(\Omega)$.

Proposition 2.10 (symbolic calculus). *Let $m(\xi)$, $m'(\xi)$ be admissible weights associated to the same weight vector $\Lambda(\xi)$ and $a_1(x,D) \in \widetilde{\mathrm{Op}}\, S_{m,\Lambda}(\Omega)$, $a_2(x,D) \in$ Op $S_{m',\Lambda}(\Omega)$. Then $a_1(x,D)a_2(x,D) = b(x,D)$, where $b(x,\xi) \in S_{mm',\Lambda}(\Omega)$, and*

$$b(x,\xi) \sim \sum_\alpha \frac{1}{\alpha} \partial_\xi^\alpha a_1(x,\xi) D_x^\alpha a_2(x,\xi)\,, \qquad D^\alpha := (-i)^{|\alpha|}\partial^\alpha\,. \tag{2.18}$$

Consider now $m(\xi) \equiv 1$, then by means of the arguments in Remark 2.7(4), $S_\Lambda(\Omega)$ is contained in the Taylor class $M_{c,0}^0(\Omega)$, for suitable $0 < c < 1$. Let us recall that a symbol $a(x,\xi)$ belongs to $M_{c,0}^0(\Omega)$, if $\xi^\gamma \partial_\xi^\gamma a(x,\xi) \in S_{c,0}^0(\Omega)$ for all multi-indices $\gamma \in \{0,1\}^n$. Then applying the arguments in Taylor [14], see also [6, Theorem 4.1, Corollary 4.2], the following property immediately follows.

Proposition 2.11. *If $a(x,\xi) \in S_\Lambda(\Omega)$, then, for any $1 < p < \infty$*

$$a(x,D) : L^p_{\mathrm{comp}}(\Omega) \mapsto L^p_{\mathrm{loc}}(\Omega).$$

3. Weighted Sobolev spaces

Consider the class of global symbols $S_{m,\Lambda}$ given by the smooth functions $a(x,\xi) \in C^\infty(\mathbb{R}^{2n})$ which satisfy

$$\sup_{x\in\mathbb{R}^n} |\partial_\xi^\alpha \partial_x^\beta a(x,\xi)| \le c_{\alpha,\beta} m(\xi)\Lambda(\xi)^{-\alpha}, \quad \xi \in \mathbb{R}^n. \tag{3.1}$$

Then the pseudodifferential operators in Op$S_{m,\Lambda}$ defined by (2.17) map continuously $\mathcal{S}(\mathbb{R}^n)$ to $\mathcal{S}(\mathbb{R}^n)$ and may be extended to bounded linear operators from $\mathcal{S}'(\mathbb{R}^n)$ into itself. Thanks again to the arguments in Remark 2.7, every pseudodifferential operator with zeroth order symbol in S_Λ, maps continuously $L^p(\mathbb{R}^n)$ to $L^p(\mathbb{R}^n)$, $1 < p < \infty$. The symbolic calculus in Proposition 2.10 is true again for $a(x,D) \in$ Op $S_{m,\Lambda}$ and $b(x,D) \in$ Op $S_{m',\Lambda}$. Moreover Proposition 2.9 assures that $m(\xi)$ admits an equivalent weight which is a symbol in $S_{m,\Lambda}$. Without loss of generality, from now on we consider $m(\xi) \in S_{m,\Lambda}$. We can then define for $1 < p < \infty$

the weighted Sobolev space of L^p type:

$$H^p_m := \{u \in \mathcal{S}'(\mathbb{R}^n), \quad \text{such that} \quad m(D)u \in L^p(\mathbb{R}^n)\}. \tag{3.2}$$

H^p_m may be equipped in natural way by the norm $\|u\|_{p,m} := \|m(D)u\|_{L^p}$ and it then realizes to be a Banach space (Hilbert space in the case $p = 2$). With standard arguments it can be proved that $\mathcal{S}(\mathbb{R}^n) \subset H^p_m \subset \mathcal{S}'(\mathbb{R}^n)$, with continuous embeddings and moreover $\mathcal{S}(\mathbb{R}^n)$ is dense in H^p_m, $1 < p < \infty$. For any open subset $\Omega \subset \mathbb{R}^n$ the following local spaces may be introduced:

$$H^p_{m,\mathrm{comp}}(\Omega) = \mathcal{E}'(\Omega) \cap H^p_m; \tag{3.3}$$

$$H^p_{m,\mathrm{loc}}(\Omega) = \{u \in \mathcal{D}'(\Omega) \text{ such that, for any } \varphi \in C^\infty_0(\Omega),\ \varphi u \in H^p_m\}. \tag{3.4}$$

It is now trivial that $C^\infty(\Omega) \subset H^p_{m,\mathrm{loc}}(\Omega)$ for any $1 < p < \infty$.

Proposition 3.1. *Consider m, m' admissible weights and $a_1(x,\xi) \in S_{m',\Lambda}$, $a_2 \in S_{m',\Lambda}(\Omega)$, then for any $p \in]1,\infty[$ we have:*

$$a_1(x,D) : H^p_m \mapsto H^p_{m/m'}; \tag{3.5}$$

$$a_2(x,D) : H^p_{m,\mathrm{comp}}(\Omega) \mapsto H^p_{m/m',\mathrm{loc}}(\Omega). \tag{3.6}$$

If moreover $a(x,D)$ is a properly supported operator in $\widetilde{\mathrm{Op}}S_{m',\Lambda}(\Omega)$ then:

$$a(x,D) : H^p_{m,\mathrm{comp}}(\Omega) \mapsto H^p_{m/m',\mathrm{comp}}(\Omega); \tag{3.7}$$

$$a(x,D) : H^p_{m,\mathrm{loc}}(\Omega) \mapsto H^p_{m/m',\mathrm{loc}}(\Omega). \tag{3.8}$$

Proof. Since m/m' is an admissible weight, the symbolic calculus in Proposition 2.10 assures that

$$[m/m'](D)a_1(x,D)u = m(D)[1/m'](D)a_1(x,D)[1/m](D)m(D)u, u \in \mathcal{S}(\mathbb{R}^n)$$

and $m(D)[1/m'](D)a_1(x,D)[1/m](D)$ is a pseudodifferential operator with symbol in S_Λ. Thanks now to the L^p continuity of $\mathrm{Op}\, S_\Lambda$, we obtain:

$$\|a_1(x,D)u\|_{m/m'} = \|\,[m/m'](D)a_1(x,D)u\|_{L^p} \le K\|m(D)u\|_{L^p} = K\|u\|_{p,m}.$$

Since $\mathcal{S}(\mathbb{R}^n)$ is dense in H^p_m, then $a_1(x,D)$ extends univocally to a bounded linear operator from H^p_m to $H^p_{m/m'}$.

By standard arguments the proof applies to (3.6)–(3.8). □

Definition 3.2 (elliptic symbols). A symbol $a(x,\xi) \in S_{m,\Lambda}(\Omega)$ is elliptic if for any compact $K \subset\subset \Omega$ there exist $c_K > 0$ and $R_K > 0$ such that:

$$|a(x,\xi)| \ge c_k m(\xi), \quad x \in K, \quad |\xi| \ge R_K. \tag{3.9}$$

Proposition 3.3 (parametrix). *Let $a(x,\xi) \in S_{m,\Lambda}(\Omega)$ be a elliptic symbol. Then a properly supported operator $b(x,D) \in \widetilde{\mathrm{Op}}S_{1/m,\Lambda}(\Omega)$ exists such that:*

$$b(x,D)a(x,D) = \mathrm{Id} + \rho(x,D), \tag{3.10}$$

where $\rho(x,\xi) \in S^{-\infty}(\Omega)$ and Id *denotes the identity operator.*

See [12, Theorem 1.3.6] for the proof of the above result.

Proposition 3.4 (Regularity of solution to elliptic equations). *Consider $p \in]1,\infty[$, the admissible weights $m(\xi)$, $m'(\xi)$, and the m'-elliptic symbol $a(x,\xi) \in S_{m',\Lambda}(\Omega)$. Then for every $u \in \mathcal{E}'(\Omega)$ such that $a(x,D)u \in H^p_{m/m',\mathrm{loc}}(\Omega)$, we have $u \in H^p_{m,\mathrm{comp}}(\Omega)$. If $a(x,D)$ is properly supported, then $u \in H^p_{m,\mathrm{loc}}(\Omega)$ for every $u \in \mathcal{D}'(\Omega)$ such that $a(x,D)u \in H^p_{m/m',\mathrm{loc}}(\Omega)$.*

Proof. Thanks to Proposition 3.3, there exists $b(x,D) \in \widetilde{\mathrm{Op}}\, S_{1/m',\Lambda}(\Omega)$, such that $b(x,D)a(x,D) = \mathrm{Id} + \rho(x,D)$, with $\rho(x,\xi) \in S^{-\infty}(\Omega)$. Since $\rho(x,D)$ is a regularizing operator and $a(x,D)u \in H^p_{m/m',\mathrm{loc}}(\Omega)$, we can conclude from (3.6) that $u = b(x,D)\,(a(x,D)u) - \rho(x,D)u \in H^p_{m,\mathrm{loc}}(\Omega)$. □

4. Microlocal properties of pseudodifferential operators with symbols in $S_{m,\Lambda}(\Omega)$

Definition 4.1. A symbol $a(x,\xi) \in S_{m,\Lambda}(\Omega)$ is microlocally elliptic in a set $X \subset \mathbb{R}^n_\xi$ at the point $x_0 \in \Omega$ if there are positive constants c_0, R_0 such that

$$|a(x_0,\xi)| \geq c_0 m(\xi), \quad \text{when} \quad \xi \in X, \quad |\xi| > R_0 \,. \tag{4.1}$$

The Λ-neighborhood of a set $X \subset \mathbb{R}^n$ with length $\varepsilon > 0$ is defined to be the open set:

$$X_{\varepsilon\Lambda} := \bigcup_{\xi^0 \in X} \{|\xi_j - \xi^0_j| < \varepsilon\lambda_j(\xi^0), \quad \text{for } j = 1,\dots,n\} \tag{4.2}$$

Moreover for $x_0 \in \Omega$ we set:

$$X(x_0) := \{x_0\} \times X, \quad X_{\varepsilon\Lambda}(x_0) := B_\varepsilon(x_0) \times X_{\varepsilon\Lambda}, \tag{4.3}$$

where $B_\varepsilon(x_0)$ is the open ball in Ω centered at x_0 with radius ε.
Noticing that $\Lambda(\xi)$ is a weight vector according to [13], the following properties of Λ-neighborhoods can be immediately deduced from [13, Lemma 1.11] (see also [8] for an explicit proof). For every $\varepsilon > 0$ a suitable ε^* (depending only on ε and Λ), satisfying $0 < \varepsilon^* < \varepsilon$, can be found in such a way that for every $X \subset \mathbb{R}^n$:

1. $(X_{\varepsilon^*\Lambda})_{\varepsilon^*\Lambda} \subset X_{\varepsilon\Lambda}$;
2. $(\mathbb{R}^n \setminus X_{\varepsilon\Lambda})_{\varepsilon^*\Lambda} \subset \mathbb{R}^n \setminus X_{\varepsilon^*\Lambda}$.

In view of [13, Lemma 1.10], one can also prove that for arbitrary $\varepsilon > 0$ and $X \subset \mathbb{R}^n$ there exists a symbol $\sigma = \sigma(\xi) \in S_\Lambda$ such that $\operatorname{supp}\sigma \subset X_{\varepsilon\Lambda}$ and $\sigma(\xi) = 1$ if $\xi \in X_{\varepsilon'\Lambda}$, for a suitable ε', $0 < \varepsilon' < \varepsilon$ depending only on ε and Λ. Moreover for every $x_0 \in \Omega$ there exists a symbol $\tau_0(x,\xi) \in S_\Lambda(\Omega)$ such that $\operatorname{supp}\tau_0 \subset X_{\varepsilon\Lambda}(x_0)$ and $\tau_0(x,\xi) = 1$, for $(x,\xi) \in X_{\varepsilon^*\Lambda}(x_0)$, with a suitable ε^* satisfying $0 < \varepsilon^* < \varepsilon$.

Proposition 4.2. *If a symbol $a(x,\xi) \in S_{m,\Lambda}(\Omega)$ is microlocally elliptic in $X \subset \mathbb{R}^n_\xi$ at the point $x_0 \in \Omega$, there exists a suitable $\varepsilon > 0$ such that (4.1) is satisfied in $X_{\varepsilon\Lambda}(x_0)$, that is for suitable constants $C, R > 0$*

$$|a(x,\xi)| \geq Cm(\xi), \quad \textit{for}\,(x,\xi) \in X_{\varepsilon\Lambda}(x_0), \quad |\xi| > R\,. \tag{4.4}$$

Proof. Let the symbol $a(x,\xi)\in S_{m,\Lambda}(\Omega)$ be microlocally elliptic in $X\subset\mathbb{R}^n$ at the point $x_0\in\Omega$ and let $\xi^0\in X$ be arbitrarily fixed. Since Ω is open, a positive ε^* can be found in such a way that $\overline{B}_{\varepsilon^*}(x_0):=\{x\in\mathbb{R}^n:|x-x_0|\le\varepsilon^*\}\subset\Omega$; for $0<\varepsilon<\varepsilon^*$ and $(x,\xi)\in X_{\varepsilon\Lambda}(x_0)$, a Taylor expansion of $a(x,\xi)$ about (x_0,ξ^0) gives

$$a(x,\xi)-a(x_0,\xi^0)=\sum_{j=1}^{n}(x^j-x_0^j)\partial_{x^j}a(x_t,\xi^t)dt+(\xi_j-\xi_j^0)\partial_{\xi_j}a(x_t,\xi^t)dt\,, \tag{4.5}$$

where it is set $(x_t,\xi^t):=((1-t)x_0+tx,(1-t)\xi^0+t\xi)$ for a suitable $0<t<1$. Since $|\xi_j^t-\xi_j^0|=|t||\xi_j-\xi_j^0|<\varepsilon\lambda_j(\xi^0)$ and $|x_t^j-x_0^j|=|t||x^j-x_0^j|<\varepsilon$, from (2.8) there exists $C^*>0$, depending only on ε^*, such that

$$|a(x,\xi)-a(x_0,\xi^0)|\le\sum_{j=1}^{n}\varepsilon C^*m(\xi^t)+\varepsilon\lambda_j(\xi^0)C^*m(\xi^t)\lambda_j^{-1}(\xi^t)\,. \tag{4.6}$$

In view of (2.3), (2.5), $\varepsilon>0$ can be chosen small enough such that

$$\frac{1}{C}\lambda_j(\xi^0)\le\lambda_j(\xi^t)\le C\lambda_j(\xi^0)\,,\quad \frac{1}{C}m(\xi^0)\le m(\xi^t)\le Cm(\xi^0)\,,\qquad 1\le j\le n\,, \tag{4.7}$$

for a suitable constant $C>1$ independent of t and ε. Then (4.6), (4.7) give

$$|a(x,\xi)-a(x_0,\xi^0)|\le\hat{C}\varepsilon m(\xi^0)\,, \tag{4.8}$$

with a suitable constant $\hat{C}>0$ independent of ε.

Let the condition (4.1) be satisfied by $a(x,\xi)$ with positive constants c_0,R_0. Provided $0<\varepsilon<\varepsilon^*$ is taken sufficiently small, one can find a positive R, depending only on R_0, such that $|\xi|>R$ and $|\xi_j-\xi_j^0|<\varepsilon\lambda_j(\xi^0)$ for all $1\le j\le n$ yield $|\xi^0|>R_0$; indeed, from (2.1)

$$|\xi-\xi^0|\le\sum_{j=1}^{n}|\xi_j-\xi_j^0|<\varepsilon\sum_{j=1}^{n}\lambda_j(\xi^0)\le nC\varepsilon^*(1+|\xi_0|)^C\,,$$

and then

$$|\xi|\le|\xi^0|+|\xi-\xi^0|\le|\xi^0|+nC\varepsilon^*(1+|\xi^0|)^C\,.$$

Hence, it is sufficient to choose R such that $R>R_0+nC\varepsilon^*(1+R_0)^C$.

Since $|\xi^0|>R_0$, the microlocal ellipticity of $a(x,\xi)$ yields

$$|a(x_0,\xi^0)|\ge c_0m(\xi^0)\,; \tag{4.9}$$

then (4.8) and (4.9) give for $(x,\xi)\in X_{\varepsilon\Lambda}(x_0)$ and $|\xi|>R$

$$|a(x,\xi)|\ge|a(x_0,\xi^0)|-|a(x,\xi)-a(x_0,\xi^0)|\ge(c_0-\hat{C}\varepsilon)m(\xi^0)\ge\frac{c_0}{2}m(\xi^0)\,, \tag{4.10}$$

up to a further shrinking of $\varepsilon>0$. From (4.10), the condition (4.4) follows at once, by using that $m(\xi)\approx m(\xi^0)$. □

Definition 4.3. We say that a symbol $a(x,\xi)\in S_{m,\Lambda}(\Omega)$ is rapidly decreasing in $\Theta\subset\Omega\times\mathbb{R}^n$ if there exists $a_0(x,\xi)\in S_{m,\Lambda}(\Omega)$ such that $a(x,\xi)\sim a_0(x,\xi)$ and $a_0(x,\xi)=0$ in Θ

Theorem 4.4. *For every symbol* $a(x,\xi) \in S_{m,\Lambda}(\Omega)$ *that is microlocally elliptic in* $X \subset \mathbb{R}^n$ *at a point* $x_0 \in \Omega$ *there exists a symbol* $b(x,\xi) \in S_{1/m,\Lambda}(\Omega)$ *such that the associated operator* $b(x,D)$ *is properly supported and*

$$b(x,D)a(x,D) = \mathrm{Id} + c(x,D), \tag{4.11}$$

where $c(x,\xi) \in S_\Lambda(\Omega)$ *is rapidly decreasing in* $X_{r\Lambda}(x_0)$ *for a suitable* $r > 0$.

Proof. We follow the same arguments used for the proof of [8, Theorem 1]. By Proposition 4.2, there exists $\varepsilon > 0$ such that $a(x,\xi)$ is microlocally elliptic at $X_{\varepsilon\Lambda}(x_0)$. Let $\tau_0(x,\xi)$ be a symbol in $S_\Lambda(\Omega)$ such that $\tau_0 \equiv 1$ on $X_{\varepsilon'\Lambda}(x_0)$, for a suitable $0 < \varepsilon' < \varepsilon$, and $\operatorname{supp}\tau_0 \subset X_{\varepsilon\Lambda}(x_0)$. We define $b_0(x,\xi)$ by setting

$$b_0(x,\xi) := \begin{cases} \dfrac{\tau_0(x,\xi)}{a(x,\xi)} & \text{for} \quad (x,\xi) \in X_{\varepsilon\Lambda}(x_0), \\ 0 & \text{otherwise}. \end{cases} \tag{4.12}$$

Since $a(x,\xi)$ satisfies (4.4), with suitable constants C, R, $b_0(x,\xi)$ is a well-defined C^∞-function on the set $\Omega \times \{|\xi| > R\}$. For $k \geq 1$, the functions $b_{-k}(x,\xi)$ are defined recursively on $\Omega \times \{|\xi| > R\}$ by

$$b_{-k}(x,\xi) := \begin{cases} -\sum\limits_{0<|\alpha|\leq k} \frac{1}{\alpha!}\partial_\xi^\alpha b_{-k+|\alpha|}(x,\xi)\dfrac{D_x^\alpha a(x,\xi)}{a(x,\xi)}, & \text{for} \quad (x,\xi) \in X_{\varepsilon\Lambda}(x_0), \\ 0 & \text{otherwise}. \end{cases} \tag{4.13}$$

The $b_{-k}(x,\xi)$ can be then extended to the whole set $\Omega \times \mathbb{R}^n$, multiplying them by a smooth cut-off function that vanishes on the set of possible zeroes of the symbol $a(x,\xi)$; for each $k \geq 0$, the extended b_{-k} is a symbol in $S_{\pi^{-k}/m,\Lambda}(\Omega)$. In view of the properties of the symbolic calculus, a symbol $b \in S_{1/m,\Lambda}(\Omega)$ can be chosen in such a way that

$$b \sim \sum_{k\geq 0} b_{-k}\,,$$

and $b(x,D)$ is a properly supported operator. By construction, the symbol of $b(x,D)a(x,D)$ is equivalent to $\tau_0(x,\xi)$; thus the symbol of $b(x,D)a(x,D) - \mathrm{Id}$ belongs to $S_\Lambda(\Omega)$ and is rapidly decreasing in $X_{r\Lambda}(x_0)$ for $0 < r \leq \varepsilon'$. □

Proposition 4.5. *For* $x_0 \in \Omega$, $X \subset \mathbb{R}^n$, $u \in \mathcal{D}'(\Omega)$, $1 < p < \infty$, *the following properties are equivalent:*

i) *there exists an operator* $a(x,D) \in \widetilde{\mathrm{Op}}S_\Lambda(\Omega)$ *whose symbol is microlocally elliptic in* X *at the point* x_0, *such that* $a(x,D)u \in H^p_{m,\mathrm{loc}}(\Omega)$*;*
ii) *there exist a symbol* $\sigma = \sigma(\xi) \in S_\Lambda$, *such that* $\operatorname{supp}\sigma \subset X_{\varepsilon\Lambda}$, $\sigma(\xi) = 1$ *when* $\xi \in X_{\varepsilon'\lambda}$ *for suitable* $0 < \varepsilon' < \varepsilon$, *and a function* $\phi \in C_0^\infty(\Omega)$, *with* $\phi(x_0) = 1$, *satisfying* $\sigma(D)(\phi u) \in H^p_m$.

Proof. i) ⇒ ii): Let the operator $a(x,D) \in \widetilde{\mathrm{Op}}S_\Lambda(\Omega)$ satisfy the assumptions in i); from Theorem 4.4 there exists $b(x,D) \in \widetilde{\mathrm{Op}}S_\Lambda(\Omega)$ satisfying

$$b(x,D)a(x,D) = \mathrm{Id} + c(x,D)\,, \tag{4.14}$$

where $c(x,\xi) \in S_\Lambda(\Omega)$ is rapidly decreasing in $X_{\varepsilon\Lambda}(x_0)$ for some $0 < \varepsilon < 1$.

Let $\sigma(\xi) \in S_\Lambda$ and $\phi(x) \in C_0^\infty(\Omega)$ satisfy the conditions in *ii*) with suitable $0 < \varepsilon' < \varepsilon$. From (4.14) we write

$$\sigma(D)(\phi u) = \sigma(D)\phi(x)b(x,D)a(x,D)u - \sigma(D)\phi(x)c(x,D)u\,. \tag{4.15}$$

Since $a(x,D)u \in H^p_{m,\mathrm{loc}}(\Omega)$, by Proposition 3.1 $\sigma(D)\phi(x)b(x,D)a(x,D)u \in H^p_m$.

As regards to the second term $\sigma(D)\phi(x)c(x,D)u$ in the right-hand side of (4.15), the operator $c(x,D)$ is known to be properly supported (since $a(x,D)$ and $b(x,D)$ are so and (4.14) holds true). Then a function $\widetilde{\phi} \in C_0^\infty(\Omega)$ can be found in such a way that

$$\phi(x)c(x,D)u = \phi(x)c(x,D)(\widetilde{\phi}u)\,. \tag{4.16}$$

Since $c(x,\xi)$ is also rapidly decreasing, there exist $c_0(x,\xi) \in S_\Lambda(\Omega)$, supported in $(\Omega \times \mathbb{R}^n) \setminus X_{\varepsilon\Lambda}(x_0)$, such that

$$\rho(x,\xi) := c(x,\xi) - c_0(x,\xi) \in S^{-\infty}(\Omega)\,. \tag{4.17}$$

Then

$$\begin{aligned}\sigma(D)(\phi(x)c(x,D)u) &= \sigma(D)(\phi(x)c(x,D)(\widetilde{\phi}u)) \\ &= \sigma(D)(\phi(x)c_0(x,D)(\widetilde{\phi}u)) + \sigma(D)(\phi(x)\rho(x,D)(\widetilde{\phi}u))\,.\end{aligned} \tag{4.18}$$

We immediately get $\sigma(D)(\phi(x)\rho(x,D)(\widetilde{\phi}u)) \in \mathcal{S}(\mathbb{R}^n) \subset H^p_m$.

Now, let us set $d(x,D) := \sigma(D)\phi(x)c_0(x,D)$; from the symbolic calculus, the symbol of $d(x,D)$ enjoys the asymptotic expansion

$$d(x,\xi) \sim \sum_\alpha \frac{1}{\alpha!}\partial_\xi^\alpha\sigma(\xi)D_x^\alpha(\phi(x)c_0(x,\xi))\,. \tag{4.19}$$

Since $\operatorname{supp}\sigma \subset X_{\varepsilon\Lambda}$, $\partial_\xi^\alpha\sigma(\xi) = 0$ as long as $\xi \notin X_{\varepsilon\Lambda}$. On the other hand, $c_0(x,\xi) = 0$ for $\xi \in X_{\varepsilon\Lambda}$ and $x \in B_\varepsilon(x_0)$ and $\phi(x) = 0$ for $x \notin B_\varepsilon(x_0)$. Hence for all $\alpha \in \mathbb{Z}^n_+$

$$\partial_\xi^\alpha\sigma(\xi)D_x^\alpha(\phi(x)c_0(x,\xi)) = 0\,, \quad \text{for } (x,\xi) \in \Omega \times \mathbb{R}^n\,; \tag{4.20}$$

thus $d(x,\xi) \in S^{-\infty}(\Omega)$ and we get $\sigma(D)(\phi(x)c_0(x,D)(\widetilde{\phi}u)) = d(x,D)(\widetilde{\phi}u) \in C^\infty(\Omega)$. On the other hand, the two operators $\sigma(D)$ and $\phi(x)c_0(x,D)$ have global symbols $\sigma(\xi)$ and $\phi(x)c_0(x,\xi)$ in S_Λ; then $d(x,\xi)$ is a global symbol in $S^{-\infty}$ and $d(x,D)(\widetilde{\phi}u) \in \mathcal{S}(\mathbb{R}^n) \subset H^p_m$.

ii) ⇒ i): Let $\sigma(\xi) \in S_\Lambda$ and $\phi(x) \in C_0^\infty(\Omega)$, satisfying the assumptions in ii), be such that $\sigma(D)(\phi u) \in H^p_m$. Since $S_\Lambda \subset S_\Lambda(\Omega)$, there exists $\widetilde{\sigma}(x,\xi) \in S_\Lambda(\Omega)$ such that $\sigma \sim \widetilde{\sigma}$ and $\widetilde{\sigma}(x,D)$ is properly supported. Let us set $a(x,D) := \widetilde{\sigma}(x,D)\phi(x) \in \widetilde{\mathrm{Op}}\,S_\Lambda(\Omega)$; we also have

$$a(x,\xi) \sim \sum_\alpha \frac{1}{\alpha!}\partial_\xi^\alpha\widetilde{\sigma}(x,\xi)D_x^\alpha\phi(x)\,. \tag{4.21}$$

In particular, $\tau(x,\xi) := a(x,\xi) - \widetilde{\sigma}(x,\xi)\phi(x) \in S_{\pi^{-1},\Lambda}(\Omega)$ and $\rho(x,\xi) := \widetilde{\sigma}(x,\xi) - \sigma(\xi) \in S^{-\infty}(\Omega)$. Hence

$$a(x,\xi) = (\sigma(\xi) + \rho(x,\xi))\phi(x) + \tau(x,\xi) = \sigma(\xi)\phi(x) + \theta(x,\xi)\,, \tag{4.22}$$

where $\theta(x,\xi) := \rho(x,\xi)\phi(x) + \tau(x,\xi) \in S_{\pi^{-1},\Lambda}(\Omega)$. This proves that $a(x,\xi)$ is microlocally elliptic in X at x_0; indeed for $\xi \in X$ (and in view of (2.1))

$$|a(x_0,\xi)| \geq |\sigma(\xi)| - |\theta(x_0,\xi)| = 1 - C_0\pi^{-1}(\xi) \geq 1 - C_1\langle\xi\rangle^{-C}\,, \tag{4.23}$$

hence $|a(x_0,\xi)| \geq \frac{1}{2}$ when $|\xi| > R$ and $R > 0$ is taken sufficiently large. Finally, one computes

$$a(x,D)u = \widetilde{\sigma}(x,D)(\phi u) = \sigma(D)(\phi u) + \rho(x,D)(\phi u) \in H^p_{m\,\mathrm{loc}}(\Omega)\,,$$

which completes the proof. □

Definition 4.6. For $X \subset \mathbb{R}^n$, $x_0 \in \Omega$ and $p \in]1,\infty[$ we say that $u \in \mathcal{D}'(\Omega)$ is microlocally H^p_m-regular in X at the point $x_0 \in \Omega$, and write $u \in mclH^p_m(X(x_0))$, if one of the equivalent properties in Proposition 4.5 is satisfied.

5. Microlocal Sobolev Continuity and Regularity

In the following, we will provide a microlocal counterpart of the properties of boundedness and regularity for pseudodifferential operators developed in Section 3. In the sequel, m, m' are two admissible weights associated to the same weight vector Λ.

Proposition 5.1. *Let $x_0 \in \Omega$, $X \subset \mathbb{R}^n$, $a(x,D) \in \widetilde{\mathrm{Op}}S_{m,\Lambda}(\Omega)$ be given. Then for $p \in]1,\infty[$ and $u \in mclH^p_{m'}(X(x_0))$ one has $a(x,D)u \in mclH^p_{m'/m}(X(x_0))$.*

Proof. From Proposition 4.5, there exists $b(x,D) \in \widetilde{\mathrm{Op}}S_\Lambda(\Omega)$, with microlocally elliptic symbol, such that $b(x,D)u \in H^p_{m',\mathrm{loc}}(\Omega)$. From Theorem 4.4 there also exists an operator $c(x,D) \in \widetilde{\mathrm{Op}}S_\Lambda(\Omega)$ such that

$$c(x,D)b(x,D) = \mathrm{Id} + \rho(x,D)\,, \tag{5.1}$$

where $\rho(x,\xi) \in S_\Lambda(\Omega)$ is rapidly decreasing in $X_{r\Lambda}(x_0)$ for some $0 < r < 1$. Let $r^* > 0$ be such that

$$(\mathbb{R}^n \setminus X_{r\Lambda})_{r^*\Lambda} \subset \mathbb{R}^n \setminus X_{r^*\Lambda}\,, \qquad 0 < r^* < r\,, \tag{5.2}$$

and take a symbol $\tau_0(x,\xi) \in S_\Lambda(\Omega)$ satisfying

$$\mathrm{supp}\,\tau_0 \subset X_{r^*\Lambda}(x_0)\,, \qquad \tau_0 \equiv 1 \text{ on } X_{r'\Lambda}(x_0)\,,$$

with a suitable $0 < r' < r^*$. Finally, let $\tau(x,\xi)$ be a symbol such that $\theta_0(x,\xi) := \tau(x,\xi) - \tau_0(x,\xi) \in S^{-\infty}(\Omega)$ and $\tau(x,D) \in \widetilde{\mathrm{Op}}S_\Lambda(\Omega)$. One can check (see [8] for details) that $\tau(x,\xi)$ is microlocally elliptic in X at x_0; in particular, $\tau(x,\xi) = \theta_0(x,\xi) \in S^{-\infty}(\Omega)$ for $(x,\xi) \notin X_{r^*\Lambda}(x_0)$.

Arguing as in the proof of [8, Theorem 2], from (5.1) we write

$$\tau(x,D)a(x,D)u = \tau(x,D)a(x,D)c(x,D)(b(x,D)u) - \tau(x,D)a(x,D)\rho(x,D)u\,. \tag{5.3}$$

Since $\tau(x,D)a(x,D)c(x,D) \in \widetilde{\mathrm{Op}}S_{m,\Lambda}(\Omega)$ and $b(x,D)u \in H^p_{m',\mathrm{loc}}(\Omega)$ then we get $\tau(x,D)a(x,D)c(x,D)(b(x,D)u) \in H^p_{m'/m,\mathrm{loc}}(\Omega)$. Moreover, it can be shown that in view of (5.2)

$$\varphi(x)\tau(x,D)a(x,D)\rho(x,D)u \in C_0^\infty(\Omega) \subset H^p_{m'/m}\,,$$

for every $\varphi \in C_0^\infty(\Omega)$, so that $\tau(x,D)a(x,D)\rho(x,D)u \in H^p_{m'/m,\mathrm{loc}}(\Omega)$ (see [8, Theorem 2]).

This proves that $\tau(x,D)a(x,D)u \in H^p_{m'/m,\mathrm{loc}}(\Omega)$ and ends the proof. □

Proposition 5.2. *For $x_0 \in \Omega$, $X \subset \mathbb{R}^n$, let the symbol of $a(x,D) \in \widetilde{\mathrm{Op}}S_{m,\Lambda}(\Omega)$ be microlocally elliptic in X at the point x_0. Then for every $p \in]1,\infty[$ and $u \in \mathcal{D}'(\Omega)$ such that $a(x,D)u \in mcl H^p_{m'/m}(X(x_0))$ one has $u \in mcl H^p_{m'}(X(x_0))$.*

Proof. From Proposition 4.5, there exists an operator $b(x,D) \in \widetilde{\mathrm{Op}}S_\Lambda(\Omega)$ microlocally elliptic in X at x_0 such that

$$b(x,D)a(x,D)u \in H^p_{m'/m,\mathrm{loc}}(\Omega)\,. \tag{5.4}$$

From Theorem 4.4 there exist $c(x,D) \in \widetilde{\mathrm{Op}}S_\Lambda(\Omega)$ and $q(x,D) \in \widetilde{\mathrm{Op}}S_{1/m,\Lambda}(\Omega)$ such that

$$c(x,D)b(x,D) = \mathrm{Id} + \rho(x,D)\,, \qquad q(x,D)a(x,D) = \mathrm{Id} + \sigma(x,D)\,, \tag{5.5}$$

with $\rho(x,\xi), \sigma(x,\xi) \in S^{-\infty}(\Omega)$ rapidly decreasing in $X_{r\Lambda}(x_0)$ for a suitable $0 < r < 1$.

Let the symbols $\tau_0(x,\xi), \tau(x,\xi) \in S_\Lambda(\Omega)$ be constructed as in the proof of Proposition 5.1. It can be proved that $\tau(x,D)u \in H^p_{m,\mathrm{loc}}(\Omega)$, by writing

$$\begin{aligned}\tau(x,D)u &= \tau(x,D)q(x,D)c(x,D)\,(b(x,D)a(x,D)u)\\ &\quad -\tau(x,D)q(x,D)\rho(x,D)a(x,D)u - \tau(x,D)\sigma(x,D)u\,,\end{aligned} \tag{5.6}$$

where the identities (5.5) have been used, and applying similar arguments as in the proof of Proposition 5.1 (see also [8, Theorem 3]). □

6. The m-filter of Sobolev singularities

Let $a(x,D)$ be a properly supported pseudo-differential operator with symbol $a(x,\xi) \in S_{m,\Lambda}(\Omega)$ and $x_0 \in \Omega$. Following [4], [13], we can define, for any $x_0 \in \Omega$,

- the m-filter of Sobolev singularities of $u \in \mathcal{D}'(\Omega)$ by

$$\mathcal{W}^p_{m,x_0}u := \{X \subset \mathbb{R}^n\,;\, u \in mcl H^p_m\left((\mathbb{R}^n \setminus X)(x_0)\right)\}\,, \quad 1 < p < \infty; \tag{6.1}$$

- the m-characteristic filter of $a(x,D) \in \widetilde{\mathrm{Op}}S_{m,\Lambda}(\Omega)$ by

$$\Sigma_{m,x_0}a(x,D) := \{X \subset \mathbb{R}^n\,,\; a(x,\xi)\,\text{is microlocally elliptic in}\ \mathbb{R}^n \setminus X\ \text{at}\ x_0\}\,. \tag{6.2}$$

It is trivial that $\mathcal{W}^p_{m,x_0}u$ and $\Sigma_{m,x_0}a(x,D)$ are filters in the sense that they are closed with respect to the intersection of a finite collection of their members and if $X \in \mathcal{W}^p_{m,x_0}u\,(\Sigma_{m,x_0}a(x,D))$ and $X \subset Y$ then $Y \in \mathcal{W}^p_{m,x_0}u\,(\Sigma_{m,x_0}a(x,D))$.

It is also straightforward to show that the results of Propositions 5.1, 5.2 can be restated as follows.

Proposition 6.1. *Assume that m, m' are two arbitrary admissible weights and let $a(x,D) \in \widetilde{\mathrm{Op}}\,S_{m,\Lambda}(\Omega)$, $x_0 \in \Omega$ and $p \in]1,\infty[$ be given. Then the following inclusions are satisfied for every $u \in \mathcal{D}'(\Omega)$:*

$$\mathcal{W}^p_{m'/m,x_0}a(x,D)u \cap \Sigma_{m,x_0}a(x,D) \subset \mathcal{W}^p_{m',x_0}u \subset \mathcal{W}^p_{m'/m,x_0}a(x,D)u\,. \tag{6.3}$$

References

[1] R. Beals *A general calculus of pseudodifferential operators*, Duke Math. J. **42** (1978), 1–42.

[2] P. Boggiatto, E. Buzano, L. Rodino *Global hypoellipticity and spectral theory*, Akademie-Verlag, Berlin 1996

[3] C. Fefferman *L^p bounds for pseudodifferential operators*, Israel J.Math. **14** (1973), 413–417.

[4] G.Garello *Inhomogeneous microlocal analysis for C^∞ and H^ν_ψ singularities*, Rend. Sem. Mat. Univ. Polit. Torino **2**, (1992), 165–181.

[5] G. Garello *Generalized Sobolev algebras and regularity of solutions of multi-quasi-elliptic semilinear equations*, Comm. in Appl. Analysis **3** n.4 (1999), 199–209

[6] G.Garello, A. Morando *L^p-bounded pseudodifferential operators and regularity for multi-quasi-elliptic equations*, Integr. equ.oper. theory **51** (2005), 501–517.

[7] G.Garello, A. Morando *L^p-microlocal regularity for pseudodifferential operators of quasi-homogeneous type*, Complex Var. Elliptic Equ. **54/8** (2009), 779–794.

[8] G. Garello, A. Morando *L^p-microlocal properties for multi-quasi-elliptic pseudodifferential operators*, Pliska Stud. Math. Bulgar. **21** (2012), 71–96.

[9] S.G. Gindikin, L.R. Volevich *The method of Newton's Polyhedron in the theory of partial differential equations*, Mathematics and its Applications (Soviet Series) Vol 86, Kluwer Academic Publishers, 1992.

[10] L. Hörmander *The Weyl calculus of pseudo-differential operators*, Comm. Pure Appl. Math. **XXXII** (1982), 359–443.

[11] L. Hörmander *The Analysis of Linear Partial Differential Operators III*, Springer Verlag, Berlin, Heidelberg 1985.

[12] F. Nicola, L. Rodino *Global Pseudo-Differential Operators on Euclidean Spaces*, Birkhäuser, 2010.

[13] L. Rodino *Microlocal analysis for spatially inhomogeneous pseudodifferential operators*, Ann. Sc. Norm. Sup. Pisa Classe di Scienze **9/2** (1982), 221–253.

[14] M.E. Taylor *Pseudodifferential Operators*, Princeton Univ. Press, 1981.

Gianluca Garello
Dipartimento di Matematica
Università di Torino
Via Carlo Alberto 10
I-10123, Torino, Italy
e-mail: gianluca.garello@unito.it

Alessandro Morando
DICATAM – Sezione di Matematica
Università di Brescia
Via Valotti 9
I-25133, Brescia, Italy
e-mail: alessandro.morando@ing.unibs.it

Fourier Analysis
Trends in Mathematics, 149–186

A Parametrix Construction for the Laplacian on $\mathbb{Q}$-rank 1 Locally Symmetric Spaces

D. Grieser and E. Hunsicker

Abstract. This paper presents the construction of parametrices for the Gauss–Bonnet and Hodge Laplace operators on noncompact manifolds modelled on $\mathbb{Q}$-rank 1 locally symmetric spaces. These operators are, up to a scalar factor, ϕ-differential operators; that is, they live in the generalised ϕ-calculus studied by the authors in a previous paper, which extends work of Melrose and Mazzeo. However, because they are not totally elliptic elements in this calculus, it is not possible to construct parametrices for these operators within the ϕ-calculus. We construct parametrices for them in this paper using a combination of the b-pseudodifferential operator calculus of R. Melrose and the ϕ-pseudodifferential operator calculus. The construction simplifies and generalizes the construction of a parametrix for the Dirac operator done by Vaillant in his thesis. In addition, we study the mapping properties of these operators and determine the appropriate Hilbert spaces between which the Gauss–Bonnet and Hodge Laplace operators are Fredholm. Finally, we establish regularity results for elements of the kernels of these operators.

Mathematics Subject Classification (2010). Primary 58AJ40; Secondary 35J75.

Keywords. Hodge Laplacian, pseudodifferential operators, symmetric spaces, noncompact spaces.

1. Introduction

Analysis of the Laplacian on locally symmetric spaces has interesting applications in analytic number theory and in the relationships between analysis and topology on singular spaces. An important tool for studying the Laplacian is the method of pseudodifferential operators. The basic philosophy of this approach is to define a space of pseudodifferential operators and one or more symbol maps on this space whose invertibility defines a subset of (totally) elliptic elements for which parametrices can be constructed within the calculus. Ideally, the operator you are

---

This work was completed with the support of Leverhulme Trust Project Assistance Grant F/00 261/Z.

interested in studying, for us the Laplacian, will be a (totally) elliptic element in your calculus, which will then provide the necessary tools for studying it.

In this paper, we are interested in the Gauss–Bonnet operator and its square, the Hodge Laplace operator, on L^2 differential forms over Riemannian manifolds modelled on $\mathbb{Q}$-rank 1 locally symmetric spaces. These spaces can be naturally compactified to manifolds with boundary. In this paper, we will refer to such manifolds with boundary as ϕ-manifolds, and to their Riemannian metrics as ϕ-cusp metrics. A ϕ-manifold, M, will be a compact manifold with boundary, $\partial M = Y$, which is the total space of a fibre bundle $Y \overset{\phi}{\to} B$. A ϕ-cusp metric will be a metric g on the interior $\mathring{M}$ that satisfies the property that near the boundary of M, with respect to some trivialization of a neighborhood U of the boundary $U \equiv [0, \epsilon) \times \partial M$, it can be written in the form

$$g = \frac{dx^2}{x^2} + \phi^* g_B + x^{2a} h. \tag{1.1}$$

Here $a \in \mathbb{N}$, x is the coordinate on $[0, \epsilon)$, h is a symmetric 2-tensor on ∂M that restricts to a metric on each fibre of ϕ and g_B is a Riemannian metric on B. Also, we assume that $\phi : Y \to B$ is a Riemannian fibration with respect to the metrics $\phi^* g_B + x^{2a} h$ on Y and g_B on B, for some (hence all) $x \in (0, \epsilon)$. We will always assume that M is orientable.

In a previous paper [4], we constructed a calculus of pseudodifferential operators adapted to the geometric setting of generalized ϕ-manifolds with general ϕ-cusp metrics (which may have a lower order perturbation from the metrics considered in this paper). This calculus, which we called the small ϕ-calculus, is a generalisation of the ϕ-calculus of Melrose and Mazzeo, [11], in that it permits a broader possible set of metric degenerations than that calculus does. In particular, the Melrose and Mazzeo calculus considers operators associated to metrics as in equation (1.1) where $a = 1$, whereas we allow a to be any natural number. Also, we allow a stack of several fibrations at the boundary with different orders of degeneration. A disappointing aspect of the small ϕ-calculus in both the Melrose–Mazzeo setting and our more general setting is that, although standard geometric operators such as Dirac operators, the Laplace–Beltrami operator and the Hodge Laplacian are elements of this calculus, they are typically not "totally elliptic" elements, which means that it is not possible to find good parametrices for them within this calculus. However, the Hodge Laplacian has an additional structure, reflecting the boundary fibration, which we can exploit to construct a parametrix in a larger ϕ-calculus.

The parametrix construction combines aspects of the original b-calculus of Melrose and the ϕ-calculus of [4]. The idea of combining elements of these two calculi to construct a parametrix was used first by Boris Vaillant in his thesis [19] for the Dirac operator on ϕ-manifolds with ϕ-cusp metrics where $a = 1$ under certain geometric conditions. The goal of this paper and the upcoming paper [6] is to simplify and generalise this construction, fully understand the space of operators in which a parametrix can be built, and characterise the conditions on ϕ-differential operators that allow this approach to work. One of the main differences between Vaillant's work and the work in this paper is that certain geometric obstructions arise in the construction of the parametrix of the (second-order) Hodge Laplacian that do not arise in the construction of the parametrix of the (first-order) Dirac operator. These obstructions have arisen in work of J. Müller [16] that studies the Hodge Laplacian on ϕ-manifolds from a perturbation theory viewpoint. Thus it is interesting to see how the obstructions also arise using a pseudodifferential approach.

Many $\mathbb{Q}$-rank 1 locally symmetric spaces compactify to ϕ-manifolds with ϕ-cusp metrics. However, a general $\mathbb{Q}$-rank 1 locally symmetric space may compactify to a manifold whose boundary is the total space of a double fibration as opposed to a single fibration. In a work in preparation, [6], we consider this more general case, and in addition generalise our results to other elliptic geometric operators on such manifolds. However, the main ideas of this general work can already be largely understood through the special case we consider in this paper, and the work of extending is quite technical. Thus the case we consider here is useful not only in its own right, but also for grasping the essential points of the general construction done in [6], and as such is a step towards the development of an extended pseudodifferential operator calculus on locally symmetric spaces of any rank and their generalisations.

Although the perspective of the b-calculus of Melrose has been taken in this paper, aspects of the construction reflect work done in other schools. The clearest similarities are to work on the setting of manifolds with edges undertaken from the perspective of the Schulze calculus, see for instance the recent work of Gil, Krainer and Mendoza in [3], [10], see also the work of Mazzeo and Vertman in [12] and the work of Dorschfeldt in [2] as well as earlier work by Schulze, [18]. At a more fundamental level, the general perspective of the need for multiple symbols when studying elliptic operators over noncompact and singular manifolds has been observed by many mathematicians in addition to Melrose and Schulze, for instance Kondratiev, [9] and Boutet de Monvel [1]. It is likely that the issues dealt with in the present paper could be addressed from any of these alternative viewpoints. To the best of our knowledge, however, the analysis of operators associated to manifolds with the particular geometry we study here has not yet been undertaken in any of these alternative approaches.

In order to state the main theorems, we need to introduce some notation. A fundamental object in the analysis of the Gauss–Bonnet and Hodge Laplace operators is the bundle $\overline{\mathcal{K}} \to B$ of fibre-harmonic forms. This is the finite-dimensional vector bundle over B whose fibre $\overline{\mathcal{K}}_{\mathbf{y}}$ over $\mathbf{y} \in B$ is the space of harmonic forms on the fibre $F_{\mathbf{y}} = \phi^{-1}(\mathbf{y})$, with respect to the metric $h_{|F_{\mathbf{y}}}$. We call a differential form on $\mathring{M}$ fibre harmonic if its restrictions to all fibres over U are harmonic, and fibre perpendicular if the restrictions are perpendicular, in the L^2 space of the fibre, to the harmonic forms. Fibre harmonic forms may be regarded as forms on $[0, \epsilon) \times B$ valued in $\overline{\mathcal{K}}$. Our operators behave differently on the subspaces of fibre harmonic forms and fibre perpendicular forms, and this difference, combined with the fact that these subspaces are not invariant under the operators, is the main difficulty in the construction. It is reflected in the structure of the Sobolev spaces on which the operators are Fredholm. We call them split Sobolev spaces. They are denoted $H^m_{\text{split}}(M, \Lambda T^*M, \text{dvol}_b)$ and defined in Section 6.1. Here the form bundle ΛT^*M is endowed with the metric induced by g, and dvol_b is a b-volume form, that is, a volume form on $\mathring{M}$ which near the boundary is $\frac{1}{x}$ times a smooth volume form on M.

The metric g induces a flat connection on the bundle $\overline{\mathcal{K}}$, and this can be used to define a Gauss–Bonnet operator D_V on $V = [0, \epsilon) \times B$ acting on forms with coefficients in $\overline{\mathcal{K}}$, see (6.3). This is an elliptic b- (or totally characteristic) operator, hence has a discrete set of critical weights associated with it, for which the operator is not Fredholm between the natural weighted Sobolev spaces. We call this set $-\operatorname{Im}\operatorname{spec}(D_V)$, see (2.10). This set also arises in the expansions of harmonic forms over M near the boundary. Recall that a smooth form u on $\mathring{M}$ is called polyhomogeneous if at the boundary it has a full asymptotic expansion of the form:

$$u \sim \sum\nolimits_{w,k} x^w (\log x)^k u_{w,k}\,, \quad x \to 0 \tag{1.2}$$

where all $u_{w,k}$ are smooth up to $x = 0$ and the sum runs over a discrete set of $w \in \mathbb{C}$ with $\operatorname{Re} w \to \infty$ and $k \leq N_w$ for each w, for some $N_w \in \mathbb{N}_0$. The pairs (w, k) that arise in the expansions of harmonic forms can be derived from the set $-\operatorname{Im}\operatorname{spec}(D_V)$. Here smoothness of $u_{w,k}$ at $x = 0$ is to be understood as smoothness of the coefficient functions when writing the form in terms of $\frac{dx}{x}, dy_j, x^a dz_k$ where the y_j are base coordinates and the z_k are fibre coordinates. This may be stated as smoothness when considering $u_{w,k}$ as a section of a rescaled bundle $\Lambda^{c\phi}T^*M$, as explained in Section 3.

Our first main result is a general parametrix construction, which is stated as Theorem 12. Using this parametrix we deduce the following results.

Theorem 1. *Let M be a ϕ-manifold endowed with a ϕ-cusp metric, g. The Gauss–Bonnet operator, $D_M = d + d^*$, is a Fredholm operator*

$$D_M : x^{\gamma+a} H^1_{\text{split}}(M, \Lambda T^*M, \text{dvol}_b) \to x^\gamma L^2(M, \Lambda T^*M, \text{dvol}_b)$$

for every $\gamma \notin -\operatorname{Im}\operatorname{spec}(D_V)$.

If $D_M u = 0$ for $u \in L^2(M, \Lambda T^*M, \text{dvol}_b)$, then u is polyhomogeneous, and in the expansion (1.2) *we have $\operatorname{Re} w > 0$ for all w, and all terms $u_{w,k}$ with $\operatorname{Re} w \leq a$ are fibre harmonic.*

In the next theorem we refer to operators D_B and Π. The operator D_B is a kind of Gauss–Bonnet operator on B with coefficients in all forms on F, see (3.9), and Π is the fibrewise orthogonal projection to the fibre-harmonic forms.

Theorem 2. *Let M be a ϕ-manifold endowed with a ϕ-cusp metric, g, and assume that $[D_B, \Pi] = 0$. Then the Hodge Laplacian, $\Delta_M = D_M^2$, is a Fredholm operator*

$$\Delta_M : x^{\gamma+2a} H^2_{\text{split}}(M, \Lambda T^*M, \text{dvol}_b) \to x^\gamma L^2(M, \Lambda T^*M, \text{dvol}_b).$$

for every $\gamma \notin -\operatorname{Im}\operatorname{spec}(D_V)$.

If $\Delta_M u = 0$ *for* $u \in L^2(M, \Lambda T^*M, \mathrm{dvol}_b)$*, then* u *is polyhomogeneous, and in the expansion* (1.2) *we have* $\mathrm{Re}\, w > 0$ *for all* w*, and all terms* $u_{w,k}$ *with* $\mathrm{Re}\, w \leq 2a$ *are fibre harmonic.*

This paper is organised as follows. In Section 2, we review the general b-calculus approach to pseudodifferential operators and give definitions, parametrix and regularity theorems in the b and small-ϕ-calculus settings. These results are useful both for comparison to Theorems 1 and 2, and also because parts of them are used in the proofs of these theorems. In Section 3, we review the geometry of fibrations and the form of the Gauss–Bonnet and Hodge Laplace operators over ϕ-manifolds with ϕ-cusp metrics. This section lays out some of the critical properties of these operators which permit the approach taken in the proofs of Theorems 1 and 2. Here also the rescaled bundle ${}^{c\phi}T^*M$ is introduced, which is needed to apply the b-calculus techniques and is used in the rest of the paper. It also explains where the condition $[D_B, \Pi] = 0$ in Theorem 2 comes from. In Section 4, we consider how to lift integral kernels of elements of the b-calculus to the ϕ-double space, and how the resulting operators combine with elements of the ϕ-calculus. A larger space of operators, the "extended" ϕ-calculus in which both ϕ-operators and lifted b-operators live and may be combined, is defined. It is also shown that the extended ϕ-calculus has a meaningful boundary symbol, and that it has properties similar to the original ϕ-calculus that will be used in the construction of parametrices. Section 5 begins with a discussion of the properties we want from our final parametrix, then contains the statement and proof of the main (parametrix) theorem of this paper, Theorem 12. Section 6 contains the definition of the spaces that arise in the Fredholm results in Theorems 1 and 2, and proofs of general Fredholm and regularity theorems as corollaries of the parametrix theorem. These theorems, when applied to the Gauss–Bonnet and Hodge Laplace operators, become Theorems 1 and 2. This paper is written in such a way that a reader interested primarily in applying the theorems, rather than in the proof details, can skip the more technical Sections 4 and 5.

2. The b- and ϕ-calculi

In the b-calculus approach to geometric pseudodifferential operators, the geometry of a singular or noncompact manifold is encoded in a Lie algebra of vector fields over a manifold with boundary that degenerate in a particular way at the boundary. The original b-setting of Melrose, [13], dealt with a Riemannian manifold $(\mathring{M}, g_b)$ which off of a compact set had an infinite cylindrical metric,

$$g_b = dr^2 + g_Y, \quad r \in [0, \infty),$$

where (Y, g_Y) was a smooth compact Riemannian manifold. Certain perturbations of such product type metrics are also considered. Under the change of coordinates

$x = e^{-r}$, the metric can be rewritten in the form:

$$g_b = \frac{dx^2}{x^2} + g_Y, \quad x \in (0, 1].$$

We compactify $\mathring{M}$ by adding a copy of Y at $x = 0$, thus obtaining a compact manifold with boundary M whose interior is $\mathring{M}$. The singular behaviour of the metric at the boundary $x = 0$ induces degenerations in the associated geometric differential operators at $x = 0$. These operators are defined on M including its boundary, and are expressible, in a smooth and non-degenerate way, in terms of vector fields tangent to the boundary. In terms of local coordinates $\{x, \mathbf{y}\} = \{x, y_1, \dots, y_n\}$ near a boundary point these vector fields are spanned by

$$x\partial_x, \partial_{y_1}, \dots, \partial_{y_n} . \tag{2.1}$$

More precisely, the geometric operators are b-differential operators. The set of b-differential operators of order m, denoted $\mathrm{Diff}_b^m(M)$, consists of order m differential operators on M which have smooth coefficients in the interior, and in coordinates near $x = 0$ can be written in the form

$$P = \sum_{j+|K| \le m} a_{j,K}(x, \mathbf{y})(x\mathcal{D}_x)^j (\mathcal{D}_\mathbf{y})^K, \tag{2.2}$$

where the functions $a_{j,K}$ are smooth up to the boundary $x = 0$, K is a multi-index, $\mathcal{D}_x = \frac{1}{i}\frac{\partial}{\partial x}$ and $(\mathcal{D}_\mathbf{y})^K = \mathcal{D}_{y_1}^{K_1} \cdots \mathcal{D}_{y_n}^{K_n}$. Similarly, for a vector bundle E over M we say $P \in \mathrm{Diff}_b^m(M, E)$ if P has, in local coordinates and with respect to a local bundle trivialization $E \cong \mathbb{R}^e$, the form (2.2) where the $a_{j,K}(x, \mathbf{y})$ are homomorphisms $\mathbb{R}^e \to \mathbb{R}^e$. Then P acts on sections of E.

In the setting of this paper, the degeneration of the ϕ-cusp metric is encoded by smooth vector fields on M that near any boundary point may be written as a smooth linear combination of vector fields of the form

$$x^{1+a}\partial_x, x^a\partial_{y_1}, \dots, x^a\partial_{y_b}, \partial_{z_1}, \dots, \partial_{z_f}, \tag{2.3}$$

where the boundary is $x = 0$, the coordinates y_i are lifted from the base B of the fibration $Y \xrightarrow{\phi} B$ and the z_j can be thought of as local fiber coordinates. Thus, a ϕ-differential operator of order m is an mth-order differential operator on M which near any boundary point can be written in the form:

$$P = \sum_{j+|K|+|L| \le m} a_{j,K,L}(x, \mathbf{y}, \mathbf{z})(x^{1+a}\mathcal{D}_x)^j (x^a\mathcal{D}_\mathbf{y})^K (\mathcal{D}_\mathbf{z})^L, \tag{2.4}$$

where the functions $a_{j,K,L}(x, \mathbf{y}, \mathbf{z})$ are smooth up to the boundary $x = 0$. We denote the algebra of ϕ-differential operators over M by $\mathrm{Diff}_\phi^*(M)$ or $\mathrm{Diff}_\phi^*(M, E)$. Then if A is a geometric differential operator of order m over M associated to the metric g, we have

$$A = x^{-ma}P, \qquad P \in \mathrm{Diff}_\phi^m(M, E)$$

where E is typically a variant of the form bundle ΛT^*M, see Section 3. Here and in the sequel $x \in C^\infty(M)$ always denotes a boundary defining function for M; locally near any boundary point it is a coordinate, and in the interior of M it is positive. It is quite reasonable to ask why we have put the factor of x^{-ma} out front, rather than building up our differential operators from vector fields of the forms $x\partial_x$, ∂_{y_i} and $x^{-a}\partial_{z_j}$. The reason is that those vector fields do not span a Lie algebra over smooth functions on M, whereas the vector fields in equation (2.3) do.

Differential operators, and the broader class of pseudodifferential operators associated to a Lie algebra of vector fields, are identified with their Schwartz kernels. These are distributions on the double space, M^2, that are singular at the diagonal. In order to separate this singularity from the degeneration at the boundary it is useful to lift these kernels to a particular model space obtained by a blow-up of M^2, where their structure can be described explicitly. The space M^2 and these model spaces are manifolds with corners. The space M^2 has two boundary hypersurfaces, $lf = \partial M \times M$ and $rf = M \times \partial M$. In the b-calculus setting, the model space M_b^2 has three boundary hypersurfaces, lf, rf and bf. In the ϕ setting, the model space M_ϕ^2 has four boundary hypersurfaces: lf, rf, bf and ff. The diagonal of M^2 lifts to M_b^2 and M_ϕ^2. Operators in the full calculi in both settings are defined by distributions on the appropriate model space that are conormal to the lifted diagonal and have polyhomogeneous expansions at each boundary face. The degree of the conormality of a kernel at the lifted diagonal determines the order of its associated operator, as in the setting of standard pseudodifferential operators. Its expansions at the various boundary faces determine how it maps between appropriate weighted Sobolev spaces.

In order to state the mapping results for elements of the full b and ϕ calculi that we will need in our parametrix construction, we need first to define weighted b and ϕ Sobolev spaces, as well as spaces of polyhomogeneous and conormal distributions. Then we will move on to a summary of the necessary results from the b and ϕ-calculi. In the interests of space, we will give only heuristic explanations, ones we find the most useful in applications, and refer the reader to the references [13], [4] and [7] for details of exactly how and why these definitions and theorems work. Note that the notation we adopt here conforms to that in [13] but not quite to that in [4]. For example, the space denoted $\Psi_b^{m,\mathcal{I}}(M, \Omega_b^{1/2})$ below is denoted $\Psi_b^{m,\mathcal{I}}(M)$ in [4].

2.1. Sobolev spaces and polyhomogeneous spaces

Throughout the paper we fix a b-volume form dvol_b on $\mathring{M}$. The index b indicates that we assume that in local coordinates near a boundary point, we can write

$$\mathrm{dvol}_b = a(x, \mathbf{y})\frac{dx}{x}dy_1, \ldots, dy_n$$

where a is smooth and positive up to the boundary $x = 0$. While this b-behavior near the boundary is important, the particular choice of dvol_b is inessential. We define Sobolev spaces always with respect to this volume form.

Elliptic b-differential operators map as Fredholm operators between appropriate b-Sobolev spaces, defined as follows. Let E be a vector bundle over M. Choose a cutoff function χ which is equal to 1 near $x = 0$ and equal to 0 for $x > \epsilon$. Then we define for $m \in \mathbb{N}_0$, $s \in \mathbb{R}$

$$\begin{aligned} x^s H_b^m(M, E, \mathrm{dvol}_b) = \{x^s u \mid u \in H^m_{\mathrm{loc}}(\mathring{M}, E), \\ (x\mathcal{D}_x)^j (\mathcal{D}_{\mathbf{y}})^K \chi u \in L^2(M, E, \mathrm{dvol}_b)\ \forall j + |K| \leq m\}. \end{aligned} \tag{2.5}$$

These spaces are independent of the choice of dvol_b and cutoff χ, local coordinates and local trivializations of E, and metric on E, and can be metrized using invertible b-pseudodifferential operators, but we will not go into these details here. This definition can be extended to $m \in \mathbb{R}$ by the usual arguments of duality and interpolation. For most applications $m \in \mathbb{N}_0$ is sufficient, but we state theorems in this greater generality. Similarly, the appropriate Sobolev spaces for ϕ-operators are expressed in terms of the vector fields in equation (2.3), so they are defined as follows:

$$\begin{aligned} x^s H_\phi^m(M, E, \mathrm{dvol}_b) = \{x^s u \mid u \in H^m_{\mathrm{loc}}(\mathring{M}, E), \\ (x^{1+a}\mathcal{D}_x)^j (x^a \mathcal{D}_{\mathbf{y}})^K (\mathcal{D}_{\mathbf{z}})^L \chi u \in L^2(M, E, \mathrm{dvol}_b)\ \forall j + |K| + |L| \leq m\}. \end{aligned} \tag{2.6}$$

Again, these can be metrized using invertible ϕ-differential operators (see [4]).

We next need to define polyhomogeneous sections, which are sections that have asymptotic expansions in terms of the boundary defining function x near ∂M. The expansions near the boundary for these sections involve powers of x and powers of $\log x$. Thus the type of these expansions can be described by index sets I, which are sets of pairs $(z, k) \in \mathbb{C} \times \mathbb{N}_0$, where the first term describes permitted powers of x and the second describes permitted powers of $\log x$. Index sets are required to satisfy $(z, k) \in I, l \leq k \Rightarrow (z, l) \in I$ and $(z + 1, k) \in I$ (this guarantees that the condition below is independent of the choice of boundary defining function x), and that for any r there is only a finite number of $(z, k) \in I$ satisfying $\mathrm{Re}\, z < r$. Polyhomogeneous sections over M are defined as follows.

Definition 1. Let I be an index set and E a vector bundle over M, a manifold with boundary. A smooth section of E over $\mathring{M}$ is said to belong to $\mathcal{A}^I_{\mathrm{phg}}(M, E)$ if for each $(z, k) \in I$ there exists a section $u_{(z,k)}$ of E, smooth up to the boundary of M, such that for all N, the difference

$$u - \sum_{\substack{(z,k)\in I \\ \mathrm{Re}(z) \leq N}} x^z (\log x)^k u_{(z,k)}$$

vanishes to Nth order at ∂M.

For an index set I, we say $I > \alpha$ if $(z, k) \in I$ implies $\mathrm{Re}\, z > \alpha$, and $I \geq \alpha$ if $(z, k) \in I$ implies $\mathrm{Re}\, z \geq \alpha$, and $k = 0$ for $\mathrm{Re}\, z = \alpha$.

When operators are composed or applied to sections, their index sets combine using two operations. The first is simple set addition, indicated by $+$. The second is an extended union:

$$I \bar{\cup} J = I \cup J \cup \{(z, k) \mid k = l_1 + l_2 + 1 \text{ where there exist } (z, l_1) \in I, (z, l_2) \in J\}.$$

For the index set $(r + \mathbb{N}_0) \times \{0\}$, where $r \in \mathbb{R}$, we use the short notation r.

Thus, $I+r$ is the index set I shifted by r. Note that for the index set $\emptyset$, we have $I\overline{\cup}\emptyset = I$, $I+\emptyset=\emptyset$ and $\emptyset+r=\emptyset$.

There is an analogous definition of polyhomogeneity for sections on manifolds with corners, for example the model double spaces M_b^2, M_ϕ^2. In this case an *index family* $\mathcal{I}$ has to be specified, which consists of an index set $\mathcal{I}_f$ for each boundary hypersurface f of the space. Also, the definition of polyhomogeneity extends to distributions on $\mathring{M}$ which are conormal with respect to a submanifold of M that meets the boundary in a product type fashion. The main example of such a submanifold is the lifted diagonal as a submanifold of M_b^2 or M_ϕ^2. Rather than go into the detail for this here, we refer the reader to [13], [4] and [7].

2.2. Mapping and regularity in the b-calculus

As mentioned above, Schwartz kernels of b-differential and b-pseudodifferential operators are best understood when considered as distributions not on the double space, M^2, but rather on a blow-up of this space, called M_b^2. When identifying an operator with its Schwartz kernel there is integration, hence a choice of measure or density involved, unless one considers the operator as acting on densities, or takes the kernel to be a density. Another possibility, which is more symmetric, is to think of both as half-densities. It is important to deal with this carefully because of the various singular coordinate changes involved in the blow-up of M^2 to M_b^2 and M_ϕ^2. We will go into this only briefly, since it is needed to define b- and ϕ-pseudodifferential operators properly, but in the end the half-densities will only act behind the scenes and not be visible.

Recall that a density on an oriented manifold M is a smooth top degree form, that is, a smooth section of the trivial one-dimensional vector bundle $\Omega = \Lambda^{\dim M}T^*M$. For us more important is the b-density bundle over an oriented manifold with boundary, M, denoted Ω_b. The volume form dvol_b is a smooth non-vanishing section of Ω_b. We can form the square root of this bundle, denoted $\Omega_b^{1/2}$. Near a boundary point its sections are of the form $u(x,\mathbf{y})\left(\frac{dx}{x}\,dy_1\ldots dy_n\right)^{1/2}$ in terms of (oriented) local coordinates $\{x,\mathbf{y}\}$, where the function u is smooth up to the boundary. Similarly, one has the b-half-density bundle on a manifold with corners.

We now define the space of b-pseudodifferential operators on a manifold with boundary, M. Recall that M_b^2 is obtained by blowing up the corner $\partial M\times\partial M$ of M^2. The front face of the blow-up is called bf, so M_b^2 has three boundary hypersurfaces, lf, rf and bf. Let $\mathcal{I}=(\mathcal{I}_{lf},\mathcal{I}_{rf},\mathcal{I}_{bf})$ be an index family for M_b^2. For $m\in\mathbb{R}$ we define the full b-calculus $\Psi_b^{m,\mathcal{I}}(M,\Omega_b^{1/2})$ as the set of $\Omega_b^{1/2}$-valued distributions on M_b^2 which are conormal with respect to the lifted diagonal of order m and polyhomogeneous at the boundary with index family $\mathcal{I}$. These operators act on b-half-densities on M. To extend this to operators acting on sections of a vector bundle E over M one proceeds in two steps. First, by tensoring kernels with sections of the $\mathrm{End}(E)$ bundle on M^2 lifted to M_b^2 we obtain operators acting on sections of $\Omega_b^{1/2} \otimes E$. Then we replace E by $E \otimes \Omega_b^{-1/2}$ in this construction and obtain operators acting on sections of E.

This space of kernels or operators is denoted by $\Psi_b^{m,\mathcal{I}}(M,E)$. For $\mathcal{I} = (\emptyset, \emptyset, 0)$, i.e., kernels vanishing to infinite order at *lf* and *rf* and smooth transversally to *bf*, we simply write $\Psi_b^m(M,E)$. This is called the small *b*-calculus. A simple calculation shows that $\mathrm{Diff}_b^m(M,E) \subset \Psi_b^m(M,E)$. The fact that here the index set at *bf* is simply 0 results from and justifies the use of *b*-half-densities instead of regular densities.

The expansions at *rf* and *bf* determine the domain of an operator in $\Psi_b^{m,\mathcal{I}}(M,E)$, and the expansions at *lf* and *bf* determine its range. This is made precise in the following theorems:

Theorem 3 (Boundedness and compactness for *b*-operators). *Let M be a compact manifold with boundary and let E be a vector bundle over M. Let $P \in \Psi_b^{m,\mathcal{I}}(M,E)$ and $\alpha, \beta \in \mathbb{R}$, $k \in \mathbb{R}$.*

1. *If*
$$\mathcal{I}_{lf} > \beta,\ \mathcal{I}_{rf} > -\alpha,\ \mathcal{I}_{bf} \geq \beta - \alpha, \tag{2.7}$$
then P is bounded as an operator
$$P : x^\alpha H_b^{k+m}(M,E,\mathrm{dvol}_b) \to x^\beta H_b^k(M,E,\mathrm{dvol}_b). \tag{2.8}$$
2. *If $m < 0$ and strict inequality holds everywhere in* (2.7) *then P, acting as in* (2.8), *is compact.*

Note that for the index set $\emptyset$, $\inf(\emptyset) = \infty$, so in particular, *b*-differential operators are bounded between equal weighted *b*-Sobolev spaces of appropriate orders. Next we can state the mapping property with respect to polyhomogeneous sections. Again the use of *b*-half-densities makes the index sets combine in a very simple manner:

Theorem 4 (Mapping of polyhomogeneous sections for *b*-operators). *Let $u \in \mathcal{A}^I_{\mathrm{phg}}(M,E)$ and $P \in \Psi_b^{m,\mathcal{J}}(M,E)$. Then if $\mathcal{J}_{rf} + I > 0$, we can define Pu and we get that $Pu \in \mathcal{A}^K_{\mathrm{phg}}(M,E)$, where $K = \mathcal{J}_{lf} \overline{\cup} (\mathcal{J}_{bf} + I)$.*

In order to state the parametrix and regularity results for *b*-elliptic operators, we need to discuss the two symbols that control ellipticity. Associated to a *b*-differential operator (2.2) is a *b*-symbol:

$${}^b\sigma_m(P)(x,\mathbf{y},\tau,\eta) = \sum\nolimits_{j+|K|=m} a_{j,K}(x,\mathbf{y})\tau^j\eta^K. \tag{2.9}$$

Then P is defined to be *b*-elliptic if ${}^b\sigma_m(P)(x,\mathbf{y},\tau,\eta)$ is invertible for $(\tau,\eta) \neq 0$. It turns out that this symbol is not enough to characterize Fredholm elements in the calculus, so we need an additional "boundary symbol" called the indicial operator. For a *b*-differential operator as above, this is the differential operator defined by

$$I(P) = \sum\nolimits_{j+|K|\leq m} a_{j,K}(0,\mathbf{y})(s\mathcal{D}_s)^j\mathcal{D}_{\mathbf{y}}^K.$$

Here $(s,\mathbf{y}) \in \mathbb{R}_+ \times \partial M$, and $I(P)$ acts on sections of the bundle $E_{|\partial M}$ pulled back to $\mathbb{R}_+ \times \partial M$. Under the Mellin transform, this becomes a holomorphic family of operators on ∂M, the indicial family:

$$I(P,\lambda) = \sum_{j+|K|\le m} a_{j,K}(0,\mathbf{y})\lambda^j \mathcal{D}_{\mathbf{y}}^K.$$

The concepts of b-principal symbol, indicial operator and indicial family can be generalized to operators in $\Psi_b^{m,\mathcal{I}}(M,E)$.

Definition 2. Let $P \in \Psi_b^{m,\mathcal{I}}(M,E)$ be b-elliptic. The sets $\mathrm{Spec}_b(P) \subset \mathbb{C} \times \mathbb{N}_0$ and $-\operatorname{Im}\operatorname{spec}(P) \subset \mathbb{R}$ are defined as:

$$\begin{aligned} \mathrm{Spec}_b(P) = \{(\lambda,k) \mid\ & I(P,\lambda) \text{ is not invertible} \\ & \text{on } C^\infty(\partial M,E) \text{ and has a pole of order } k+1 \text{ at } \lambda\}, \end{aligned}$$

$$-\operatorname{Im}\operatorname{spec}(P) = \{-\operatorname{Im}(\lambda) \mid I(P,\lambda) \text{ is not invertible on } C^\infty(\partial M,E)\}. \tag{2.10}$$

These sets are central in describing the mapping properties of the operator P. The set $-\operatorname{Im}\operatorname{spec}(P)$ is a discrete subset of $\mathbb{R}$.

We can now state the parametrix theorem for b-operators. In this paper we are not interested in the precise index sets, so we state it in the following form (cf. Proposition 5.59 in [13]).

Theorem 5 (Parametrix in the b-calculus). *Let $P \in \mathrm{Diff}_b^m(M,E)$ be b-elliptic. Then for each $\alpha \notin -\operatorname{Im}\operatorname{spec}(P)$ there is an index family $\mathcal{E}(\alpha)$ for M_b^2 determined by $\mathrm{Spec}_b(P)$ and satisfying*

$$\mathcal{E}(\alpha)_{lf} > \alpha, \quad \mathcal{E}(\alpha)_{rf} > -\alpha, \quad \mathcal{E}(\alpha)_{bf} \ge 0$$

and parametrices

$$Q_{b,r,\alpha},\ Q_{b,l,\alpha} \in \Psi_b^{-m,\mathcal{E}(\alpha)}(M,E),$$

such that

$$P \circ Q_{b,r,\alpha} = \mathrm{Id} - R_{b,r,\alpha}, \quad Q_{b,l,\alpha} \circ P = \mathrm{Id} - R_{b,l,\alpha},$$

where the remainders satisfy

$$R_{b,r,\alpha} \in x^\infty \Psi_b^{-\infty,\mathcal{E}(\alpha)}(M,E), \quad R_{b,l,\alpha} \in \Psi_b^{-\infty,\mathcal{E}(\alpha)}(M,E)x^\infty.$$

Note that the x^∞ factor on the left means that the kernel of $R_{b,r,\alpha}$ actually vanishes to infinite order at lf and bf, while the x^∞ factor on the right means that the kernel of $R_{b,l,\alpha}$ vanishes to infinite order at rf and bf.

Combined with Theorems 3 and 4, and a similar mapping result for the remainders, this gives (cf. Theorem 5.60 and Prop. 5.61 in [13]):

Theorem 6 (Fredholmness and regularity of elliptic b-operators). *Let $P \in \mathrm{Diff}_b^m(M,E)$ be b-elliptic. Then P is Fredholm as a map*

$$P : x^\alpha H_b^{k+m}(M,E,\mathrm{dvol}_b) \to x^\alpha H_b^k(M,E,\mathrm{dvol}_b)$$

for any $\alpha \notin -\operatorname{Im}\operatorname{spec}(P)$ and any $k \in \mathbb{R}$.

Further, if $u \in x^\alpha H_b^k(M, E, \mathrm{dvol}_b)$ *for some* $\alpha, k \in \mathbb{R}$, *then* $Pu \in \mathcal{A}^I_{\mathrm{phg}}(M, E)$ *implies that* $u \in \mathcal{A}^J_{\mathrm{phg}}(M, E)$, *where* $J = I \overline{\cup} K$ *for some index set* $K > \alpha$ *determined by* $\mathrm{Spec}_b(P)$.

In particular, if u has only Sobolev regularity, but is mapped by an elliptic b-differential operator to a section with an asymptotic expansion at ∂M, for instance if u is in the kernel of P, then u must also have an expansion at ∂M.

For simplicity, the theorems in this section have been formulated for compact manifolds with boundary such as our ϕ-manifold M. However, it is easy to extend them to non-compact manifolds with compact boundary such as $V = B \times [0, \epsilon)$, which we will consider later, under suitable support assumptions. Specifically, Theorem 3 holds if the Schwartz kernel of P is compactly supported, and Theorem 4 holds for compactly supported sections. For Definition 2 and Theorem 5 it is sufficient to require that P is b-elliptic (that is, ${}^b\sigma_m(P)$ is invertible) near the boundary, then of course the parametrix will only be valid near the boundary.

2.3. Mapping and regularity in the ϕ-calculus

The small ϕ-calculus, which contains parametrices for fully elliptic ϕ-operators, is simpler than the b-calculus, since one does not need to worry about complicated spectral or index sets. The price for this is the strong requirement of full ellipticity. Since the Gauss–Bonnet and Hodge Lapace operators are not fully elliptic, we need a bigger calculus, called the full ϕ-calculus. Operators in the full ϕ-calculus are again degenerate operators on a compact manifold with boundary, but now the boundary is assumed to be the total space of a fibre bundle $F \to \partial M \overset{\phi}{\to} B$. We always extend this fibration to a neighborhood of the boundary. We fix an order of degeneracy $a \in \mathbb{N}$. Recall that the degeneracy of ϕ-differential operators is described by the vector fields (2.3), where we always use coordinates $(x, y_1, \dots, y_b, z_1, \dots, z_f)$ adapted to the fibration. Sometimes we will write ϕ-differential operators (2.4) as

$$P = \sum\nolimits_{j+|K|\le m} P^{x,\mathbf{y}}_{j,K}\,(x^{1+a}\mathcal{D}_x)^j (x^a \mathcal{D}_y)^K \tag{2.11}$$

with $P^{x,\mathbf{y}}_{j,K}$ differential operators on the fibre $F_{x,\mathbf{y}}$, of order $\le m - j - |K|$.

Once again, we study these operators by considering distributions that live on a blown up double space, M^2_ϕ, that has four boundary hypersurfaces, *lf*, *rf*, *bf* and *ff*. It is a degree a blowup of the b-double space, M^2_b, and *ff* is the new boundary hypersurface created by this blowup. As for the b-calculus, we fix an index family $\mathcal{I} = (\mathcal{I}_{lf}, \mathcal{I}_{rf}, \mathcal{I}_{bf}, \mathcal{I}_{ff})$ for M^2_ϕ and define the full calculus $\Psi^{m,\mathcal{I}}_\phi(M, \Omega_b^{1/2})$ as the space of distributions on M^2_ϕ which are conormal with respect to the lifted diagonal of order m and polyhomogeneous at the boundary with index family $\mathcal{I}$. However, as is explained in [4, Section 3.4], for the ϕ-calculus it is natural to take these distributions valued in a half-density bundle denoted $\Omega^{1/2}_{b\phi}$ rather than in $\Omega^{1/2}_b$. Sections of $\Omega_{b\phi}$ behave like b-densities at lf, rf, bf, but like ϕ-densities at ff, that is, they are $x^{-a(b+1)}$ times a b-density there, so that the total exponent of x^{-1} is the same as the sum of all the exponents of x in the vector fields (2.3).

With this normalization, operators in $\Psi_\phi^{m,\mathcal{I}}(M,\Omega_b^{1/2})$ act naturally on b-half-densities on M. The extension to $\Psi_\phi^{m,\mathcal{I}}(M,E)$ for a vector bundle E over M works as before, and the normalization also gives $\mathrm{Diff}_\phi^m(M,E) \subset \Psi_\phi^m(M,E) := \Psi_\phi^{m,(\emptyset,\emptyset,\emptyset,0)}(M,E)$. The latter space is called the small ϕ-calculus.

We first note mapping properties of elements of the full ϕ-calculus on Sobolev and polyhomogeneous spaces.

Theorem 7 (Boundedness and compactness for ϕ-operators). *Let M be a ϕ-manifold and let E be a vector bundle over M. Let $P \in \Psi_\phi^{m,\mathcal{I}}(M,E)$ and $\alpha, \beta, k \in \mathbb{R}$.*

1. *If*

$$\mathcal{I}_{lf} > \beta,\ \mathcal{I}_{rf} > -\alpha,\ \mathcal{I}_{bf} \geq \beta - \alpha,\ \mathcal{I}_{ff} \geq \beta - \alpha \tag{2.12}$$

and

strict inequality holds in (2.12) *for at least one of bf, ff* (2.13)

then P is bounded as an operator

$$x^\alpha H_\phi^{k+m}(M,E,\mathrm{dvol}_b) \to x^\beta H_\phi^k(M,E,\mathrm{dvol}_b). \tag{2.14}$$

2. *If $m < 0$ and strict inequality holds everywhere in* (2.12) *then P, acting as in* (2.14), *is compact.*

Note that although the differentiations defining the ϕ-Sobolev spaces are those arising from the ϕ-vector fields (2.3), the volume form in (2.14) is still the b-volume form. This is not surprising as this volume form relates to the conditions on the weights, so the expansions at lf and rf, and elements of the full ϕ-calculus that vanish near ff are actually in the full b-calculus.

Theorem 8 (Mapping of polyhomogeneous sections for ϕ-operators). *Let $u \in \mathcal{A}_{\mathrm{phg}}^I(M,E)$ and $P \in \Psi_\phi^{m,\mathcal{J}}(M,E)$. Then if $\mathcal{J}_{rf} + I > 0$, we can define Pu and we get that $Pu \in \mathcal{A}_{\mathrm{phg}}^K(M,E)$, where $K = \mathcal{J}_{lf} \overline{\cup} (\mathcal{J}_{bf} + I) \overline{\cup} (\mathcal{J}_{ff} + I)$.*

We will also use the following result from [4] that says how elements of the full ϕ-calculus compose.

Theorem 9 (Composition in the full ϕ-calculus). *If $P \in \Psi_\phi^{m,\mathcal{I}}(M,E)$ and $Q \in \Psi_\phi^{m',\mathcal{J}}(M,E)$ and if $I_{rf} + J_{lf} > 0$, then $PQ \in \Psi_\phi^{m+m',\mathcal{K}}(M,E)$, where, with $A = a(b+1)$,*

$$\begin{aligned}
\mathcal{K}_{lf} &= \mathcal{I}_{lf} \overline{\cup} (\mathcal{I}_{bf} + \mathcal{J}_{lf}) \overline{\cup} (\mathcal{I}_{ff} + \mathcal{J}_{lf}),\\
\mathcal{K}_{rf} &= \mathcal{J}_{rf} \overline{\cup} (\mathcal{I}_{rf} + \mathcal{J}_{bf}) \overline{\cup} (\mathcal{I}_{rf} + \mathcal{J}_{ff}),\\
\mathcal{K}_{bf} &= (\mathcal{I}_{lf} + \mathcal{J}_{rf}) \overline{\cup} (\mathcal{I}_{bf} + \mathcal{J}_{bf}) \overline{\cup} (\mathcal{I}_{ff} + \mathcal{J}_{bf}) \overline{\cup} (\mathcal{I}_{bf} + \mathcal{J}_{ff}),\\
\mathcal{K}_{ff} &= (\mathcal{I}_{lf} + \mathcal{J}_{rf} + A) \overline{\cup} (\mathcal{I}_{bf} + \mathcal{J}_{bf} + A) \overline{\cup} (\mathcal{I}_{ff} + \mathcal{J}_{ff}).
\end{aligned}$$

In particular, if one factor is in the small ϕ-calculus and the other factor has index family $\mathcal{I}$ then the composition is defined and also has index family $\mathcal{I}$.

We now discuss some results in the small ϕ-calculus. To an operator $P \in \Psi_\phi^*(M,E)$ we may associate a ϕ-principal symbol. In particular, when $P \in$

$\mathrm{Diff}^m_\phi(M,E)$ is given by the sum in equation (2.4), the ϕ-principal symbol is given by:

$$^\phi\sigma_m(P)(x,\mathbf{y},\mathbf{z},\tau,\eta,\zeta) = \sum_{j+|K|+|L|=m} a_{j,K,L}(x,\mathbf{y},\mathbf{z})\tau^j\eta^K\zeta^L. \tag{2.15}$$

We say that P is ϕ-elliptic if its ϕ-principal symbol is invertible for $(\tau,\eta,\zeta)\neq 0$.

As in the b-calculus we need a second symbol to fully characterize Fredholm and regularity properties of ϕ-operators. This is the normal operator, which for a ϕ-differential operator P as in (2.11) is given by

$$N(P) = \sum_{j+|K|\le m} P^{0,\mathbf{y}}_{j,K}\mathcal{D}^j_T\mathcal{D}^K_Y.$$

This is a family of operators parametrized by $\mathbf{y}\in B$, acting on sections of the bundle $E_{|\partial M}$ pulled back to $\mathbb{R}_T\times\mathbb{R}^b_Y\times F_\mathbf{y}$ which are rapidly decaying as $|(T,Y)|\to\infty$. More generally, for operators $P\in\Psi^m_\phi(M,E)$ the normal operator may be defined using the restriction of the Schwartz kernel to *ff*. The *normal family* carries the same information as the normal operator. It is a family of operators on $F_\mathbf{y}$ parametrized by $(\tau,\eta)\in\mathbb{R}\times\mathbb{R}^b$ and by $\mathbf{y}\in B$. For ϕ-differential operators it is obtained by replacing $\mathcal{D}_T$ by τ and $\mathcal{D}_Y$ by η in the normal operator, so

$$\hat{N}(P)(\mathbf{y},\tau,\eta) = \sum_{j+|K|\le m} P^{0,\mathbf{y}}_{j,K}\tau^j\eta^K. \tag{2.16}$$

For computations the following characterization is useful (see [4, Equation (29)]): Fix $\tau\in\mathbb{R},\eta\in\mathbb{R}^b$ and $\mathbf{y}_0\in B$, in coordinates $\mathbf{y}$. Let $g(x,\mathbf{y}) = -\tau a^{-1}x^{-a}+\eta(\mathbf{y}-\mathbf{y}_0)x^{-a}$. Then for a section $u\in C^\infty(F_{\mathbf{y}_0},E)$ we have

$$\hat{N}(P)(\mathbf{y},\tau,\eta)u = \left(e^{-ig}Pe^{ig}u\right)_{|\mathbf{y}=\mathbf{y}_0,x=0} \tag{2.17}$$

If P is ϕ-elliptic and $N(P)$ is invertible (which is equivalent to invertibility of $\hat{N}(P)(\mathbf{y},\tau,\eta)$ for each $\mathbf{y},\tau,\eta$), then we say P is *fully elliptic*. We have (Theorem 9 from [4], but see also [11]):

Theorem 10 (Parametrices for elliptic ϕ-operators in the small ϕ-calculus). *Let $P\in\Psi^m_\phi(M,E)$ be a ϕ-elliptic ϕ-operator over a ϕ-manifold, M. Then there exists an operator $Q\in\Psi^{-m}_\phi(M,E)$ such that $PQ=I+R_1$ and $QP=I+R_2$, where $R_i\in\Psi^{-\infty}_\phi(M,E)$. If P is fully elliptic then Q can be chosen so that $R_i\in x^\infty\Psi^{-\infty}_\phi(M,E)$.*

This implies the following Fredholm and regularity result.

Theorem 11 (Fredholmness and regularity of fully elliptic ϕ-operators). *Let $P\in\Psi^m_\phi(M,E)$ be a fully elliptic ϕ-operator on a ϕ-manifold. Then P is Fredholm as a map $P: x^cH^k_\phi(M,E,\mathrm{dvol}_b)\to x^cH^{k-m}_\phi(M,E,\mathrm{dvol}_b)$ for any $c,k\in\mathbb{R}$. If u is a tempered distribution and $Pu\in x^\alpha H^k_\phi(M,E,\mathrm{dvol}_b)$, then $u\in x^\alpha H^{k+m}_\phi(M,E,\mathrm{dvol}_b)$. If u is in any weighted Sobolev space and $Pu\in\mathcal{A}^J_{\mathrm{phg}}(M,E)$ then in fact $u\in\mathcal{A}^J_{\mathrm{phg}}(M,E)$.*

3. Structure of the Laplacian

Assume that M is a ϕ-manifold with boundary $\partial M = Y \overset{\phi}{\to} B$ and ϕ-cusp metric g. In a neighborhood of the boundary we can extend the boundary fibration map, ϕ, to the interior using the product structure:

$$U \cong Y \times [0, \epsilon) \overset{\Phi=\phi\times\mathrm{id}}{\longrightarrow} V \cong B \times [0, \epsilon) \overset{\Phi_2}{\to} [0, \epsilon). \tag{3.1}$$

In order to understand the structure of the Gauss–Bonnet and Hodge Laplace operators it is useful to recall a few facts about Riemannian fibrations.

First, let us recall which structures are associated to a fibration $\phi : Y \to B$. The tangent bundle TY has the tangent spaces to the fibres as a natural subbundle, called the vertical tangent subbundle. The cotangent bundle T^*Y has the forms annihilating vertical vectors as a natural subbundle, called the horizontal cotangent subbundle. Let $\mathbf{y} = (y_1, \ldots, y_b)$ be local coordinates on B and lift them to a partial set of local coordinate functions on Y. Then the exterior derivatives $dy_1, \ldots, dy_b$ span the horizontal cotangent subbundle. Now supplement these coordinate functions to a complete set of local coordinates for Y and call the additional coordinates $\mathbf{z} = (z_1, \ldots, z_f)$. We may think of the z_j as fibre local coordinates. Choosing the z_j is equivalent to choosing a trivialization of the fibration ϕ locally on Y. The set of coordinates $y_1, \ldots, y_b, z_1, \ldots, z_f$ defines vector fields $\partial_{y_1}, \ldots, \partial_{y_b}, \partial_{z_1}, \ldots, \partial_{z_f}$. The vertical tangent subbundle is locally spanned by the vector fields $\partial_{z_1}, \ldots, \partial_{z_f}$.

Now consider the additional structures defined by Riemannian metrics g_Y and g_B on Y and B, respectively, and recall what it means for ϕ to be a Riemannian fibration. The metric g_Y defines the horizontal tangent subbundle (or horizontal distribution) of TY, as the orthogonal complement of the vertical tangent subbundle. For any $p \in Y$ the differential of the fibration $\phi : Y \to B$ restricts to an isomorphism of the horizontal subspace of T_pY to $T_{\phi(p)}B$, and the assumption that ϕ is a Riemannian fibration for the metrics g_Y, g_B means that this isomorphism is an isometry for each p. Equivalently, $g_Y = \phi^* g_B + h$ where the symmetric two-tensor h vanishes on the horizontal subspace. Dually, the dual metric on T^*Y defines the vertical cotangent subbundle as orthogonal complement of the horizontal cotangent subbundle, and the natural map $T^*Y \to T^*F$, given by restricting a form to TF, restricts to an isomorphism of the vertical cotangent space and T^*F. If ϕ is Riemannian then this isomorphism is an isometry.

In coordinates, the horizontal tangent subspace will usually *not* be spanned by the coordinate vector fields $\partial_{y_1}, \ldots, \partial_{y_b}$. A choice of coordinates $\mathbf{y}, \mathbf{z}$ for which this does happen is possible if and only if the horizontal distribution is integrable, that is, if there is a local foliation by submanifolds that are orthogonal and transversal to the fibers. By Frobenius' theorem this is equivalent to the vanishing of the curvature, denoted R, of the horizontal distribution. Note that in this case, the requirement that ϕ is a Riemannian fibration means that the horizontal submanifolds are also totally geodesic. In contrast, the fibres of ϕ will not generally be totally geodesic. This happens if and only if their second fundamental forms, denoted $\mathbb{II}$, vanish.

In coordinates $\mathbf{y}, \mathbf{z}$ the matrix of g_Y has the form

$$g_Y = \begin{pmatrix} A_{00}(\mathbf{y}) & A_{01}(\mathbf{y}, \mathbf{z}) \\ A_{10}(\mathbf{y}, \mathbf{z}) & A_{11}(\mathbf{y}, \mathbf{z}) \end{pmatrix} \tag{3.2}$$

where A_{00} depends only on $\mathbf{y}$ because ϕ is Riemannian. The submatrix A_{00} encodes g_B, and A_{01}, A_{10}, A_{11} encode h. If the curvature of the horizontal distribution vanishes then coordinates can be chosen so that the A_{01} and A_{10} terms vanish, and if in addition the second fundamental form of the fibres vanishes then the A_{11} term is a function only of the $\mathbf{z}$ (fibre) variables. So the vanishing of both terms means that the Riemannian fibration is locally trivial in a metric sense.

Now consider the metric g in (1.1), where we fix a product decomposition (3.1). Choose local coordinates $\mathbf{y}, \mathbf{z}$ associated to ϕ. We use the above discussion with h replaced by $x^{2a}h$. This would introduce factors x^{2a} in front of the A_{01}, A_{10}, A_{11} terms of (3.2). However, it is advantageous to write g in terms of the rescaled local coordinate vector fields

$$x\partial_x, \{\partial_{y_i}\}_{i=1}^b, \{x^{-a}\partial_{z_j}\}_{j=1}^f; \tag{3.3}$$

then the metric g has the form

$$g = \begin{pmatrix} 1 & 0 & 0 \\ 0 & A_{00}(\mathbf{y}) & x^a A_{01}(\mathbf{y}, \mathbf{z}) \\ 0 & x^a A_{10}(\mathbf{y}, \mathbf{z}) & A_{11}(\mathbf{y}, \mathbf{z}) \end{pmatrix}. \tag{3.4}$$

The fact that this is smooth and non-degenerate up to $x = 0$ motivates the introduction of the vector bundle ${}^{c\phi}TM$ over M, defined by the requirements that the vector fields (3.3) form a local basis of sections of ${}^{c\phi}TM$ over any coordinate patch in U and that sections of ${}^{c\phi}TM$ over $\mathring{M}$ are smooth vector fields on $\mathring{M}$. There is a canonical identification of ${}^{c\phi}TM$ and TM over $\mathring{M}$, but not over M, since $x^{-a}\partial_{z_j}$ is not defined as a section of TM at $x = 0$ and $x\partial_x$ vanishes there. Now (3.4) shows that g, which was defined only in the interior $x > 0$, defines a metric on the bundle ${}^{c\phi}TM$ over all of M, i.e., in $x \geq 0$. The dual bundle ${}^{c\phi}T^*M$ is locally spanned by $\frac{dx}{x}, dy_1, \dots, dy_b, x^a dz_1, \dots, x^a dz_f$, over U.

Similarly, over $V = B \times [0, \epsilon)$ we have the bundle ${}^bTV$ locally spanned by $x\partial_x, \partial_{y_1}, \dots, \partial_{y_b}$. The differential of Φ maps ${}^{c\phi}TU \to {}^bTV$ by sending $x^{-a}\partial_{z_j}$ to zero. From $V = B \times [0, \epsilon)$ we have ${}^bT^*V = \operatorname{span}\{\frac{dx}{x}\} \oplus T^*B$.

We apply the discussion above about Riemannian fibrations to the fibration $\Phi : U \to V$ and the bundles ${}^{c\phi}T^*U$, ${}^bT^*V$. The inclusion $TF \subset TU$ is replaced by $x^{-a}TF \subset {}^{c\phi}TU$ with the dual restriction map ${}^{c\phi}T^*U \to x^aT^*F$. We obtain the orthogonal decomposition of vector bundles

$${}^{c\phi}T^*U = \operatorname{span}\{\tfrac{dx}{x}\} \oplus \mathcal{H} \oplus \mathcal{V}. \tag{3.5}$$

Here $\operatorname{span}\{\frac{dx}{x}\} \oplus \mathcal{H}$ is the image of $\Phi^* : {}^bT^*V \to {}^{c\phi}T^*U$ and $\mathcal{V}$ is its orthogonal complement with respect to the dual of the metric (3.4). The map ${}^{c\phi}T^*U \to$

x^aT^*F restricts to an isomorphism $\mathcal{V} \to x^aT^*F$, which is an isometry since Φ is Riemannian. Under the identification of ${}^{c\phi}T^*U$ and T^*U over the interior, (3.5) is the horizontal/vertical decomposition with respect to the singular metric g; the use of the rescaled bundles allows us to extend this decomposition smoothly to the boundary.

The Gauss–Bonnet operator D_M and Hodge Laplacian Δ_M for the metric g are first defined over $\mathring{M}$ only, but when considered as operators acting on sections of the bundle $E = \Lambda^{c\phi}T^*M$ they are elliptic ϕ-cusp-operators on M, that is,

$$D_M = x^{-a}P,\ P \in \mathrm{Diff}^1_\phi(M, \Lambda^{c\phi}T^*M), \quad \Delta_M = x^{-2a}T,\ T \in \mathrm{Diff}^2_\phi(M, \Lambda^{c\phi}T^*M)$$

with P, T ϕ-elliptic. This follows from general principles as shown in [14] or from concrete calculation, see [8]. Concretely, this means that if we write forms on M near the boundary as

$$\sum_{K,L} \left(a_{K,L} + \tfrac{dx}{x} \wedge b_{K,L}\right), \quad a_{K,L} = \alpha_{K,L}\, d\mathbf{y}^K \wedge (x^a d\mathbf{z})^L, \quad b_{K,L} = \beta_{K,L}\, d\mathbf{y}^K \wedge (x^a d\mathbf{z})^L \tag{3.6}$$

for multiindices K, L and then consider how D_M, Δ_M act on the coefficients $\alpha_{K,L}$, $\beta_{K,L}$, then P, T are expressible in terms of the ϕ-vector fields (2.3), with coefficients smooth up to the boundary.

In addition, we can write D_M and Δ_M in terms of the vertical/horizontal decomposition, and this will be essential for our analysis. The decomposition (3.5) of ${}^{c\phi}T^*U$ induces a decomposition of $\Lambda^{c\phi}T^*U$. It is useful to write the exterior derivative d_U on U in terms of this orthogonal decomposition, rather than in coordinates $x, \mathbf{y}, \mathbf{z}$, since then its adjoint is easy to compute. We obtain (see [8]):

$$\begin{aligned} d_U - x^{-a}&d_F \\ &+ d_D\,|_{\mathcal{V}} + \Phi^* d_B \quad \mathrm{I\!I\!I} \\ &+ x^a R \\ &+ (\Phi_0 \circ \Phi)^*(d_x) - A. \end{aligned} \tag{3.7}$$

In this decomposition, the term d_F is the exterior derivative on the fibres, under the isomorphism $\mathcal{V} \to x^aT^*F$. The term $d_B\,|_{\mathcal{V}}$ represents the action of the derivatives in the B directions on the $\mathcal{V}$-components of a form and the term Φ^*d_B is the pullback to U of the B differential on $V = B \times [0, \epsilon)$. Here we identify $\mathcal{H}$ with T^*B via the differential of Φ. The terms $\mathrm{I\!I\!I}$ and R are the second fundamental form and curvature operators for the metric g. R is independent of x and $\mathrm{I\!I\!I}$ depends smoothly on x. The operator d_x is pulled back from $[0, \epsilon)$ via $\Phi_0 \circ \Phi : U \to [0, \epsilon)$ in (3.1). It first acts by $x\partial_x$ then wedges with $\frac{dx}{x}$, so it is a b-operator. The 0th-order differential operator A acts on each summand in (3.6) by

$$A(a_{K,L} + \tfrac{dx}{x} \wedge b_{K,L}) = a|L|\,\tfrac{dx}{x} \wedge a_{K,L}. \tag{3.8}$$

Note that in terms of the decomposition of forms by subbundle degrees, we get that d_F increases the $\mathcal{V}$ degree by 1, the various d_B and $\mathrm{I\!I\!I}$ terms increase the $\mathcal{H}$ degree by 1 and R increases the $\mathcal{H}$ degree by 2 and decreases the $\mathcal{V}$ degree by 1.

The d_x and A terms increase the span$\{\frac{dx}{x}\}$ degree by 1. We define the following combination to simplify notation:

$$d_B - \mathrm{III} := d_B\mid_{\mathcal{V}} + \Phi^* d_B - \mathrm{III}.$$

Passing to the Gauss–Bonnet operator, we fix the following notation:

$$\begin{aligned} D_U &= d_U + d_U^*,\ D_B = (d_B - \mathrm{III}) + (d_B - \mathrm{III})^*,\ D_F = d_F + d_F^*, \\ \mathbf{R} &= R + R^*,\ xD_x = (\Phi_0 \circ \Phi)^* d_x + ((\Phi_0 \circ \Phi)^* d_x)^*,\ \mathbf{A} = A + A^*. \end{aligned} \tag{3.9}$$

Then we get

$$D_U = x^{-a} D_F + D_B + x^a \mathbf{R} + xD_x - \mathbf{A}. \tag{3.10}$$

3.1. The "split" property for the Gauss–Bonnet operator and the Hodge Laplacian

In this section, we identify some of the properties of the Gauss–Bonnet operator, D_U, and the Hodge Laplacian, Δ_U, that will permit us to construct parametrices for them using a combination of b- and ϕ-calculus operators, and that form the model for more general "split elliptic" operators in [6]. When multiplied with x^a and x^{2a}, respectively, both operators are elliptic as ϕ-operators in the standard sense, so we will focus on additional properties they have in the neighbourhood U of the boundary.

We calculate the normal operator of $x^a D_U$. First, we consider the normal family of $x^a d_U$. Recall that the normal family of a ϕ-operator in $\mathrm{Diff}^m_\phi(M, \Lambda^{c\phi}T^*M)$ is a family of operators at the boundary acting on $C^\infty(F_{\mathbf{y}}; \Lambda^{c\phi}T^*M)$ and parametrized by $(\mathbf{y}, \tau, \eta) \in B \times \mathbb{R}^{b+1}$. Consider the bundle ${}^{c\phi}T^*M$ at a boundary point $(0, \mathbf{y}, \mathbf{z})$. Since the metric (3.4) is block diagonal at $x = 0$, the decomposition (3.5) reduces naturally to ${}^{c\phi}T^*_{(0,\mathbf{y},\mathbf{z})}M = W^*_{\mathbf{y}} \oplus x^a T^*_{\mathbf{z}} F_{\mathbf{y}}$ where $W_{\mathbf{y}} = {}^bT_{(0,\mathbf{y})}V$. Since $W_{\mathbf{y}}$ is a vector space, we have $W^*_{\mathbf{y}} = T^*_w W_{\mathbf{y}}$ for any $w \in W_{\mathbf{y}}$. If we identify $x^a T^*_{\mathbf{z}} F_{\mathbf{y}}$ with $T^*_{\mathbf{z}} F_{\mathbf{y}}$ and change the order, then we get a natural identification, for any $w \in W_{\mathbf{y}}$,

$${}^{c\phi}T^*_{(0,\mathbf{y},\mathbf{z})}M = T^*_{(\mathbf{z},w)}(F_{\mathbf{y}} \times W_{\mathbf{y}}).$$

When calculating $\hat{N}(x^a d_U)$ from (3.7), the zero-order terms III, R, A drop out because of the additional factor x^a. Using (2.17) one easily computes

$$\hat{N}(x^a d_U)(\mathbf{y}, \tau, \eta)u = d_F u + i(\tau \tfrac{dx}{x} + \eta\, d\mathbf{y}) \wedge u.$$

When interpreting τ, η as coordinates on ${}^bT^*_{(0,\mathbf{y})}V = W^*_{\mathbf{y}}$, the second term is simply the symbol of the exterior derivative operator $d_{W_{\mathbf{y}}}$ on the vector space $W_{\mathbf{y}}$, so $N(x^a d_U)(\mathbf{y}) = d_{F_{\mathbf{y}}} + d_{W_{\mathbf{y}}}$. Taking adjoints, where the scalar product on $W_{\mathbf{y}}$ is given by $g_{\mathbf{y}} := \frac{dx^2}{x^2} + g_B(\mathbf{y})$, we get

$$N(x^a D_U)(\mathbf{y}) = D_{F_{\mathbf{y}}} + D_{W_{\mathbf{y}}} \tag{3.11}$$

acting on $\Lambda T^*(F_{\mathbf{y}} \times W_{\mathbf{y}})$, which is the Gauss–Bonnet operator on $F_{\mathbf{y}} \times W_{\mathbf{y}}$ with metric $h_{|F_{\mathbf{y}}} + g_{\mathbf{y}}$. Note that this shows that the space $\mathbb{R}^{b+1}$ in the local definition of the normal family is, invariantly, $W_{\mathbf{y}} = {}^bT_{(0,\mathbf{y})}V$.

Define the vector bundle

$$\mathcal{K} = \text{the bundle of } F\text{-harmonic forms over } V. \tag{3.12}$$

More precisely, for any $\mathbf{y} \in B$ let $\tilde{\mathcal{K}}_{\mathbf{y}} = \ker D_F \subset C^\infty(F_{\mathbf{y}}, \Lambda x^a T^* F_{\mathbf{y}})$ be the space of harmonic forms on $F_{\mathbf{y}}$ for the metric $g_{|F_{\mathbf{y}}} = x^{2a} h_{|F_{\mathbf{y}}}$, and let $\mathcal{K}_{(x,\mathbf{y})} = (\Lambda^b T^*_{(x,\mathbf{y})} V) \otimes \tilde{\mathcal{K}}_{\mathbf{y}} \subset C^\infty(F_{(x,\mathbf{y})}, \Lambda^{c\phi} T^* U)$, using the isometric identification of $x^a T^* F$ with $\mathcal{V}$. By the Hodge theorem, $\dim \tilde{\mathcal{K}}_{\mathbf{y}}$ is finite and independent of $\mathbf{y}$, so $\mathcal{K}$ is a vector bundle over V. Note that we keep the x factors in the bundle. In Section 6.2 we explain how $\mathcal{K}$ relates to the bundle $\overline{\mathcal{K}}$ mentioned in the introduction. It is important to distinguish these bundles carefully in order to get the weight conditions in the main theorems right. We also define the infinite-dimensional vector bundle

$$\mathcal{C} = \mathcal{K}^\perp = \text{forms on } F \text{ that are perpendicular to the fibre harmonic forms,} \tag{3.13}$$

with respect to the L^2 scalar product on the fibres. We have $C^\infty(U, \Lambda^{c\phi} T^* M) = C^\infty(V, \mathcal{K}) \oplus C^\infty(V, \mathcal{C})$. We refer to sections of $\mathcal{C}$ as "fibre-perpendicular forms" over V.

Now we can define the following families of operators pointwise over V:

$$\Pi = \text{orthogonal projection onto } \mathcal{K} \tag{3.14}$$

$$\Pi^\perp := I - \Pi = \text{orthogonal projection onto } \mathcal{C} \tag{3.15}$$

Observe that the family $\Pi(x, \mathbf{y})$ of projection operators is smooth in the variable y since the family of metrics on F is smooth, and constant in x, since these metrics are independent of x. These families of projection operators on $C^\infty(F, \Lambda^{c\phi} T^* U)$ together define operators, also denoted Π and $\Pi^\perp$, on $C^\infty(U, \Lambda^{c\phi} T^* U)$.

Note that the normal operator $N(x^a D_M) = D_F + D_W$ commutes with Π. In particular, it maps $C^\infty(V, \mathcal{C}) \to C^\infty(V, \mathcal{C})$. Its restriction to this subspace is invertible since its square is $(\Delta_F + \Delta_W)_{|C^\infty(V,\mathcal{C})} \geq \Delta_{F|C^\infty(V,\mathcal{C})} \geq \min_{\mathbf{y} \in B} \lambda_{\mathbf{y}} > 0$, where $\lambda_{\mathbf{y}}$ is the smallest positive eigenvalue of $\Delta_{F_{\mathbf{y}}}$. We can decompose the restriction D_U of D_M to U using these projections as the sum of four pieces:

$$(D_U)_{00} := \Pi D_U \Pi, \qquad (D_U)_{01} := \Pi D_U \Pi^\perp,$$
$$(D_U)_{10} := \Pi^\perp D_U \Pi, \qquad (D_U)_{11} := \Pi^\perp D_U \Pi^\perp.$$

In order to construct the parametrix for D_M, we need to study the properties of each of these four pieces. It is convenient to organise these in a matrix:

$$D_U = \begin{pmatrix} (D_U)_{00} & (D_U)_{01} \\ (D_U)_{10} & (D_U)_{11} \end{pmatrix} =: x^{-a} \begin{pmatrix} x^a P_{00} & x^a P_{01} \\ x^a P_{10} & P_{11} \end{pmatrix}. \tag{3.16}$$

In [8], it is proved that the operator $\mathfrak{d} := \Pi(d_B - \mathbb{III})\Pi$ is the differential on forms over B with values in $\mathcal{K}$, with respect to the flat connection on $\mathcal{K}$ induced by the metric g. When we project D_U using Π we obtain from (3.10):

$$\Pi D_U \Pi = P_{00} = \mathfrak{d} + \mathfrak{d}^* + x D_x - \mathbf{A} + x^a \Pi \mathbf{R} \Pi, \tag{3.17}$$

which is a b-operator acting on sections of $\mathcal{K}$. The fact that $\eth$ is the differential on $\mathcal{K}$-valued differential forms over B implies that P_{00} is b-elliptic. The operators P_{01}, P_{10} and P_{11} have smooth coefficients up to $x = 0$ and are ϕ-operators in a sense made precise in the next section.

The Hodge Laplacian is the square of the Gauss–Bonnet operator, so we can understand it also in terms of its $\mathcal{K}$ and $\mathcal{C}$ components. Using (3.16) we obtain

$$\begin{aligned}\Delta_U &= x^{-2a}\begin{pmatrix} x^{2a}(P_{00})^2 + x^{2a}P_{01}P_{10} & x^{2a}P_{01}x^{-a}P_{11} + x^{2a}P_{00}P_{01} \\ x^a P_{11}P_{10} + x^{2a}P_{10}P_{00} & x^a P_{11}x^{-a}P_{11} + x^{2a}P_{10}P_{01}\end{pmatrix} \\ &=: x^{-2a}\begin{pmatrix} x^{2a}T_{00} & x^a T_{01}' + x^{2a}T_{01} \\ x^a T_{10}' + x^{2a}T_{10} & T_{11}\end{pmatrix}.\end{aligned} \tag{3.18}$$

Squaring (3.10) we see that $T_{00} = \Pi D_U^2 \Pi$ is a second-order b-operator acting on sections of $\mathcal{K}$, which is b-elliptic. The term $x^{2a}P_{10}P_{01}$ as well as $x^a P_{11}x^{-a} - P_{11}$ vanish at the boundary, so the normal family of the lower right term is the same as the normal family of P_{11}^2.

The critical difference between analysis of the Gauss–Bonnet and Hodge Laplace operators comes from the fact that the off-diagonal terms in $x^{2a}\Delta_M$ generally vanish like x^a. It turns out that our parametrix construction requires the greater order of vanishing x^{2a}, so this introduces a restriction on the metrics we consider in this paper. Since xD_x and $\mathbf{A}$ commute with Π, we have from (3.10)

$$P_{01} = \Pi\,[D_B + x^a\mathbf{R}]\,\Pi^\perp, \quad P_{10} = \Pi^\perp\,[D_B + x^a\mathbf{R}]\,\Pi,$$

so we see that the problematic term in each one is the one involving D_B. The term vanishes if this operator commutes with Π. Also note that, in this case, we have $T_{00} = (P_{00})^2 + O(x^{2a})$. The commutation condition is satisfied in the setting of $\mathbb{Q}$-rank 1 locally symmetric spaces. However, we are also interested in understanding how to study manifolds that generalise symmetric spaces. This question of how better to characterise boundary fibre bundles with this property was studied by J. Mueller in [16], and he is taking it up again together with the authors of this paper in current work.

4. The extended ϕ-calculus

In this section we consider how to lift fiber-harmonic b-kernels to the ϕ-double space. The motivation for this is as follows. Philosophically, we will create a parametrix for the Gauss–Bonnet and Hodge Laplace operators over a ϕ manifold by using the ϕ parametrix for the fiber-perpendicular part of the operator and the b-parametrix for the fiber-harmonic part of the operator near the boundary, and the standard parametrix on the interior. If the off-diagonal terms of our operators with respect to the fiber-harmonic, fiber-perpendicular splitting vanish in a neighborhood of the boundary, then in fact this simple diagonal parametrix suffices as a parametrix for the operator. However, when there are off-diagonal terms, we will need to improve this parametrix. In order to do this, and to understand the result we obtain from this, we need to understand how the b-parametrix composes with various parts of our ϕ-operator. We do this by reinterpreting (lifting) the b-parametrix as a ϕ-operator and then using the composition theorem for ϕ-operators, Theorem 9. Since the fibration ϕ is only defined near the boundary, these considerations take place on the neighborhood $U \equiv \partial M \times [0, \epsilon)$ and its projection $V = B \times [0, \epsilon)$.

In order to analyze the lift of b-operators on V to ϕ-operators on U, let us recall the spaces on which the Schwartz kernels of these operators live, i.e., are conormal. The boundary fibration determines the blown-up double space where the Schwartz kernels of ϕ-operators on U live. We denote this space, constructed in [4], by U^2_ϕ. It is constructed as an a-quasihomogeneous blow-up from the b-double space U^2_b. We also have the b-double space V^2_b, and we can do the corresponding a-quasihomogeneous blow-up of it, with respect to the trivial fibration $B \to B$ of the boundary of V, to obtain V^2_ϕ. It is useful to put the various double spaces in a commutative diagram:

$$\begin{array}{ccc} U^2_\phi & \xrightarrow{\tilde{\tilde{\Phi}}^2} & V^2_\phi \\ \big\downarrow{\scriptstyle \beta_{\phi,Y}} & & \big\downarrow{\scriptstyle \beta_{\phi,B}} \\ U^2_b & \xrightarrow{\tilde{\Phi}^2} & V^2_b \\ \big\downarrow{\scriptstyle \beta_{b,Y}} & & \big\downarrow{\scriptstyle \beta_{b,B}} \\ U^2 & \xrightarrow{\Phi^2} & V^2. \end{array} \tag{4.1}$$

The rows in this diagram represent fiber bundles with fibres F^2. The columns are sequences of blow-down maps. The diagram embodies the fact that in both the b- and ϕ-blowups of the U double space, the fibres F^2 are carried along like parameters. In the parametrix construction we construct a local parametrix for the fiber harmonic part of the operator on V^2_b (with coefficients in the harmonic form bundle). To combine this with the other pieces of the parametrix, we need to lift this vertically under $\beta_{\phi,B}$, then horizontally by $\tilde{\tilde{\Phi}}^2$ (the other order would work the same, of course), to get a kernel on U^2_ϕ. For notational simplicity in the rest of this section we will suppress the bundle coefficients E.

We first consider the lift of kernels under the vertical map $\beta_{\phi,B}$ in (4.1).

Proposition 1. *If $T \in \Psi^{m,\mathcal{I}}_b(V;\mathcal{K})$ then its kernel, K_T, lifts by the map $\beta_{\phi,B}$ to give the kernel of an operator $\tilde{T} \in \Psi^{m,\mathcal{J}}_\phi(V,\mathcal{K}) + \Psi^{-\infty,\mathcal{J}'}_\phi(V,\mathcal{K})$, where $\mathcal{J}$, $\mathcal{J}'$ agree with $\mathcal{I}$ at lf, rf, bf and*

$$\mathcal{J}_{ff} = \mathcal{I}_{bf} + a(-m), \qquad \mathcal{J}'_{ff} = \mathcal{I}_{bf} + a((-m)\overline{\cup}(b+1)), \tag{4.2}$$

where $(-m)$ denotes the index set $(-m + \mathbb{N}_0) \times \{0\}$ for any $m \in \mathbb{R}$ and $aJ = \{(az, k) : (z, k) \in J\}$.

The proof of this proposition will be given in [6]. However, for intuition we can make the following remarks.

1. Note that the shift by $-am$ in the first term at the front face in (4.2) is easily explained in the case of differential operators. For example, $x^m \partial_x^m \in \Psi_b^m$, but only after multiplying it with x^{am} does it become an operator in Ψ_ϕ^m. However, we will only use the proposition for lifting the parametrix, i.e., for negative m.
2. The second summand of $\tilde{T}$ may have log terms in its expansion at the front face even if T doesn't (encoded by the sign $\overline{\cup}$ in the term $(-m)\overline{\cup}(b+1)$ in (4.2)); these come from log terms in the expansion of the kernel of T at the diagonal.

We now consider the lift of a kernel under the horizontal map $\tilde{\tilde{\Phi}}^2$ in (4.1), more precisely, the lift of the kernel of an operator in $\Psi_\phi^m(V,\mathcal{K})$ to a kernel on U_ϕ^2. Consider kernels of operators in $\Psi_\phi^m(U)$. These are distributions on U_ϕ^2 with a conormal singularity along the lifted diagonal. We can think of these as kernels of the form $K(x,x',y,y',z,z')$ which are conormal to the diagonal $x = x', y = y', z = z'$ uniformly (in an appropriate rescaled sense given by the blowup) down to $x = x' = 0$.

By comparison, the kernel of an operator in $\Psi_\phi^m(V,\mathcal{K})$ is a finite sum of terms of the form $K_B(x,x',y,y') \otimes L(z,z')$, where K_B is a distribution on V_ϕ^2 with a conormal singularity along the lifted diagonal. Again, we can think of this as being conormal to the set $x = x', y = y'$ uniformly up to the boundary in the same rescaled sense as above. The functions $L \in \mathcal{K} \otimes \mathcal{K}$ in this sum are smooth since harmonic forms on F are smooth. To lift kernels of operators in $\Psi_\phi^m(V,\mathcal{K})$ horizontally, we simply consider them as distributions on U_ϕ^2. Note that these lifts are not the kernels of pseudodifferential operators. This is because pseudodifferential operators are singular only along the diagonal, $x = x', y = y', z = z'$, whereas these horizontally lifted distributions are singular on the larger set $x = x', y = y'$, which we call the *fibre diagonal.* When we have lifted the kernel of an element in $\Psi_\phi^m(V,\mathcal{K})$, we will also generally want to glue it to an interior kernel, so we will multiply the result by a smooth cutoff function supported compactly in U_ϕ^2 and equal to 1 near *bf* and *ff*. We define a space of operators on U in which these cutoff lifts live as follows.

Definition 3. Let $\Psi_{\phi,F}^m(U)$ be the space of operators which are ϕ-operators on V valued in smoothing operators on F, and with kernels compactly supported in U_ϕ^2. That is, the kernels of operators in $\Psi_{\phi,F}^m(U)$ are conormal of degree m with respect to the fibre diagonal, smoothly up to the front face, and vanish to infinite order at the other faces.

We make some notes about this space.

1. Note that this space does not require that the smoothing operators on F vanish on fibre-perpendicular sections. Thus this space contains more than just the horizontal lifts of operators in $\Psi_\phi^m(V,\mathcal{K})$.

2. We may generalise this space in a natural way to include operators with polyhomogeneous expansions given by an index family $\mathcal{G}$ at the boundary hypersurfaces in U^2_ϕ, which we will denote by $\Psi^{m,\mathcal{G}}_{\phi,F}(U)$.
3. In Equation (3.14), we defined the projection operator Π and we used it in our decomposition of the Gauss–Bonnet operator. Note that this is the lift of the identity operator in $\Psi^0_\phi(V,\mathcal{K})$ to an operator in $\Psi^0_{\phi,F}(U)$. It is not a pseudifferential operator on U.

Now consider the operator $\Pi^\perp$. This is not smoothing in the F factor, so it is not an element of $\Psi^0_{\phi,F}(U)$. Rather, it is the difference of the identity element in $\Psi^0_\phi(U)$ and $\Pi \in \Psi^0_{\phi,F}(U)$. Thus when we consider the compositions $\Pi x^a D_U \Pi$, $\Pi^\perp x^a D_U \Pi$ and $\Pi x^a D_U \Pi^\perp$, we will also arrive at such sums. This motivates the following definition.

Definition 4. Define the extended calculus as the sum of two pieces:

$$\Psi^{m,\mathcal{G}}_{\phi,\mathrm{ext}}(M) = \Psi^{m,\mathcal{G}}_{\phi}(M) + \Psi^{m,\mathcal{G}}_{\phi,F}(U). \tag{4.3}$$

As noted above, elements in this calculus need not be pseudodifferential operators since their Schwartz kernels may have singularities outside the diagonal. It should be noted, however, that this is a minor extension which serves only to formulate the parametrix construction in a simple way. The final parametrix that we consider is actually a pseudodifferential operator in the interior of M. This follows from ellipticity by the classicial pseudodifferential calculus.

There is no interior symbol map for operators in the extended calculus. However, the normal operator is still defined for elements in the subspace $\Psi^m_{\phi,\mathrm{ext}}(M)$. Recall that the normal operator of the ϕ-calculus takes values in the space of suspended pseudodifferential ϕ-operators, which we denote by $\Psi^m_{\mathrm{sus}-\phi}(\partial M)$. These are pseudodifferential operators on $\mathbb{R}_T \times \mathbb{R}^b_Y \times F$ (locally near a point of B) which are translation invariant in T, Y, so that they are given by a convolution kernel $K(T, Y, z, z')$ with a conormal singularity at $T = 0, Y = 0, z = z'$, and such that K decays rapidly as $(T, Y) \to \infty$. By adding terms that satisfy the same condition except that they have a conormal singularity on $T = 0, Y = 0$, we obtain the extended suspended calculus, denoted $\Psi^m_{\mathrm{sus}-\phi,\mathrm{ext}}(\partial M)$.

The following proposition shows that $\Psi^m_{\phi,\mathrm{ext}}(M)$, $\Psi^m_{\mathrm{sus}-\phi,\mathrm{ext}}(\partial M)$ behave much like the standard ϕ-pseudodifferential calculus and suspended calculus, respectively. As before, we suppress the bundle E in the notation.

Proposition 2.

1. *$\Psi^*_{\phi,\mathrm{ext}}(M)$ and $\Psi^*_{\mathrm{sus}-\phi,\mathrm{ext}}(\partial M)$ are closed under composition.*
2. *$\Psi^m_{\phi,\mathrm{ext}}(M) \circ \Psi^{-\infty}_\phi(M) \subset \Psi^{-\infty}_\phi(M)$, $\Psi^{-\infty}_\phi(M) \circ \Psi^m_{\phi,\mathrm{ext}}(M) \subset \Psi^{-\infty}_\phi(M)$ and similar results hold for $\Psi^*_{\mathrm{sus}-\phi,\mathrm{ext}}$.*
3. *Let $P \in \mathrm{Diff}^m_\phi(M)$ be ϕ-elliptic and such that $N(P)$ is diagonal with respect to the splitting $C^\infty(\mathbb{R}^{b+1} \times F) = C^\infty(\mathbb{R}^{b+1}, \mathcal{K}) \oplus C^\infty(\mathbb{R}^{b+1}, \mathcal{C})$ and so that its*

$C^\infty(\mathbb{R}^{b+1},\mathcal{C}) \to C^\infty(\mathbb{R}^{b+1},\mathcal{C})$ part is invertible. Let

$$B = \begin{cases} \left[N(P)_{|C^\infty(\mathbb{R}^{b+1},\mathcal{C})\to C^\infty(\mathbb{R}^{b+1},\mathcal{C})}\right]^{-1} & \text{on } C^\infty(\mathbb{R}^{b+1},\mathcal{C}) \\ 0 & \text{on } C^\infty(\mathbb{R}^{b+1},\mathcal{K}). \end{cases} \tag{4.4}$$

Then $B \in \Psi^{-m}_{\mathrm{sus}-\phi,\mathrm{ext}}(\partial M)$.

4. *The short exact sequence for the normal operator of the ϕ-calculus extends to a short exact sequence, for each m,*

$$0 \longrightarrow x\Psi^m_{\phi,\mathrm{ext}}(M) \hookrightarrow \Psi^m_{\phi,\mathrm{ext}}(M) \xrightarrow{N} \Psi^m_{\mathrm{sus}-\phi,\mathrm{ext}}(\partial M) \longrightarrow 0.$$

5. *Any $A \in \Psi^m_{\phi,\mathrm{ext}}(M)$ defines a bounded operator, for all k,*

$$H^{k+m}_\phi(M,\mathrm{dvol}_b) \to H^k_\phi(M,\mathrm{dvol}_b).$$

Proof. 1. This follows essentially the same lines as the proof that $\Psi^*_\phi(M)$ is closed under composition, see Theorem 8 in [4].

2. This follows from $\Psi^{-\infty}_{\phi,\mathrm{ext}}(M) = \Psi^{-\infty}_\phi(M)$.

3. We show this by a variation on the well-known argument that the inverse of an invertible pseudodifferential operator is pseudodifferential again. Since P is ϕ-elliptic, $N := N(P)$ is an elliptic element of $\Psi^m_{\mathrm{sus}-\phi}(\partial M)$. Hence it has a parametrix $C \in \Psi^{-m}_{\mathrm{sus}-\phi}(\partial M)$ so that $NC = I+R$, $CN = I+R'$ with $R, R' \in \Psi^{-\infty}_{\mathrm{sus}-\phi}(\partial M)$. We multiply these identities from the left and right with B respectively and solve for B. This gives $B = BNC - BR = BNC - (CNB - R'B)R$. Now use $NB = BN = \Pi^\perp = \mathrm{Id} - \Pi$ to obtain $B = C - \Pi C - S$ where $S = -C\Pi^\perp R + R'BR \in \Psi^{-\infty}_{\mathrm{sus}-\phi}(\partial M)$ by part 1 and standard facts, so $B \in -\Pi C + \Psi^{-m}_{\mathrm{sus}-\phi}(\partial M)$. The result follows.

4. This is obvious; compare Lemma 4 in [4].

5. By composing with invertible fully elliptic ϕ-operators one reduces to the case $k = m = 0$. Since ϕ-operators of order zero are bounded on L^2, we only need to check L^2-boundedness for $P \in \Psi^0_{\phi,F}(U)$. Using a partition of unity we may assume that the kernel of P is supported near the fibre over a point of $B \times B$. Then by fixing $z, z' \in F$ we may regard P as a family of ϕ-operators on V parametrized by $z, z' \in F$. Each of these is bounded on $L^2(V)$, uniformly in z, z'. The usual argument showing that Hilbert–Schmidt operators are bounded in L^2 then shows that P is bounded on $L^2(U)$. □

Because our aim in this paper is to present a clear construction, not to obtain the sharpest possible results (for example with respect to identifying the precise index sets that arise), we will work with the following spaces of b and ϕ-pseudodifferential operators in the construction. In these spaces, we require only certain leading order behaviour of the kernel expansions at the various faces in the blown-up double spaces. This makes the construction easier to follow, but of course, it only tells us that the resulting operator has expansions at the various faces satisfying certain bounds on the exponents in the leading terms.

Definition 5. For $k, \alpha \in \mathbb{R}$ define the space of b-pseudodifferential operators of order k and weight α as

$$\Psi_b^{k,\alpha}(M) = \bigcup_{\mathcal{E}} \Psi_b^{k,\mathcal{E}}(M),$$

where the union is over those index families $\mathcal{E}$ for M_b^2 satisfying $\mathcal{E}_{lf} > \alpha$, $\mathcal{E}_{rf} > -\alpha$, $\mathcal{E}_{bf} \geq 0$. Similarly, let

$$\Psi_\phi^{k,\alpha}(M) = \bigcup_{\mathcal{E}} \Psi_\phi^{k,\mathcal{E}}(M),$$

where the union is over those index families $\mathcal{E}$ satisfying $\mathcal{E}_{lf} > \alpha$, $\mathcal{E}_{rf} > -\alpha$, $\mathcal{E}_{bf} \geq 0$, $\mathcal{E}_{ff} > 0$. We define the extended version of this space, $\Psi_{\phi,\mathrm{ext}}^{k,\alpha}(M)$, in the analogous way.

As motivation for this simplification of the full calculi, note that, by Theorems 3 and 7, operators in $\Psi_{b/\phi}^{k,\alpha}(M)$ are bounded $x^\alpha L^2 \to x^\alpha H_{b/\phi}^m$. Also, the adjoint of an operator in $\Psi_{b/\phi}^{k,\alpha}(M)$ is in $\Psi_{b/\phi}^{k,-\alpha}(M)$.

The properties of these spaces that we need in the parametrix construction are collected in the following proposition. Because these properties are only relevant near the boundary of M when we construct the parametrix, in this proposition and for Section 5, we use $\Psi_b^{k,\alpha}$ to denote $\Psi_b^{k,\alpha}(V, \mathcal{K})$ and $\Psi_\phi^{k,\alpha}$ to denote $\Psi_\phi^{k,\alpha}(U)$. Also, write $\Psi_{b,\mathrm{ext}}^{k,\alpha} := \Psi_{b,\mathrm{ext}}^{k,\alpha}(U)$, defined in an analogous way as $\Psi_{\phi,\mathrm{ext}}^{k,\alpha}(U)$, and $\Psi_{\phi,\mathrm{ext}}^{k,\alpha} := \Psi_{\phi,\mathrm{ext}}^{k,\alpha}(U)$.

Proposition 3. *Let $k, l, \alpha, c \in \mathbb{R}$. Then we have*

a) $\Psi_b^{k,\alpha} \circ \Psi_b^{l,\alpha} \subset \Psi_b^{k+l,\alpha}$,
b) $\Psi_\phi^{k,\alpha} \circ \Psi_\phi^{l,\alpha} \subset \Psi_\phi^{k+l,\alpha}$,
c) $\Psi_b^{k,\alpha} \subset \Psi_\phi^{k,\alpha}$ *via lifting if* $k < 0$,
d) $x^{-c}\Psi_b^{k,\alpha}x^c = \Psi_b^{k,\alpha-c}$, $x^{-c}\Psi_\phi^{k,\alpha}x^c = \Psi_\phi^{k,\alpha-c}$,
e) $x^\infty\Psi_b^{k,\alpha} = x^\infty\Psi_\phi^{k,\alpha}$, *where* $x^\infty\Psi_{b/\phi}^{k,\alpha} := \bigcap_{c\in\mathbb{R}} x^c\Psi_{b/\phi}^{k,\alpha}$,
f) $\Psi_b^{k,\alpha}x^c\Psi_\phi^{l,\alpha} \subset \Psi_{b,\mathrm{ext}}^{-\infty,\alpha} + x^c\Psi_{b\phi}^{k+l}$ *and* $\Psi_\phi^{k,\alpha}x^c\Psi_\phi^{l,\alpha} \subset \Psi_{b,\mathrm{ext}}^{-\infty,\alpha} + x^c\Psi_{b\phi}^{k+l}$ *for* $c \geq 0$ *and* $k \leq 0$, *where* $\Psi_{b\phi}^m := \bigcap_{\alpha\in\mathbb{R}} \Psi_\phi^{m,\alpha}$ *is the space of ϕ-operators having empty index sets at lf, rf, index set ≥ 0 at bf and > 0 at ff.*
g) *These properties are all still true if we replace* $\Psi_\phi^{m,\alpha}$ *with* $\Psi_{\phi,\mathrm{ext}}^{m,\alpha}$.

Concerning f), note that $\Psi_b^{k,\alpha}x^c\Psi_\phi^{l,\alpha} \subset \Psi_\phi^{k+l,\alpha}$ follows from b) and c), but the factor x^c is lost in this crude argument; f) is a refinement needed below. The main point of the proposition is to record the behavior of kernels and compositions as b- or ϕ-kernels and their expansions at the left and right faces. We will not make use of the information about their orders.

Proof. Parts a) and b) follow immediately from the composition theorems for the b- and ϕ-calculus. Part c) follows from the lifting theorem. For d) and e) observe that multiplying a b-operator by x^c from the left or right means multiplying its Schwartz kernel by x^c or $(x')^c$, respectively. Therefore, its index sets at lf, bf, respectively at bf, rf, are simply shifted up by c. This is similar for ϕ-operators. This implies d), and e) follows from the fact that functions vanishing to infinite order at a front face of a blow-up are still smooth after collapsing that front face.

Now we prove f). Let $S \in \Psi_\phi^{l,\alpha}$. By introducing a cutoff function supported in a neighborhood of $\mathrm{Diag}_\phi \cup ff$ and equal to one in a smaller such neighborhood, where Diag_ϕ is the diagonal in M_ϕ^2, we can write $S = S_1 + S_2$ where $S_1 \in \Psi_b^{-\infty,\alpha}$ (since a ϕ-operator whose kernel vanishes near ff is a b-operator) and $S_2 \in \Psi_{b\phi}^l$. Thus

$$\Psi_\phi^{l,\alpha} \subset \Psi_{b,\mathrm{ext}}^{-\infty,\alpha} + \Psi_b^l, \tag{4.5}$$

so $\Psi_b^{k,\alpha} x^c \Psi_\phi^{l,\alpha} \subset \Psi_{b,\mathrm{ext}}^{-\infty,\alpha} + \Psi_{b,\mathrm{ext}}^{k,\alpha} x^c \Psi_{b\phi}^l$ by a). Next, from $x^c \Psi_{b\phi}^l = \Psi_{b\phi}^l x^c$ and b), c) we get $\Psi_b^{k,\alpha} x^c \Psi_{b\phi}^l \subset \Psi_\phi^{k+l,\alpha} x^c$. Finally, using (4.5) again, and $c \geq 0$, we see that this is contained in $\Psi_{b,\mathrm{ext}}^{-\infty,\alpha} + \Psi_{b\phi}^{k+l} x^c$. This gives the first claim in f). For the second claim use (4.5) to obtain $\Psi_\phi^{k,\alpha} \subset \Psi_b^{-\infty,\alpha} + \Psi_{b\phi}^k$. The first term is handled by the first containment in f), and the second yields $\Psi_{b\phi}^k x^c \Psi_\phi^{l,\alpha} = x^c \Psi_{b\phi}^k \Psi_\phi^{l,\alpha} \subset x^c \Psi_\phi^{k+l,\alpha}$ by b). Property g) is proved by essentially the same arguments as a) through f). □

5. Statement and proof of main parametrix theorem

Before we construct parametrices, we should consider for a moment what sort of parametrices we want. Recall that a right parametrix for P is an operator Q so that $PQ = \mathrm{Id} - R$, where the remainder R is 'good' in a suitable sense, and similar is true for a left parametrix. Parametrices satisfying more stringent notions of 'good' are harder to construct in general than those satisfying weaker notions, but also lead to stronger corollaries. For example, the classical pseudodifferential parametrix construction for an elliptic operator, P, over a smooth manifold yields a remainder R which is smoothing. This may be used to prove that solutions of $Pu = 0$ are smooth in the interior of M. If M is a compact manifold without boundary, such a remainder is also a compact operator, which shows that P is a Fredholm operator between Sobolev spaces. However, in our setting a smoothing remainder need not be compact on the natural weighted Sobolev spaces on which it is bounded. Thus in order to get a Fredholm result we need to improve the parametrix. If we would further like to get full asymptotic expansions at the boundary for solutions u of $Pu = 0$, this requires an even more refined parametrix. More precisely, in the ϕ-calculus setting this requires the Schwartz kernel of the left remainder R to be smooth in the interior of M_ϕ^2 and to vanish to infinite order at the faces bf, ff and also at rf. The expansion of R at lf then determines the expansion of u. This is the type of parametrix we will construct here. As usual we will first construct a right parametrix, where the remainder has the same properties with lf and rf interchanged, and then use adjoints to get a left parametrix.

The regularity and vanishing properties of the remainder kernel correspond to the two parts of the parametrix construction, which, however, we carry out in the opposite order.

I. First we find a right parametrix 'at the boundary'. This yields a right parametrix Q_∂ with a remainder R_∂ that vanishes to infinite order at bf, ff and lf.

II. We combine Q_∂ with a 'small' right ϕ-parametrix of P, obtained by inverting the principal symbol, to improve the error to be smoothing in addition. This comes at the cost of losing some control of the $\Pi, \Pi^\perp$ splitting, but this is no problem for the Fredholm and regularity results we want.

5.1. Statement of the main theorem

Recall from equation (3.16) that the Gauss–Bonnet operator for a ϕ-cusp metric near the boundary may be written as

$$D_U = x^{-a}\begin{pmatrix} x^a P_{00} & x^a P_{01} \\ x^a P_{10} & P_{11} \end{pmatrix}$$

with respect to the decomposition $C^\infty(V, \mathcal{K}) \oplus C^\infty(V, \mathcal{C})$ into the space of fibre-harmonic forms and its orthogonal complement. Also, recall from (3.18) and the discussion after it that the Hodge Laplacian may be written, under the condition that $[D_B, \Pi] = 0$, as

$$\Delta_U = x^{-2a}\begin{pmatrix} x^{2a} T_{00} & x^{2a} T_{01} \\ x^{2a} T_{10} & T_{11} \end{pmatrix}.$$

If we can obtain a parametrix for $x^a D_U$ and for $x^{2a}\Delta_U$, we will be able to get parametrices for D_U and Δ_U by composing with appropriate powers of x. Therefore, in this section we will consider the more general situation that P is a Π-split operator, which we define as follows:

Definition 6. Let P be a ϕ-elliptic differential ϕ-operator of order m

$$P : C^\infty(M, E) \to C^\infty(M, E),$$

$\mathcal{K}$ be a finite-dimensional sub-bundle of the infinite-dimensional $C^\infty(F, E)$ bundle over V, with orthogonal complement $\mathcal{C}$, and let $\Pi, \Pi^\perp$ be the projections on the first and second factor in the orthogonal decomposition

$$C^\infty(U, E) = C^\infty(V, C^\infty(F, E)) = C^\infty(V, \mathcal{K}) \oplus C^\infty(V, \mathcal{C}).$$

Write P in terms of this decomposition as

$$P = \begin{pmatrix} x^{am} P_{00} & x^{am} P_{01} \\ x^{am} P_{10} & P_{11} \end{pmatrix}, \tag{5.1}$$

where $x^{am} P_{00} = \Pi P \Pi$ etc. We say that P is Π-*split* if the following conditions are satisfied.

1. $P_{00} \in \mathrm{Diff}_b^m(V, \mathcal{K})$ is an elliptic b-operator.
2. There is $\tilde{P} \in \mathrm{Diff}_\phi^m(U, E)$ such that $P_{01} = \Pi \tilde{P} \Pi^\perp$ and $P_{10} = \Pi^\perp \tilde{P} \Pi$.
3. $N(P_{11})$ is invertible on forms over $\mathbb{R}^{b+1}$ with values in $\mathcal{C}$.

Note that condition 2 implies that $N(P)$ is diagonal with respect to the decomposition $C^\infty(\mathbb{R}^{b+1} \times F, E) = C^\infty(\mathbb{R}^{b+1}, \mathcal{K}) \oplus C^\infty(\mathbb{R}^{b+1}, \mathcal{C})$, so condition 3 makes sense. Also, the conditions imply that $P_{01}, P_{10}, P_{11} \in \Psi^m_{\phi,\mathrm{ext}}(U)$.

As discussed in Sections 3 and 4, these conditions are satisfied by $x^a D_M$ and $x^{2a}\Delta_M$.

As in the b-calculus, we will not generally be able to construct a parametrix for such operators that gives Fredholm results for all weighted L^2 spaces. Rather, we will find a family of parametrices corresponding to a dense set of weights. In fact, these weights come from the same source as the weights in the b setting, as will become clear in the construction. For any admissible weight α, the corresponding parametrix will allow us to prove that

$$P : x^\alpha H^m_{\mathrm{split}} \to x^\alpha L^2$$

is Fredholm. The Sobolev space H^m_{split} that appears on the left side of this map is introduced in Section 6.1.

The following theorem is the main technical result of this paper. We use the short notation $\Psi_b^{k,\alpha}$, $\Psi^{k,\alpha}_{\phi,\mathrm{ext}}$ and $\Psi^k_{b\phi,\mathrm{ext}}$ for $\Psi_b^{k,\alpha}(V,\mathcal{K})$, $\Psi^{k,\alpha}_{\phi,\mathrm{ext}}(U,E)$ and $\Psi^k_{b\phi,\mathrm{ext}}(U,E)$, respectively. These spaces were defined in Definition 5 and Proposition 3.

Theorem 12. *Let P be a Π-split differential operator of order m as in Definition 6. Let $\alpha \in \mathbb{R}$ and assume*

$$\alpha - am \notin -\operatorname{Im}\operatorname{spec}(P_{00})\,. \tag{5.2}$$

Then there are right and left parametrices which in the interior of M are pseudodifferential operators of order $-m$ and over U are in the spaces

$$Q_{r,\alpha}, Q_{l,\alpha} \in \begin{pmatrix} x^{-am}\Psi_b^{-m,\alpha} + \Psi^{-m}_{b\phi,\mathrm{ext}} & 0 \\ 0 & \Psi^{-m}_{\phi,\mathrm{ext}} + \Psi^{-m,\alpha}_{\phi,\mathrm{ext}} x^{am} \end{pmatrix} + \begin{pmatrix} 0 & x^{-am}\Psi^{-m,\alpha}_{\phi,\mathrm{ext}} x^{am} \\ \Psi^{-m,\alpha}_{\phi,\mathrm{ext}} & 0 \end{pmatrix}$$

with

$$PQ_{r,\alpha} = \mathrm{Id} - R_{r,\alpha}, \quad Q_{l,\alpha}P = \mathrm{Id} - R_{l,\alpha},$$

where the remainders are smooth in the interior and over U satisfy

$$R_{r,\alpha} \in x^\infty \Psi_\phi^{-\infty,\alpha}(\Pi + x^{am}\Pi^\perp), \tag{5.3}$$

$$R_{l,\alpha} \in (x^{-am}\Pi + \Pi^\perp)\Psi_\phi^{-\infty,\alpha} x^\infty. \tag{5.4}$$

The notation in (5.4) means that $R_{l,\alpha}$ is a sum of the form $x^{-am}\Pi R' x^\infty + \Pi^\perp R'' x^\infty$ with $R', R'' \in \Psi_\phi^{-\infty,\alpha}$, and the notation in (5.3) is the same. It will be clear from the construction that the index sets for the various parts of the parametrix could be described more precisely if required and will be computed from $\mathrm{Spec}_b(P_{00})$. The remainder of this section is devoted to the proof of Theorem 12.

5.2. Construction of right parametrix

In order to make this section easier to read, we leave out the subscript ext in the notation for spaces of pseudodifferential operators. Thus, Ψ^*_ϕ is to be understood as $\Psi^*_{\phi,\mathrm{ext}}$ throughout this section.

Assume that P is Π-split. Decompose P into its diagonal and off-diagonal parts:

$$P = P_d + P_o, \quad P_d = \begin{pmatrix} x^{am}P_{00} & 0 \\ 0 & P_{11} \end{pmatrix}, \quad P_o = \begin{pmatrix} 0 & x^{am}P_{01} \\ x^{am}P_{10} & 0 \end{pmatrix}.$$

The construction now proceeds in five steps. The first four steps are needed to construct the boundary parametrix, and in the fifth step, we combine this with the interior parametrix to make the remainder smoothing. We first give a short description of each step.

Step 1: First, we find a parametrix, Q_d, for the diagonal part, P_d, of P. Since P_{00} is a b-elliptic b-operator, it has a parametrix in the b-calculus, which we set as the upper left piece in Q_d. For the lower right part we use the invertibility of $N(P_{11})$ and Proposition 2.3. The diagonal remainder, $R_d := P_dQ_d - I$ vanishes to infinite order in x at all boundary faces, but Q_d is not a parametrix for the full operator P.

Step 2: Using the off-diagonal terms of P we modify Q_d to Q_2 so that PQ_2 yields a remainder R_2 that vanishes to some order at *bf* and *ff*.

Step 3: By constructing formal solutions of the equation $Pu = f$, we construct a right parametrix Q_3 with a remainder R_3 that vanishes to infinite order at *lf*, but still has non-trivial index sets at *bf* and *ff*. However, the remainder does vanish to some order at these faces, so we can correct this in the next step.

Step 4: We use a Neumann series to get a right parametrix Q_∂ with a remainder R_∂ that vanishes to infinite order at *bf*, *ff* and *lf*. This finishes the construction of the boundary parametrix.

Step 5: By combining Q_∂ with a 'small' right ϕ-parametrix of P, we get a remainder that in addition is smoothing.

We now provide the details.

Step 1: Here we use the assumption that the normal operator of P is diagonal. We only use the $\Pi^\perp$ part of the normal operator, since in its Π part the normal operator forgets important information. For example, if the Π part is of the form $x^{1+a}\partial_x + x^a h$ (with h a zero-order operator) then the normal operator will be $x^{1+a}\partial_x$, while when we write this as $x^a(x\partial_x + h)$, then we see that h is included in the indicial operator of $P_{00} = x\partial_x + h$, so it is essential for the correct b-parametrix. Therefore, in the Π-part we use the b-parametrix.

In order to construct the right parametrix, it is useful to write $\Pi P \Pi$ as $P'_{00}x^{am}$ instead of $x^{am}P_{00}$. Thus, define the b-operator

$$P'_{00} = x^{am}P_{00}x^{-am}$$

and choose a right parametrix Q_{00} for P'_{00} in the b-calculus, corresponding to the given weight α, see Theorem 5. This is possible under the condition $\alpha \notin$

$-\operatorname{Im}\operatorname{spec} P'_{00}$. Since $-\operatorname{Im}\operatorname{spec} x^{am}P_{00}x^{-am} = -\operatorname{Im}\operatorname{spec} P_{00} + am$ this condition is equivalent to condition (5.2). We obtain

$$x^{am}P_{00}x^{-am}Q_{00} = \mathrm{Id} - R_{00}, \quad Q_{00} \in \Psi_b^{-m,\alpha},\ R_{00} \in x^\infty\Psi_b^{0,\alpha}.$$

(The remainder actually is in $x^\infty\Psi_b^{-\infty,\alpha}$, but this is of no use here.) We now invert $N(P_{11})$ on $\mathcal{C}$ and define its inverse B as in (4.4). Then by Proposition 2.3, $B \in \Psi^{-m}_{\mathrm{sus}-\phi}(\partial M)$, and by Proposition 2.4 we may extend B to the interior of M^2_ϕ and obtain a parametrix Q'_{11}:

$$P_{11}Q'_{11} = \mathrm{Id} - R'_{11}, \quad Q'_{11} \in \Psi_\phi^{-m},\ R'_{11} \in x\Psi^0_\phi.$$

Using Proposition 2.5 and the standard Neumann series argument this can be improved to

$$P_{11}Q_{11} = \mathrm{Id} - R_{11}, \quad Q_{11} \in \Psi_\phi^{-m},\ R_{11} \in x^\infty\Psi^0_\phi.$$

Now we have $P_dQ_d = \mathrm{Id} - R_d$ with

$$Q_d = \begin{pmatrix} x^{-am}Q_{00} & 0 \\ 0 & Q_{11} \end{pmatrix} \in \begin{pmatrix} x^{-am}\Psi_b^{-m,\alpha} & 0 \\ 0 & \Psi_\phi^{-m} \end{pmatrix}, \tag{5.5}$$

$$R_d = \begin{pmatrix} R_{00} & 0 \\ 0 & R_{11} \end{pmatrix} \in \begin{pmatrix} x^\infty\Psi_\phi^{0,\alpha} & 0 \\ 0 & x^\infty\Psi^0_\phi \end{pmatrix}. \tag{5.6}$$

Step 2: In this step we make use of the fact that the x^{am} factors in P_o mean than it is of lower order near the boundary than P_d. The idea comes from the formal inversion

$$P^{-1} = P_d^{-1}(\mathrm{Id} + P_oP_d^{-1})^{-1} = P_d^{-1} - P_d^{-1}P_oP_d^{-1} + \cdots.$$

Since P_d may not be invertible we use its parametrix Q_d instead. The first two terms suffice for our purposes. Thus, we set

$$Q_2 = Q_d + Q_o, \quad \text{where} \quad Q_o := -Q_dP_oQ_d.$$

Then $P_dQ_d = \mathrm{Id} - R_d$ implies

$$PQ_2 = \mathrm{Id} - R_2, \quad \text{where } R_2 = R_d - R_o + (P_oQ_d)^2, \quad \text{and } R_o := R_dP_oQ_d.$$

To analyze Q_2 and R_2 we use Proposition 3 to get $\Psi^m_\phi\Psi_b^{-m,\alpha} \subset \Psi_\phi^{0,\alpha}$. Together with $x^{am}\Psi^0_\phi = \Psi^0_\phi x^{am}$ we get

$$P_oQ_d \in \begin{pmatrix} 0 & \Psi^0_\phi x^{am} \\ \Psi_\phi^{0,\alpha} & 0 \end{pmatrix}. \tag{5.7}$$

This gives

$$Q_o = Q_dP_oQ_d \in \begin{pmatrix} 0 & x^{-am}\Psi_\phi^{-m,\alpha}x^{am} \\ \Psi_\phi^{-m,\alpha} & 0 \end{pmatrix}. \tag{5.8}$$

Now we analyze R_2. From (5.6) and (5.7) we get

$$R_o = R_d P_o Q_d \in \begin{pmatrix} 0 & x^\infty \Psi_\phi^{0,\alpha} x^{am} \\ x^\infty \Psi_\phi^{0,\alpha} & 0 \end{pmatrix}, \quad (P_o Q_d)^2 \in \begin{pmatrix} x^{am}\Psi_\phi^{0,\alpha} & 0 \\ 0 & \Psi_\phi^{0,\alpha} x^{am} \end{pmatrix}. \tag{5.9}$$

The overall x^{am} factor in $(P_o Q_d)^2$ (as opposed to x^{am} in one entry only in (5.7)) is the reason that Q_2 is better than Q_d. Note that this factor increases the order of vanishing of the Schwartz kernel at *bf* and *ff*, no matter whether it is on the left or on the right. However, on the left it increases the order at *lf* but not at *rf*, while on the right it increases the order at *rf* but not at *lf*. Thus in the second expression in equation (5.9), we cannot move both factors of x^{am} to the same side of the pseudodifferential term.

Step 3: The parts R_d and $R_d P_o Q_d$ in R_2 already have the infinite-order vanishing at *lf*, *bf*, *ff* that we need. However, the term $(P_o Q_d)^2$ does not, see (5.9). The standard procedure to improve remainder terms is a Neumann series argument as in Step 4, which replaces a remainder R by arbitrarily high powers R^N. However, as can be seen from the Composition Theorem 9, taking powers of R does not improve order of vanishing at *lf*, and also does not improve order of vanishing at *bf*, *ff* unless R already vanishes at *lf* to arbitrarily high order. There is a standard remedy for these problems. By constructing formal solutions u of the equation $Pu = f$, where f arises from the *lf* expansion of the remainder, combining these solutions into a Schwartz kernel and subtracting the resulting operator from the parametrix, one gets an improved remainder, which vanishes to infinite order at *lf*. Taking powers of this remainder then yields arbitrarily high orders of vanishing at *bf* and *ff* also. This standard procedure is explained in [13, Section 5.20]. The setting there is for a *b*-elliptic *b*-operator P, and we need to adapt this to our situation. We obtain the following proposition. Its proof will appear in [6].

Proposition 4. *Let P be Π-split. Suppose the Schwartz kernel of $R' \in \Psi_b^{-\infty,\alpha}$ is supported near lf and satisfies, with index sets listed in the order lf, bf,*

$$\begin{aligned} &\Pi R' && \text{has index sets } > \alpha + am, \ \geq am, \\ &\Pi^\perp R' && \text{has index sets } > \alpha, \qquad\quad \geq am. \end{aligned}$$

Then there are operators $Q' \in \Psi_b^{-\infty,\alpha}$, $R'' \in \Psi_b^{-\infty,\infty}$ with Schwartz kernels supported near lf and satisfying

$$\begin{aligned} &\Pi Q' && \text{has index sets } > \alpha, \ \geq 0, \\ &\Pi^\perp Q' && \text{has index sets } > \alpha, \ \geq am, \\ &R'' && \text{has index sets } \emptyset, \quad \geq am, \end{aligned}$$

$$PQ' = R' + R''.$$

We apply the proposition with $R' = \chi R_2$ (understood in the sense of kernels), where χ is a smooth cut-off function on the blown up double space supported and equal to one near *lf*. Then (5.6), (5.9) imply that R' satisfies the conditions in Proposition 4. With Q', R'' from the proposition we set

$$Q_3 = Q_2 + Q', \quad R_3 = (1-\chi)R_2 - R''.$$

Then $PQ_3 = \mathrm{Id} - R_3$ since $PQ_2 = \mathrm{Id} - R_2$, $PQ' = \chi R_2 + R''$. Note from (5.6), (5.9) and Proposition 4 that R_3 is a ϕ-operator of weight α having index sets $\emptyset$ at *lf*, $\geq am$ at *bf* and $> am$ at *ff*, and so that $R_3\Pi^\perp$ has an extra am degrees of vanishing at *rf*. Equivalently

$$R_3 \in \Psi_R := \begin{pmatrix} x^{am}\Psi^{0,\alpha}_{\phi,lf} & \Psi^{0,\alpha}_{\phi,lf}x^{am} \\ \Psi^{0,\alpha}_{\phi,lf} & \Psi^{0,\alpha}_{\phi,lf}x^{am} \end{pmatrix}, \tag{5.10}$$

where $\Psi^{0,\alpha}_{\phi,lf}$ is the space of those elements of $\Psi^{0,\alpha}_{\phi}$ that vanish to infinite order at *lf*.

Step 4: In order to improve the remainder further, we use the standard Neumann series argument, i.e., we multiply $PQ_3 = \mathrm{Id} - R_3$ by $\mathrm{Id} + R_3 + R_3^2 + \cdots$ from the right and sum asymptotically. We need to check that the asymptotic sum makes sense. Clearly $\Psi^{0,\alpha}_{\phi,lf}x^{am} \subset x^{am}\Psi^{0,\alpha}_{\phi,lf}$, and this implies that

$$R_3^2 \in \begin{pmatrix} x^{am}\Psi^{0,\alpha}_{\phi,lf} & x^{am}\Psi^{0,\alpha}_{\phi,lf}x^{am} \\ x^{am}\Psi^{0,\alpha}_{\phi,lf} & \Psi^{0,\alpha}_{\phi,lf}x^{am} \end{pmatrix}$$

and then inductively that $R_3^{2N} \in x^{(N-1)am}\Psi_R$ for all $N \in \mathbb{N}$. Therefore, the index sets at *bf*, *ff* of R_3^N wander off to infinity as $N \to \infty$. In addition, the index sets of R_3^N at *rf* stabilize by a simple argument as in [13, Section 5.22]; hence the asymptotic sum $R_3' = \sum_{N=1}^{\infty} R_3^N$ makes sense and yields an operator in Ψ_R, see (5.10). Setting $Q_\partial = Q_3R_3'$ we obtain $PQ_\partial = \mathrm{Id} - R_\partial$ where R_∂ is in the intersection of all the spaces $x^{(N-1)am}\Psi_R$; hence

$$R_\partial \in x^\infty\Psi^{0,\alpha}_{\phi}(\Pi + x^{am}\Pi^\perp). \tag{5.11}$$

We now analyze $Q_\partial = Q_3R_3'$. Recall $Q_3 = Q_d + Q_o + Q'$ and write $R_3' = \tilde{R}_d + \tilde{R}_o$, where $\tilde{R}_d$ is diagonal and $\tilde{R}_o$ is off-diagonal with respect to the $\Pi, \Pi^\perp$-splitting. We analyze each term in the product Q_3R_3' separately, using (5.6), (5.8) and (5.10). The $\Pi\Pi$ parts of $Q_d\tilde{R}_d$, $Q_o\tilde{R}_o$ are in $x^{-am}\Psi_b^{-m,\alpha}x^{am}\Psi^{0,\alpha}_{\phi,lf}$ and $x^{-am}\Psi_\phi^{-m,\alpha}x^{am}\Psi^{0,\alpha}_{\phi,lf}$ respectively, hence by Proposition 3f) are contained in $x^{-am}\Psi_b^{-\infty,\alpha} + \Psi_{b\phi}^{-m}$. The $\Pi^\perp\Pi^\perp$ terms are unproblematic, and we get

$$Q_d\tilde{R}_d + Q_o\tilde{R}_o \in \begin{pmatrix} x^{-am}\Psi_b^{-\infty,\alpha} + \Psi_{b\phi}^{-m} & 0 \\ 0 & \Psi_\phi^{-m,\alpha}x^{am} \end{pmatrix}. \tag{5.12}$$

Next, by a similar argument, we get

$$Q_d\tilde{R}_o + Q_o\tilde{R}_d \in \begin{pmatrix} 0 & x^{-am}\Psi_\phi^{-m,\alpha}x^{am} \\ \Psi_\phi^{-m,\alpha} & 0 \end{pmatrix}. \tag{5.13}$$

Finally, Proposition 4 gives in particular

$$Q' \in \begin{pmatrix} \Psi_{b,rf}^{-\infty,\alpha} & \Psi_{b,rf}^{-\infty,\alpha} \\ \Psi_{b,rf}^{-\infty,\alpha} x^{am} & \Psi_{b,rf}^{-\infty,\alpha} x^{am} \end{pmatrix}, \tag{5.14}$$

where $\Psi_{\phi,rf}^{-\infty,\alpha}$ are those operators in $\Psi_{\phi}^{-\infty,\alpha}$ whose kernels vanish at rf to infinite order. Then with $R_3' \in \Psi_R$ from Equation 5.10 we get

$$Q'R_3' \in \begin{pmatrix} \Psi_{\phi}^{-\infty,\alpha} & \Psi_{\phi}^{-\infty,\alpha} x^{am} \\ \Psi_{\phi}^{-\infty,\alpha} & \Psi_{\phi}^{-\infty,\alpha} x^{am} \end{pmatrix} = \Psi_{\phi}^{-\infty,\alpha}(\Pi + x^{am}\Pi^{\perp}). \tag{5.15}$$

Step 5: Now we improve the parametrix further using a small ϕ-parametrix of P. That is, since P is ϕ-elliptic, we may find a parametrix by Theorem 10

$$PQ^{\sigma} = \mathrm{Id} - R^{\sigma}, \quad Q^{\sigma} \in \Psi_{\phi}^{-m}, \quad R^{\sigma} \in \Psi_{\phi}^{-\infty}.$$

Then we define

$$Q_r = Q_{\partial} + Q^{\sigma} R_{\partial}, \quad R_r = R^{\sigma} R_{\partial},$$

so $PQ_r = \mathrm{Id} - R_r$.

5.3. Construction of left parametrix and proof of Theorem 12

The rough idea for constructing a left parametrix is as follows. Let P^* be the formal adjoint of P. Then P^* is also Π-split, so we may construct a right parametrix Q_r' as above but for P^*, and obtain a remainder R_r', so $P^*Q_r' = \mathrm{Id} - R_r'$. Taking adjoints we obtain $Q_l P = \mathrm{Id} - R_l$ where $Q_l = (Q_r')^*$, $R_l = (R_r')^*$.

To show that the formal adjoint of a Π-split operator is again Π-split, consider what taking the formal adjoint of P means. It means that the Schwartz kernel of P is reflected across the diagonal, so coordinates are switched $x \leftrightarrow x', y \leftrightarrow y', z \leftrightarrow z'$. Since Π is an orthogonal projection, the 2×2 matrices representing the $\Pi, \Pi^{\perp}$ decomposition are simply flipped, and adjoints of its parts are taken.

The $\Pi\Pi$ part of P^* is $(P_{00})^* x^{am}$. Now $\mathrm{Im}\,\mathrm{spec}(P_{00})^* = -\mathrm{Im}\,\mathrm{spec}(P_{00})$, so by assumption (5.2) we have $am - \alpha \notin -\mathrm{Im}\,\mathrm{spec}(P_{00})^*$. Therefore, there is a right b-parametrix for $(P_{00})^*$ for the weight $am - \alpha$, and we can construct a right parametrix Q_r' as above for P^* and for this weight. Let $Q_l = (Q_r')^*$.

Proof of Theorem 12. We collect all terms in $Q_r = Q_{\partial} + Q^{\sigma} R_{\partial}$. First, Q_{∂} is the sum of Q_d in (5.6), Q_o in (5.8) and the terms in (5.12), (5.13), (5.14) and (5.15), and each one is in the space given for $Q_D + Q_O$. Here the $\Pi\Pi$ term in $Q'R_3$ is split up as in (4.5). Then from (5.11) we have $Q^{\sigma} R_{\partial} \in R_{\partial} \in x^{\infty}\Psi_{\phi}^{-m,\alpha}(\Pi + x^{am}\Pi^{\perp})$, which is also of the form given in the theorem. In the same way we get the claim for $R_r = R^{\sigma} R_{\partial}$. Looking at the left parametrix constructed above we see

$$(x^{-am}\Psi_b^{-m,am-\alpha})^* = \Psi_b^{-m,\alpha-am} x^{-am} = x^{-am}\Psi_b^{-m,\alpha},$$

and similar calculations show that the left parametrix we obtain is precisely of the same type as the right parametrix. The left remainder is in

$$R_l \in \left(x^\infty \Psi_\phi^{-\infty,am-\alpha}(\Pi + x^{am}\Pi^\perp)\right)^* = (\Pi + x^{am}\Pi^\perp)\Psi_\phi^{-\infty,\alpha-am} x^\infty$$
$$= (x^{-am}\Pi + \Pi^\perp)\Psi_\phi^{-\infty,\alpha} x^\infty. \qquad \square$$

6. Proofs of Fredholm and regularity results

In this section, we first define the Sobolev spaces which reflect the different regularity of fibre harmonic and fibre perpendicular forms, and then prove Fredholm and regularity results.

6.1. Split Sobolev spaces

When dealing with the Gauss–Bonnet and Hodge Laplace operators, we naturally encounter Sobolev spaces which encode different sorts of regularity in the fibre-harmonic and fibre-perpendicular parts of sections near the boundary. Recall the definition of the bundle $\mathcal{K}$ in Section 3.

Definition 7. Let M be a ϕ-manifold. For any $k \in \mathbb{R}$ define

$$\Pi H_b^k(M, E, \mathrm{dvol}_b) = \{\mu \in H_{\mathrm{loc}}^k(\mathring{M}, E, \mathrm{dvol}_b) : \mu_{|U} \in H_b^k(V, \mathcal{K}, \mathrm{dvol}_b)\}$$

and let $\Pi^\perp H_\phi^k(U, E, \mathrm{dvol}_b)$ denote the image of the map $\Pi^\perp$. Then define the *split Sobolev spaces* for any $k \in \mathbb{R}$ by

$$H_{\mathrm{split}}^k(M, E, \mathrm{dvol}_b) = x^{-ak}\Pi H_b^k(M, E, \mathrm{dvol}_b) + \chi\,\Pi^\perp H_\phi^k(U, E, \mathrm{dvol}_b),$$

where $\chi \in C_0^\infty(U)$ equals one near ∂M.

In the sequel we will often write H_ϕ^k, H_{split}^k for

$$H_\phi^k(M, E, \mathrm{dvol}_b), \quad H_{\mathrm{split}}^k(M, E, \mathrm{dvol}_b)$$

etc. That these are the natural Sobolev spaces for Π-split operators can be seen in the special case where the operator P in (5.1) is diagonal, i.e., $P_{01} = P_{10} = 0$. Then P_{00} maps between b-Sobolev spaces and P_{11} between ϕ-Sobolev spaces. We can notice three important things about these spaces.

1. These spaces are complete inner product spaces under an inner product that depends on the metric on M and E, on the volume form dvol_b and on the cut-off function χ. The topology induced by this inner product is as usual independent of these choices.
2. $H_{\mathrm{split}}^0 = L^2$ and there are continuous inclusions

$$H_\phi^k \subset H_{\mathrm{split}}^k \subset x^{-ak} H_b^k \text{ for } k \geq 0. \tag{6.1}$$

This follows from $H_\phi^k \subset x^{-ak} H_b^k$, which is obvious from the definitions, by applying it to the Π and $\Pi^\perp$ parts separately.

3. Unlike usual Sobolev spaces, the split Sobolev spaces do not in general form a scale of spaces. That is, if $\mathcal{K} \neq 0$, then

$$H^{k+1}_{\text{split}} \not\subset H^{k}_{\text{split}}, \tag{6.2}$$

since higher regularity comes with potentially greater rate of blow-up of the fibre harmonic part of a section. This is the reason that our Fredholm theorem below holds only as a map $H^m_{\text{split}} \to L^2$ and not, as might be expected, $H^{m+k}_{\text{split}} \to H^k_{\text{split}}$ for all $k \in \mathbb{R}$.

6.2. Proof of main theorems

We first prove Fredholm and regularity theorems for general Π-split operators P. Then we apply the results to deduce Theorems 1 and 2.

Theorem 13. *Let P be a Π-split operator of order m. Let $\alpha \in \mathbb{R}$. If $\alpha - am \notin -\operatorname{Im}\operatorname{spec}(P_{00})$ then P is Fredholm as an operator*

$$x^\alpha H^m_{\text{split}}(M, E, \mathrm{dvol}_b) \to x^\alpha L^2(M, E, \mathrm{dvol}_b).$$

Also, if $\alpha \notin -\operatorname{Im}\operatorname{spec}(P_{00})$ then P is Fredholm

$$x^\alpha L^2(M, E, \mathrm{dvol}_b) \to x^\alpha H^{-m}_{\text{split}}(M, E, \mathrm{dvol}_b).$$

Note that the conditions on α are the natural ones when only considering the $\Pi\,\Pi$ part of the operator. In the first statement, $x^{am}P_{00}$ is to map $x^{\alpha-am}H^m_b \to x^\alpha L^2$, so P_{00} is to map $x^{\alpha-am}H^m_b \to x^{\alpha-am}L^2$, which is Fredholm if $\alpha - am \notin -\operatorname{Im}\operatorname{spec}(P_{00})$. In the second statement, $x^{am}P_{00}$ is to map $x^\alpha L^2 \to x^{\alpha+am}H^{-m}_b$, so P_{00} is to map $x^\alpha L^2 \to x^\alpha H^{-m}_b$, which is Fredholm if $\alpha \notin -\operatorname{Im}\operatorname{spec}(P_{00})$.

Proof. Conjugating by a power of x, we may assume $\alpha = 0$. We leave out the index α for parametrices and remainders. First, we show that P is bounded $H^m_{\text{split}} \to L^2$. We look at each term in (5.1) separately. Clearly, $x^{am}P_{00} = P'_{00}x^{am} : x^{-am}\Pi H^m_b \to L^2$ and $P_{11} : \Pi^\perp H^m_\phi \to L^2$, so it remains to check the cross terms. Boundedness $x^{am}P_{01} : H^m_\phi \to L^2$ follows from $P_{01} \in \Psi^m_{\phi,\text{ext}}$, and boundedness $x^{am}P_{10} : x^{-am}H^m_b \to L^2$ follows from $P_{10} = \Pi^\perp \tilde{P}\Pi$ with $\tilde{P} \in \operatorname{Diff}^m_\phi \subset \operatorname{Diff}^m_b$.

Next, we check boundedness of the left and right parametrix $Q_l, Q_r : L^2 \to H^m_{\text{split}}$. First, consider the $\Pi\,\Pi$ terms. Clearly, any element of $x^{-am}\Psi_b^{-m,0}$ is bounded $L^2 \to x^{-am}H^m_b$. Also, an element of $\Psi^{-m}_{b\phi}$ is bounded $L^2 \to H^m_\phi$, so also into $x^{-am}H^m_b$. Next, the $\Pi^\perp\Pi^\perp$ terms are in $\Psi_\phi^{-m,0}$, hence bounded $L^2 \to H^m_\phi$. The $\Pi\,\Pi^\perp$ term is in $x^{-am}\Psi_\phi^{-m,0}x^{am}$, and we need to show that it is bounded $L^2 \to x^{-am}H^m_b$. Split the operator as in (4.5). Then the first part is in $x^{-am}\Psi_{b,\text{ext}}^{-\infty,0}$, hence ok, and the second part is in $\Psi_\phi^{-m,0}$, hence maps $L^2 \to H^m_\phi$ and therefore into $x^{-am}H^m_b$. Finally, the $\Pi^\perp\Pi$ part of the parametrix is in $\Psi_\phi^{-m,0}$, hence bounded $L^2 \to H^m_\phi$ as required.

It remains to prove compactness of the remainders. First, compactness of $R_r : L^2 \to L^2$ follows from Theorem 7.2. To prove the compactness of $R_l : H^m_{\text{split}} \to H^m_{\text{split}}$, we first map

$$H^m_{\text{split}} \hookrightarrow x^{-am} H^m_b \xrightarrow{x^{am}} H^m_b .$$

We now consider ΠR_l and $\Pi^\perp R_l$ separately. Using (4.5) again, we split ΠR_l as a sum of $R'_l \in x^{-am}\Psi_b^{-\infty,0} x^\infty$ and $R''_l \in x^\infty \Psi_{b\phi}^{-\infty,0}$. Since elements of $\Psi_b^{-\infty,0}x^\infty$ are compact as operators $H^m_b \to H^m_b$ by Theorem 3, the operator $R'_l : H^m_b \to x^{-am}H^m_b$ is compact. For the remaining terms R''_l and $\Pi^\perp R_l$, we first map $H^m_b \hookrightarrow H^m_\phi$. Both of these operators are compact $H^m_\phi \to H^m_\phi$ by Theorem 7; hence the inclusion $H^m_\phi \to H^m_{\text{split}}$ completes the proof.

Finally, to prove the last statement, apply the first statement to P^* and the weight $-\alpha$. This is possible under the condition $-\alpha \notin -\operatorname{Im}\operatorname{spec}((P_{00})^*) = \operatorname{Im}\operatorname{spec}(P_{00})$. Then take adjoints and identify the dual spaces of $x^{-\alpha}L^2$, $x^{-\alpha}H^m_{\text{split}}$ with $x^\alpha L^2$, $x^\alpha H^{-m}_{\text{split}}$ via the L^2 scalar product. □

Theorem 14. *Suppose $u \in x^\alpha H^m_{\text{split}}(M, E, \text{dvol}_b)$ for some $\alpha \in \mathbb{R}$ and $Pu = 0$, where the operator P is Π-split of order m. Then $\Pi u \in x^{-am}\mathcal{A}^K_{\text{phg}}(U, E)$ and $\Pi^\perp u \in \mathcal{A}^K_{\text{phg}}(U, E)$ for some index set $K > \alpha$ determined by* $\operatorname{Spec}_b(P_{00})$.

Also, if $u \in x^\alpha L^2(M, E, \text{dvol}_b)$ and $Pu = 0$ then $\Pi u \in \mathcal{A}^K_{\text{phg}}(U, E)$, $\Pi^\perp u \in x^{am}\mathcal{A}^K_{\text{phg}}(U, E)$ with $K > \alpha$ determined by $\operatorname{Spec}_b(P_{00})$.

In addition, u is smooth in the interior of M by elliptic regularity.

Proof. First, assume $\alpha - am \notin -\operatorname{Im}\operatorname{spec}(P_{00})$. The identity $Q_{l,\alpha}P = \operatorname{Id} - R_{l,\alpha}$ holds on $x^\alpha H^m_{\text{split}}$, since it holds on C_0^∞ and both sides are continuous in $x^\alpha H^m_{\text{split}}$ by the proof of Theorem 13. Hence it may be applied to u. From $Pu = 0$ it follows that $u = R_{l,\alpha}u$. Now $\Psi_\phi^{-\infty,\alpha}x^\infty = \Psi_b^{-\infty,\alpha}x^\infty$ as in Proposition 3e), and the result follows from (5.4) and the mapping properties of b-operators. If $\alpha - am \in -\operatorname{Im}\operatorname{spec}(P_{00})$ then apply this argument with α replaced by $\alpha - \epsilon$ where $\epsilon > 0$ is such that $\alpha - \epsilon \notin -\operatorname{Im}\operatorname{spec}(P_{00})$ and the resulting index set K has no powers in the interval $(\alpha - \epsilon, \alpha)$. Then the expansion of u has powers with real parts $\geq \alpha$, but those terms with equality must vanish since they are not in $x^\alpha H^m_{\text{split}}$.

For the second statement, observe that $Q_{l,\alpha+am}$ is bounded $x^\alpha H^{-m}_{\text{split}} \to x^\alpha L^2$. This follows from the same arguments as in the proof of Theorem 13. Therefore, the identity $Q_{l,\alpha+am}P = \operatorname{Id} - R_{l,\alpha+am}$ holds on $x^\alpha L^2$. Therefore $u = R_{l,\alpha+am}u$, and the claim follows as before. □

Before we prove Theorems 1 and 2 we need to define the operator D_V used there. The operator P_{00} in (3.17) acts on sections of $\mathcal{K}$, defined after (3.12), while the theorems are stated in terms of sections of $\overline{\mathcal{K}}$. Because of $\mathcal{K} = (\Lambda^b T^*V) \otimes \tilde{\mathcal{K}}$ we may consider P_{00} as acting on sections of $\tilde{\mathcal{K}}$ instead. Recall that elements of $\tilde{\mathcal{K}}$ are sections of $\Lambda x^a T^*F$ over F. There is an obvious bundle map which identifies

$\Lambda x^a T^*F$ with ΛT^*F. It sends $\mathcal{K}$ to the space $\overline{\mathcal{K}}$ of harmonic forms on F for the metric $h_{|F}$. Under this identification, P_{00} turns into the operator

$$D_V = \eth + \eth^* + xD_x + x^a\Pi\mathbf{R}\Pi \tag{6.3}$$

since the term A in (3.8) arose from the commutators $[x\partial_x, x^{a|L|}]$ when introducing the scaling factors x^a, and in $\overline{\mathcal{K}}$ these factors are eliminated again.

Proof of Theorems 1 *and* 2. We first prove the Fredholm theorem for D_M. Recall that $D_M = x^{-a}P$ where P is Π-split of order 1, for the bundle $E = \Lambda^{c\phi}T^*M$ and $\mathcal{K} \subset C^\infty(F, E)$ as defined after (3.12). Theorem 13 shows that

$$P : x^\alpha H^1_{\text{split}}(M, E, \text{dvol}_b) \to x^\alpha L^2(M, E, \text{dvol}_b)$$

is Fredholm if $\alpha - a \notin -\operatorname{Im}\operatorname{spec}(P_{00})$. Over the interior of M, we may identify the bundles ΛT^*M, with metric g, with the bundle E, with regular metric on all of M. Therefore, we get that

$$D_M : x^\alpha H^1_{\text{split}}(M, \Lambda T^*M, \text{dvol}_b) \to x^{\alpha-a}L^2(M, \Lambda T^*M, \text{dvol}_b)$$

is Fredholm. Setting $\gamma = \alpha - a$ gives the claim.

The Fredholm claim for Δ_M is proved in the same way, using the condition $[D_B, \Pi] = 0$ to get that $\Delta_M = x^{-2a}T$ for a Π-split operator T of order 2 (see (3.18) and the remarks following it). These remarks also show that the condition implies $T_{00} = (P_{00})^2 + O(x^{2a})$, so the indicial families of T_{00} and $(P_{00})^2$ coincide, implying that $-\operatorname{Im}\operatorname{spec}(T_{00}) = -\operatorname{Im}\operatorname{spec}(P_{00}) = -\operatorname{Im}\operatorname{spec}(D_V)$. Therefore, the condition on γ is the same as for D_M.

Finally, the polyhomogeneity claims in Theorems 1 and 2 follow from the second statement in Theorem 14, using that $x^w(\log x)^k$ is in $L^2(M, E, \text{dvol}_b)$ if and only if $\operatorname{Re} w > 0$. □

References

[1] L. Boutet de Monvel, *Comportement d'un opérateur pseudo-différentiel sur une variété à bord. I. La propriété de transmission*, J. Analyse Math. 17 (1966) 241–253.

[2] C. Dorschfeldt, *Algebras of Pseudodifferential Operators Near Edge and Corner Singularities*, Mathematical Research vol. 102. Wiley-VCH, Berlin (1998).

[3] J. Gil, T. Krainer and G. Mendoza, *On the Closure of Elliptic Wedge Operators*, J. Geom. Anal. 23 (2013), no. 4, 2035–2062.

[4] D. Grieser, E. Hunsicker, *Pseudodifferential operator calculus for generalized Q-rank 1 locally symmetric spaces, I*, Journal of Functional Analysis 257 (2009) 3748–3801.

[5] D. Grieser, E. Hunsicker, *Techniques for the b-calculus*, in preparation.

[6] D. Grieser, E. Hunsicker, *Pseudodifferential operator calculus for generalized Q-rank 1 locally symmetric spaces, II*, in preparation.

[7] D. Grieser, *Basics of the b-calculus*, in J.B. Gil et al. (eds.), *Approaches to Singular Analysis*, 30–84, Operator Theory: Advances and Applications, 125. Advances in Partial Differential Equations, Birkhäuser, Basel, 2001.

[8] T. Hausel, E. Hunsicker, and R. Mazzeo, *Hodge cohomology of gravitational instantons*, Duke Mathematical Journal **122** (2004), no. 3, 485–548.

[9] V.A. Kondratiev, *Boundary value problems for elliptic equations in domains with conical or angular points*, (Russian) Trudy Moskov. Mat. Obsc. 16 (1967), 209–292.

[10] T. Krainer and G. Mendoza, *Boundary value problems for first order elliptic wedge operators*, preprint 2013, arXiv:1307.2398.

[11] R. Mazzeo and R. Melrose, *Pseudodifferential operators on manifolds with fibred boundaries* in *"Mikio Sato: a great Japanese mathematician of the twentieth century"*, Asian J. Math. **2** (1998) no. 4, 833–866.

[12] R. Mazzeo, B. Vertman, *Elliptic theory of differential edge operators*, II: *boundary value problems*, preprint 2013, arXiv:1307:2266

[13] R. Melrose, *The Atiyah–Patodi–Singer index theorem*, A.K. Peters, Newton (1991).

[14] R. Melrose, *Pseudodifferential operators, corners and singular limits*, Proc. Int. Congr. Math., Kyoto/Japan 1990, Vol. I, 217–234 (1991).

[15] R. Melrose, *Differential analysis on manifolds with corners*, in preparation, partially available at http://www-math.mit.edu/~rbm/book.html.

[16] J. Müller, *A Hodge-type theorem for manifolds with fibered cusp metrics*, Geom. Funct. Anal. 21 (2011), no. 2, 443–482.

[17] W. Müller, *Manifolds with cusps of rank* 1, Lecture Notes in Math., vol. 1244, Springer-Verlag, New York (1987).

[18] B.-W. Schulze, *Pseudo-differential Operators on Manifolds with Singularities*, North–Holland, Amsterdam (1991).

[19] B. Vaillant, *Index and spectral theory for manifolds with generalized fibred cusps*, Ph.D. thesis, Univ. of Bonn, 2001. arXiv:math-DG/0102072.

D. Grieser
Institut für Mathematik
Carl von Ossietzky Universität Oldenburg
D-26111 Oldenburg, Germany
e-mail: daniel.grieser@uni-oldenburg.de

E. Hunsicker
Department of Mathematical Sciences
Loughborough University
Loughborough, LE11 3TU, UK
e-mail: E.Hunsicker@lboro.ac.uk

Fourier Analysis
Trends in Mathematics, 187–209

A Class of Elliptic Operators on a Manifold with Edge and Boundary

N. Habal, W. Rungrottheera and B.-W. Schulze

Abstract. On a compact manifold with edge and boundary we construct new classes of elliptic elements in the edge calculus of boundary value problems. The operators will be obtained without extra trace and potential conditions. Boundary value problems here refer to the case with or without the transmission property at the smooth part of the boundary.

Mathematics Subject Classification (2010). Primary 35J70; Secondary 35S35, 47G30, 58J40.

Keywords. Edge boundary problems, edge pseudo-differential calculus.

1. Introduction

One of the major tasks of the analysis on a manifold with singularities is the study of elliptic elements within a corresponding operator algebra with symbolic structure and to establish parametrices. In the case of a manifold with smooth boundary or edge the operators A have a principal symbolic hierarchy

$$\sigma(A) := (\sigma_0(A), \sigma_1(A))$$

where $\sigma_0(A)$ is the scalar (usual or edge-degenerate) principal interior symbol and $\sigma_1(A)$ the so-called principal edge symbol which is operator-valued and lives on the cotangent bundle of the edge minus zero section, cf., e.g., [6], [17]. Recall that already the case of operators on a manifold with smooth boundary with the transmission property at the boundary is interesting and has many applications, cf. Boutet de Monvel [2], Grubb [9], or Rempel and Schulze [15]. Also the case without the transmission property belongs to the important aspects, cf. the book of Eskin [7], the articles of Rempel and Schulze [14], or Schulze and Seiler [21]. The general theory on singular manifolds gives rise to a number of challenges, for instance, to establish an index theory along the lines of Gelfand's program, originally initiated for classical boundary value problems on a smooth manifold, cf. [8]. An essential specific point concerns a topological obstruction for the existence of

Shapiro–Lopatinskij elliptic boundary conditions, cf. Atiyah and Bott [1], and the paper [18]. Another problem, as soon as the obstruction vanishes, is that we should be able to construct sufficiently many explicit examples of elliptic operators with Shapiro–Lopatinskij ellipticity (called σ_∂-ellipticity, i.e., for an edge regarded as a boundary we also write σ_∂ rather than σ_1). Although this is not difficult for many standard operators, e.g., the Laplacian with Dirichlet or Neumann conditions, for general operators that are σ_0-elliptic the construction of extra boundary conditions to obtain a σ_∂-elliptic boundary problem can be very hard, since we need to know not only the Fredholm index of $\sigma_\partial(A)$ but the kernels and cokernels in order to pass to bijective block matrices $\sigma_\partial(\mathcal{A})$ of boundary symbols for a future boundary value problem (BVP) $\mathcal{A}$ with $\sigma_\partial(A)$ as the upper left corner of $\sigma_\partial(\mathcal{A})$. Concerning the "abstract" K-theoretic background of such constructions, cf. [2], or [12, Section 3.3.4]. Now in the case of a manifold with edge and boundary the operators A have a triple

$$\sigma(A) := (\sigma_0(A), \sigma_\partial(A), \sigma_1(A)) \tag{1.1}$$

of principal symbols, and besides the construction of operators $\mathcal{A}$ that are σ_∂-elliptic we should be able to also find a way for constructing σ_1-elliptic examples. In Dines [4], there have been obtained remarkable classes of elliptic edge and corner operators, based on specific trace and potential operators with respect to embedded manifolds in an ambient manifold.

The program of the present paper is to establish large new classes of elliptic elements in the edge algebra of boundary value problems with the symbolic structure (1.1). The idea here is completely different from [4]. We employ the result of [21] which states that truncated operators on a manifold with boundary belong to the edge calculus. In addition we also conclude ellipticity. Note that truncation operations play a role in analysing mixed and transmission problems, see [7] or [12].

Here we extend the truncation process to manifolds with edge and boundary, where we mainly consider zero-order operators, and we reach in that way operators of higher singularity orders, cf. Theorems 2.5 and 2.7. The edge algebra of BVPs in general is outlined in [13, Chapter 6] in connection with the investigation of crack problems and in [12, Chapters 6, 7] for mixed and transmission problems. Concerning aspects of the corner pseudo-differential calculus, cf. [3], [5], [10], [19], [20], and the references there.

2. Operators on a manifold with edge

2.1. Boundary problems as particular edge problems

By $L^\mu_{(\mathrm{cl})}(X)$ for a C^∞ manifold X we denote the space of pseudo-differential operators on X of order $\mu \in \mathbb{R}$, classical or non-classical where (cl) means that a consideration is valid both in the classical and the general case. We employ the standard symbol classes

$$S^\mu(U \times \mathbb{R}^m) \subset C^\infty(U \times \mathbb{R}^m)$$

for $U \subseteq \mathbb{R}^n$ open, defined by the symbolic estimates

$$|D_x^\alpha D_\xi^\beta a(x,\xi)| \le c\langle\xi\rangle^{\mu-|\beta|}$$

for all $(x,\xi) \in K \times \mathbb{R}^m, K \Subset U$, and $\alpha \in \mathbb{N}^n, \beta \in \mathbb{N}^m$, for constants $c = c(\alpha,\beta,K) > 0$. We call a symbol classical, i.e., write $a(x,\xi) \in S^\mu_{\mathrm{cl}}(U \times \mathbb{R}^m)$, if there is a sequence of homogeneous components $a_{(\mu-j)}(x,\xi) \in C^\infty(U \times (\mathbb{R}^m \setminus \{0\}))$, with $a_{(\mu-j)}(x,\delta\xi) = \delta^{\mu-j}a_{(\mu-j)}(x,\xi), \delta \in \mathbb{R}_+, j \in \mathbb{N}$, such that the symbol is an asymptotic sum (up to an element in $S^{-\infty}(U \times \mathbb{R}^m) = \bigcap_{\mu\in\mathbb{R}} S^\mu(U \times \mathbb{R}^m)$)

$$a(x,\xi) \sim \sum_{j=0}^{\infty} \chi(\xi)a_{(\mu-j)}(x,\xi) \tag{2.1}$$

for any excision function χ in $\mathbb{R}^m$ (i.e., $\chi = 0$ for $|\xi| < c_0$, $\chi = 1$ for $|\xi| > c_1$ for some $0 < c_0 < c_1$).

The definition of the space $L^\mu_{(\mathrm{cl})}(X;\mathbb{R}^l)$ of parameter-dependent pseudo-differential operators with parameter $\lambda \in \mathbb{R}^l$ is based on local symbols $a(x,\xi,\lambda) \in S^\mu_{(\mathrm{cl})}(U \times \mathbb{R}^{m+l}_{\xi,\lambda})$, combined with a system of charts on X and a subordinate partition of unity. Details on basic constructions of that kind may be found in any textbook on pseudo-differential operators. On X we fix a Riemannian metric; then we can identify the space $L^{-\infty}(X) = \bigcap_{\mu\in\mathbb{R}} L^\mu(X)$ with $C^\infty(X \times X)$ via $\int_X c(x,x')u(x')dx'$, $c(x,x') \in C^\infty(X \times X)$, with dx' being the measure associated with the Riemannian metric. Under some general topological condition on X (say, X is a countable union of compact sets) the space $C^\infty(X \times X)$ is Fréchet and this also turns $L^{-\infty}(X)$ to a Fréchet space. We set

$$L^{-\infty}(X;\mathbb{R}^l) := \mathcal{S}(\mathbb{R}^l, L^{-\infty}(X)).$$

It will be necessary to employ operator-valued symbols and associated operators referring to Hilbert (or Fréchet) spaces with group action.

A Hilbert space H is said to be endowed with a group action $\kappa = \{\kappa_\delta\}_{\delta\in\mathbb{R}_+}$ if $\kappa_\delta : H \to H$ is an isomorphism for every $\delta \in \mathbb{R}_+, \kappa_\delta\kappa_\nu = \kappa_{\delta\nu}$ for every $\delta,\nu \in \mathbb{R}_+$, and if $\delta \to \kappa_\delta h$ defines an element of $C(\mathbb{R}_+, H)$ for every $h \in H$. A similar notion makes sense for a Fréchet space E written as a projective limit of Hilbert spaces $\varprojlim_{k\in\mathbb{N}} E^k$ with continuous embeddings $E^k \hookrightarrow E^0, k \in \mathbb{N}$, and a group action κ on E^0 such that $\kappa|_{E^k}$ is a group action on E^k for every k. Now if H and $\widetilde{H}$ are Hilbert spaces with group actions κ and $\tilde{\kappa}$, respectively, then $S^\mu(U \times \mathbb{R}^m; H, \widetilde{H}), U \subseteq \mathbb{R}^n$ open, is defined to be the set of all $a(x,\xi) \in C^\infty(U \times \mathbb{R}^m, \mathcal{L}(H,\widetilde{H}))$ such that

$$\|\tilde{\kappa}^{-1}_{\langle\xi\rangle}\{D_x^\alpha D_\xi^\beta a(x,\xi)\}\kappa_{\langle\xi\rangle}\|_{\mathcal{L}(H,\widetilde{H})} \le c\,\langle\xi\rangle^{\mu-|\beta|}$$

for all $(x,\xi) \in K \times \mathbb{R}^m, K \Subset U$, and $\alpha \in \mathbb{N}^n, \beta \in \mathbb{N}^m$, for constants $c = c(\alpha,\beta,K) > 0$. The notion of classical operator-valued symbols is based on twisted homogeneity in the following sense. By

$$S^{(\mu)}(U \times (\mathbb{R}^m \setminus \{0\}); H, \widetilde{H})$$

we denote the set of all $a_{(\mu)}(x,\xi) \in C^\infty(U \times (\mathbb{R}^m \setminus \{0\}), \mathcal{L}(H, \widetilde{H}))$ such that

$$a_{(\mu)}(x, \delta\xi) = \delta^\mu \tilde{\kappa}_\delta a_{(\mu)}(x,\xi)\kappa_\delta^{-1}$$

for all $\delta \in \mathbb{R}_+$. Analogously as in the case of scalar symbols we define the space $S^\mu_{\mathrm{cl}}(U \times \mathbb{R}^m; H, \widetilde{H})$ of classical symbols in terms of asymptotic expansions (2.1) for $a_{(\mu-j)} \in S^{(\mu-j)}(U \times (\mathbb{R}^m \setminus \{0\}); H, \widetilde{H}), j \in \mathbb{N}$. More details on operator-valued symbols, also the variant with Fréchet spaces rather than Hilbert spaces with group action, may be found in [17].

An example in that sense are so-called Green symbols of type 0 in Boutet de Monvel's calculus of boundary value problems, with the corresponding Fréchet space

$$\mathcal{S}(\overline{\mathbb{R}}_+) := \mathcal{S}(\mathbb{R})|_{\mathbb{R}_+} = \varprojlim_{k\in\mathbb{N}} H^{k,k}(\mathbb{R}_+)$$

for the standard Sobolev spaces $H^k(\mathbb{R}), H^k(\mathbb{R}_+) := H^k(\mathbb{R})|_{\mathbb{R}_+}$, and $H^{k,k}(\mathbb{R}_+) := \langle r\rangle^{-k} H^k(\mathbb{R}_+)$, where r is the variable of $\mathbb{R}_+$. The group action in this case is defined as

$$(\kappa_\delta u)(r) = \delta^{1/2}u(\delta r), \delta \in \mathbb{R}_+, \tag{2.2}$$

where the above-mentioned E^0 is just $L^2(\mathbb{R}_+) = H^0(\mathbb{R}_+)$ with (2.2) being unitary. There is then the space of all operator functions

$$g(y,\eta) \in C^\infty(\Omega \times \mathbb{R}^q, \mathcal{L}(L^2(\mathbb{R}_+), L^2(\mathbb{R}_+)))$$

that represent symbols

$$g(y,\eta), g^*(y,\eta) \in S^\mu_{\mathrm{cl}}(\Omega \times \mathbb{R}^q; L^2(\mathbb{R}_+), \mathcal{S}(\overline{\mathbb{R}}_+));$$

here $*$ denotes the (y,η)-wise adjoint. The variables (y,r) here refer to $\Omega \times \overline{\mathbb{R}}_+$, corresponding to a chart on a manifold X with boundary $Y = \partial X, n = \dim X$, where $\Omega \subseteq \mathbb{R}^q$ open belongs to chart on Y, and $x \in X$ is splitted into $x = (y,r), q = n-1$. There are also so-called trace and potential symbols

$$t(y,\eta) \in S^\mu_{\mathrm{cl}}(\Omega \times \mathbb{R}^q; L^2(\mathbb{R}_+), \mathbb{C}), \;\; k(y,\eta) \in S^\mu_{\mathrm{cl}}(\Omega \times \mathbb{R}^q; \mathbb{C}, L^2(\mathbb{R}_+))$$

where we ask

$$k(y,\eta) \in S^\mu_{\mathrm{cl}}(\Omega \times \mathbb{R}^q; \mathbb{C}, \mathcal{S}(\overline{\mathbb{R}}_+))$$

based on the group action id on $\mathbb{C}$ and (2.2) on $\mathcal{S}(\overline{\mathbb{R}}_+)$ while the trace symbol t is defined by requiring t^* to be a potential symbol.

Now let $a(x,\xi) \in S^\mu_{\mathrm{cl}}(\Omega \times \overline{\mathbb{R}}_+ \times \mathbb{R}^n_\xi)$ $(= S^\mu_{\mathrm{cl}}(\Omega \times \mathbb{R} \times \mathbb{R}^n_\xi)|_{\Omega\times\overline{\mathbb{R}}_+\times\mathbb{R}^n})$ be a symbol. Then it is an important aspect in BVPs to assess the nature of truncated operators

$$\mathrm{op}^+(a)(y,\eta) := \mathrm{r}^+\mathrm{op}(a)(y,\eta)\mathrm{e}^+$$

with e^+ being the operator of extension of distributions (say, in $H^s(\mathbb{R}_+), s > -\frac{1}{2}$) by zero to the negative half-axis, $\mathrm{r}^+u := u|_{\mathbb{R}_+}$ for some $u \in \mathcal{D}'(\mathbb{R})$, and

$$\mathrm{op}(a)(y,\eta)v(r) := \iint e^{i(r-r')\varrho}a(y,r,\eta,\varrho)v(r')dr'đ\varrho, \tag{2.3}$$

$đ\varrho = (2\pi)^{-1}d\varrho$. Here $\xi = (\eta, \varrho)$ is the splitting of covariables according to $x = (y, r)$. Note that

$$\mathrm{op}^+(a)(y,\eta) : H^s(\mathbb{R}_+) \to H^{s-\mu}(\mathbb{R}_+)$$

is not always continuous for all $s > -\frac{1}{2}$. A sufficient condition for that is the transmission property of the symbol at the boundary $r = 0$, cf. [2], [9], [15]. Then, as soon as the symbol a is independent of r for large r, we have

$$\mathrm{op}^+(a)(y,\eta) \in S^\mu(\Omega \times \mathbb{R}^q; H^s(\mathbb{R}_+), H^{s-\mu}(\mathbb{R}_+)),$$

$s > -\frac{1}{2}$, and $\mathrm{op}^+(a)(y,\eta) \in S^\mu(\Omega \times \mathbb{R}^q; \mathcal{S}(\overline{\mathbb{R}}_+), \mathcal{S}(\overline{\mathbb{R}}_+))$. This gives us many examples for operator-valued symbols.

On a manifold X with boundary, not necessarily compact, there is Boutet de Monvel's algebra of pseudo-differential BVPs with the transmission property, furnished by spaces $\mathcal{B}^{\mu,d}(X)$ of operators of order $\mu \in \mathbb{Z}$ and type $d \in \mathbb{N}$. The operators in $\mathcal{B}^{\mu,d}(X)$ consist of 2×2 block matrices, locally (up to smoothing operators) near the boundary described by $\mathrm{Op}_y(\boldsymbol{a})$ for a symbol

$$\boldsymbol{a}(y,\eta) = \begin{pmatrix} \mathrm{op}^+(a)(y,\eta) + g(y,\eta) & k(y,\eta) \\ t(y,\eta) & q(y,\eta) \end{pmatrix}$$

belonging to $S^\mu(\Omega \times \mathbb{R}^q; H^s(\mathbb{R}_+) \oplus \mathbb{C}^{j_-}, H^{s-\mu}(\mathbb{R}_+) \oplus \mathbb{C}^{j_+})$. Here $\mathrm{op}^+(a)(y,\eta)$ is as before, where the symbol $a(x,\xi)$ has the transmission property at the boundary, $g(y,\eta)$ is a Green symbol, now of type $d \in \mathbb{N}$, which means

$$g(y,\eta) = \sum_{j=0}^{d} g_j(y,\eta)\partial_r^j$$

for $g_j(y,\eta)$ of the above-mentioned kind, of order $\mu - j$. Moreover, the components of $t(y,\eta)$ have the form $\sum_{j=0}^d t_j(y,\eta)\partial_r^j$, where the $t_j(y,\eta)$ are trace symbols as above, of order $\mu - j$, the components of $k(y,\eta)$ are potential symbols while $q(y,\eta)$ is a $j_+ \times j_-$-matrix of symbols in $S^\mu_{\mathrm{cl}}(\Omega \times \mathbb{R}^q)$.

In order to formulate continuity properties of $\mathrm{Op}_y(\cdot)$ for some symbol in $S^\mu(\Omega \times \mathbb{R}^q; H, \tilde{H})$ we briefly recall the notion of abstract edge spaces.

By $\mathcal{W}^s(\mathbb{R}^q, H)$ for a Hilbert space H with group action κ we denote the completion of $\mathcal{S}(\mathbb{R}^q, H)$ with respect to the norm

$$\|u\|_{\mathcal{W}^s(\mathbb{R}^q,H)} = \| \langle\eta\rangle^s \kappa^{-1}_{\langle\eta\rangle} \hat{u}(\eta) \|_{L^2(\mathbb{R}^q,H)}$$

where $\hat{u}(\eta) = (F_{y\to\eta}u)(\eta) = \int e^{-iy\eta}u(y)dy$ is the Fourier transform. Note that $\mathcal{W}^\infty(\mathbb{R}^q, H) = \bigcap_{s\in\mathbb{R}} \mathcal{W}^s(\mathbb{R}^q, H)$ does not depend on the choice of κ; therefore, an equivalent semi-norm system in $\mathcal{W}^\infty(\mathbb{R}^q, H)$ is $u \to \| \langle\eta\rangle^s \hat{u}(\eta)\|_{L^2(\mathbb{R}^q,H)}$, $s \in \mathbb{R}$.

Then, for instance, if $a(y,\eta) \in S^\mu(\mathbb{R}^q \times \mathbb{R}^q; H, \widetilde{H})$ satisfies suitable growth conditions for $|y| \to \infty$, the operator $\mathrm{Op}(a) : C_0^\infty(\mathbb{R}^q, H) \to C^\infty(\mathbb{R}^q, \widetilde{H})$ extends to continuous operators

$$\mathrm{Op}(a) : \mathcal{W}^s(\mathbb{R}^q, H) \to \mathcal{W}^{s-\mu}(\mathbb{R}^q, \widetilde{H})$$

for all $s \in \mathbb{R}$. Moreover, over an open set $\Omega \subseteq \mathbb{R}^q$ we have obvious analogues of the standard comp/loc-Sobolev spaces, namely, $\mathcal{W}^s_{\mathrm{comp}}(\Omega, H)$ and $\mathcal{W}^s_{\mathrm{loc}}(\Omega, H)$, respectively, and then, for any $a \in S^\mu(\Omega \times \mathbb{R}^q; H, \widetilde{H})$ we have continuous operators

$$\mathrm{Op}(a) : \mathcal{W}^s_{\mathrm{comp}}(\Omega, H) \to \mathcal{W}^{s-\mu}_{\mathrm{loc}}(\Omega, \widetilde{H}) \tag{2.4}$$

for all $s \in \mathbb{R}$.

Note that in the case $H = L^2(\mathbb{R}^m)$ the group action $(\kappa_\delta u)(x) := \delta^{m/2} u(\delta x)$, $\delta \in \mathbb{R}_+$, is unitary, and then

$$\mathcal{W}^0(\mathbb{R}^q, L^2(\mathbb{R}^m)) = L^2(\mathbb{R}^q \times \mathbb{R}^m).$$

More generally, if κ_δ is a unitary action on a Hilbert space H then

$$\mathcal{W}^0(\mathbb{R}^q, H) = L^2(\mathbb{R}^q, H). \tag{2.5}$$

2.2. Tools from the edge pseudo-differential calculus

Let us now pass to a few elements of the edge pseudo-differential calculus which is a generalisation of the calculus of BVPs with the transmission property. First the edge calculus contains BVPs without the transmission property at the boundary as a special case. This in turn is a subcalculus of the edge algebra where the boundary is interpreted as the edge and a collar neighbourhood of the boundary as local wedge. However, the symbols in the edge calculus are much more general than those belonging to BVPs. They are edge-degenerate, i.e., have the form

$$p(r, y, \varrho, \eta) = \tilde{p}(r, y, r\varrho, r\eta)$$

where $\tilde{p}(r, y, \tilde{\varrho}, \tilde{\eta})$ is a symbol in the "ordinary" sense in the covariables $(\tilde{\varrho}, \tilde{\eta})$, smooth in r up to the boundary. We interchanged here the order in the variables, i.e., write (r, y) instead of (y, r) which comes from the formalism in the edge calculus. Finally, in the general edge algebra we replace $\overline{\mathbb{R}}_+ \times \Omega \ni (r, y)$ by $N^\Delta \times \Omega$ for a cone

$$N^\Delta := (\overline{\mathbb{R}}_+ \times N)/(\{0\} \times N) \tag{2.6}$$

for a smooth closed manifold N. Below we often consider the open stretched cone $N^\wedge := \mathbb{R}_+ \times N$ in the variables (r, x). In the case $\dim N = 0$ we recover again the case of a manifold with boundary.

A manifold M with (smooth) edge Y is characterised by the properties that $M \setminus Y$ is a C^∞ manifold and M close to Y is modelled on wedges $N^\Delta \times \Omega$ where Ω plays the role of a chart on Y. The nature of transition maps $N^\Delta \times \Omega \to N^\Delta \times \widetilde{\Omega}$ between different such local models is characterised by smoothness of the map $\mathbb{R}_+ \times N \times \Omega \to \mathbb{R}_+ \times N \times \widetilde{\Omega}$ which is required to be the restriction of a diffeomorphism $\overline{\mathbb{R}}_+ \times N \times \Omega \to \overline{\mathbb{R}}_+ \times N \times \widetilde{\Omega}$. All this is required in the frame isomorphisms between (in this case trivial) bundles with fibres N^Δ, $N^\wedge$ or $\overline{\mathbb{R}}_+ \times N$ over the corresponding Ω and $\widetilde{\Omega}$.

For purposes below we formulate the local description of M near the edge in a more precise manner in terms of singular charts χ^Δ close to the edge. Every point $y \in Y$ has a neighbourhood V in M such there is a homeomorphism

$$\chi^{\Delta} : V \to N^{\Delta} \times \Omega \tag{2.7}$$

which restricts to diffeomorphisms

$$\chi^{\wedge} : V \setminus Y \to N^{\wedge} \times \Omega, \quad \chi^{0} : V \cap Y \to \Omega. \tag{2.8}$$

For reasons that become clear below we introduce so-called weight data $\boldsymbol{g} := (\gamma, \gamma - \mu, \Theta)$ where $\gamma \in \mathbb{R}$ is a weight in future weighted Sobolev spaces while $\Theta = (-(k+1), 0]$ for some $k \in \mathbb{N} \cup \{+\infty\}$ is an interval that plays a role in formulating asymptotic properties under the operations. Now the pseudo-differential calculus on M locally near Y is determined by a space

$$\mathcal{R}^{\mu}(\Omega \times \mathbb{R}^{q}, \boldsymbol{g}) \tag{2.9}$$

of operator-valued amplitude functions of the following structure. Let σ, σ' and $\omega, \omega', \omega''$ be cut-off functions on the r half-axis, i.e., real-valued elements of $C_0^{\infty}(\overline{\mathbb{R}}_+)$ that are $\equiv 1$ close to $r = 0$. Here we require $\omega'' \prec \omega \prec \omega'$ where $\omega \prec \omega'$ means that $\omega' \equiv 1$ on $\operatorname{supp} \omega$. While the cut-off functions σ, σ' only depend on r for a cut-off function ω we write

$$\omega_{\eta}(r) := \omega(r[\eta])$$

where $\eta \to [\eta]$ is some fixed strictly positive C^{∞} function in $\mathbb{R}^q$ such that $[\eta] = |\eta|$ for large $|\eta|$. Below we need the notation $\omega_{|\eta|}(r) := \omega(r|\eta|)$. We define

$$\begin{aligned} a(y,\eta) := \sigma r^{-\mu}\{\omega_{\eta} \mathrm{op}_M^{\gamma - n/2}(h)(y,\eta)\omega'_{\eta} + (1-\omega_{\eta})\mathrm{Op}_r(p)(y,\eta)(1-\omega''_{\eta})\}\sigma' \\ + \varphi \mathrm{Op}_r(p_{\mathrm{int}})(y,\eta)\varphi' + m(y,\eta) + g(y,\eta), \end{aligned} \tag{2.10}$$

$n = \dim N$, with the following ingredients. The main part is an operator function $p(r, y, \varrho, \eta) = \tilde{p}(r, y, r\varrho, r\eta)$ where

$$\tilde{p}(r, y, \tilde{\varrho}, \tilde{\eta}) \in C^{\infty}(\overline{\mathbb{R}}_+ \times \Omega, L^{\mu}_{\mathrm{cl}}(N; \mathbb{R}^{1+q}_{\tilde{\varrho},\tilde{\eta}}))$$

(we tacitly employ the canonical Fréchet space structure of $L^{\mu}_{\mathrm{cl}}(\cdot\,; \mathbb{R}^l)$). The pseudo-differential operator $\mathrm{Op}_r(\cdot)$ with respect to (r, ϱ) within the amplitude function is as in (2.3) (the notation Op or op only indicates different contexts). In order to define h in (2.10) we recall that in the edge calculus by $M^{\mu}_{\mathcal{O}}(N; \mathbb{R}^l_{\lambda})$ we denote the subspace of all $f(w, \lambda) \in \mathcal{A}(\mathbb{C}_w, L^{\mu}_{\mathrm{cl}}(N; \mathbb{R}^l_{\lambda}))$ such that $f(\beta + i\varrho, \lambda) \in L^{\mu}_{\mathrm{cl}}(N; \mathbb{R}_{\varrho} \times \mathbb{R}^l_{\lambda})$ for every $\beta \in \mathbb{R}$, uniformly in finite β-intervals. Here $\mathcal{A}(G, E)$ for an open $G \subseteq \mathbb{C}$ and a Fréchet space E is the space of all holomorphic functions over G with values in E. The space is Fréchet in the topology of uniform convergence on compact subsets. Now h in (2.10) defined by

$$h(r, y, w, \eta) = \tilde{h}(r, y, w, r\eta)$$

for some

$$\tilde{h}(r, y, w, \tilde{\eta}) \in C^{\infty}(\overline{\mathbb{R}}_+ \times \Omega, M^{\mu}_{\mathcal{O}}(N; \mathbb{R}^q_{\tilde{\eta}})).$$

The notation $\mathrm{op}_M^\delta(\cdot)$ occurring in (2.10) means a pseudo-differential action on $\mathbb{R}_+$, based on the weighted Mellin transform. The standard Mellin transform

$$Mu(w) = \int_0^\infty r^w u(r) \frac{dr}{r}$$

for $u \in C_0^\infty(\mathbb{R}_+)$ is an entire function in $\mathbb{C} \ni w$, and we define

$$M_\delta u(w) := Mu(w)|_{\Gamma_{\frac{1}{2}-\delta}}$$

where $\Gamma_\beta := \{w \in \mathbb{C} : \mathrm{Re}\, w = \beta\}$ for any real β. Now

$$\mathrm{op}_M^\delta(f) := M_\delta^{-1} f M_\delta$$

is the pseudo-differential operator based on M_δ with the amplitude function $f(r,w)$. In our case f may depend on additional variables, and f is operator-valued. Thus we explained the first part on the right of (2.10). The next summand is motivated by remainders when we change the cut-off functions σ, σ'. We assume $\varphi, \varphi' \in C_0^\infty(\mathbb{R}_+)$ and

$$p_{\mathrm{int}}(r, y, \varrho, \eta) \in C^\infty(\mathbb{R}_+ \times \Omega, L_{\mathrm{cl}}^\mu(N; \mathbb{R}_{\varrho,\eta}^{1+q})).$$

The remaining $m + g$ terms on the right of (2.10) are responsible for asymptotic properties in the edge calculus. First $g(y, \eta)$ is an analogue of the above-mentioned Green symbols. In contrast to the mapping properties with $\mathcal{S}(\overline{\mathbb{R}}_+)$ in the image, in particular, Taylor asymptotics for $r \to 0$, we now admit arbitrary so-called discrete asymptotics of certain types $\mathcal{P}$. Those are represented by sequences

$$\mathcal{P} := \{(p_j, m_j)\}_{j=0,\ldots,J} \subset \mathbb{C} \times \mathbb{N}$$

for some $J = J(\mathcal{P}) \in \mathbb{N} \cup \{\infty\}$. We consider $\mathcal{P}$ in connection with a so-called weight interval $\Theta = (-(k+1), 0]$, $k \in \mathbb{N} \cup \{\infty\}$, on the left of the weight line $\Gamma_{\frac{n+1}{2}-\gamma}$. More precisely, we assume

$$\pi_{\mathbb{C}}\mathcal{P} := \{p_j\}_{j=0,\ldots,J} \subset \left\{\frac{n+1}{2} - \gamma - (k+1) < \mathrm{Re}\, w < \frac{n+1}{2} - \gamma\right\}$$

and J to be finite for finite k, and $\mathrm{Re}\, p_j \to -\infty$ as $j \to \infty$ as soon as $\pi_{\mathbb{C}}\mathcal{P}$ is infinite in the case $k = +\infty$. Such an asymptotic type is said to be associated with the weight data (γ, Θ).

We define the space $\mathcal{H}^{s,\gamma}(N^\wedge)$ for $s \in \mathbb{N}, \gamma \in \mathbb{R}$, to be the set of all $u(r,x) \in r^{-n/2}L^2(\mathbb{R}_+ \times N)$ such that $(r\partial_r)^j D_x^m u(r,x) \in r^{-n/2}L^2(\mathbb{R}_+ \times N)$ for all $j + m \leq s$, where D_x^m is any differential operator with smooth coefficients on N of order m. For $-s \in \mathbb{N}$ we obtain $\mathcal{H}^{s,\gamma}(N^\wedge)$ by duality with respect to the scalar product of $\mathcal{H}^{0,0}(N^\wedge) = r^{-n/2}L^2(\mathbb{R}_+ \times N)$. For real s we define $\mathcal{H}^{s,\gamma}(N^\wedge)$ by (complex) interpolation, cf. [13, Theorem 2.1.16].

Moreover, let $H_{\mathrm{cone}}^s(N^\wedge)$ defined to be the set of all $u(r,x) \in H_{\mathrm{loc}}^s(N^\wedge)$ such that for any cut-off function $\omega(r)$ and every coordinate neighbourhood U on N and a diffeomorphism $\chi : U \to V$ to some $V \subset S^n$ and the associated map

$(1 \times \chi) : \mathbb{R}_+ \times U \to \Gamma$, $(1 \times \chi)(r, x) := r\chi(x), \Gamma := \{\tilde{x} \in \mathbb{R}^{1+n} : \tilde{x}/|\tilde{x}| \in V\}$, we have

$$((1 \times \chi)^{-1})^*((1 - \omega)\varphi u) \in H^s(\mathbb{R}^{1+n}_{\tilde{x}})$$

for every $\varphi \in C_0^\infty(U)$, with upper $*$ being the function pull-back. For any fixed cut-off function ω we now form

$$\mathcal{K}^{s,\gamma}(N^\wedge) := \{\omega u + (1 - \omega)v : u \in \mathcal{H}^{s,\gamma}(N^\wedge), v \in H^s_{\text{cone}}(N^\wedge)\}.$$

For $s = \gamma = 0$ we have

$$\mathcal{H}^{0,0}(N^\wedge) = \mathcal{K}^{0,0}(N^\wedge) = r^{-n/2}L^2(\mathbb{R}_+ \times N),$$

where $L^2(\mathbb{R}_+ \times N)$ refers to the measure $drdx$.
Moreover, for finite k and $\Theta = (-(k+1), 0]$ we form

$$\mathcal{K}^{s,\gamma}_\Theta(N^\wedge) := \varprojlim_{l \in \mathbb{N}} \mathcal{K}^{s,\gamma+(k+1)-(1+l)^{-1}}(N^\wedge)$$

which corresponds to the space of flat functions of flatness $k + 1 - 0$ with respect to the weight γ. In order to form spaces with asymptotics of type $\mathcal{P}$ we set

$$\mathcal{K}^{s,\gamma}_\mathcal{P}(N^\wedge) := \mathcal{K}^{s,\gamma}_\Theta(N^\wedge) + \mathcal{E}_\mathcal{P}(N^\wedge)$$

where

$$\mathcal{E}_\mathcal{P}(N^\wedge) := \Big\{\omega(r) \sum_{j=0}^{J} \sum_{l=0}^{m_j} c_{jl}(x) r^{-p_j} \log^l r : c_{jl} \in C^\infty(N)\Big\}.$$

Let

$$\mathcal{K}^{\infty,\gamma}_\mathcal{P}(N^\wedge) := \varprojlim_{j \in \mathbb{N}} \mathcal{K}^{j,\gamma}_\mathcal{P}(N^\wedge).$$

Moreover, we set $\mathcal{K}^{s,\gamma;e}(N^\wedge) := \langle r \rangle^{-e} \mathcal{K}^{s,\gamma}(N^\wedge)$ for $s, \gamma, e \in \mathbb{R}$. If $\mathcal{P}$ is an asymptotic type, associated with the weight data (γ, Θ) we define

$$\mathcal{S}^\gamma_\mathcal{P}(N^\wedge) := \varprojlim_{l \in \mathbb{N}} \langle r \rangle^{-l} \mathcal{K}^{l,\gamma}_\mathcal{P}(N^\wedge)$$

in the Fréchet topology of the projective limit. The definition implies

$$(1 - \omega)\mathcal{S}^\gamma_\mathcal{P}(N^\wedge) = (1 - \omega)\mathcal{S}(\overline{\mathbb{R}}_+, C^\infty(N))$$

for any cut-off function ω. Observe that

$$\begin{aligned} &\sigma r^{-\mu}\{\omega_\eta \mathrm{op}_M^{\gamma-n/2}(h)(y,\eta)\omega'_\eta + (1-\omega_\eta)\mathrm{Op}_r(p)(y,\eta)(1-\omega''_\eta)\}\sigma' + \varphi \mathrm{Op}_r(p_{\text{int}})(y,\eta)\varphi' \\ &\quad \in S^\mu(\Omega \times \mathbb{R}^q; \mathcal{K}^{s,\gamma}(N^\wedge), \mathcal{K}^{s-\mu,\gamma-\mu}(N^\wedge)) \qquad (2.11) \end{aligned}$$

for every $s \in \mathbb{R}$. The Hilbert spaces $\mathcal{K}^{s,\gamma}(N^\wedge)$ are endowed with the group action

$$(\kappa_\delta u)(r, x) = \delta^{\frac{n+1}{2}} u(\delta r, x), \delta \in \mathbb{R}_+.$$

In future we omit the second term in (2.11), since it only plays the role to keep the expression independent of the choice of σ, σ', modulo so-called Green symbols to be defined below. However, by choosing σ, σ' in such a way that $\sigma \succ \varphi$ and $\sigma' \succ \varphi'$ we may rephrase the second summand as a part of the first term in (2.11).

By $\mathcal{R}^\nu_G(\Omega\times\mathbb{R}^q,\boldsymbol{g})$ for weight data $\boldsymbol{g}:=(\gamma,\gamma-\mu,\Theta),\gamma,\mu\in\mathbb{R},\Theta=(-(k+1),0],k\in\mathbb{N}\cup\{+\infty\}$, we denote the space of so-called Green symbols $g(y,\eta)$ of order $\nu\in\mathbb{R}$, defined by the relations,

$$g(y,\eta)\in S^\nu_{\mathrm{cl}}(\Omega\times\mathbb{R}^q;\mathcal{K}^{s,\gamma;e}(N^\wedge),\mathcal{S}^{\gamma-\mu}_{\mathcal{P}}(N^\wedge)),$$
$$g^*(y,\eta)\in S^\nu_{\mathrm{cl}}(\Omega\times\mathbb{R}^q;\mathcal{K}^{s,-\gamma+\mu;e}(X^\wedge),\mathcal{S}^{-\gamma}_{\mathcal{Q}}(N^\wedge))$$

for all $s,e\in\mathbb{R}$, for g-dependent asymptotic types $\mathcal{P}$ and $\mathcal{Q}$ associated with $(\gamma-\mu,\Theta)$ and $(-\gamma,\Theta)$, respectively. The (y,η) wise formal adjoint g^* refers to the scalar product of $\mathcal{K}^{0,0}(N^\wedge)=r^{-n/2}L^2(\mathbb{R}_+\times N)$.

Now we also define $m(y,\eta)$. To this end we fix a sequence

$$\mathcal{R}:=\{(r_j,n_j)\}_{j\in\mathbb{Z}}\subset\mathbb{C}\times\mathbb{N}$$

such that $\pi_{\mathbb{C}}\mathcal{R}:=\{r_j\}_{j\in\mathbb{Z}}$ intersects every strip $\{c\le\operatorname{Re}w\le c'\},c<c'$, in a finite set. Such $\mathcal{R}$ are called Mellin asymptotic types. Then $M^{-\infty}_{\mathcal{R}}(N)$ is defined as the space of all $f\in\mathcal{A}(\mathbb{C}\setminus\pi_{\mathbb{C}}\mathcal{R},L^{-\infty}(N))$ such that for any $\pi_{\mathbb{C}}\mathcal{R}$-excision function χ (i.e., $\chi(w)=0$ for $\operatorname{dist}(\pi_{\mathbb{C}}\mathcal{R},w)<c_0,\operatorname{dist}(\pi_{\mathbb{C}}\mathcal{R},w)>c_1$ for any $0<c_0<c_1$) we have

$$(\chi(w)f(w))|_{\Gamma_\beta}\in\mathcal{S}(\Gamma_\beta,L^{-\infty}(N))$$

for every $\beta\in\mathbb{R}$, uniformly in compact β-intervals. In addition we require that f is meromorphic with poles at r_j of multiplicity n_j+1 and Laurent coefficients at $(w-r_j)^{-(k+1)}$ in $L^{-\infty}(N)$ and being of finite rank, $0\le k\le n_j$.

The spaces $M^{-\infty}_{\mathcal{R}}(N)$ are Fréchet in a natural way. We set

$$M^\mu_{\mathcal{R}}(N):=M^\mu_{\mathcal{O}}(N)+M^{-\infty}_{\mathcal{R}}(N)$$

in the Fréchet topology of the non-direct sum.

Given such an $f\in M^{-\infty}_{\mathcal{R}}(N)$ and a weight $\gamma\in\mathbb{R}$ then

$$r^{-\mu}\omega\operatorname{op}^{\gamma-n/2}_M(f)\omega':\mathcal{K}^{s,\gamma}(N^\wedge)\to\mathcal{K}^{\infty,\gamma-\mu}(N^\wedge)$$

is continuous for any cut-off functions ω,ω', provided that $\Gamma_{\frac{n+1}{2}-\gamma}\cap\pi_{\mathbb{C}}\mathcal{R}=\emptyset$. Moreover, for every asymptotic type $\mathcal{P}$ associated with the weight data (γ,Θ) there is a $\mathcal{Q}$ associated with $(\gamma-\mu,\Theta)$ such that

$$r^{-\mu}\omega\operatorname{op}^{\gamma-n/2}_M(f)\omega':\mathcal{K}^{s,\gamma}_{\mathcal{P}}(N^\wedge)\to\mathcal{K}^{\infty,\gamma-\mu}_{\mathcal{Q}}(N^\wedge)$$

is continuous for all $s\in\mathbb{R}$. These operators are now the raw material for building up operator-valued symbols taking values in such Mellin operators. For $k\in\mathbb{N}$ in the weight interval $\Theta=(-(k+1),0]$ we form

$$m(y,\eta):=r^{-\mu}\omega_\eta\sum_{j=0}^{k}r^j\sum_{|\alpha|\le j}\operatorname{op}^{\gamma_{j\alpha}-n/2}_M(f_{j\alpha})(y)\eta^\alpha\omega'_\eta \tag{2.12}$$

for $f_{j\alpha}(y,w)\in C^\infty(\Omega,M^{-\infty}_{\mathcal{R}_{j\alpha}}(N))$ and $\Gamma_{\frac{n+1}{2}-\gamma_{j\alpha}}\cap\pi_{\mathbb{C}}\mathcal{R}_{j\alpha}=\emptyset,\gamma-j\le\gamma_{j\alpha}\le\gamma$ for all j,α. Let $\mathcal{R}^\mu_{M+G}(\Omega\times\mathbb{R}^q,\boldsymbol{g})$, $\boldsymbol{g}:=(\gamma,\gamma-\mu,(-(k+1),0])$, denote the set of all $(m+g)(y,\eta)$ for operator functions $m(y,\eta)$ as in (2.12) and $g(y,\eta)\in\mathcal{R}^\mu_G(\Omega\times\mathbb{R}^q,\boldsymbol{g})$.

Then we have (y,η)-dependent families of continuous operators in the above spaces, more precisely, we have

$$m(y,\eta) \in S^{\mu}_{\mathrm{cl}}(\Omega \times \mathbb{R}^q; \mathcal{K}^{s,\gamma}(N^\wedge), \mathcal{K}^{\infty,\gamma-\mu}(N^\wedge)),$$
$$m(y,\eta) \in S^{\mu}_{\mathrm{cl}}(\Omega \times \mathbb{R}^q; \mathcal{K}^{s,\gamma}_{\mathcal{P}}(N^\wedge), \mathcal{K}^{\infty,\gamma-\mu}_{\mathcal{Q}}(N^\wedge))$$

for all $s \in \mathbb{R}$ and every $\mathcal{P}$ associated with (γ,Θ) for some resulting $\mathcal{Q}$ associated with $(\gamma-\mu,\Theta)$.

For any $a(y,\eta) \in \mathcal{R}^\mu(\Omega \times \mathbb{R}^q, \boldsymbol{g})$ we define the (twisted) homogeneous principal edge symbol of order μ by

$$\sigma_1(a)(y,\eta) = r^{-\mu}\{\omega_{|\eta|}\mathrm{op}_M^{\gamma-n/2}(h_0)(y,\eta)\omega'_{|\eta|} + (1-\omega_{|\eta|})\mathrm{Op}_r(p_0)(y,\eta)(1-\omega''_{|\eta|})\} + \sigma_1(m)(y,\eta) + \sigma_1(g)(y,\eta) \quad (2.13)$$

where $h_0(r,y,w,\eta) := \tilde{h}(0,y,w,r\eta), p_0(r,y,\varrho,\eta) := \tilde{p}(0,y,r\varrho,r\eta)$.

Moreover, $\sigma_1(m)(y,\eta)$ and $\sigma_1(g)(y,\eta)$ are the (twisted) homogeneous components of order μ of the classical symbols m and g, respectively. For $\sigma_1(m)(y,\eta)$ we have an explicit expression, namely,

$$\sigma_1(m)(y,\eta) := r^{-\mu}\omega_{|\eta|}\sum_{j=0}^{k} r^j \sum_{|\alpha|=j} \mathrm{op}_M^{\gamma_{j\alpha}-n/2}(f_{j\alpha})(y)\eta^\alpha\omega'_{|\eta|}.$$

From (2.11) and $(m+g)(y,\eta) \in S^{\mu}_{\mathrm{cl}}(\Omega \times \mathbb{R}^q; \mathcal{K}^{s,\gamma}(N^\wedge), \mathcal{K}^{s-\mu,\gamma-\mu}(N^\wedge))$ we see that

$$\mathcal{R}^\mu(\Omega \times \mathbb{R}^q, \boldsymbol{g}) \subset S^\mu(\Omega \times \mathbb{R}^q; \mathcal{K}^{s,\gamma}(N^\wedge), \mathcal{K}^{s-\mu,\gamma-\mu}(N^\wedge)),$$

$s \in \mathbb{R}$. Then, according to (2.4) for $a(y,\eta) \in \mathcal{R}^\mu(\Omega \times \mathbb{R}^q, \boldsymbol{g})$ we obtain continuous operators

$$\mathrm{Op}(a) : \mathcal{W}^s_{\mathrm{comp}}(\Omega, \mathcal{K}^{s,\gamma}(N^\wedge)) \to \mathcal{W}^{s-\mu}_{\mathrm{loc}}(\Omega, \mathcal{K}^{s-\mu,\gamma-\mu}(N^\wedge)),$$

$s \in \mathbb{R}$. Later on we mainly consider the case $\mu = 0$ and $\gamma = 0$. Then we have

$$\mathrm{Op}(a) : \mathcal{W}^0_{\mathrm{comp}}(\Omega, \mathcal{K}^{0,0}(N^\wedge)) \to \mathcal{W}^0_{\mathrm{loc}}(\Omega, \mathcal{K}^{0,0}(N^\wedge)),$$

where comp and loc refer to the y-variable. Observe that

$$\mathcal{W}^0(\mathbb{R}^q, \mathcal{K}^{0,0}(N^\wedge)) = r^{-n/2}L^2(\mathbb{R}_+ \times N \times \mathbb{R}^q), \quad (2.14)$$

cf. also the relation (2.5).

Globally on our (first compact) manifold M with edge Y we have the space $H^{0,0}(M) \subset L^2_{\mathrm{loc}}(M \setminus Y)$ which is locally near Y described by (2.14). Depending on fixed Riemannian metrics over N far from Y we have a scalar product in $H^{0,0}(M)$. The first upper subscript means Sobolev smoothness 0, the second one the weight 0 in $\mathcal{K}^{0,0}(N^\wedge)$ on the left of (2.14).

We also form subspaces $H^{\infty,0}_{\mathcal{P}}(M) \subset C^\infty(M \setminus Y)$ that are locally near Y described by

$$\mathcal{W}^\infty(\mathbb{R}^q, \mathcal{K}^{\infty,0}_{\mathcal{P}}(N^\wedge)),$$

for some asymptotic type $\mathcal{P}$ associated with the weight data $(0,\Theta)$. These spaces are Fréchet in a natural way.

By $L^{-\infty}(M,\boldsymbol{g})$ on a compact manifold M with edge and $\boldsymbol{g}=(0,0,\Theta),\Theta=(-(k+1),0]$, we denote the set of all $C\in\mathcal{L}(H^{0,0}(M),H^{0,0}(M))$ that induce continuous operators

$$C:H^{0,0}(M)\to H^{\infty,0}_{\mathcal{P}}(M),\; C^*:H^{0,0}(M)\to H^{\infty,0}_{\mathcal{Q}}(M) \tag{2.15}$$

for certain C-dependent asymptotic types $\mathcal{P}$ and $\mathcal{Q}$, both associated with the weight data $(0,\Theta)$. The subspace

$$L^{-\infty}(M,\boldsymbol{g})_{\mathcal{P},\mathcal{Q}} \tag{2.16}$$

of operators for fixed $\mathcal{P},\mathcal{Q}$ is a Fréchet space; thus $L^{-\infty}(M,\boldsymbol{g})$ is a union of Fréchet spaces.

On a compact manifold M with edge Y we now define the algebra

$$L^0(M,\boldsymbol{g}) \tag{2.17}$$

of edge pseudo-differential operators of order 0 referring to the weight data $\boldsymbol{g}=(0,0,(-(k+1),0])$. The space (2.17) is characterised as the subspace of all $A\in L^0_{\mathrm{cl}}(M\setminus Y)$ such that $A-C$ for some $C\in L^{-\infty}(M,\boldsymbol{g})$ is of the form

$$\sigma\Big\{\sum_{j=1}^{N}\varphi_j(\chi_j^\wedge)_*^{-1}\mathrm{Op}_y(a_j)\varphi_j'\Big\}\sigma'+(1-\sigma)A_{\mathrm{int}}(1-\sigma''). \tag{2.18}$$

Here $\sigma''\prec\sigma\prec\sigma'$ are cut-off functions on M, and

$$\chi_j^\Delta:V_j\to N^\Delta\times\mathbb{R}^q$$

is a system of singular charts which induce charts

$$\chi_j^\wedge:V_j\setminus Y\to N^\wedge\times\mathbb{R}^q,\;\chi_j^0:V_j\cap Y\to\mathbb{R}^q,$$

cf. (2.7), (2.8), and $\{\varphi_1,\dots,\varphi_N\}$ is a partition of unity subordinate to the open covering $\{V_1\cap Y,\dots,V_N\cap Y\}$ of Y, $\varphi_j'\in C_0^\infty(V_j\cap Y)$ with $\varphi_j'\succ\varphi_j$ for all j. Moreover, $a_j(y,\eta)\in\mathcal{R}^0(\mathbb{R}^q\times\mathbb{R}^q,\boldsymbol{g}),j=1,\dots,N$, and $A_{\mathrm{int}}\in L^0_{\mathrm{cl}}(M\setminus Y)$.

There is a straightforward extension of (2.17) to the case of a non-compact manifold with edge. It suffices to replace the finite sum in (2.18) by a locally finite sum referring to a corresponding countable system of singular charts and to smoothing operators defined in terms of comp/loc-analogues of the spaces occurring in (2.15).

By virtue of $L^0(M,\boldsymbol{g})\subset L^0_{\mathrm{cl}}(M\setminus Y)$ an $A\in L^0(M,\boldsymbol{g})$ has a zero-order homogeneous principal symbol $\sigma_0(A)$ which is a function in $C^\infty(T^*(M\setminus Y)\setminus 0)$. In the splitting $(r,x,y)\in\mathbb{R}_+\times N\times\Omega$ of variables locally near the edge Y and the corresponding covariables (ϱ,ξ,η) we can write

$$\sigma_0(A)(r,x,y,\varrho,\xi,\eta)=\tilde{\sigma}_0(A)(r,x,y,r\varrho,\xi,r\eta) \tag{2.19}$$

for a $\tilde{\sigma}_0(A)(r,x,y,\tilde{\varrho},\xi,\tilde{\eta})$ homogeneous of order zero in $(\tilde{\varrho},\xi,\tilde{\eta})\neq 0$ and smooth up to $r=0$. We also call $\tilde{\sigma}_0(A)(r,x,y,\varrho,\xi,\eta)$ the reduced interior principal symbol. In addition an $A\in L^0(M,\boldsymbol{g})$ has a (twisted) homogeneous principal edge symbol

$\sigma_1(A)(y,\eta)$, living on $T^*Y \setminus 0$, which is an operator function

$$\sigma_1(A)(y,\eta) : \mathcal{K}^{0,0}(N^\wedge) \to \mathcal{K}^{0,0}(N^\wedge).$$

The definition of $\sigma_1(A)$ immediately follows from the local expressions (2.13), formulated on the level of amplitude functions $a(y,\eta)$, here applied to $\mu = \gamma = 0$. In our calculus we need two generalisations of the edge operator space $L^0(M, \boldsymbol{g})$, namely, a parameter-dependent analogue

$$L^0(M, \boldsymbol{g}; \mathbb{R}^l) \tag{2.20}$$

with parameters $\lambda \in \mathbb{R}^l$ which is defined by amplitude functions $a(y,\eta,\lambda)$ where everywhere the covariable $\eta \in \mathbb{R}^q$ is simply replaced by $(\eta,\lambda) \in \mathbb{R}^{q+l}$. Details of this calculus are developed in [5]. Let us recall some notation. In order to formulate the class of smoothing operators we set

$$L^{-\infty}(M, \boldsymbol{g}; \mathbb{R}^l) := \mathcal{S}(\mathbb{R}^l, L^{-\infty}(M, \boldsymbol{g})) \tag{2.21}$$

which is nothing else than $\bigcup_{\mathcal{P},\mathcal{Q}} \mathcal{S}(\mathbb{R}^l, L^{-\infty}(M, \boldsymbol{g})_{\mathcal{P},\mathcal{Q}})$, cf. the notation (2.16). Another slight generalisation concerns $d_+ \times d_-$ block matrices of operators in (2.20). Setting $\boldsymbol{v} := (d_-, d_+)$ the corresponding space will be denoted by

$$L^0(M, \boldsymbol{g}; \boldsymbol{v}; \mathbb{R}^l). \tag{2.22}$$

The components of the principal symbol of an A in (2.20) can be computed in terms of homogeneities where λ is treated as an additional covariable. We obtain in this way the parameter-dependent principal symbol

$$\sigma(A(\lambda)) = (\sigma_0(A(\lambda)), \sigma_1(A(\lambda))).$$

Moreover, for A in (2.22) we may compute the symbols separately for every entry, and since

$$L^0(M, \boldsymbol{g}; \boldsymbol{v}; \mathbb{R}^l) = L^0(M, \boldsymbol{g}; \mathbb{R}^l) \otimes \mathbb{C}^{d_+} \otimes \mathbb{C}^{d_-}$$

we obtain corresponding $d_+ \times d_-$ block matrices. In our application we mainly need the case $\boldsymbol{v} = (2, 2)$.

Theorem 2.1. *$A \in L^0(M, \boldsymbol{g}; \boldsymbol{v}; \mathbb{R}^l), B \in L^0(M, \boldsymbol{g}; \boldsymbol{w}; \mathbb{R}^l)$ for $\boldsymbol{v} = (d_0, d_+), \boldsymbol{w} = (d_-, d_0)$ implies $AB \in L^0(M, \boldsymbol{g}; \boldsymbol{v} \circ \boldsymbol{w}; \mathbb{R}^l)$ for $\boldsymbol{v} \circ \boldsymbol{w} = (d_-, d_+)$, and we have*

$$\sigma(AB) = \sigma(A)\sigma(B)$$

for componentwise composition.

The proof of Theorem 2.1 follows in a similar manner as the corresponding composition result for edge operators without parameters, see [5, Theorem 2.5.24, Proposition 2.6.11]. The only change in the parameter-dependent case is to interpret the parameter as an extra edge covariable and to interpret homogeneity for σ_0 and twisted homogeneity for σ_1 including the parameter.

In the following definition we assume $\boldsymbol{v} := (d, d)$.

Definition 2.2.

(i) An $A \in L^0(M, \boldsymbol{g}; \boldsymbol{v}; \mathbb{R}^l)$ is called (parameter-dependent) elliptic, if $\det \sigma_0(A(\lambda))$ never vanishes on $T^*(M \setminus Y) \times \mathbb{R}^l \setminus 0$, the reduced symbol of $\det \sigma_0(A(\lambda))$ does not vanish for $(\varrho, \xi, \eta, \lambda) \neq 0$, and if in addition the (parameter-dependent) principal edge symbol

$$\sigma_1(A)(y, \eta, \lambda) : \mathcal{K}^{0,0}(N^\wedge, \mathbb{C}^d) \to \mathcal{K}^{0,0}(N^\wedge, \mathbb{C}^d) \tag{2.23}$$

is bijective for all y and $(\eta, \lambda) \neq 0$.

(ii) A $P \in L^0(M, \boldsymbol{g}; \boldsymbol{v}; \mathbb{R}^l)$ is called a (parameter-dependent) parametrix of $A \in L^0(M, \boldsymbol{g}; \boldsymbol{v}; \mathbb{R}^l)$ if

$$PA - I, AP - I \in L^{-\infty}(M, \boldsymbol{g}; \boldsymbol{v}; \mathbb{R}^l).$$

Remark 2.3. The ellipticity of operators in the edge calculus usually refers to additional edge conditions. We apply here the simpler variant requiring the bijectivity of (2.23) without further entries in a corresponding block matrix family of operators, since our application is just of this kind.

In the construction of parametrices of elliptic elements it is known that also the subordinate conormal symbols have to be inverted. Those are meromorphic operator functions, in general depending on the variable $y \in Y$. The arising Mellin asymptotic types then become y-dependent which requires an extension of the calculus to continuous or variable discrete asymptotics. In order to avoid this voluminous point, cf., e.g., [12], [17], here we assume that the inverted conormal symbols represent constant (in y) discrete asymptotics. Under this additional condition on the operator A the first part of the following theorem holds. The second part is true in general.

Theorem 2.4.

(i) *An elliptic* $A \in L^0(M, \boldsymbol{g}; \boldsymbol{v}; \mathbb{R}^l)$ *has a parametrix* $P \in L^0(M, \boldsymbol{g}; \boldsymbol{v}; \mathbb{R}^l)$.

(ii) *An elliptic* $A \in L^0(M, \boldsymbol{g}; \boldsymbol{v}; \mathbb{R}^l)$ *induces a family of Fredholm operators*

$$A(\lambda) : H^{0,0}(M, \mathbb{C}^d) \to H^{0,0}(M, \mathbb{C}^d). \tag{2.24}$$

For $l > 0$ *the operators* (2.24) *are of index* 0 *and become isomorphisms for* $|\lambda| \geq C$ *for a suitable* $C > 0$.

For Theorem 2.4 we refer to [5, Theorems 3.1.10 and 3.3.1]. Note that the Fredholm property between edge Sobolev spaces is λ-wise, while remainders in this construction belong to the space (2.21) which causes invertibility for large $|\lambda|$. Thus the Fredholm operators are of index zero for all λ.

2.3. The truncation of operators on manifolds with boundary

Let X be a smooth compact manifold with smooth boundary $\partial X = Y$. We assume that X is embedded in a smooth closed manifold N where N is represented as $N = X_+ \cup X_-$ for $X_+ := X$ with X_- being another smooth compact manifold with boundary Y, such that $X_+ \cap X_- = Y$. On N we fix a Riemannian metric that is close to Y the product metric $(-1, 1) \times Y$ on a tubular neighbourhood of Y in N, for some Riemannian metric on Y. Consider the spaces $L^2(N)$ as well as $L^2(X_\pm)$ with the corresponding scalar products. In the following for convenience we assume that X_- is a diffeomorphic to X_+ (in the category of smooth manifolds with boundary). We fix a diffeomorphism

$$\varepsilon : X_+ \to X_-$$

such that $\varepsilon(y) = y$ for $y \in Y$, the common boundary of $X_\pm$. Then the function pull-back ε^* induces an isomorphism

$$\varepsilon^* : L^2(X_-) \to L^2(X_+). \tag{2.25}$$

Let $\mathrm{r}^\pm : L^2(N) \to L^2(X_\pm)$ be operators of restrictions of functions to the $\pm$-sides, and denote by $\mathrm{e}^\pm : L^2(X_\pm) \to L^2(N)$ the operators that extend functions by zero to the respective opposite side.

Let $A \in L^0_{\mathrm{cl}}(N; \mathbb{R}^l)$ be a classical parameter-dependent pseudo-differential operator of order 0 on N and consider A as a family of continuous maps

$$A(\lambda) : L^2(N) \to L^2(N),$$

$\lambda \in \mathbb{R}^l$. Then, composing A with the isomorphisms

$$\begin{pmatrix} \mathrm{r}^+ \\ \mathrm{r}^- \end{pmatrix} : L^2(N) \longrightarrow \begin{matrix} L^2(X_+) \\ \oplus \\ L^2(X_-) \end{matrix}, \quad \begin{pmatrix} \mathrm{e}^+ & \mathrm{e}^- \end{pmatrix} : \begin{matrix} L^2(X_+) \\ \oplus \\ L^2(X_-) \end{matrix} \longrightarrow L^2(N) \tag{2.26}$$

we obtain a family of continuous maps

$$\begin{pmatrix} \mathrm{r}^+ \\ \mathrm{r}^- \end{pmatrix} A \begin{pmatrix} \mathrm{e}^+ & \mathrm{e}^- \end{pmatrix} = \begin{pmatrix} \mathrm{r}^+ A\mathrm{e}^+ & \mathrm{r}^+ A\mathrm{e}^- \\ \mathrm{r}^- A\mathrm{e}^+ & \mathrm{r}^- A\mathrm{e}^- \end{pmatrix} : \begin{matrix} L^2(X_+) \\ \oplus \\ L^2(X_-) \end{matrix} \longrightarrow \begin{matrix} L^2(X_+) \\ \oplus \\ L^2(X_-) \end{matrix}.$$

Moreover, applying the reflection isomorphism (2.25) yields operators

$$\begin{aligned} \mathbb{A} &:= \begin{pmatrix} 1 & 0 \\ 0 & \varepsilon^* \end{pmatrix} \begin{pmatrix} \mathrm{r}^+ A\mathrm{e}^+ & \mathrm{r}^+ A\mathrm{e}^- \\ \mathrm{r}^- A\mathrm{e}^+ & \mathrm{r}^- A\mathrm{e}^- \end{pmatrix} \begin{pmatrix} 1 & 0 \\ 0 & (\varepsilon^*)^{-1} \end{pmatrix} \\ &= \begin{pmatrix} \mathrm{r}^+ A\mathrm{e}^+ & \mathrm{r}^+ A\mathrm{e}^-(\varepsilon^*)^{-1} \\ \varepsilon^*\mathrm{r}^- A\mathrm{e}^+ & \varepsilon^*\mathrm{r}^- A\mathrm{e}^-(\varepsilon^*)^{-1} \end{pmatrix} : L^2(X, \mathbb{C}^2) \to L^2(X, \mathbb{C}^2). \end{aligned} \tag{2.27}$$

We now apply a known result from [21] which tells us that for every fixed λ the operator $\mathbb{A}$ is a (2×2 system of) pseudo-differential boundary value problem(s) over X with respect to Y, in general, without the transmission property at the boundary. At the same time the operator $\mathbb{A}$ represents an edge problem in the edge calculus of [17] where the boundary is regarded as the edge. Clearly, if the original operator A has the transmission property at the boundary, then $\mathbb{A}$ belongs to Boutet de Monvel's calculus, cf. [2], [15]. The operators

$$\mathrm{r}^+ A\mathrm{e}^-(\varepsilon^*)^{-1},\ \varepsilon^*\mathrm{r}^- A\mathrm{e}^+ \tag{2.28}$$

are then Green operators of type 0. In the general case, i.e., for arbitrary A the operators (2.28) belong to $L^0_{M+G}(X, \boldsymbol{g}; \boldsymbol{v})$, $\boldsymbol{g} = (0, 0, (-(k+1), 0])$, $\boldsymbol{v} = (2, 2)$, the substructure of smoothing Mellin plus Green operators. Recall that Eskin in his book [7] for truncated operators on the half-axis observed smoothing Mellin operators plus Hilbert–Schmidt operators in such a connection. In addition the operators $\mathrm{r}^+ A \mathrm{e}^+$ (and also $\varepsilon^* \mathrm{r}^- A \mathrm{e}^- (\varepsilon^*)^{-1}$) have been recognized as Mellin pseudo-differential operators on the half-axis, also modulo certain more regular terms, later on charaterized as smoothing Mellin plus Green operators in the cone calculus over $\mathbb{R}_+$, cf. [17]. Let us now formulate as a first consequence the following result. Assume for the moment $l = 0$.

Theorem 2.5.

(i) *The map* $\Phi : L^0_{\mathrm{cl}}(N) \to L^0(X, \boldsymbol{g}; \boldsymbol{v})$, $\boldsymbol{g} = (0, 0, (-(k+1), 0])$, $\boldsymbol{v} = (2, 2)$, $A \to \mathbb{A}$, *is an algebra homomorphism where for* $A, B \in L^0_{\mathrm{cl}}(N)$

$$\sigma_i(\Phi(AB)) = \sigma_i(\Phi(A))\sigma_i(\Phi(B)),\ i = 0, 1. \tag{2.29}$$

(ii) *Applying* Φ *to an elliptic* A *and denoting by* $P \in L^0_{\mathrm{cl}}(N)$ *a parametrix of* A, *then* $\Phi(P) \in L^0(X, \boldsymbol{g}; \boldsymbol{v}), \boldsymbol{v} = (2, 2)$, *is a parametrix of* $\Phi(A)$ *in the edge calculus.*

(iii) *The (standard) ellipticity of* $A \in L^0_{\mathrm{cl}}(N)$ *entails the ellipticity of* $\Phi(A)$ *in the edge calculus with respect to* (σ_0, σ_1) *without any additional boundary (or edge) conditions.*

Proof. (i) The fact that $\mathbb{A}$ belongs to the space $L^0(X, \boldsymbol{g}; \boldsymbol{v})$ for $A \in L^0_{\mathrm{cl}}(N)$ is a consequence of [21, Section 2.5]. From the construction it is also clear that for another $B \in L^0_{\mathrm{cl}}(N)$ and the associated $\mathbb{B} = \Phi(B) \in L^0(X, \boldsymbol{g}; \boldsymbol{v})$ we have $\Phi : AB \to \mathbb{A}\mathbb{B}$. Thus, since in the edge algebra $L^0(X, \boldsymbol{g}; \boldsymbol{v})$ the composition is compatible with the multiplication of the associated symbols we have

$$\sigma_i(\mathbb{A}\mathbb{B}) = \sigma_i(\mathbb{A})\sigma_i(\mathbb{B})$$

for $i = 0, 1$, which corresponds to the relations (2.29).

(ii) If A is elliptic and P a parametrix as in the assumption then we can form the operators $\mathbb{A}$ and $\mathbb{P}$ where $\sigma_i(\mathbb{A})^{-1} = \sigma_i(\mathbb{P})$ for $i = 0, 1$. By virtue of (i) we have

$$\sigma_i(\mathbb{P}\mathbb{A} - \mathbb{I}) = \sigma_i(\mathbb{A}\mathbb{P} - \mathbb{I}) = 0$$

for $i = 0, 1$. Thus $\mathbb{P} = \Phi(P)$ is a parametrix of $\mathbb{A} = \Phi(A)$ in the sense of

$$\mathbb{P}\mathbb{A} - \mathbb{I}, \mathbb{A}\mathbb{P} - \mathbb{I} \in L^{-1}(X, \boldsymbol{g}; \boldsymbol{v})$$

for $L^{-1}(X, \boldsymbol{g}; \boldsymbol{v}) := \{\mathbb{G} \in L^0(X, \boldsymbol{g}; \boldsymbol{v}) : \sigma_i(\mathbb{G}) = 0 \text{ for } i = 0, 1\}$. We may content ourselves with this conclusion, but we can also pass to an improved parametrix by a formal Neumann series expansion. Then the remainder would belong to $L^{-\infty}(X, \boldsymbol{g}; \boldsymbol{v})$.

(iii) If $A \in L^0_{\mathrm{cl}}(N)$ is elliptic we have a parametrix $P \in L^0_{\mathrm{cl}}(N)$ such that $AP - 1, PA - 1 \in L^{-\infty}(N)$. It follows that $\Phi(P) = \mathbb{P} \in L^{-\infty}(X, \boldsymbol{g}; \boldsymbol{v})$ is a parametrix of $\Phi(A) = \mathbb{A} \in L^{-\infty}(X, \boldsymbol{g}; \boldsymbol{v})$. Thus, because of (i) we have $\sigma_i(\mathbb{P}) = \sigma_i(\mathbb{A})^{-1}$, $i = 0, 1$, i.e., $\mathbb{A}$ is elliptic. □

Let us now give some more comment on some analytic details connected with the conclusion (ii)⇒(iii). If $A \in L^0_{\mathrm{cl}}(N)$ is elliptic then also $\mathrm{r}^+ A \mathrm{e}^+$ and $\mathrm{r}^- A \mathrm{e}^-$ are elliptic over X_+ and X_-, respectively. Now the ellipticity under the edge quantisation remains preserved, see the proof of [21, Lemma 1.3.1]. In fact, let us consider, for instance, $\mathrm{r}^+ A \mathrm{e}^+$. The ellipticity of the upper left corner $\mathbb{A}_{11}$ in (2.27) far from the boundary is evident. It remains to argue for the symbols σ_0 and $\tilde{\sigma}_0$ in the edge sense close to $r = 0$. Let $a^+_{(0)}(x, \xi)$ denote the standard homogeneous principal symbol of $\mathrm{r}^+ A \mathrm{e}^+$. Then close to $\partial X_+ = Y$ in the splitting of variables $x = (r, y)$ and covariables $\xi = (\varrho, \eta)$ we have

$$a^+_{(0)}(r, y, \varrho, \eta) \neq 0 \ \text{ for } \ (\varrho, \eta) \neq 0, \text{ up to} \, r = 0.$$

Now in the interpretation of X_+ as a manifold with edge Y the interior symbol in the edge algebra sense means

$$\sigma_0(\mathbb{A}^+)(r, y, r\varrho, r\eta) = a^+_{(0)}(r, y, r\varrho, r\eta)$$

which is non-vanishing for $(\varrho, \eta) \neq 0$ and $r \in \mathbb{R}_+$, while

$$\tilde{\sigma}_0(\mathbb{A}^+)(r, y, \tilde{\varrho}, \tilde{\eta}) = a^+_{(0)}(r, y, \tilde{\varrho}, \tilde{\eta}) \neq 0$$

for $(\tilde{\varrho}, \tilde{\eta}) \neq 0$, up to $r = 0$, cf. also notation (2.19). Analogous arguments work for $\mathbb{A}^-$, the lower right corner of $\mathbb{A}$. The other entries of $\mathbb{A}$ are in the class of smoothing Mellin plus Green operators; those do not participate in the σ_0-ellipticity.
For the σ_1-symbolic component we first consider the homogeneous principal symbol of $A \in L^0_{\mathrm{cl}}(N)$ in a tubular neighbourhood of $Y \subset N$, denoted by $a_{(0)}(r, y, \varrho, \eta)$ which is compatible with the former notation over X_+. From the ellipticity of A, i.e., $a_{(0)} \neq 0$ for $(\varrho, \eta) \neq 0$ we obtain that

$$b(y, \eta) := \mathrm{op}_r(a_{(0)}|_{r=0})(y, \eta) : L^2(\mathbb{R}) \to L^2(\mathbb{R}) \tag{2.30}$$

is an isomorphism for every $\eta \neq 0$. In fact, we have $\mathrm{op}_r(a_{(0)}|_{r=0})^{-1}(y, \eta) = \mathrm{op}_r((a_{(0)}|_{r=0})^{-1})(y, \eta)$. But then, denoting by $(\mathrm{e}^+ \ \ \mathrm{e}^-)$ and $\begin{pmatrix} \mathrm{r}^+ \\ \mathrm{r}^- \end{pmatrix}$ the analogues of the above extension and restriction operators on $\mathbb{R}$ with respect to $\mathbb{R} = \overline{\mathbb{R}}_- \cup \overline{\mathbb{R}}_+$, we have analogues of the isomorphisms (2.26), now for $\mathbb{R}$ and $\mathbb{R}_\pm$ instead of N and $X_\pm$. The reflection diffeomorphism $\varepsilon : \mathbb{R}_+ \to \mathbb{R}_-$ gives rise to a 2×2 block matrix isomorphism

$$\boldsymbol{b}(y, \eta) = \begin{pmatrix} \mathrm{r}^+ b(y, \eta) \mathrm{e}^+ & \mathrm{r}^+ b(y, \eta) \mathrm{e}^- (\varepsilon^*)^{-1} \\ \varepsilon^* \mathrm{r}^- b(y, \eta) \mathrm{e}^+ & \varepsilon^* \mathrm{r}^- b(y, \eta) \mathrm{e}^- (\varepsilon^*)^{-1} \end{pmatrix} : L^2(\mathbb{R}_+, \mathbb{C}^2) \to L^2(\mathbb{R}_+, \mathbb{C}^2) \tag{2.31}$$

which is in one-to-one correspondence to (2.30). It is now justified to set $\boldsymbol{b}(y, \eta) := \sigma_1(\mathbb{A})(y, \eta)$ and call it the boundary symbol of $\mathbb{A}$. We proved the bijectivity of (2.31) for $\eta \neq 0$ which is just the σ_1-ellipticity of $\mathbb{A}$.

Remark 2.6. An inspection of the proof of Theorem 2.5 from the paper [21] (which employs many subtle details of the edge pseudo-differential calculus) shows that Theorem 2.5 can be extended to the parameter-dependent case. In other words for $A(\lambda) \in L^0_{\mathrm{cl}}(N; \mathbb{R}^l)$ the λ-wise application of Φ gives us a map into the parameter-dependent edge algebra

$$\Phi : L^0_{\mathrm{cl}}(N; \mathbb{R}^l) \to L^0(X, \boldsymbol{g}; \boldsymbol{v}; \mathbb{R}^l),$$

and the assertions (i), (ii), (iii) remain valid in analogous form in the respective parameter-dependent theories.

Let us illustrate the process of obtaining Shapiro–Lopatinskij elliptic boundary conditions on X_+ by starting with an elliptic differential operator D on M of order $\mu \in \mathbb{N}$. The continuous operator $D : H^\mu(M) \to L^2(M)$ may be composed with an $R \in L^{-\mu}_{\mathrm{cl}}(M)$ which reduces orders in the sense of isomorphisms $R : H^s(M) \to H^{s+\mu}(M)$ for any s. Such operators are necessarily elliptic, and then

$$A := DR : L^2(M) \to L^2(M)$$

is elliptic as well and of order 0. Now Theorem 2.5 (iii) gives us a Shapiro–Lopatinskij elliptic boundary value problem $\mathbb{A} = \Phi(A)$ on X_+. As in the general case, no elliptic boundary or potential conditions are required for the Fredholm property of $\mathbb{A}$. Nevertheless it may happen that D itself, restricted to X_+, does not admit Shapiro–Lopatinskij elliptic boundary conditions at Y. This is the case for Dirac operators on even-dimensional manifolds, i.e., we then have the situation of non-vanishing of the topological obstruction for the existence of such boundary value problems, see Atiyah and Bott [1]. There are order reductions R where the composition DR preserves vanishing or non-vanishing of the obstruction. It suffices to take an R where the homogeneous principal symbol only depends on $|\xi|$. Let R be of that kind. Then for $A = DR$ we cannot pose Shapiro–Lopatinskij elliptic boundary conditions on X_+, in contrast to $\mathbb{A}$, as we just explained. Recall that operators with non-vanishing obstruction also form a calculus. In the case of pseudo-differential operators with the transmission property at $Y = \partial X_+$ such an algebra has been introduced in [18] while [22] treats the case without the transmission property at the boundary. We observed before that examples in the present context are obtained by starting with $A = DR$ where the principal symbol of R only depends on $|\xi|$.

Let us note that it is also desirable to study higher-order operators from the point of view of truncations and within the higher corner calculus. An expectation is that the order-reducing corner operators constructed in [16] may be useful in this program. In any case those reductions of orders also represent new classes of elliptic elements.

2.4. Truncation of operators on manifolds with edge and boundary

The following considerations refer to a compact manifold M with edge singularities, cf. the considerations at the beginning of Subsection 2.2. From the definition it follows that M is a stratified space, here a disjoint union of strata

$$M = s_0(M) \cup s_1(M) \tag{2.32}$$

with $s_j(M), j = 0, 1$, being smooth manifolds, $\dim s_1(M) < \dim s_0(M) =: \dim M$ where $s_1(M)$ is the edge and the main stratum $s_0(M) := M \setminus s_1(M)$, and $s_1(M)$ has a neighbourhood V in M with the structure of a (locally trivial) N^Δ-bundle over $s_1(M)$ for some closed and compact manifold N. We assume that N can be represented as $N = X_+ \cup X_-$ as explained at the beginning of Subsection 2.3.

Recall that the edge calculus of operators over M of order and weights zero is represented by a space $L^0(M, \boldsymbol{g})$ for $\boldsymbol{g} = (0, 0, \Theta), \Theta = (-(k+1), 0], k \in \mathbb{N} \cup \{+\infty\}$. Parallel to the stratification (2.32) the operators A have a principal symbolic hierarchy

$$\sigma(A) = (\sigma_0(A), \sigma_1(A))$$

already used in Section 2. Let us also recall that $\sigma_0(A)$ in the splitting of variables $(t, x, z) \in \mathbb{R}_+ \times N \times \mathbb{R}^d$, $d = \dim s_1(M)$, with the covariables (τ, ξ, ζ) is edge-degenerate, i.e., it can be written in the form

$$\sigma_0(A)(t, x, z, \tau, \xi, \zeta) = \tilde{\sigma}_0(A)(t, x, z, t\tau, \xi, t\zeta)$$

for the reduced symbol $\tilde{\sigma}_0(A)(t, x, z, \tilde{\tau}, \xi, \tilde{\zeta})$ that is homogeneous of order zero in $(\tilde{\tau}, \xi, \tilde{\zeta}) \neq 0$ and smooth up to $t = 0$. We realise A in the space $H^{0,0}(M)$, locally close to the edge, i.e., near $t = 0$ defined by

$$t^{-n/2} L^2(\mathbb{R}_+ \times N \times \mathbb{R}^d) = \mathcal{W}^0(\mathbb{R}^d, \mathcal{K}^{0,0}(N^\wedge)).$$

and outside the edge by $L^2_{\mathrm{loc}}(s_0(M))$. By virtue of the assumptions on the subdivision of $N = X_+ \cup X_-$ we also have the subdivision

$$\mathbb{R}_+ \times N \times \mathbb{R}^d = \mathbb{R}_+ \times X_+ \times \mathbb{R}^d \cup \mathbb{R}_+ \times X_- \times \mathbb{R}^d$$

and a direct sum

$$t^{-n/2} L^2(\mathbb{R}_+ \times N \times \mathbb{R}^d) = t^{-n/2} L^2(\mathbb{R}_+ \times X_+ \times \mathbb{R}^d) \oplus t^{-n/2} L^2(\mathbb{R}_+ \times X_- \times \mathbb{R}^d).$$

We assume altogether that M is the double of a manifold M_+ with edge and boundary, i.e., there is a smooth submanifold $s_1(M_+) \subset M_+$ such that $M_+ \setminus s_1(M_+)$ is a (non-compact) smooth manifold $s_0(M_+)$ with boundary, and $s_1(M_+)$ has a neighbourhood $V_+ \subset M_+$ with the structure of (locally trivial) X_+-bundle over $s_1(M_+)$. In the following for convenience we assume that bundles of that kind are trivial. The simple generalisation to the case of non-trivial bundles is straightforward. Now, taking another copy M_- we can pass to $M = M_+ \cup M_-$, the double of M_+ by gluing together $M_+ \setminus s_1(M_+)$ and $M_- \setminus s_1(M_-)$ (where $s_1(M_+) = s_1(M_-)$) to a compact manifold with edge, $s_1(M) = s_1(M_+)$.

We easily obtain a direct decomposition

$$H^{0,0}(M) = H^{0,0}(M_+) \oplus H^{0,0}(M_-)$$

for $H^{0,0}(M_\pm) := \mathrm{r}^\pm H^{0,0}(M)$ with the restriction operators $\mathrm{r}^\pm$ to the $\pm$ sides of M. From the known edge pseudo-differential calculus any $A \in L^0(M, \boldsymbol{g})$ induces a continuous operator

$$A : H^{0,0}(M) \to H^{0,0}(M). \tag{2.33}$$

Let

$$\mathrm{e}^\pm : H^{0,0}(M_\pm) \to H^{0,0}(M)$$

be the operators that extend functions in $H^{0,0}(M_\pm)$ by zero to the opposite sides $M_\mp$. Moreover, let

$$\varepsilon : M_+ \to M_-$$

be defined by identifying any point in M_+ with the corresponding point in M_-. Clearly we then have $\varepsilon(m) = m$ for all $m \in \partial(M_+ \setminus s_1(M_+)) \cup \partial(M_- \setminus s_1(M_-))$. From (2.33) we pass to the operator

$$\begin{pmatrix} \mathrm{r}^+ \\ \mathrm{r}^- \end{pmatrix} A \begin{pmatrix} \mathrm{e}^+ & \mathrm{e}^- \end{pmatrix} : \begin{matrix} H^{0,0}(M_+) \\ \oplus \\ H^{0,0}(M_-) \end{matrix} \longrightarrow \begin{matrix} H^{0,0}(M_+) \\ \oplus \\ H^{0,0}(M_-) \end{matrix}.$$

Using the canonical isomorphism

$$\varepsilon^* : H^{0,0}(M_-) \to H^{0,0}(M_+)$$

we then obtain an analogue of (2.27), namely,

$$\mathbb{A} = \begin{pmatrix} \mathrm{r}^+ A \mathrm{e}^+ & \mathrm{r}^+ A \mathrm{e}^- (\varepsilon^*)^{-1} \\ \varepsilon^* \mathrm{r}^- A \mathrm{e}^+ & \varepsilon^* \mathrm{r}^- A \mathrm{e}^- (\varepsilon^*)^{-1} \end{pmatrix} : H^{0,0}(M_+, \mathbb{C}^2) \to H^{0,0}(M_+, \mathbb{C}^2)$$

for $H^{0,0}(M_+, \mathbb{C}^2) = H^{0,0}(M_+) \otimes \mathbb{C}^2$.

In Theorem 2.5 we obtained an algebra homomorphism by the correspondence $A \to \mathbb{A}$, including compatibility with the respective principal symbolic structure. Now we immediately conclude such a result for the operator spaces

$$\mathcal{L}(H^{0,0}(M_+, \mathbb{C}^2), H^{0,0}(M_+, \mathbb{C}^2)) \cap L^0(M_+ \setminus s_1(M), \boldsymbol{g}; \boldsymbol{v}), \tag{2.34}$$

$\boldsymbol{v} = (2, 2)$. The space $M_+ \setminus s_1(M)$ is interpreted as a non-compact manifold with edge $\partial(M_+ \setminus s_1(M))$. Thus the edge algebra $L^0(M_+ \setminus s_1(M), \boldsymbol{g}; \boldsymbol{v})$ makes sense, and the truncation in this case refers to the decomposition of the smooth manifold $M \setminus s_1(M)$ into submanifolds with boundary

$$M \setminus s_1(M) = (M_+ \setminus s_1(M)) \cup (M_- \setminus s_1(M)).$$

There is an analogue of the edge algebra for a manifold W with edge and boundary, cf. [12, Chapter 7] or [13, Chapter 4] for operators with the transmission property at the smooth part of the boundary. A similar structure is possible in general, i.e., without the transmission property. The operator space (here of order zero) is denoted by

$$L^0(W, \boldsymbol{g}; \boldsymbol{w}) \tag{2.35}$$

with dimension data $\boldsymbol{w} = (j_-, j_+; d_-, d_+)$ indicating the numbers of components j_- and j_+ of potential and trace conditions along ∂W, and d_- and d_+ of potential and trace conditions along the edge of W. The principal symbolic hierarchy of operators in (2.35) now has the form

$$\sigma(\cdot) = (\sigma_0(\cdot), \sigma_\partial(\cdot), \sigma_1(\cdot))$$

where we have an additional boundary symbol $\sigma_\partial(\cdot)$ compared with the case without boundary.

For part (ii) of the following theorem we make a similar assumption as in the analogous situation of Theorem 2.4, but the assertion remains true in an enlarged edge calculus with continuous asymptotics.

Theorem 2.7.

(i) *The map* $\Phi_{\text{edge}} : L^0(M, \boldsymbol{g}) \to L^0(M_+, \boldsymbol{g}; \boldsymbol{v}), \boldsymbol{v} = (2,2), A \to \mathbb{A}$, *is an algebra homomorphism where for* $A, B \in L^0(M, \boldsymbol{g})$

$$\sigma_i(\Phi_{\text{edge}}(AB)) = \sigma_i(\Phi_{\text{edge}}(A))\sigma_i(\Phi_{\text{edge}}(B)),\ i = 0, 1, \tag{2.36}$$

$$\sigma_\partial(\Phi_{\text{edge}}(AB)) = \sigma_\partial(\Phi_{\text{edge}}(A))\sigma_\partial(\Phi_{\text{edge}}(B)). \tag{2.37}$$

(ii) *Applying* Φ_{edge} *to a* (σ_0, σ_1)*-elliptic* $A \in L^0(M, \boldsymbol{g})$ *and if* $P \in L^0(M, \boldsymbol{g})$ *is a parametrix of* A *in* $L^0(M, \boldsymbol{g})$, *then* $\Phi_{\text{edge}}(P)$ *is a parametrix of* $\Phi_{\text{edge}}(A)$ *in* $L^0(M_+, \boldsymbol{g}; \boldsymbol{v})$.

(iii) *The* (σ_0, σ_1)*-ellipticity of* $A \in L^0(M, \boldsymbol{g})$ *entails the* $(\sigma_0, \sigma_\partial, \sigma_1)$ *ellipticity of* $\Phi_{\text{edge}}(A) \in L^0(M_+, \boldsymbol{g}; \boldsymbol{v})$ *without any additional boundary or edge conditions.*

Proof. (i) From the definition of the map Φ_{edge} it is clear that for $A, B \in L^0(M, \boldsymbol{g})$ and $\mathbb{A} := \Phi_{\text{edge}}(A)$, $\mathbb{B} := \Phi_{\text{edge}}(B)$ we have $\Phi_{\text{edge}}(AB) = \mathbb{AB}$. In fact, in the composition up to the reflection map we have a factor

$$\begin{pmatrix} \mathrm{e}^+ & \mathrm{e}^- \end{pmatrix} \begin{pmatrix} \mathrm{r}^+ \\ \mathrm{r}^- \end{pmatrix} = \mathrm{id}_{H^{0,0}(M)}.$$

Since the symbols σ_0 and σ_1 in $L^0(M_+, \boldsymbol{g}; \boldsymbol{v})$ are compatible under multiplication of operators we obtain the composition rule (2.36). For (2.37) it suffices to note that also the boundary symbols are multiplicative under compositions of operators.

(ii) The proof for the parametrix can be given in an analogous manner as in Theorem 2.5.

(iii) The ellipticity of A in $L^0(M, \boldsymbol{g})$ is a consequence of (i) and (ii), similarly as in the proof of Theorem 2.5. □

Let us give some additional comment on the ellipticity of the new boundary symbol that appears after applying Φ_{edge} to an elliptic $A \in L^0(M, \boldsymbol{g})$. Locally near the smooth part of the boundary of M_+ (or M_-) and close to $t = 0$, the edge, we may refer to the variables $(t, x, z, \tau, \xi, \zeta)$ in the splitting $x = (r, y)$ and covariables $\xi = (\varrho, \eta)$. Then, similarly as (2.30) we have isomorphisms

$$b(t, y, z, \tau, \eta, \zeta) := \mathrm{Op}_r(\sigma_0(A)|_{r=0})(t, y, z, \tau, \eta, \zeta) : L^2(\mathbb{R}_r) \to L^2(\mathbb{R}_r)$$

for $(\tau, \eta, \zeta) \neq 0$, as soon as $\sigma_0(A)$ does not vanish for $(\varrho, \xi, \zeta) \neq 0$, and

$$\sigma_\partial(\mathbb{A})(t, y, z, \tau, \eta, \zeta) : L^2(\mathbb{R}_+, \mathbb{C}^2) \to L^2(\mathbb{R}_+, \mathbb{C}^2)$$

is obtained as a 2×2 block matrix family of isomorphisms, by first writing $L^2(\mathbb{R}) = \mathrm{e}^+ L^2(\mathbb{R}_+) \oplus \mathrm{e}^- L^2(\mathbb{R}_-)$ and then applying the reflection isomorphism ε. Isomorphism just means that the Shapiro–Lopatinskij condition holds without additional entries of trace and potential type. A similar conclusion applies far from edge Z; here the arguments are exactly as in the proof of Theorem 2.5 (this part of the proof does not refer to any compactness of the underlying manifold).

References

[1] M.F. Atiyah and R. Bott, *The index problem for manifolds with boundary.* Coll. Differential Analysis, Tata Institute Bombay, Oxford University Press, Oxford, 1964, pp. 175–186.

[2] L. Boutet de Monvel, *Boundary problems for pseudo-differential operators.* Acta Math. **126** (1971), 11–51.

[3] D. Calvo and B.-W. Schulze, *Edge symbolic structures of second generation.* Math. Nachr. **282** (2009), 348–367.

[4] N. Dines, *Elliptic operators on corner manifolds.* Ph.D. thesis, University of Potsdam, 2006.

[5] Ch. Dorschfeldt, *Algebras of pseudo-differential operators near edge and corner singularities*, Math. Res. **102**, Akademie Verlag, Berlin, 1998.

[6] Ju.V. Egorov and B.-W. Schulze, *Pseudo-differential operators, singularities, applications.* Oper. Theory: Adv. Appl. **93**, Birkhäuser Verlag, Basel, 1997.

[7] G.I. Eskin, *Boundary value problems for elliptic pseudodifferential equations.* Transl. of Nauka, Moskva, 1973, Math. Monographs, Amer. Math. Soc. **52**, Providence, Rhode Island 1980.

[8] I.M. Gelfand, *On elliptic equations.* Uspechi Math. Nauk. **15** (1960), 121–132. (Russian)

[9] G. Grubb, *Functional calculus of pseudo-differential boundary problems.* Second Edition, Birkhäuser Verlag, Boston, 1996.

[10] N. Habal and B.-W. Schulze, *Holomorphic corner symbols.* J. Pseudo-Differ. Oper. Appl. **2** (2011), 419–465.

[11] N. Habal and B.-W. Schulze, *Mellin quantisation in corner operators.* Dedicated to the 70th anniversary of V. Rabinovich. Oper. Theory: Adv. Appl. **228**, Springer, Basel, 2013, pp. 151–172.

[12] G. Harutyunyan and B.-W. Schulze, *Elliptic mixed, transmission and singular crack problems.* European Mathematical Soc., Zürich, 2008.

[13] D. Kapanadze and B.-W. Schulze, *Crack theory and edge singularities.* Kluwer Academic Publ., Dordrecht, 2003.

[14] S. Rempel and B.-W. Schulze, *Parametrices and boundary symbolic calculus for elliptic boundary problems without transmission property.* Math. Nachr. **105** (1982), 45–149.

[15] S. Rempel and B.-W. Schulze, *Index theory of elliptic boundary problems.* Akademie-Verlag, Berlin, 1982.

[16] W. Rungrottheera, *Parameter-dependent corner operators.* Asian-Eur. J. Math. **6** (2013), 1–29.

[17] B.-W. Schulze, *Boundary value problems and singular pseudo-differential operators.* J. Wiley, Chichester, 1998.

[18] B.-W. Schulze, *An algebra of boundary value problems not requiring Shapiro–Lopatinskij conditions.* J. Funct. Anal. **179** (2001), 374–408.

[19] B.-W. Schulze, *Operators with symbol hierarchies and iterated asymptotics.* Publ. Res. Inst. Math. Sci., Kyoto **38** (2002), 735–802.

[20] B.-W. Schulze, *The iterative structure of the corner calculus.* Oper. Theory: Adv. Appl. **213**, Pseudo-Differential Operators: Application and Computations (L. Rodino et al. eds.), Birkhäuser Verlag, Basel, 2011, pp. 79–103.

[21] B.-W. Schulze and J. Seiler, *The edge algebra structure of boundary value problems.* Ann. Glob. Anal. Geom. **22** (2002), 197–265.

[22] B.-W. Schulze and J. Seiler, *Pseudodifferential boundary value problems with global projection conditions.* J. Funct. Anal. **206** (2004), 449–498.

N. Habal, W. Rungrottheera and B.-W. Schulze
University of Potsdam
Am Neuen Palais 10
D-14469 Potsdam, Germany
e-mail: habal@uni-potsdam.de
rungrott@uni-potsdam.de
schulze@math.uni-potsdam.de

Fourier Analysis
Trends in Mathematics, 211–233

A Representation of the Fundamental Solution for the Fokker–Planck Equation and Its Application

Chisato Iwasaki

Abstract. The aim of this paper is to construct the fundamental solution for the Fokker–Planck equation by calculus of pseudo-differential operators. Eigenfunction expansion of the Fokker–Planck operator is obtained by this exact construction.

Mathematics Subject Classification (2010). Primary 35Q84; Secondary 35P10.

Keywords. Pseudo-differential operators, the fundamental solution, Fokker–Planck equations.

1. Introduction

A. Melin [8] characterized the symbol of differential operators which satisfy a sub-elliptic estimate. C. Iwasaki and N. Iwasaki [6], [7] constructed the fundamental solution for the heat equations corresponding to degenerate operators which A. Melin characterized. They showed that the fundamental solution is obtained as a pseudo-differential operator of the Weyl symbol and they gave the exact expression of its main part represented in terms of derivatives of the second order of the principal symbol and the sub-principal symbol of degenerate operators. If the symbol of operators are quadratic polynomials with respect of x and ξ, the main part of the fundamental solution coincides with the fundamental solution itself (see also O. Calin, D.-C. Chang, K. Furutani, C. Iwasaki [2]). This fact is useful to show that the spectral zeta function for sub-Laplacians on nilmanifolds has only one singularity (see W. Bauer, K. Furutani, C. Iwasaki [1]).

In this paper, we apply this expression of the fundamental solution to the Fokker–Planck equation. We obtain the exact symbol of the fundamental solution

This work was supported by the Grant-in-aid for Scientific Research (C) No. 24540189 of JSPS (Japan Society for the Promotion of Science).

and the eigenfunction expansion to the Fokker–Planck operator as its application. The conclusions are different corresponding to the potentials of operators. It is shown that the generalized eigenfunctions appear under some potentials.

The plan of this paper is as follows: In Section 2 we introduce the Fokker–Planck operator which will be studied in this paper and give some remarks. Section 3 is devoted to the exact symbol of the fundamental solution for the heat equation in case the symbol is a polynomial. In the next section we apply this expression to the Fokker–Planck equation and obtain one of the main theorems of this paper. The following section is devoted to its proof. In Section 6 a key proposition is stated which is useful if one gets the eigenfunction expansion by the symbol of the fundamental solution as a pseudo-differential operator of Weyl symbol. The last section is devoted to obtain the eigenfunction expansion of the Fokker–Planck operator, applying the key proposition.

2. Fokker–Planck equation

We call the following operator K the Fokker–Planck (Kramers) operator on $(x, v) \in \mathbb{R}^n \times \mathbb{R}^n$ (see H. Riskin [9], Helffer and Nier [3]):

$$\begin{aligned} K &= v \cdot \partial_x - \big(\partial_x V(x)\big) \cdot \partial_v - \triangle_v + \frac{|v|^2}{4} - \frac{n}{2} \\ &= X_0 - \triangle_v + \frac{|v|^2}{4} - \frac{n}{2}. \end{aligned}$$

We note that the operator K is hypo-elliptic but not self-adjoint. $\Phi(x, v)$ is a classical Hamiltonian

$$\Phi(x, v) = \frac{|v|^2}{2} + V(x)$$

and X_0 is the corresponding Hamiltonian vector field

$$X_0 = v \cdot \partial_x - \big(\partial_x V(x)\big) \cdot \partial_v.$$

There are many studies about relations between the Fokker–Planck operators and the Witten Laplacian

$$\triangle_{\Phi/2} = -\triangle + \frac{1}{4}|\nabla\Phi|^2 - \frac{1}{2}\triangle\Phi$$

(see [3]).

Helffer and Nier [3] showed that the following inequality holds under some assumption for $V(x)$

$$\|\Lambda_x^{2/3} u\|^2 + \|\Lambda_v^2 u\|^2 \leq C(\|Ku\|^2 + \|u\|^2), \tag{2.1}$$

where $\Lambda_x = (-\triangle_x + x^2/4)^{1/2}, \Lambda_v = (-\triangle_v + v^2/4)^{1/2}$.

In F. Hérau and F. Nier [5] the following estimate was studied

$$\|u(t) - u^*\|_{L^2} \leq Ce^{-\tau t} \tag{2.2}$$

for the following solution $u(t)$ under some assumptions on $V(x)$:

$$\frac{d}{dt}u(t) + Ku(t) = 0, \quad t > 0, \quad u(0) = u_0.$$

We call an operator $E(t)$ the fundamental solution of the Fokker–Planck equation if $E(t)$ satisfies

$$\frac{d}{dt}E(t) + KE(t) = 0, \quad t > 0, \quad E(0) = I.$$

Our aim is to give the precise expression of the fundamental solution $E(t)$ as a pseudo-differential operator of Weyl symbol and to get the eigenfunction expansion of K as its application. The estimates (2.1) is obtained by the construction of the fundamental solution as we will mention in Section 4. Because we can choose $\int_0^T E(t)dt \quad (T > 0)$ as a parametrix for K. We have also (2.2) by the eigenfunction expansion. We note that K may have the generalized eigenvalues if $V(x)$ is chosen suitably, which was avoided in [9].

3. Expression of the fundamental solution

A pseudo-differential operator P on $\mathbb{R}^d$ of a Weyl symbol $p(x,\xi)$ is defined by:

$$Pu(x) = p(x,D)u(x) = \int_{\mathbb{R}^d\times\mathbb{R}^d} e^{i(x-y)\cdot\xi} p(\frac{x+y}{2},\xi)u(y)d\xi dy$$

We will construct the following fundamental solution $E(t)$ for a heat equation for an operator P whose the principal symbol is nonnegative:

$$\left(\frac{d}{dt} + P\right)E(t) = 0, \quad E(0) = I.$$

The following Theorem 3.1 has been obtained (see [7], [2]).

Theorem 3.1. *Suppose $p(x,\xi)$ is a quadratic polynomial with respect $X^t = (x,\xi) \in \mathbb{R}^d \times \mathbb{R}^d$*

$$p = \frac{1}{2}\langle X, HX\rangle + i\langle X, p_0\rangle + b$$

with a nonnegative real symmetric $2d \times 2d$ matrix H, $p_0 \in \mathbb{C}^{2d}$ and $b \in \mathbb{C}$. Then $E(t)$ is constructed as pseudo-differential operator whose symbol $e(t;x,\xi)$ is given by

$$e(t;x,\xi) = \frac{e^{-bt}}{\sqrt{\det\cosh(At/2)}} \exp\Big[-i\Big\{\langle J\tanh(At/2)X, X\rangle + t\langle J\tanh(At/2)(At/2)^{-1}X, Jp_0\rangle + \frac{t^2}{4}\langle JG(At/2)Jp_0, Jp_0\rangle\Big\}\Big],$$

where

$$G(x) = (1 - x^{-1}\tanh x)/x$$

and $A = iJH$ *is a* $2d \times 2d$ matrix*:*

$$J = \begin{pmatrix} 0 & I_d \\ -I_d & 0 \end{pmatrix}, \quad I_d \textit{ is the } d \times d \textit{ identity matrix.}$$

In our case $d = 2n$, $X =^t (x, v, \xi, \rho)$

$$\sigma(K) = p(x, v, \xi, \rho) = i\Big(\langle v, \xi\rangle - \langle \partial_x V(x), \rho\rangle\Big) + |\rho|^2 + \frac{|v|^2}{4} - \frac{n}{2}.$$

4. The exact form of the fundamental solution to the Fokker–Planck equation

Suppose $V(x) = a \cdot x + \frac{1}{2}\langle \varepsilon x, x\rangle$ with a symmetric matrix ε. We have the following theorem by Theorem 3.1, setting

$$H = \begin{pmatrix} 0 & 0 & 0 & -i\varepsilon \\ 0 & \frac{1}{2}I_n & iI_n & 0 \\ 0 & iI_n & 0 & 0 \\ -i\varepsilon & 0 & 0 & 2I_n \end{pmatrix}, \quad p_0 = \begin{pmatrix} 0 \\ 0 \\ 0 \\ -a \end{pmatrix}, \quad b = -\frac{n}{2}$$

and A is a $4n \times 4n$ matrix of the form

$$A = \begin{pmatrix} 0 & -I_n & 0 & 0 \\ \varepsilon & 0 & 0 & 2iI_n \\ 0 & 0 & 0 & -\varepsilon \\ 0 & -\frac{i}{2}I_n & I_n & 0 \end{pmatrix}.$$

We have the following result for the symbol $e(t; x, v, \xi, \rho)$ of the fundamental solution $E(t)$ for the Fokker–Planck equation.

Theorem 4.1. (I) *If* $V(x) = a \cdot x$, *we have*

$$e(t; x, v, \xi, \rho) = \Big(\frac{1 + e^{-t}}{2}\Big)^{-n} \exp\Big[- 2\tanh(t/2)\left\{|\rho|^2 + |v|^2/4 + i(\langle v, \xi\rangle - \langle a, \rho\rangle)\right\} - 2\left(t/2 - \tanh(t/2)\right)\left(|\xi|^2 + |a|^2/4\right)\Big].$$

(II) *Assume* $n = 1$ *and* $\varepsilon = \mu \neq 0$. *Let* δ *be the following constant depending on* μ*:*

$$\delta = \begin{cases} \sqrt{1 - 4\mu}, & \textit{for } -\infty < \mu \le \frac{1}{4}, \\ i\sqrt{4\mu - 1}, & \textit{for } \mu > \frac{1}{4}. \end{cases}$$

(II-1) *If* $\mu \neq \frac{1}{4}$, *then*

$$e(t; x, v, \xi, \rho) = C_\delta(t) \exp\big\{-2\phi(t)\big\},$$

where $\phi(t)$ *and* $C_\delta(t)$ *are given by the following formula:*

$$\begin{aligned} C_\delta(t) &= 4\Big\{1 + e^{-t} + 2e^{-t/2}\cosh(\delta t/2)\Big\}^{-1}, \\ \phi(t) &= F_1(t)\left(|\rho|^2 + |v|^2/4\right) + F_2(t)\left(|\xi|^2 + |\mu x + a|^2/4\right) \\ &\quad + iF_3(t)\left(\langle v, \xi\rangle - \langle \mu x + a, \rho\rangle\right), \end{aligned}$$

where

$$\begin{aligned} F_1(t) &= \frac{1}{\delta}\Big(\lambda_1 g(\lambda_1 t/2) - \lambda_2 g(\lambda_2 t/2)\Big), \\ F_2(t) &= \frac{1}{\delta}\Big(\frac{1}{\lambda_2} g(\lambda_2 t/2) - \frac{1}{\lambda_1} g(\lambda_1 t/2)\Big), \\ F_3(t) &= \frac{1}{\delta}\Big(g(\lambda_1 t/2) - g(\lambda_2 t/2)\Big) \end{aligned}$$

with $g(x) = \tanh x$ *and*

$$\lambda_1 = \frac{1}{2}(1+\delta), \quad \lambda_2 = \frac{1}{2}(1-\delta).$$

(II-2) *If* $\mu = \frac{1}{4}$, *then*

$$\begin{aligned} &e(t; x, v, \xi, \rho) \\ &= 4(1+e^{-t/2})^{-2}\exp\Big[-2\Big(\tanh(t/4) + \frac{t/4}{\cosh^2(t/4)}\Big)\left(|\rho|^2 + |v/2|^2\right) \\ &\qquad - i\frac{t}{\cosh^2(t/4)}\left(\langle v, \xi\rangle - \langle x/4 + a, \rho\rangle\right) \\ &\qquad - 8\Big(\tanh(t/4) - \frac{t/4}{\cosh^2(t/4)}\Big)\left(|\xi|^2 + |x/4 + a|^2/4\right)\Big]. \end{aligned}$$

Remark 4.2. We note that

$$\begin{aligned} t/2 - \tanh(t/2) &= \frac{1}{24}t^3 + O(t^5), \\ F_2(t) &= \frac{1}{24}t^3 + O(t^5), \\ 4\tanh(t/4) - \frac{t}{\cosh^2(t/4)} &= \frac{1}{24}t^3 + O(t^5). \end{aligned}$$

By the above behavior of the symbol of the fundamental solution it is easy to see that the estimate (2.1) for $\mu \neq 0$ holds and that if $\varepsilon = 0$ we have the following estimate

$$\|(-\Delta_x)^{2/3}u\|^2 + \|\Lambda_v^2 u\|^2 \leq C(\|Ku\|^2 + \|u\|^2).$$

Remark 4.3. Since $xg(x)$ is an increasing function and $\dfrac{g(x)}{x}$ is a decreasing functions, $F_1(t)$ and $F_2(t)$ are non-negative functions. $F_3(t)$ is a real function.

5. Proof of Theorem 4.1

For the proof of (I) we apply Theorem 3.1 to the following the $2n \times 2n$ matrix

$$H = \begin{pmatrix} \frac{1}{2}I_n & 0 \\ 0 & 2I_n \end{pmatrix}, \quad p_0 = \begin{pmatrix} \xi \\ -a \end{pmatrix}, \quad b = -\frac{n}{2}.$$

Then $2n \times 2n$ matrix A is of the form

$$A = \begin{pmatrix} 0 & 2iI_n \\ -\frac{i}{2}I_n & 0 \end{pmatrix}.$$

Let $e_j = {}^t(0, 0, \ldots, 1, 0, \ldots, 0)$ be n-vector and $2n$-vectors E_j^+, E_j^- be as follow:

$$E_j^+ = \begin{pmatrix} ie_j \\ \frac{1}{2}e_j \end{pmatrix}, \quad E_j^- = \begin{pmatrix} -ie_j \\ \frac{1}{2}e_j \end{pmatrix}.$$

Then the following equalities hold:

$$AE_j^+ = E_j^+, \quad AE_j^- = -E_j^-, \quad i\langle JE_j^+, E_k^-\rangle = \delta_{jk}.$$

For any $2n$ vector $\begin{pmatrix} v \\ \rho \end{pmatrix}$ we have

$$\begin{pmatrix} v \\ \rho \end{pmatrix} = \sum_{j=1}^{n} \left(\rho_j - \frac{i}{2}v_j\right) E_j^+ + \sum_{j=1}^{n} \left(\rho_j + \frac{i}{2}v_j\right) E_j^-.$$

So if $h(x)$ is an odd function, then

$$i\left\langle Jh(At/2)\begin{pmatrix} v \\ \rho \end{pmatrix}, \begin{pmatrix} v \\ \rho \end{pmatrix}\right\rangle = 2h(t/2)(|\rho|^2 + |v/2|^2).$$

On the other hand if $h(x)$ is an even function, then

$$i\left\langle Jh(At/2)\begin{pmatrix} v \\ \rho \end{pmatrix}, \begin{pmatrix} a \\ \xi \end{pmatrix}\right\rangle = ih(t/2)(\langle \rho, a\rangle - \langle \xi, v\rangle).$$

So we obtain the assertion (I), noting

$$\frac{t^2}{4}G(t/2) = \frac{t}{2} - \tanh(t/2),$$

$$(\det[\cosh(At/2)])^{-1} = (\cosh(t/2))^{-n} = e^{-nt/2}\left(\frac{1+e^{-t}}{2}\right)^{-n}.$$

(II) We note that

$$F_1(t) = \frac{1}{2}\{g(\lambda_1 t/2) + g(\lambda_2 t/2)\} + \frac{1}{2}F_3(t) \tag{5.1}$$

$$\mu F_2(t) = \frac{1}{2}\{g(\lambda_1 t/2) + g(\lambda_2 t/2)\} - \frac{1}{2}F_3(t) \tag{5.2}$$

by the definition. It is sufficient to show (II-1) because if $\mu = 1/4$, we have $\lambda_1 = \lambda_2 = 1/2$, $\delta = 0$ and

$$\lim_{\delta \to 0} F_3(t) = \lim_{\delta \to 0} \frac{\Big(\tanh((1+\delta)t/4) - \tanh((1-\delta)t/4)\Big)}{\delta} = \frac{t/2}{\cosh^2(t/4)}.$$

By the above equation and (5.1), (5.2) we get the assertion for $\mu = 1/4$.

In case $\mu \neq 1/4$ the proof for (II-2) is complete by Theorem 3.1 and the following Lemma 5.1.

Lemma 5.1.

(1) *If* $U ={}^t (x, v, \xi, \rho)$, *then*

$$\begin{aligned} i\langle J \tanh(At/2)U, U\rangle &= 2F_1(t)(|\rho|^2 + |v|^2/4) + 2F_2(t)(|\xi|^2 + |\mu x|^2/4) \\ &\quad + 2iF_3(t)\big(\langle v, \xi\rangle - \langle \mu x, \rho\rangle\big). \end{aligned}$$

(2) *If* $W ={}^t (0, -a, 0, 0)$, *then*

$$i\frac{t^2}{4}\langle JG(At/2)W, W\rangle = \frac{a^2}{2}F_2(t).$$

(3) *If* $U ={}^t (x, v, \xi, \rho)$, $W ={}^t (0, -a, 0, 0)$, *then*

$$it\langle J \tanh(At/2)(At/2)^{-1}U, W\rangle = 2a\left(\frac{\mu x}{2}F_2(t) - i\rho F_3(t)\right).$$

(4)
$$\frac{1}{\sqrt{\det \cosh(At/2)}} = e^{-t/2}C_\delta(t).$$

Proof. (1) If $\mu \neq 1/4$, then eigenvalues $\pm\lambda_1, \pm\lambda_2$ of A are distinct. The corresponding eigenvectors are as follow:

$$w_j = {}^t(-1/\lambda_j, 1, i\mu/(2\lambda_j), -i/2)/\left(\sqrt{1 - \frac{\mu}{\lambda_j^2}}\right), \quad j = 1, 2,$$

$$\tilde{w}_j = {}^t(1/\lambda_j, 1, i\mu/(2\lambda_j), i/2)/\left(\sqrt{1 - \frac{\mu}{\lambda_j^2}}\right), \quad j = 1, 2,$$

$$Aw_j = \lambda_j w_j, \quad A\tilde{w}_j = -\lambda_j \tilde{w}_j.$$

Note that λ_j are the solutions of $\lambda^2 - \lambda + \mu = 0$ and

$$i\langle Jw_j, w_k\rangle = 0 \quad i\langle Jw_j, \tilde{w}_k\rangle = \delta_{jk}, \quad j, k = 1, 2.$$

Then for any $U \in \mathbb{R}^4$ we can write

$$U = a_1 w_1 + a_2 w_2 + \tilde{a}_1 \tilde{w}_1 + \tilde{a}_2 \tilde{w}_2,$$

where

$$a_j = -i\langle J\tilde{w}_j, U\rangle, \tilde{a}_j = i\langle Jw_j, U\rangle, \quad j = 1, 2.$$

For any analytic function $h(x)$ we have

$$i\langle Jh(A)U,W\rangle = h(\lambda_1)\langle J\tilde{w}_1,U\rangle\langle Jw_1,W\rangle + h(\lambda_2)\langle J\tilde{w}_2,U\rangle\langle Jw_2,W\rangle$$
$$- h(-\lambda_1)\langle Jw_1,U\rangle\langle J\tilde{w}_1,W\rangle - h(-\lambda_2)\langle Jw_2,U\rangle\langle J\tilde{w}_2,W\rangle.$$

Especially we have the following results (i) and (ii).

(i) If $h(x)$ is an odd function, we have

$$i\langle Jh(A)U,U\rangle = 2\sum_{j=1}^{2} h(\lambda_j)\langle J\tilde{w}_j,U\rangle\langle Jw_j,U\rangle. \tag{5.3}$$

(ii) If $h(x)$ is an even function, we have

$$i\langle Jh(A)U,W\rangle = \sum_{j=1}^{2} h(\lambda_j)\Big\{\langle J\tilde{w}_j,U\rangle\langle Jw_j,W\rangle - \langle Jw_j,U\rangle\langle J\tilde{w}_j,W\rangle\Big\}. \tag{5.4}$$

By the following equations

$$\langle J\tilde{w}_j,U\rangle = \left(\frac{i}{2}\frac{\mu}{\lambda_j}x + \frac{i}{2}v - \frac{\xi}{\lambda_j} - \rho\right)\Big/\left(\sqrt{1-\frac{\mu}{\lambda_j^2}}\right),$$
$$\langle Jw_j,U\rangle = \left(\frac{i}{2}\frac{\mu}{\lambda_j}x - \frac{i}{2}v + \frac{\xi}{\lambda_j} - \rho\right)\Big/\left(\sqrt{1-\frac{\mu}{\lambda_j^2}}\right)$$

and noting that

$$1\Big/\left(1-\frac{\mu}{\lambda_1^2}\right) = \frac{\lambda_1}{\delta},\quad 1\Big/\left(1-\frac{\mu}{\lambda_2^2}\right) = -\frac{\lambda_2}{\delta},$$

we get

$$\langle J\tilde{w}_1,U\rangle\langle Jw_1,U\rangle = \frac{\lambda_1}{\delta}\left\{\left(\frac{i}{2}\frac{\mu}{\lambda_1}x-\rho\right)^2 - \left(\frac{i}{2}v-\frac{\xi}{\lambda_1}\right)^2\right\}$$
$$= \frac{\lambda_1}{\delta}(|\rho|^2+|v|^2/4) - \frac{1}{\lambda_1\delta}(|\xi|^2+|\mu x|^2/4) + \frac{i}{\delta}(\langle v,\xi\rangle - \langle \mu x,\rho\rangle)$$

and

$$\langle J\tilde{w}_2,U\rangle\langle Jw_2,U\rangle = -\frac{\lambda_2}{\delta}\left\{\left(\frac{i}{2}\frac{\mu}{\lambda_2}x-\rho\right)^2 - \left(\frac{i}{2}v-\frac{\xi}{\lambda_2}\right)^2\right\}$$
$$= -\frac{\lambda_2}{\delta}(|\rho|^2+|v|^2/4) + \frac{1}{\lambda_2\delta}(|\xi|^2+|\mu x|^2/4) - \frac{i}{\delta}(\langle v,\xi\rangle - \langle \mu x,\rho\rangle).$$

We obtain the following equation by (5.3)

$$i\langle J\tanh(At/2)U,U\rangle = 2\sum_{j=1}^{2}\tanh(\lambda_j t/2)\langle J\tilde{w}_j,U\rangle\langle Jw_j,U\rangle.$$

The proof of the assertion (1) is complete.

(2) Noting

$$\langle Jw_j, W\rangle = ia/\left(2\sqrt{1-\frac{\mu}{\lambda_j^2}}\right), \quad \langle J\tilde{w}_j, W\rangle = -ia/\left(2\sqrt{1-\frac{\mu}{\lambda_j^2}}\right),$$

we have by (5.3)

$$i\frac{t^2}{4}\langle JG(At/2)W, W\rangle = \frac{a^2t^2}{8\delta}\Big\{\lambda_1 G(\lambda_1 t/2) - \lambda_2 G(\lambda_2 t/2)\Big\}.$$

Set $G(x) = (1 - G_1(x))/x$ with $G_1(x) = g(x)/x$. Then the following equation holds:

$$\begin{aligned} i\frac{t^2}{4}\langle JG(At/2)W, W\rangle &= \frac{a^2t}{4\delta}\Big\{-G_1(\lambda_1 t/2) + G_1(\lambda_2 t/2)\Big\}. \\ &= \frac{a^2t}{4\delta}\left\{-\frac{2g(\lambda_1 t/2)}{\lambda_1 t} + \frac{2g(\lambda_2 t/2)}{\lambda_2 t}\right\} \\ &= \frac{a^2}{2}F_2(t). \end{aligned}$$

(3) We have by (5.4)

$$\begin{aligned} it\langle J\tanh(At/2)(At/2)^{-1}U, W\rangle &= \frac{iat}{2\delta}\left[-\frac{4\rho}{t}\{\tanh(\lambda_1 t/2) - \tanh(\lambda_2 t/2)\}\right. \\ &\quad \left. + i\mu x\Big\{\frac{\tanh(\lambda_1 t/2)}{\lambda_1 t/2} - \frac{\tanh(\lambda_2 t/2)}{\lambda_2 t/2}\Big\}\right] \\ &= 2a\left(\frac{\mu x}{2}F_2(t) - i\rho F_3(t)\right). \end{aligned}$$

(4) Eigenvalues of A are $\pm\lambda_1$ and $\pm\lambda_2$. So we have

$$\det\cosh(At/2) = \cosh^2(\lambda_1 t/2)\cosh^2(\lambda_2 t/2),$$

$$\begin{aligned} \sqrt{\det\cosh(At/2)} &= \cosh(\lambda_1 t/2)\cosh(\lambda_2 t/2) \\ &= e^{\lambda_1 t/2}e^{\lambda_2 t/2}\Big(\frac{1+e^{-\lambda_1 t}}{2}\Big)\Big(\frac{1+e^{-\lambda_2 t}}{2}\Big) \\ &= \frac{1}{4}e^{t/2}\Big(1 + e^{-\lambda_1 t} + e^{-\lambda_2 t} + e^{-t}\Big) \\ &= \frac{1}{4}e^{t/2}\Big(1 + e^{-t} + 2e^{-t/2}\cosh(\delta t/2)\Big). \end{aligned}$$

Therefore we obtain

$$\frac{1}{\sqrt{\det\cosh(At/2)}} = C_\delta(t)e^{-t/2}.$$

□

6. The key proposition for the eigenfunction expansion

In the rest of this paper we assume that $n = 1$ and $\mu > 0$. It is easy to see for $\mu > 0$

$$\begin{aligned} e_\infty(x, v, \xi, \rho) &= \lim_{t\to\infty} e(t; x, v, \xi, \rho) \\ &= 4\exp\Big[-2(|\rho|^2 + |v|^2/4) - \frac{2}{\mu}(|\xi|^2 + |\mu x + a|^2/4)\Big] \end{aligned}$$

and we have

$$\begin{aligned} &(2\pi)^{-2}\int_{\mathbb{R}^2} e^{i(x-x')\cdot\xi + i(v-v')\cdot\rho} e_\infty\left(\frac{x+x'}{2}, \frac{v+v'}{2}, \xi, \rho\right) d\xi d\rho \\ &= \frac{\sqrt{\mu}}{2\pi}\exp\Big[-\frac{1}{4}\left\{v^2 + v'^2 + \mu(x + a/\mu)^2 + \mu(x' + a/\mu)^2\right\}\Big]. \end{aligned}$$

We show a method of obtaining eigenfunctions due to the fundamental solution represented by pseudo-differetial operator of Weyl symbols. The following proposition is the key of a proof of an expansipn of the kernel of the fundamental solution obtained as a pseudo-differential operator.

Proposition 6.1. *If the symbol of a pseudo-differential operator P is of the form*

$$p(x, \xi) = k(x, \xi) g(x, \xi),$$

then the kernel of the operator P is

$$(2\pi)^{-d}\int_{\mathbb{R}^d} e^{i(x-x')\cdot\xi} p\left(\frac{x+x'}{2}, \xi\right) d\xi = k\left(\frac{r}{2}, -i\partial_q\right)\tilde{g}\left(\frac{r}{2}, q\right)\Big|_{r=x+x', q=x-x'},$$

where

$$\tilde{g}\left(\frac{r}{2}, q\right) = (2\pi)^{-d}\int_{\mathbb{R}^d} e^{iq\cdot\xi} g\left(\frac{r}{2}, \xi\right) d\xi.$$

Proof. Formally if the symbol $k(x, \xi) = \sum_{j=0}^\infty a_j(x)\xi^j$ has the expansion, we have with $r = x + x', q = x - x'$

$$\begin{aligned} &(2\pi)^{-d}\int_{\mathbb{R}^d} e^{i(x-x')\cdot\xi} p\left(\frac{x+x'}{2}, \xi\right) d\xi \\ &= (2\pi)^{-d}\int_{\mathbb{R}^d}\sum_{j=0}^\infty a_j\left(\frac{r}{2}\right)\xi^j e^{iq\cdot\xi} g\left(\frac{r}{2}, \xi\right) d\xi \\ &= \sum_{j=0}^\infty a_j\left(\frac{r}{2}\right)(-i\partial_q)^j (2\pi)^{-d}\int_{\mathbb{R}^d} e^{iq\cdot\xi} g\left(\frac{r}{2}, \xi\right) d\xi \\ &= k\left(\frac{r}{2}, -i\partial_q\right)\tilde{g}\left(\frac{r}{2}, q\right)\Big|_{r=x+x', q=x-x'}. \end{aligned}$$

□

7. The eigenfunction expansion for K

Apply Proposition 6.1 for $d = 2$, $g = e_\infty(x, v, \xi, \rho)$, $k = e(t; x, v, \xi, \rho)/e_\infty(x, v, \xi, \rho)$. In our case we have

$$\tilde{e}_\infty\left(\frac{x+x'}{2}, \frac{v+v'}{2}, x-x', v-v'\right) = \psi(x, v)\psi(x', v'),$$

where

$$\psi(x, v) = \left(\frac{\sqrt{\mu}}{2\pi}\right)^{1/2} \exp\left[-\frac{1}{4}\left\{v^2 + \mu\left(x + \frac{a}{\mu}\right)^2\right\}\right] \tag{7.1}$$

by the argument of the previous section.

Set $z_j = e^{-\lambda_j t}$ $j = 1, 2$. Then

$$\frac{C_\delta(t)}{4} = \frac{1}{(1+z_1)(1+z_2)}.$$

The methods to obtain the eigenfunction expansion are similar for any positive μ. But the symbol $k(t; x, v, \xi, \rho)$ for $\mu = 1/4$ is different from that for $\mu \neq 1/4$. In fact there are generalized eigenfunctions in case $\mu = 1/4$ as discussed later.

7.1. K in case of positive $\mu \neq \frac{1}{4}$

Set

$$h(z) = \frac{z}{1+z},$$

Then we have

$$\frac{1}{2}\left\{\tanh(\lambda_j t/2) - 1\right\} = -h(z_j), \quad j = 1, 2.$$

In our case the symbol $k(t; x, v, \xi, \rho)$ is obtained as

$$k(t;x,v,\xi,\rho) - \frac{1}{(1+z_1)(1+z_2)} \exp\left\{\frac{1}{\delta}h(z_1)\tilde{k}_{(\lambda_1)}(x,v,\xi,\rho) - \frac{1}{\delta}h(z_2)\tilde{k}_{(\lambda_2)}(x,v,\xi,\rho)\right\},$$

where

$$\begin{aligned}
&\tilde{k}_{(\lambda_j)}(x, v, \xi, \rho) \\
&= 4\left[\lambda_j\left(\rho^2 + \frac{v^2}{4}\right) + i\left(\langle v, \xi\rangle - \langle \mu x + a, \rho\rangle\right) - \frac{1}{\lambda_j}\left\{\xi^2 + \frac{1}{4}(\mu x + a)^2\right\}\right] \\
&= 4\left\{\sqrt{\lambda_j}\left(i\rho + \frac{v}{2}\right) + \frac{1}{\sqrt{\lambda_j}}\left(i\xi + \frac{1}{2}(\mu x + a)\right)\right\} \\
&\quad \times \left\{\sqrt{\lambda_j}\left(-i\rho + \frac{v}{2}\right) - \frac{1}{\sqrt{\lambda_j}}\left(-i\xi + \frac{1}{2}(\mu x + a)\right)\right\}.
\end{aligned}$$

So we have with $r_1 = x + x', r_2 = v + v', q_1 = x - x', q_2 = v - v'$

$$\begin{aligned}
&\tilde{k}_{(\lambda_j)}\left(\frac{r_1}{2}, \frac{r_2}{2}, -i\partial_{q_1}, -i\partial_{q_2}\right) \\
&\quad = 4\left\{\sqrt{\lambda_j}\,\left(\partial_{q_2} + \frac{r_2}{4}\right) + \frac{1}{\sqrt{\lambda_j}}\left(\partial_{q_1} + \frac{1}{4}\mu r_1 + \frac{a}{2}\right)\right\} \\
&\qquad \times \left\{\sqrt{\lambda_j}\,\left(-\partial_{q_2} + \frac{r_2}{4}\right) - \frac{1}{\sqrt{\lambda_j}}\left(-\partial_{q_1} + \frac{1}{4}\mu r_1 + \frac{a}{2}\right)\right\} \\
&\quad = \left\{\sqrt{\lambda_j}(b_v + b_{v'}^*) + \frac{1}{\sqrt{\lambda_j}}(a_x + a_{x'}^*)\right\}\left\{\sqrt{\lambda_1}(b_v^* + b_{v'}) - \frac{1}{\sqrt{\lambda_j}}(a_x^* + a_{x'})\right\},
\end{aligned}$$

where $a_x, b_v, a_{x'}, b_{v'}$ are differential operators of the first order:

$$b_v = \partial_v + \frac{v}{2}, \qquad a_x = \partial_x + \frac{\mu x}{2} + \frac{a}{2},$$
$$b_{v'} = \partial_{v'} + \frac{v'}{2}, \quad a_{x'} = \partial_{x'} + \frac{\mu x'}{2} + \frac{a}{2}.$$

Definition 7.1. For $\lambda = \lambda_1, \lambda_2$ we define differential operators:

$$\begin{aligned}
M_\lambda &= \frac{1}{\sqrt{\delta}}\left(\sqrt{\lambda} b_v - \frac{1}{\sqrt{\lambda}} a_x\right), & N_\lambda &= \frac{1}{\sqrt{\delta}}\left(\sqrt{\lambda} b_v + \frac{1}{\sqrt{\lambda}} a_x\right), \\
M'_\lambda &= \frac{1}{\sqrt{\delta}}\left(\sqrt{\lambda} b_{v'} - \frac{1}{\sqrt{\lambda}} a_{x'}\right), & N'_\lambda &= \frac{1}{\sqrt{\delta}}\left(\sqrt{\lambda} b_{v'} + \frac{1}{\sqrt{\lambda}} a_{x'}\right), \\
S &= \left(N_{\lambda_1} + (N'_{\lambda_1})^*\right)(M^*_{\lambda_1} + M'_{\lambda_1}), & \tilde{S} &= -\left(N_{\lambda_2} + (N'_{\lambda_2})^*\right)(M^*_{\lambda_2} + M'_{\lambda_2}).
\end{aligned}$$

Using the Definition 7.1, we can write

$$\frac{1}{\delta}\tilde{k}_{(\lambda_1)}\left(\frac{r_1}{2}, \frac{r_2}{2}, -i\partial_{q_1}, -i\partial_{q_2}\right) = S$$

and

$$\frac{1}{\delta}\tilde{k}_{(\lambda_2)}\left(\frac{r_1}{2}, \frac{r_2}{2}, -i\partial_{q_1}, -i\partial_{q_2}\right) = -\tilde{S}.$$

So we have by Proposition 6.1

$$e(t; x, v, x', v') = \frac{1}{(1 + z_1)(1 + z_2)} \exp\left\{h(z_1)S + h(z_2)\tilde{S}\right\}\left(\psi(x, v)\psi(x', v')\right).$$

It is clear that

$$[a_x, a_x^*] = \mu, \quad [b_v, b_v^*] = 1.$$

So we have

$$\begin{aligned}
&[N_{\lambda_1}, M^*_{\lambda_1}] = [M_{\lambda_1}, N^*_{\lambda_1}] = 1, \\
&[N_{\lambda_2}, M^*_{\lambda_2}] = [M_{\lambda_2}, N^*_{\lambda_2}] = -1, \\
&[N_{\lambda_1}, M^*_{\lambda_2}] = [N_{\lambda_2}, M^*_{\lambda_1}] = [M_{\lambda_1}, N^*_{\lambda_2}] = [M_{\lambda_2}, N^*_{\lambda_1}] = 0.
\end{aligned}$$

Then $M^*_{\lambda_1} + M'_{\lambda_1}$, $N_{\lambda_1} + (N'_{\lambda_1})^*$, $M^*_{\lambda_2} + M'_{\lambda_2}$, $N_{\lambda_2} + (N'_{\lambda_2})^*$ are commuting. Also S and $\tilde{S}$ are commuting. So we have

$$e(t;x,v,x',v') = \sum_{j=0}^{\infty}\sum_{k=0}^{\infty} \frac{1}{j!k!} C_j(S)C_k(\tilde{S})\Big(\psi(x,v)\psi(x',v')\Big) z_1^j z_2^k,$$

where the operators $C_j(P)$ are defined as

$$\frac{1}{(1+z)} \exp\left(\frac{z}{1+z}P\right) = \sum_{j=0}^{\infty} \frac{C_j(P)}{j!} z^j. \tag{7.2}$$

Definition 7.2. For a pair of nonnegative integers m and k we set

$$\psi_{m,k}(x,v) = \frac{(M^*_{\lambda_1})^m}{\sqrt{m!}} \frac{(-M^*_{\lambda_2})^k}{\sqrt{k!}} \psi(x,v), \quad \psi^{m,k}(x,v) = \frac{(N^*_{\lambda_1})^m}{\sqrt{m!}} \frac{(N^*_{\lambda_2})^k}{\sqrt{k!}} \psi(x,v).$$

The following expansion of the kernel of the fundamental solution is obtained. The assertion of (2) and (3) Theorem 7.3 means that $\psi_{m,k}(x,v)$ are the orthonormal eigenfunctions whose eigenvalues are $\lambda_1 m + \lambda_2 k$.

Theorem 7.3.

(1) *We have the eigenfunction expansion for the kernel of the heat operator:*

$$e(t;x,v,x',v') = \sum_{n_1,n_2=0}^{\infty} e^{-(\lambda_1 n_1 + \lambda_2 n_2)t} \psi_{n_1,n_2}(x,v)\psi^{n_1,n_2}(x',v').$$

(2) *For any* $(m,k) \in \mathbb{N}_0 \times \mathbb{N}_0$ *it holds that*

$$K\psi_{m,k}(x,v) = (\lambda_1 m + \lambda_2 k)\psi_{m,k}(x,v).$$

(3) *For any* $(m,k),(n,\ell) \in \mathbb{N}_0 \times \mathbb{N}_0$ *we have*

$$\langle \psi_{m,k}, \psi^{n,\ell} \rangle = \delta_{m,n}\delta_{k,\ell}.$$

For the proof of Theorem 7.3 we prepare two propositions.

Proposition 7.4. *The operator* K *is written by* M_λ *and* N_λ *as follows:*

$$K = \lambda_1 M^*_{\lambda_1} N_{\lambda_1} - \lambda_2 M^*_{\lambda_2} N_{\lambda_2}.$$

Proof. It is easy to see

$$K = b_v^* b_v + \big(b_v^* a_x - a_x^* b_v\big).$$

We can show the following equation by the direct calculation.

$$b_v^* b_v + \big(b_v^* a_x - a_x^* b_v\big) = \lambda_1 M^*_{\lambda_1} N_{\lambda_1} - \lambda_2 M^*_{\lambda_2} N_{\lambda_2}.$$ □

Proposition 7.5.

(1) $\psi_{m,k}(x,v)$ *and* $\psi^{m,k}(x,v)$ *have the following properties:*

$$a_x\psi(x,v) = b_v\psi(x,v) = 0,$$
$$M_{\lambda_1}\psi(x,v) = M_{\lambda_2}\psi(x,v) = N_{\lambda_1}\psi(x,v) = N_{\lambda_2}\psi(x,v) = 0.$$

Moreover

$$N_{\lambda_1}\psi_{m,\ell}(x,v)=\sqrt{m}\psi_{m-1,\ell}(x,v),\qquad N_{\lambda_2}\psi_{m,\ell}(x,v)=\sqrt{\ell}\psi_{m,\ell-1}(x,v),$$
$$M_{\lambda_1}\psi^{m,\ell}(x,v)=\sqrt{m}\psi^{m-1,\ell}(x,v),\quad -M_{\lambda_2}\psi^{m,\ell}(x,v)=\sqrt{\ell}\psi^{m,\ell-1}(x,v).$$

(2) *For any nonnegative integers k we have*

$$N^k_{\lambda_1}(M^*_{\lambda_1})^k\psi(x,v)=N^k_{\lambda_2}(-M^*_{\lambda_2})^k\psi(x,v)=k!\psi(x,v)\qquad (k\in\mathbb{N}_0),$$
$$M^k_{\lambda_1}(N^*_{\lambda_1})^k\psi(x,v)=(-M_{\lambda_2})^k(N^*_{\lambda_2})^k\psi(x,v)=k!\psi(x,v)\quad (k\in\mathbb{N}_0).$$

(3) $C_\ell(S)$ *defined by* (7.2) *satisfies*

$$C_\ell(S)\Big(\psi_{0,m}(x,v)\psi^{0,r}(x',v')\Big)=\ell!\psi_{\ell,m}(x,v)\psi^{\ell,r}(x',v')\quad (\ell,m,r\in\mathbb{N}_0).$$

(4) $C_\ell(\tilde{S})$ *defined by* (7.2) *satisfies*

$$C_\ell(\tilde{S})\Big(\psi_{m,0}(x,v)\psi^{r,0}(x',v')\Big)=\ell!\psi_{m,\ell}(x,v)\psi^{r,\ell}(x',v')\quad (\ell,m,r\in\mathbb{N}_0).$$

Proof. (1) By the following equations the assertions are obtained.

$$\begin{aligned}N_{\lambda_1}\psi_{m,\ell}(x,v)&=\left\{[N_{\lambda_1},(M^*_{\lambda_1})^m]\psi_{0,\ell}(x,v)+(M^*_{\lambda_1})^m N_{\lambda_1}\psi_{0,\ell}(x,v)\right\}/\sqrt{m!}\\&=m(M^*_{\lambda_1})^{m-1}\psi_{0,\ell}(x,v)/\sqrt{m!}\\&=\sqrt{m}\psi_{m-1,\ell}(x,v).\end{aligned}$$

$$\begin{aligned}N_{\lambda_2}\psi_{m,\ell}(x,v)&=\left\{[N_{\lambda_2},(-M^*_{\lambda_2})^\ell]\psi_{m,0}(x,v)+(-M^*_{\lambda_2})^\ell N_{\lambda_2}\psi_{m,0}(x,v)\right\}/\sqrt{\ell!}\\&=\ell(-M^*_{\lambda_2})^{\ell-1}\psi_{m,0}(x,v)/\sqrt{\ell!}\\&=\sqrt{\ell}\psi_{m,\ell-1}(x,v).\end{aligned}$$

$$\begin{aligned}M_{\lambda_1}\psi^{m,\ell}(x,v)&=\left\{[M_{\lambda_1},(N^*_{\lambda_1})^m]\psi^{0,\ell}(x,v)+(N^*_{\lambda_1})^m M_{\lambda_1}\psi^{0,\ell}(x,v)\right\}/\sqrt{m!}\\&=m(N^*_{\lambda_1})^{m-1}\psi^{0,\ell}(x,v)/\sqrt{m!}\\&=\sqrt{m}\psi^{m-1,\ell}(x,v).\end{aligned}$$

$$\begin{aligned}M_{\lambda_2}\psi^{m,\ell}(x,v)&=\left\{[M_{\lambda_2},(N^*_{\lambda_2})^\ell]\psi^{m,0}(x,v)+(N^*_{\lambda_2})^\ell M_{\lambda_2}\psi^{m,0}(x,v)\right\}/\sqrt{\ell!}\\&=-\ell(N^*_{\lambda_2})^{\ell-1}\psi^{m,0}(x,v)/\sqrt{\ell!}\\&=-\sqrt{\ell}\psi^{m,\ell-1}(x,v).\end{aligned}$$

(2) By (1) we have

$$\begin{aligned}\frac{N^k_{\lambda_1}(M^*_{\lambda_1})^k}{k!}\psi(x,v)&=N^k_{\lambda_1}\psi_{k,0}(x,v)/\sqrt{k!}=\sqrt{k}N^{k-1}_{\lambda_1}\psi_{k-1,0}(x,v)/\sqrt{k!}\\&=N^{k-1}_{\lambda_1}\psi_{k-1,0}(x,v)/\sqrt{(k-1)!}=\frac{N^{k-1}_{\lambda_1}(M^*_{\lambda_1})^{k-1}}{(k-1)!}\psi(x,v).\end{aligned}$$

We get the assertion by the induction with respect to k. Similarly we obtain the assertion for

$$\frac{(-M_{\lambda_2})^k (N_{\lambda_2}^*)^k}{k!}\psi(x,v).$$

(3) The following equations hold:

$$\begin{aligned}
&S\Big(\psi_{m,k}(x,v)\psi^{n,\ell}(x',v')\Big)\\
&= \sqrt{(m+1)(n+1)}\psi_{m+1,k}(x,v)\psi^{n+1,\ell}(x',v') + (m+n+1)\psi_{m,k}(x,v)\psi^{n,\ell}(x',v')\\
&\quad + \sqrt{mn}\psi_{m-1,k}(x,v)\psi^{n-1,\ell}(x',v'),
\end{aligned}$$

$$\begin{aligned}
&\tilde{S}\Big(\psi_{m,k}(x,v)\psi^{n,\ell}(x',v')\Big)\\
&= \sqrt{(k+1)(\ell+1)}\psi_{m,k+1}(x,v)\psi^{n,\ell+1}(x',v') + (k+\ell+1)\psi_{m,k}(x,v)\psi^{n,\ell}(x',v')\\
&\quad + \sqrt{k\ell}\psi_{m,k-1}(x,v)\psi^{n,\ell-1}(x',v'),
\end{aligned}$$

which will be shown as follows:

We note that

$$\begin{aligned}
&(M_{\lambda_1}^* + M_{\lambda_1}')(\psi_{m,k}(x,v)\psi^{n,\ell}(x',v'))\\
&\quad = \sqrt{m+1}\psi_{m+1,k}(x,v)\psi^{n,\ell}(x',v') + \sqrt{n}\psi_{m,k}(x,v)\psi^{n-1,\ell}(x',v')
\end{aligned}$$

and

$$\begin{aligned}
&(N_{\lambda_1} + (N_{\lambda_1}')^*)(\psi_{m,k}(x,v)\psi^{n,\ell}(x',v'))\\
&\quad = \sqrt{m}\psi_{m-1,k}(x,v)\psi^{n,\ell}(x',v') + \sqrt{n+1}\psi_{m,k}(x,v)\psi^{n+1,\ell}(x',v').
\end{aligned}$$

So by the above equations the assertion for $S = (N_{\lambda_1} + (N_{\lambda_1}')^*)(M_{\lambda_1}^* + M_{\lambda_1}')$ is obtained. Similarly we can show the assertion for $\tilde{S}$.

We will show the assertion (3) by the induction with respect to ℓ. The assertion holds for $\ell = 0$ because $C_0(P) = 1$ for all P. Note that for any operator P the operators $C_j(P)$ satisfies the following equation (7.3) for which we give a proof in Proposition 7.10 below in the more general form.

$$C_{\ell+1}(P) = PC_\ell(P) - (2\ell+1)C_\ell(P) - \ell^2 C_{\ell-1}(P), (\ell \geq 0), \tag{7.3}$$

$$C_0(P) = 1. \tag{7.4}$$

Assume that (3) holds for all $\ell \leq j$. Then by the equation (7.3) we have

$$\begin{aligned}
&C_{j+1}(S)\Big(\psi_{0,m}(x,v)\psi^{0,r}(x',v')\Big)\\
&= SC_j(S)\Big(\psi_{0,m}(x,v)\psi^{0,r}(x',v')\Big) - (2j+1)C_j(S)\Big(\psi_{0,m}(x,v)\psi^{0,r}(x',v')\Big)\\
&\quad - j^2 C_{j-1}(S)\Big(\psi_{0,m}(x,v)\psi^{0,r}(x',v')\Big)\\
&= j!\Big\{S(\psi_{j,m}(x,v)\psi^{j,r}(x',v')) - (2j+1)\psi_{j,m}(x,v)\psi^{j,r}(x',v')\\
&\quad - j\psi_{j-1,m}(x,v)\psi^{j-1,r}(x',v')\Big\}.
\end{aligned}$$

On the other hand it holds by the equation for S

$$\begin{aligned}
&S(\psi_{j,m}(x,v)\psi^{j,r}(x',v')) \\
&\quad = (j+1)\psi_{j+1,m}(x,v)\psi^{j+1,r}(x',v') + (2j+1)\psi_{j,m}(x,v)\psi^{j,r}(x',v') \\
&\quad\quad + j\psi_{j-1,m}(x,v)\psi^{j-1,r}(x',v').
\end{aligned}$$

So we get (3) for $j+1$.
(4) is proved similarly. □

Now we give a proof of Theorem 7.3.

(1) By Proposition 7.5 we obtain the following equation:

$$\begin{aligned}
C_j(S)C_k(\tilde{S})\Big(\psi(x,v)\psi(x',v')\Big) &= k!C_j(S)\Big(\psi_{0,k}(x,v)\psi^{0,k}(x',v')\Big) \\
&= k!j!\psi_{j,k}(x,v)\psi^{j,k}(x',v').
\end{aligned}$$

Finally we have

$$\begin{aligned}
&e(t;x,v,x',v') \\
&\quad = \sum_{j=0}^{\infty}\sum_{k=0}^{\infty} \frac{1}{j!k!} C_j(S)C_k(\tilde{S})\Big(\psi(x,v)\psi(x',v')\Big) z_1^j z_2^k \\
&\quad = \sum_{j=0}^{\infty}\sum_{k=0}^{\infty} \psi_{j,k}(x,v)\psi^{j,k}(x',v') z_1^j z_2^k \\
&\quad = \sum_{j=0}^{\infty}\sum_{k=0}^{\infty} e^{-(\lambda_1 j + \lambda_2 k)t} \psi_{j,k}(x,v)\psi^{j,k}(x',v').
\end{aligned}$$

(2) By Proposition 7.4 and Proposition 7.5 we have

$$\begin{aligned}
K\psi_{m,k} &= \lambda_1\sqrt{m}M^*_{\lambda_1}\psi_{m-1,k} - \lambda_2\sqrt{k}M^*_{\lambda_2}\psi_{m,k-1} \\
&= \lambda_1 m\psi_{m,k} + \lambda_2 k\psi_{m,k} \\
&= (\lambda_1 m + \lambda_2 k)\psi_{m,k}.
\end{aligned}$$

(3) Using Proposition 7.5, we have for $m \geq n$

$$\begin{aligned}
\sqrt{m!n!}\langle\psi_{m,0},\psi^{n,0}\rangle &= \int \psi(x,v)\ M^{m-n}_{\lambda_1} M^n_{\lambda_1}(N^*_{\lambda_1})^n\psi(x,v)dxdv \\
&= \int \psi(x,v)\ M^{m-n}_{\lambda_1} n!\psi(x,v)dxdv \\
&= n!\delta_{m,n}\int |\psi(x,v)|^2 dxdv = n!\delta_{m,n}.
\end{aligned}$$

If $\ell \ge k$, we have

$$\begin{aligned}\sqrt{m!k!n!\ell!}\langle \psi_{m,k}, \psi^{n,\ell}\rangle &= \int (M^*_{\lambda_1})^m(-M^*_{\lambda_2})^k\psi(x,v)\ (N^*_{\lambda_1})^n(N^*_{\lambda_2})^\ell\psi(x,v)dxdv \\ &= \int (M^*_{\lambda_1})^m N^{\ell-k}_{\lambda_2} N^k_{\lambda_2}(-M^*_{\lambda_2})^k\psi(x,v)\ (N^*_{\lambda_1})^n\psi(x,v)dxdv \\ &= k!\delta_{k,\ell}\int (M^*_{\lambda_1})^m\psi(x,v)\ (N^*_{\lambda_1})^n\psi(x,v)dxdv \\ &= k!\sqrt{m!n!}\delta_{k,\ell}\langle\psi_{m,0},\psi^{n,0}\rangle = k!m!\delta_{k,\ell}\delta_{m,n}.\end{aligned}$$

7.2. K in the case $\mu = \frac{1}{4}$

In the case $\mu = \frac{1}{4}$ we have by (7.1)

$$\psi(x,v) = \frac{1}{2\sqrt{\pi}}\exp\left\{-\left(\frac{v^2}{4} + \frac{(x+4a)^2}{16}\right)\right\}$$

and with $z = e^{-t/2}$ the symbol $k(t;x,v,\xi,\rho)$ is of the form

$$\begin{aligned}k(t;x,v,\xi,\rho) = \frac{1}{(1+z)^2}\exp\Bigg\{&2\left(\frac{2z}{1+z} - \frac{tz}{(1+z)^2}\right)\left(\rho^2 + \frac{v^2}{4}\right) \\ &- i\frac{4tz}{(1+z)^2}\Big(\langle v,\xi\rangle - \langle x/4 + a,\rho\rangle\Big) \\ &+ 2\left(\frac{2z}{1+z} + \frac{tz}{(1+z)^2}\right)\left(4\xi^2 + \frac{(x+4a)^2}{16}\right)\Bigg\}.\end{aligned}$$

So we have with $r_1 = x + x', r_2 = v + v', q_1 = x - x', q_2 = v - v'$

$$\begin{aligned}&k\left(t;\frac{r_1}{2},\frac{r_2}{2},-i\partial_{q_1},-i\partial_{q_2}\right) \\ &= \frac{1}{(1+z)^2}\exp\Bigg[\frac{4z}{1+z}\Bigg\{\left(2\partial_{q_1} + \frac{r_1}{8} + a\right)\left(-2\partial_{q_1} + \frac{r_1}{8} + a\right) \\ &+ \left(\partial_{q_2} + \frac{r_2}{4}\right)\left(-\partial_{q_2} + \frac{r_2}{4}\right)\Bigg\} + \frac{2tz}{(1+z)^2}\Bigg\{\left(2\partial_{q_1} + \frac{r_1}{8} + a\right) \\ &+ \left(\partial_{q_2} + \frac{r_2}{4}\right)\Bigg\}\Bigg\{\left(-2\partial_{q_1} + \frac{r_1}{8} + a\right) - \left(-\partial_{q_2} + \frac{r_2}{4}\right)\Bigg\}\Bigg] \\ &= \frac{1}{(1+z)^2}\exp\Bigg[\frac{z}{1+z}\{(b_v + b^*_{v'})(b^*_v + b_{v'}) + (a_x + a^*_{x'})(a^*_x + a_{x'})\} \\ &- \frac{tz}{2(1+z)^2}\{(b_v + b^*_{v'} + a_x + a^*_{x'})(b^*_v + b_{v'} - a^*_x - a_{x'})\}\Bigg],\end{aligned}$$

where

$$b_v = \partial_v + \frac{v}{2},\quad a_x = 2\partial_x + \frac{x}{4} + a,\quad b_{v'} = \partial_{v'} + \frac{v'}{2},\quad a_{x'} = 2\partial_{x'} + \frac{x'}{4} + a.$$

So we have

$$e(t;x,v,x',v') = \frac{1}{(1+z)^2}\exp\left[\frac{z}{1+z}\Big\{(b_v+b_{v'}^*)(b_v^*+b_{v'})+(a_x+a_{x'}^*)(a_x^*+a_{x'})\Big\}\right.$$
$$\left.-\frac{tz}{2(1+z)^2}(b_v+b_{v'}^*+a_x+a_{x'}^*)(b_v^*+b_{v'}-a_x^*-a_{x'})\right]\,(\psi(x,v)\psi(x',v')).$$

Definition 7.6.

$$M=\frac{1}{\sqrt{2}}(b_v-a_x),\quad N=\frac{1}{\sqrt{2}}(b_v+a_x),\quad M'=\frac{1}{\sqrt{2}}(b_{v'}-a_{x'}),\quad N'=\frac{1}{\sqrt{2}}(b_{v'}+a_{x'}),$$

$$Q=(N+(N')^*)(M^*+M'),\quad S_M=(M+(M')^*)(M^*+M')$$
$$S_N=(N+(N')^*)(N^*+N'),\qquad S=S_M+S_N.$$

Using Definition 7.6 we can write

$$e(t;x,v,x',v')=\frac{1}{(1+z)^2}\exp\left(\frac{z}{1+z}S-\frac{tz}{(1+z)^2}Q\right)\Big(\psi(x,v)\psi(x',v')\Big).$$

It is clear that

$$[a_x,a_x^*]=1,\quad [b_v,b_v^*]=1.$$

So we have

$$[M,M^*]=[N,N^*]=1,\quad [M,N^*]=[N,M^*]=0,$$
$$[M^*+M',N+(N')^*]=[M+(M')^*,M^*+M']=[N+(N')^*,N^*+N']=0.$$

It is easy to see that S_M, S_N and Q are commuting. Then we have

$$\frac{1}{(1+z)^2}\exp\Big(\frac{z}{1+z}(S_M+S_N)\Big)\exp\Big(-\frac{tz}{(1+z)^2}Q\Big)$$
$$=\sum_{k=0}^{\infty}\frac{(-t)^k}{k!}\sum_{j=0}^{\infty}\sum_{\ell=0}^{j}\frac{1}{(j-\ell)!\ell!}C_\ell(S_N,k)C_{j-\ell}(S_M,k)Q^k z^{j+k},$$

where operators $C_j(P,k)$ are defined as

$$\frac{1}{(1+z)^{k+1}}\exp\Big(\frac{z}{1+z}P\Big)=\sum_{j=0}^{\infty}\frac{C_j(P,k)}{j!}z^j. \tag{7.5}$$

Definition 7.7. For a pair of nonnegative integers m and k we set

$$\psi_{m,k}(x,v)=\frac{(M^*)^m}{\sqrt{m!}}\frac{(N^*)^k}{\sqrt{k!}}\psi(x,v),$$
$$\varphi_{m,k}(x,v)=\psi_{m-k,k}(x,v),\quad (0\le k\le m).$$

We obtain the following expansion of the kernel of the fundamental solution. By the assertion (2) and (3) of Theorem 7.8 for any nonnegative integer n the set $\{\varphi_{n,k}(x,v)\}_{k=0}^{n}$ are the set of orthogonal generalized eigenfunctions whose eigenvalue is $n/2$.

Theorem 7.8.

(1) *We have the eigenfunction expansion for the kernel of the heat operator:*

$$e(t;x,v,x',v') = \sum_{n=0}^{\infty} e^{-nt/2}\left(\sum_{k=0}^{n}\frac{(-t)^k}{k!}\sum_{\ell=0}^{n-k}\sqrt{\frac{(n-\ell)!(k+\ell)!}{\ell!(n-k-\ell)!}}\varphi_{n,\ell}(x,v)\varphi_{n,k+\ell}(x',v')\right).$$

(2) *For any* $(m,k)\,(m\in\mathbb{N}_0, 0\le k\le m)$ *it holds that*

$$K\varphi_{m,k}(x,v) = (m/2)\,\varphi_{m,k}(x,v) + \sqrt{k(m-k+1)}\varphi_{m,k-1}(x,v).$$

(3) *For any* $(m,k)\,(m\in\mathbb{N}_0, 0\le k\le m)$ *and* $(n,\ell)\,(n\in\mathbb{N}_0, 0\le \ell\le n)$

$$\langle\varphi_{m,k},\varphi_{n,\ell}\rangle = \delta_{m,n}\delta_{k,\ell}.$$

For the proof of Theorem 7.8 we prepare the following two propositions.

Proposition 7.9. *K has the following representation by the operators M and N:*

$$K = \frac{1}{2}(M^*M + N^*N + 2M^*N).$$

Proof.

$$\begin{aligned} K &= b_v^* b_v + \frac{1}{2}(b_v^* a_x - a_x^* b_v)\\ &= \frac{1}{4}\Big\{2(M^*+N^*)(M+N) + (M^*+N^*)(N-M) - (N^*-M^*)(M+N)\Big\}\\ &= \frac{1}{2}(M^*M + N^*N + 2M^*N). \end{aligned}$$

□

Proposition 7.10.

(1) *For any* $(m,\ell)\in\mathbb{N}_0\times\mathbb{N}_0$ *we have*

$$a_x\psi(x,v) = b_v\psi(x,v) = 0, M\psi(x,v) = N\psi(x,v) = 0.$$

Moreover

$$M\psi_{m,\ell}(x,v) = \sqrt{m}\psi_{m-1,\ell}(x,v), \quad N\psi_{m,\ell}(x,v) = \sqrt{\ell}\psi_{m,\ell-1}(x,v)$$

(2) *For nonnegative integers k we have*

$$M^k(M^*)^k\psi(x,v) = k!\psi(x,v), N^k(N^*)^k\psi(x,v) = k!\psi(x,v) \quad (k\in\mathbb{N}_0).$$

(3) *For nonnegative integers k we have*

$$Q^k\Big(\psi(x,v)\psi(x',v')\Big) = k!\psi_{k,0}(x,v)\psi_{0,k}(x',v') \quad (k\in\mathbb{N}_0).$$

(4) *We have for any* $\ell,k,m\in\mathbb{N}_0$

$$C_\ell(S_M,k)\Big(\psi_{k,m}(x,v)\psi_{0,r}(x',v')\Big) = \sqrt{\frac{(k+\ell)!\ell!}{k!}}\psi_{k+\ell,m}(x,v)\psi_{\ell,r}(x',v').$$

(5) *We have for any* $\ell,k,m\in\mathbb{N}_0$

$$C_\ell(S_N,k)\Big(\psi_{m,0}(x,v)\psi_{r,k}(x',v')\Big) = \sqrt{\frac{(k+\ell)!\ell!}{k!}}\psi_{m,\ell}(x,v)\psi_{r,k+\ell}(x',v').$$

Proof. (1) The proof of the assertion for the operators M and N are similar.

$$M\psi_{m,\ell}(x,v) = \{[M,(M^*)^m]\psi_{0,\ell}(x,v) + (M^*)^m M\psi_{0,\ell}(x,v)\}/\sqrt{m!}$$
$$= m(M^*)^{m-1}\psi_{0,\ell}(x,v)/\sqrt{m!} = \sqrt{m}\psi_{m-1,\ell}(x,v).$$

(2) By (1) we have

$$\frac{M^k(M^*)^k}{k!}\psi(x,v) = M^k\psi_{k,0}(x,v)/\sqrt{k!} = M^{k-1}\psi_{k-1,0}(x,v)/\sqrt{(k-1)!}$$
$$= \frac{M^{k-1}(M^*)^{k-1}}{(k-1)!}\psi(x,v).$$

We get the assertion by the induction. The proof of the assertion for the operator N is similar.

(3) We will show by the induction with respect to k. We note that

$$(M^* + M')(\psi_{m,k}(x,v)\psi_{n,\ell}(x',v'))$$
$$= \sqrt{m+1}\psi_{m+1,k}(x,v)\psi_{n,\ell}(x',v') + \sqrt{n}\psi_{m,k}(x,v)\psi_{n-1,\ell}(x',v')$$

and

$$(N + (N')^*)(\psi_{m,k}(x,v)\psi_{n,\ell}(x',v'))$$
$$= \sqrt{k}\psi_{m,k-1}(x,v)\psi_{n,\ell}(x',v') + \sqrt{\ell+1}\psi_{m,k}(x,v)\psi_{n,\ell+1}(x',v').$$

Applying these indentities, we have

$$Q(\psi_{m,0}(x,v)\psi_{0,\ell}(x',v')) = \sqrt{(m+1)(\ell+1)}\psi_{m+1,0}(x,v)\psi_{0,\ell+1}(x',v').$$

So we have

$$Q^k\Big(\psi(x,v)\psi(x',v')\Big) = k!\psi_{k,0}(x,v)\psi_{0,k}(x',v').$$

(4) The following equations hold:

$$S_M\Big(\psi_{m,k}(x,v)\psi_{n,\ell}(x',v')\Big)$$
$$= \sqrt{(m+1)(n+1)}\psi_{m+1,k}(x,v)\psi_{n+1,\ell}(x',v')$$
$$+ (m+n+1)\psi_{m,k}(x,v)\psi_{n,\ell}(x',v') + \sqrt{mn}\psi_{m-1,k}(x,v)\psi_{n-1,\ell}(x',v'),$$

$$S_N\Big(\psi_{m,k}(x,v)\psi_{n,\ell}(x',v')\Big)$$
$$= \sqrt{(k+1)(\ell+1)}\psi_{m,k+1}(x,v)\psi_{n,\ell+1}(x',v')$$
$$+ (k+\ell+1)\psi_{m,k}(x,v)\psi_{n,\ell}(x',v') + \sqrt{k\ell}\psi_{m,k-1}(x,v)\psi_{n,\ell-1}(x',v'),$$

which will be shown as follows: By (1) we have

$$(M + (M')^*)(\psi_{m,k}(x,v)\psi_{n,\ell}(x',v'))$$
$$= \sqrt{m}\psi_{m-1,k}(x,v)\psi_{n,\ell}(x',v') + \sqrt{n+1}\psi_{m,k}(x,v)\psi_{n+1,\ell}(x',v').$$

The proof for $S_M = (M + (M')^*)(M^* + M')$ is complete by the above equation. Similarly we get the result for S_N.

We will show the assertion (4) by the induction with respect to ℓ. The assertion holds for $\ell = 0$ because $C_0(P, k) = 1$ for all P. Note that for any operator P the operators $C_j(P, k)$ defined by (7.5) satisfy the following equation of which we give a proof below.

$$C_{\ell+1}(P,k) = PC_\ell(P,k) - (2\ell + k + 1)C_\ell(P,k) - \ell(\ell + k)C_{\ell-1}(P,k), \tag{7.6}$$

$$C_0(P,k) = 1. \tag{7.7}$$

Assume that (4) holds for all $\ell \leq j$. Then by the equation (7.6) we have

$$\begin{aligned}
&C_{j+1}(S_M,k)\Big(\psi_{k,m}(x,v)\psi_{0,r}(x',v')\Big)\\
&= S_M C_j(S_M,k)\Big(\psi_{k,m}(x,v)\psi_{0,r}(x',v')\Big)\\
&\quad - (2j+k+1)C_j(S_M,k)\Big(\psi_{k,m}(x,v)\psi_{0,r}(x',v')\Big)\\
&\quad - j(j+k)C_{j-1}(S_M,k)\Big(\psi_{k,m}(x,v)\psi_{0,r}(x',v')\Big)\\
&= \sqrt{\frac{(k+j)!j!}{k!}} S_M\left(\psi_{k+j,m}(x,v)\psi_{j,r}(x',v')\right)\\
&\quad - (2j+k+1)\sqrt{\frac{(k+j)!j!}{k!}}\psi_{k+j,m}(x,v)\psi_{j,r}(x',v')\\
&\quad\quad j(j+k)\sqrt{\frac{(k+j-1)!(j-1)!}{k!}}\psi_{k+j-1,m}(x,v)\psi_{j-1,r}(x',v').
\end{aligned}$$

On the other hand it holds by the argument for S_M

$$\begin{aligned}
S_M\left(\psi_{k+j,m}(x,v)\psi_{j,r}(x',v')\right) &= \sqrt{(j+k+1)(j+1)}\psi_{k+j+1,m}(x,v)\psi_{j+1,r}(x',v')\\
&\quad + (2j+k+1)\psi_{k+j,m}(x,v)\psi_{j,r}(x',v')\\
&\quad + \sqrt{j(j+k)}\psi_{k+j-1,m}(x,v)\psi_{j-1,r}(x',v').
\end{aligned}$$

So we get (4) for $j+1$.

We will give a proof of (7.6). Set an analytic function $f(z)$ as follows:

$$f(z) = \frac{1}{(1+z)^{k+1}} \exp\left(\frac{z}{1+z}P\right).$$

Then the following equations hold.

$$f'(z)(1+z)^2 + (k+1)(1+z)f(z) = f(z)P, \quad f(0) = 1.$$

Differentiating the above equation ℓ times, we get

$$\begin{aligned}
&(1+z)^2 f^{(\ell+1)}(z) + 2\ell(1+z)f^{(\ell)}(z) + \ell(\ell-1)f^{(\ell-1)}(z)\\
&\qquad + (k+1)\{(1+z)f^{(\ell)}(z) + \ell f^{(\ell-1)}(z)\} = f^{(\ell)}(z)P.
\end{aligned}$$

Putting $z = 0$ we get (7.6) for $C_\ell(P,k) = f^{(\ell)}(0)$.

(5) is proved by the similar way to (4). □

Finally we will give the proof of Theorem 7.8.

(1) By Proposition 7.10 we obtain the following equation:

$$\begin{aligned}
&C_\ell(S_N,k)C_{j-\ell}(S_M,k)Q^k\Big(\psi(x,v)\psi(x',v')\Big)\\
&\quad= k!C_\ell(S_N,k)C_{j-\ell}(S_M,k)\Big(\psi_{k,0}(x,v)\psi_{0,k}(x',v')\Big)\\
&\quad= k!\sqrt{\frac{(j-\ell)!(j-\ell+k)!}{k!}}C_\ell(S_N,k)\Big(\psi_{k+j-\ell,0}(x,v)\psi_{j-\ell,k}(x',v')\Big)\\
&\quad= k!\sqrt{\frac{(j-\ell)!(j-\ell+k)!}{k!}}\sqrt{\frac{\ell!(\ell+k)!}{k!}}\psi_{k+j-\ell,\ell}(x,v)\psi_{j-\ell,k+\ell}(x',v')\\
&\quad= \sqrt{(j-\ell)!(j-\ell+k)!\ell!(\ell+k)!}\psi_{k+j-\ell,\ell}(x,v)\psi_{j-\ell,k+\ell}(x',v').
\end{aligned}$$

Finally we have for $z=e^{-t/2}$

$$\begin{aligned}
e(t;x,v,x',v') &= \sum_{k=0}^\infty \frac{(-tz)^k}{k!}\frac{1}{(1+z)^{2k+2}}\exp\Big(\frac{zS}{1+z}\Big)Q^k\Big(\psi(x,v)\psi(x',v')\Big)\\
&= \sum_{k=0}^\infty \frac{(-tz)^k}{k!}\sum_{j=0}^\infty z^j\sum_{\ell=0}^j \frac{1}{(j-\ell)!\ell!}C_\ell(S_N,k)C_{j-\ell}(S_M,k)Q^k\Big(\psi(x,v)\psi(x',v')\Big)\\
&= \sum_{k=0}^\infty \frac{(-tz)^k}{k!}\sum_{j=0}^\infty z^j\sum_{\ell=0}^j \sqrt{\frac{(j+k-\ell)!(\ell+k)!}{(j-\ell)!\ell!}}\psi_{k+j-\ell,\ell}(x,v)\psi_{j-\ell,k+\ell}(x',v')\\
&= \sum_{n=0}^\infty z^n\sum_{k=0}^n \frac{(-t)^k}{k!}\sum_{\ell=0}^{n-k}\sqrt{\frac{(n-\ell)!(\ell+k)!}{(n-k-\ell)!\ell!}}\psi_{n-\ell,\ell}(x,v)\psi_{n-k-\ell,k+\ell}(x',v')\\
&= \sum_{n=0}^\infty z^n\sum_{k=0}^n \frac{(-t)^k}{k!}\sum_{\ell=0}^{n-k}\sqrt{\frac{(n-\ell)!(\ell+k)!}{(n-k-\ell)!\ell!}}\varphi_{n,\ell}(x,v)\varphi_{n,k+\ell}(x',v').
\end{aligned}$$

(2) By Proposition 7.9 and Proposition 7.10 we have

$$\begin{aligned}
K\psi_{m,k} &= \frac{1}{2}\{M^*\sqrt{m}\psi_{m-1,k}+N^*\sqrt{k}\psi_{m,k-1}+2M^*\sqrt{k}\psi_{m,k-1}\}\\
&= \frac{1}{2}\{m\psi_{m,k}+k\psi_{m,k}+2\sqrt{k(m+1)}\psi_{m+1,k-1}\}\\
&= \frac{1}{2}(m+k)\psi_{m,k}+\sqrt{k(m+1)}\psi_{m+1,k-1}.
\end{aligned}$$

The above equation means

$$K\varphi_{m,k}(x,v) = \frac{m}{2}\varphi_{m,k}(x,v)+\sqrt{k(m-k+1)}\varphi_{m,k-1}(x,v).$$

(3) Using Proposition 7.10, we have for $m \geq n$

$$\sqrt{m!n!}\langle \psi_{m,0}, \psi_{n,0}\rangle = \int \psi(x,v)\ M^{m-n}M^n(M^*)^n\psi(x,v)dxdv$$
$$= \int \psi(x,v)\ M^{m-n}n!\psi(x,v)dxdv = n!\delta_{m,n}\int |\psi(x,v)|^2dxdv = n!\delta_{m,n}.$$

If $\ell \geq k$, we have

$$\sqrt{m!k!n!\ell!}\langle \psi_{m,k}, \psi_{n,\ell}\rangle = \int (M^*)^m(N^*)^k\psi(x,v)\ (M^*)^n(N^*)^\ell\psi(x,v)dxdv$$
$$= \int (M^*)^m N^{\ell-k}N^k(N^*)^k\psi(x,v)\ (M^*)^n\psi(x,v)dxdv$$
$$= k!\delta_{k,\ell}\int (M^*)^m\psi(x,v)\ (M^*)^n\psi(x,v)dxdv$$
$$= k!\sqrt{m!n!}\delta_{k,\ell}\langle \psi_{m,0}, \psi_{n,0}\rangle = k!n!\delta_{k,\ell}\delta_{m,n}.$$

References

[1] W. Bauer, K. Furutani, C. Iwasaki: *Spectral zeta function of the sub-Laplacian on two step nilmanifolds*, J. Math. Pures Appl. **97** (2011), 242–261.

[2] O. Calin, D.-C. Chang, K. Furutani, C. Iwasaki: *Heat Kernels for Elliptic and Sub-elliptic Operators, methods and techniques*, Applied and Harmonic Analysis, vol. XVIII (2010), Birkhäuser.

[3] B. Helffer and F. Nier, *Hypoellipticity and spectral theory for Fokker–Plancks operators and Witten Laplacians*, Lecture Notes in Math. **1862** (2005), Springer.

[4] L. Hörmander. *The Weyl calculus of pseudo-differential operators*, Comm. Pure Appl. Math. **32** (1979), 359–443.

[5] F. Hérau and F. Nier, *Isotropic hypoellipticity and trend to equilibrium for the Fokker–Planck equation with a high-degree potential*, Arch. Rational. Mech. Anal. **171** (2004), 151–218.

[6] C. Iwasaki and N. Iwasaki. *Parametrix for a degenerate parabolic equation*, Proc. Japan Acad. **55** (1979), 237–240.

[7] C. Iwasaki and N. Iwasaki. *Parametrix for a Degenerate Parabolic Equation and its Application to the Asymptotic Behavior of Spectral Functions for Stationary Problems*, Publ. Res. Inst. Math. Sci. **17** (1981), 557–655.

[8] A. Melin. *Lower bounds for pseudo-differential operators*, Ark.Mat. **9** (1971), 117–140.

[9] H. Risken, *The Fokker–Planck equation, Methods of solution and applications*, second edition (1989), Springer.

Chisato Iwasaki
Department of Mathematics
University of Hyogo
2167 Shosha Himeji
671-2201, Japan
e-mail: iwasaki@sci.u-hyogo.ac.jp

Fourier Analysis
Trends in Mathematics, 235–251

The Fourier Transform and Convolutions Generated by a Differential Operator with Boundary Condition on a Segment

Baltabek Kanguzhin and Niyaz Tokmagambetov

Abstract. We introduce the concepts of the Fourier transform and convolution generated by an arbitrary restriction of the differentiation operator in the space $L_2(0, b)$. In contrast to the classical convolution, the introduced convolution explicitly depends on the boundary condition that defines the domain of the operator L. The convolution is closely connected to the inverse operator or to the resolvent. So, we first find a representation for the resolvent, and then introduce the required convolution.

Mathematics Subject Classification (2010). 34B10, 34L10, 47G30, 47E05.

Keywords. Fourier transform, convolution, differential operator, non-local boundary condition, resolvent, spectrum, coefficient functional, basis.

1. Introduction

The standard Fourier transform is a unitary transform in the Hilbert space $L_2(-\infty, +\infty)$ and it is generated by the operator of differentiation $(-i\frac{d}{dx})$, because the system of exponents $\{\exp(i\lambda x), \lambda \in R\}$ is a system of "eigenfunctions" corresponding to its continuous spectrum. And an application of the classical Fourier transform in the pseudo-differential operators on the torus can be found in the monograph [1] (also see [2]). The Fourier transform is closely connected to the bilinear, commutative, associative convolution without annihilators. An important fact is that the convolution with the fundamental solution allows us to find solutions of the inhomogeneous differential equation, which commutes with differentiation. Corresponding constructions can be generalized to arbitrary self-adjoint operators. Instead of the differential operator $(-i\frac{d}{dx})$ in the space $L_2(-\infty, +\infty)$,

---

This work was supported partially by fundamental research programs and projects of the MES RK, Grant 0732/GF, 2012–2014.

consider an operator L in the Hilbert space $L_2(0,b)$, where $b<\infty$, which is generated by the differential operator $(-i\frac{d}{dx})$ and a boundary condition. We introduce the concepts of the Fourier transform and convolution generated by an arbitrary restriction of the differentiation operator in the space $L_2(0,b)$. In contrast to the classical convolution, the introduced convolution explicitly depends on the boundary condition that defines the domain of the operator L.

As noted above, the convolution is closely connected to the inverse operator or to the resolvent. So, we first find a representation for the resolvent, and then introduce the required convolution.

2. Resolvent and spectrum of the operator L

Without loss of generality, we assume that the origin belongs to the resolvent set of the operator L, that is, there is an inverse operator L^{-1}. By M. Otelbaev's theorem [3] such operators are parameterized by a "boundary" function $\sigma(x)$ from the space $L_2(0,b)$.

Theorem 2.1. *Let the action of the linear operator L in $L_2(0,b)$ be defined by formula $Ly=-i\frac{dy}{dx}$ with some (fixed) boundary condition. Suppose there exists the inverse operator L^{-1} in $L_2(0,b)$. Then there is a unique function $\sigma(x)\in L_2(0,b)$ such that the domain of operator L is*

$$D(L)=\left\{y\in W_2^1[0,b]:y(0)-\int_0^b\Big(-i\frac{dy}{dx}\Big)\overline{\sigma(x)}dx=0\right\}.$$

Proof. Let us consider equation $Ly=f$, where $f\in L_2(0,b)$. Since there is the bounded inverse operator L^{-1}. We have $y=L^{-1}f$. Each solution of the differential equation $-i\frac{dy}{dx}=f$ has the form $y=c+K^{-1}f$, where c is an arbitrary constant and the operator K corresponds to the Cauchy problem with zero condition at zero:

$$Ky=-i\frac{dy}{dx}, D(K)=\{y\in W_2^1[0,b]:y(0)=0\}.$$

Therefore, the constant $c=L^{-1}f-K^{-1}f$ is dependent on f and represents the value of a bounded linear functional on the Hilbert space $L_2(0,b)$. Then $c=c(f)$, and by the Riesz theorem on bounded linear functionals on $L_2(0,b)$, we have

$$c=\int_0^b f(x)\overline{\sigma(x)}dx, \sigma(x)\in L_2(0,b).$$

The element $\sigma(x)$ is uniquely determined. So, the solutions of the operator equation $Ly=f$ have the form $y=\int_0^b f(x)\overline{\sigma(x)}dx+(K^{-1}f)(x)$. In the last equation, we substitute $x=0$. As a result, we get

$$y(0)=\int_0^b f(x)\overline{\sigma(x)}dx.$$

The converse assertion is also true and can be verified directly.

If the function $y(x)$ from $W_2^1[0,b]$ satisfies condition

$$U(y) = 0,$$

then it will belong to the domain D_L of L, where

$$U(y) := y(0) - \int_0^b \Big(- i \frac{dy}{dx} \Big) \overline{\sigma(x)} dx.$$

Let us denote by $\Delta(\lambda)$ the entire function

$$\Delta(\lambda) = 1 - \lambda \int_0^b \exp(i\lambda x) \overline{\sigma(x)} dx.$$

Then the resolvent of the operator L is

$$\begin{aligned} (L - \lambda I)^{-1} f = {} & i \int_0^x \exp(i\lambda(x-\xi)) f(\xi) d\xi \\ & + \frac{\exp(i\lambda x)}{\Delta(\lambda)} \Big(\int_0^b f(x) \overline{\sigma(x)} dx + \lambda i \int_0^b \overline{\sigma(x)} dx \int_0^x \exp(i\lambda(x-\xi)) f(\xi) d\xi \Big). \end{aligned} \tag{2.1}$$

Indeed, denoting the right-hand side of this equality by $y(x)$, we find it is a direct consequence of

$$y'(x) = i f(x) + i\lambda y(x).$$

Let us calculate

$$\begin{aligned} U(y) - y(0) \quad & \int_0^b \Big(-i \frac{dy}{dx} \Big) \overline{\sigma(x)} dx \\ & = y(0) - \lambda \int_0^b y(x) \overline{\sigma(x)} dx - \int_0^b f(x) \overline{\sigma(x)} dx \\ & = \frac{1}{\Delta(\lambda)} \Big(\int_0^b f(x) \overline{\sigma(x)} dx + \lambda i \int_0^b \overline{\sigma(x)} dx \int_0^x \exp(i\lambda(x-\xi)) f(\xi) d\xi \Big) \\ & \quad - \lambda \int_0^b \overline{\sigma(x)} dx \Big[i \int_0^x \exp(i\lambda(x-\xi)) f(\xi) d\xi + \frac{\exp(i\lambda x)}{\Delta(\lambda)} \Big(\int_0^b f(x) \overline{\sigma(x)} dx \\ & \quad + \lambda i \int_0^b \overline{\sigma(x)} dx \int_0^x \exp(i\lambda(x-\xi)) f(\xi) d\xi \Big) \Big] - \int_0^b f(x) \overline{\sigma(x)} dx \\ & = \frac{1}{\Delta(\lambda)} \int_0^b f(x) \overline{\sigma(x)} dx + \frac{i\lambda}{\Delta(\lambda)} \int_0^b \overline{\sigma(x)} dx \int_0^x \exp(i\lambda(x-\xi)) f(\xi) d\xi \\ & \quad - \frac{i\lambda}{\Delta(\lambda)} \int_0^b \overline{\sigma(x)} dx \int_0^x \exp(i\lambda(x-\xi)) f(\xi) d\xi \\ & \quad + \frac{i\lambda^2}{\Delta(\lambda)} \int_0^b \overline{\sigma(x)} dx \int_0^x \exp(i\lambda(x-\xi)) f(\xi) d\xi \int_0^b \exp(i\lambda\mu) \overline{\sigma(\mu)} d\mu \\ & \quad - \frac{\lambda}{\Delta(\lambda)} \int_0^b \exp(i\lambda x) \overline{\sigma(x)} dx \int_0^b f(\xi) \overline{\sigma(\xi)} d\xi \end{aligned}$$

$$- \frac{i\lambda^2}{\Delta(\lambda)} \int_0^b \exp(i\lambda x)\overline{\sigma(x)}dx \int_0^b \overline{\sigma(\mu)}d\mu \int_0^\mu \exp(i\lambda(\mu-\xi))f(\xi)d\xi$$
$$- \frac{1}{\Delta(\lambda)} \int_0^b f(x)\overline{\sigma(x)}dx + \frac{\lambda}{\Delta(\lambda)} \int_0^b f(\xi)\overline{\sigma(\xi)}d\xi \int_0^b \exp(i\lambda x)\overline{\sigma(x)}dx = 0.$$

The proof is complete. □

From the representation of the resolvent by definition of the spectrum of operator L we get the following theorem.

Theorem 2.2. *The set of zeros with multiplicities of the entire function $\Delta(\lambda)$ is exactly the same as the spectrum of the operator L.*

Since $\Delta(\lambda)$ is the entire function in λ, we have that the spectrum of the operator L consists of isolated eigenvalues of finite multiplicity, and limit points of the spectrum can only be infinity. From the Paley–Wiener theorem it immediately follows that:

Theorem 2.3. *The operator L has either countable number of eigenvalues, or they are absent. The spectrum is empty if and only if there exists constant $c \in [0, b]$ such that $\sigma(x) = i$ for $0 \leq x \leq c$ and $\sigma(x) = 0$ for $c < x \leq b$.*

The proof immediately follows from the representation

$$\Delta(\lambda) = \exp(i\lambda c) - \int_0^c \exp(i\lambda x)\overline{(\sigma(x) - i)}dx - \lambda \int_c^b \exp(i\lambda x)\overline{\sigma(x)}dx$$

and from the Paley–Wiener theorem, since the presence of the integral term in the right side of this equation leads to the existence of the growing product $\exp(i\lambda c) \cdot \Delta(\lambda)$.

In what follows, suppose that the indicator diagram of the entire function $\Delta(\lambda)$ is the segment $[0, ib]$. Then the spectrum of the operator L is a countable set.

In order for the indicator diagram $\Delta(\lambda)$ to be represented by the interval $[0, ib]$, it is necessary and sufficient that

$$\min(\mathrm{supp}(\sigma(x) - i)) = 0, \;\; \max(\mathrm{supp}(\sigma(x))) = b, \tag{2.2}$$

where $\mathrm{supp}(g)$ is the support of g.

The following theorem is proved just as in the work E.Titmarsh [4].

Theorem 2.4. *Let the condition (2.2) hold. Then the number of zeros $N(r)$ of the function $\Delta(\lambda)$, which satisfy the inequality $|\lambda| \leq r$, satisfies the limit inequality*

$$\lim_{r\to\infty} \frac{N(r)}{r} = \frac{b}{\pi}.$$

In the work of M. Cartwright [5] it is shown that if $\sigma(x)$ satisfies (2.2) and it is a function of a bounded variation, then all the zeros of the function $\Delta(\lambda)$ are in a horizontal strip and the value of $(N(r) - \frac{br}{\pi})$ is bounded uniformly by r.

3. Convolution generated by operator L

To obtain a convolution, we rewrite the resolvent $(L-\lambda I)^{-1}$ in the form

$$\begin{aligned}(L-\lambda I)^{-1}f(x) &= i\int_0^x \frac{\exp(i\lambda(x-\xi))}{\Delta(\lambda)}f(\xi)d\xi \\ &\quad - \int_0^b \overline{\sigma(\mu)}d\mu\frac{\partial}{\partial\mu}\Big(\int_\mu^x \frac{\exp(i\lambda(x-\xi+\mu))}{\Delta(\lambda)}f(\xi)d\xi\Big).\end{aligned} \tag{3.1}$$

Indeed, from the previous view of the resolvent (2.1), we have the chain of equalities:

$$\begin{aligned}&(L-\lambda I)^{-1}f(x)\\ &= i\int_0^x \exp(i\lambda(x-\xi))f(\xi)d\xi \\ &\quad + \frac{\exp(i\lambda x)}{\Delta(\lambda)}\Big(\int_0^b f(x)\overline{\sigma(x)}dx + i\lambda\int_0^b\overline{\sigma(x)}dx\int_0^x \exp(i\lambda(x-\xi))f(\xi)d\xi\Big)\\ &= i\int_0^x \frac{\exp(i\lambda(x-\xi))}{\Delta(\lambda)}f(\xi)d\xi - i\lambda\int_0^b\overline{\sigma(\mu)}d\mu\int_0^x\frac{\exp(i\lambda(x-\xi+\mu))}{\Delta(\lambda)}f(\xi)d\xi\\ &\quad + \int_0^b\frac{\exp(i\lambda x)}{\Delta(\lambda)}f(\xi)\overline{\sigma(\xi)}d\xi + i\lambda\int_0^b\overline{\sigma(\mu)}d\mu\int_0^\mu\frac{\exp(i\lambda(x-\xi+\mu))}{\Delta(\lambda)}f(\xi)d\xi\\ &= i\int_0^x \frac{\exp(i\lambda(x-\xi))}{\Delta(\lambda)}f(\xi)d\xi - i\lambda\int_0^b\overline{\sigma(\mu)}d\mu\int_\mu^x\frac{\exp(i\lambda(x-\xi+\mu))}{\Delta(\lambda)}f(\xi)d\xi\\ &\quad + \int_0^b\frac{\exp(i\lambda x)}{\Delta(\lambda)}f(\xi)\overline{\sigma(\xi)}d\xi\\ &= i\int_0^x \frac{\exp(i\lambda(x-\xi))}{\Delta(\lambda)}f(\xi)d\xi - \int_0^b\overline{\sigma(\mu)}d\mu\frac{\partial}{\partial\mu}\Big(\int_\mu^x\frac{\exp(i\lambda(x-\xi+\mu))}{\Delta(\lambda)}f(\xi)d\xi\Big).\end{aligned}$$

Lemma 3.1. *The convolution, defined by the formula*

$$(g*f)(x) := i\int_0^x g(x-\xi)f(\xi)d\xi - \int_0^b\overline{\sigma(\mu)}d\mu\frac{\partial}{\partial\mu}\Big(\int_\mu^x g(x-\xi+\mu)f(\xi)d\xi\Big)$$

for $g, f \in W_2^1[0,b]$ is bilinear, commutative and associative.

Proof. Let us introduce an operation $\circ$ as

$$(g\circ f)(x,t) := \int_t^x g(x-\xi+t)f(\xi)d\xi,$$

then we can rewrite the expression $g*f$ in the form

$$\begin{aligned}(g*f) &= i(g\circ f)(x,0) - \int_0^b\frac{\partial}{\partial\mu}(g\circ f)(x,\mu)\overline{\sigma(\mu)}d\mu\\ &= i\Big[(g\circ f)(x,0) - \int_0^b\Big(-i\frac{\partial}{\partial\mu}(g\circ f)(x,\mu)\Big)\overline{\sigma_0(\mu)}d\mu\Big] = iU_t(g\circ f)(x,t),\end{aligned}$$

where

$$U(y) = y(0) - \int_0^b \Big(- i\frac{\partial y(\mu)}{\partial \mu}\Big)\overline{\sigma(\mu)}d\mu.$$

Hence, bilinearity is obvious. It is easy to see that the operation $\circ$ is commutative. Indeed

$$(g \circ f)(x,t) = \int_t^x g(x+t-\xi)f(\xi)d\xi = \begin{pmatrix} \xi' = x+t-\xi \\ d\xi = -d\xi' \\ x \to t \\ t \to x \end{pmatrix}$$

$$= -\int_x^t g(\xi')f(x+t-\xi')d\xi' = \int_t^x f(x+t-\xi)g(\xi)d\xi = (f \circ g)(x,t),$$

i.e.,

$$(g \circ f)(x,t) = (f \circ g)(x,t).$$

Then the operation $*$ is commutative too,

$$\begin{aligned}(g * f)(x) &= i(g \circ f)(x,0) - \int_0^b \frac{\partial}{\partial \mu}(g \circ f)(x,\mu)\overline{\sigma(\mu)}d\mu \\ &= i(f \circ g)(x,0) - \int_0^b \frac{\partial}{\partial \mu}(f \circ g)(x,\mu)\overline{\sigma(\mu)}d\mu = (f * g)(x).\end{aligned}$$ □

Remark 3.2. The convolution is expressed in terms of boundary conditions and has the form

$$(f * g)(x) = iU_\mu\left\{\int_\mu^x f(\xi)g(x+\mu-\xi)d\xi\right\}, \tag{3.2}$$

where

$$U(y) := y(0) - \int_0^b \Big(- i\frac{dy}{dx}\Big)\overline{\sigma(x)}dx.$$

Remark 3.3. From (3.1) it is easy to see that the resolvent in terms of convolution is represented in the form

$$(L - \lambda I)^{-1} f = \frac{\exp(i\lambda x)}{\Delta(\lambda)} * f(x). \tag{3.3}$$

Lemma 3.4. *For any f from the domain D_L and for arbitrary g from $W_2^1[0,b]$, the equality*

$$\frac{d}{dx}(f * g) = \frac{df}{dx} * g$$

holds true.

Proof. Since the equality $U(f) = 0$ holds for $f \in D_L$, we get

$$\begin{aligned}\frac{d}{dx}(f * g) &= if(0)g(x) + i\int_0^x \frac{df(x-\xi)}{dx}g(\xi)d\xi \\ &\quad - \int_0^b \overline{\sigma(\mu)}d\mu\frac{\partial}{\partial \mu}\Big(f(\mu)g(x) + \int_\mu^x \frac{\partial}{\partial x}f(x-\xi+\mu)g(\xi)d\xi\Big)\end{aligned}$$

$$= \frac{df}{dx} * g + if(0)g(x) - \int_0^b \overline{\sigma(x)} \frac{\partial}{\partial \mu} f(\mu) g(x) d\mu$$

$$= \frac{df}{dx} * g + ig(x)U(f) = \frac{df}{dx} * g. \qquad \square$$

Lemma 3.5. *The convolution defined in Lemma 3.1 is without annihilators, i.e., if for arbitrary $g \in W_2^1[0,b]$ and for some $f \in L_2(0,b)$ the equality $(g * f)(x) = 0$ holds, then f is identically equal to zero.*

Proof. Let $(g * f)(x) = 0$, $f \in L_2(0,b)$, then we take

$$g := \left. \frac{\exp(i\lambda x)}{\Delta(\lambda)} \right|_{\lambda=0} = 1.$$

By the definition of our convolution, the expression $1 * f = 0$ denotes the equality $L^{-1}f = 0$. Denoting $L^{-1}f$ by y, we obtain respectively $Ly = f$, but $y = 0$, hence, as the L a linear operator, it follows that $f = 0$. $\square$

Lemma 3.6. *If the function f is from the domain D_L of the operator L, then for each $g \in L_2(0,b)$ the convolution $f * g$ will also belong to the domain D_L.*

Proof. Let

$$f := \left. \frac{\exp(i\lambda_0 x)}{\Delta(\lambda_0)} \right|_{\lambda_0=0} = 1.$$

Let us denote by y the function $y - 1 * y$, by construction of the convolution this expression means that $y = L^{-1}g$, i.e., $Ly - g$, which implies $y \in D_L$. Now we fix any λ such that $\Delta(\lambda) \neq 0$. Let $f := \frac{\exp(i\lambda x)}{\Delta(\lambda)}$. If by y we denote $y = 1 * g$, then this expression by the convolution construction mean that $y = L^{-1}g$, i.e., $Ly = g$, and implies $y \in D_L$. Now we fix any λ, such that $\Delta(\lambda) \neq 0$. Let $f := \frac{\exp(i\lambda x)}{\Delta(\lambda)}$. Let y denote the convolution $y = \frac{\exp(i\lambda x)}{\Delta(\lambda)} * g$, then from the definition of the convolution, we get $y = (L - \lambda I)^{-1} g$, i.e., $Ly = \lambda y + g$ for any λ. Since zeros of the function $\Delta(\lambda)$ are countable set, then there is sequence $\{\lambda_n\}$ such that $\Delta(\lambda_n) \neq 0$ and the system $\exp(i\lambda_n x)$ is basis in the space $W_2^1[0,b]$. For any natural number n it is λ_n such that the convolution $\exp(i\lambda_n x) * g$ will belong to the domain D_L. From $f(x) = \sum_{n=0}^{\infty} c_n \exp(i\lambda_n x)$ and from the bilinearity of the convolution, the convolution $f * g$ belongs to the domain D_L. $\square$

Lemma 3.7. *For arbitrary λ and β such that $\lambda \neq \beta$ we have the equality*

$$\exp(i\lambda x) * \exp(i\beta x) = \frac{\exp(i\beta x)\Delta(\lambda) - \exp(i\lambda x)\Delta(\beta)}{\beta - \lambda}.$$

Proof. We write by definition

$$\exp(i\lambda x) * \exp(i\beta x)$$
$$= i \int_0^x \exp(i\lambda(x - \xi)) \exp(i\beta\xi) d\xi$$

$$
\begin{aligned}
&-\int_0^b \overline{\sigma(\mu)}d\mu\frac{\partial}{\partial\mu}\left(\int_\mu^x \exp(i\lambda(x-\xi+\mu))\exp(i\beta\xi)d\xi\right)\\
&=\frac{i\exp(i\lambda x)(\exp(i(\beta-\lambda)x)-1)}{i(\beta-\lambda)}\\
&-\int_0^b \overline{\sigma(\mu)}d\mu\frac{\partial}{\partial\mu}\left(\exp(i\lambda(x+\mu))\frac{\exp(i(\beta-\lambda)x)-\exp(i(\beta-\lambda)\mu)}{i(\beta-\lambda)}\right)\\
&=\frac{\exp(i\beta x)-\exp(i\lambda x)}{\beta-\lambda}-\\
&-\int_0^b \overline{\sigma(\mu)}d\mu\frac{\partial}{\partial\mu}\left[\frac{\exp(i\lambda\mu)\exp(i\beta x)-\exp(i\lambda x)\exp(i\beta\mu)}{i(\beta-\lambda)}\right]\\
&=\frac{\exp(i\beta x)-\exp(i\lambda x)}{\beta-\lambda}-\\
&-\int_0^b \frac{i\lambda\exp(i\lambda\mu)\exp(i\beta x)-i\beta\exp(i\lambda x)\exp(i\beta\mu)}{i(\beta-\lambda)}\overline{\sigma(\mu)}d\mu\\
&=\frac{1}{\beta-\lambda}\left[\exp(i\beta x)-\exp(i\lambda x)-\lambda\exp(i\beta x)\int_0^b \exp(i\lambda\mu)\overline{\sigma(\mu)}d\mu\right.\\
&\left.+\beta\exp(i\lambda x)\int_0^b \exp(i\beta\mu)\overline{\sigma(\mu)}d\mu\right]=\frac{\exp(i\beta x)\Delta(\lambda)-\exp(i\lambda x)\Delta(\beta)}{\beta-\lambda}. \quad \square
\end{aligned}
$$

4. Fourier transform and convolution

For any function f from the space $L_2(0,b)$, let us associate the expansion

$$f \sim \sum_{\lambda_n\in\sigma(L)} P_n f,$$

where the orthogonal projection is given by

$$P_n f = -\frac{1}{2\pi i}\oint_{|\lambda-\lambda_n|=\delta}(L-\lambda I)^{-1} f d\lambda.$$

By using (3.3) we obtain

$$
\begin{aligned}
P_n f &= \operatorname{res}_{\lambda_n}\frac{\exp(i\lambda x)}{\Delta(\lambda)} * f\\
&=\left[\frac{1}{(m_n-1)!}\lim_{\lambda\to\lambda_n}\frac{\partial^{m_n-1}}{\partial\lambda^{m_n-1}}\Big(\frac{(\lambda-\lambda_n)^{m_n}}{\Delta(\lambda)}\exp(i\lambda x)\Big)\right] * f(x)\\
&=\left[\sum_{j=0}^{m_n-1}\frac{d_{j,n}}{j!}\Big(\frac{(ix)^{m_n-1-j}}{(m_n-1-j)!}\exp(i\lambda_n x)\Big)\right] * f(x).
\end{aligned}
$$

Let us define

$$u_{m_n-1,n} := \sum_{j=0}^{m_n-1} \frac{d_{j,n}}{j!} \left(\frac{(ix)^{m_n-1-j}}{(m_n-1-j)!} \exp(i\lambda_n x) \right), \tag{4.1}$$

where $\{d_{j,n}\}$ are the expansion coefficients in the Taylor series of the function $\frac{(\lambda-\lambda_n)^{m_n}}{\Delta(\lambda)}$ in front of powers of $(\lambda - \lambda_n)$. Then

$$P_n f = u_{m_n-1,n} * f. \tag{4.2}$$

Let us introduce a system of root functions of the operator L corresponding to the eigenvalue λ_n by the expression

$$\begin{aligned} u_{s,n} &:= L u_{s+1,n} - \lambda_n u_{s+1,n} \\ &= \sum_{j=0}^{s} \frac{d_{j,n}}{j!} \frac{(ix)^{s-j}}{(s-j)!} \exp(i\lambda_n x), \quad s = 0, \dots, m_n - 2. \end{aligned} \tag{4.3}$$

Lemma 4.1. *The system of root functions $\{u_{k,n}, k = 0, \dots, m_n - 1\}$ is linearly independent.*

Proof. Consider the linear combination

$$\alpha_0 u_{0,n} + \cdots + \alpha_{m_n-2} u_{m_n-2,n} + \alpha_{m_n-1} u_{m_n-1,n} = 0. \tag{4.4}$$

It is easy to see that

$$(L - \lambda_n I)^{m_n-1} u_{s,n} = 0. s = 0, \dots, m_n - 2. \tag{4.5}$$

By applying to (4.4) the operator $(L - \lambda_n I)^{m_n-1}$ on both sides, we get

$$\alpha_{m_n-1} (L - \lambda_n I)^{m_n-1} u_{m_n-1,n} = 0. \tag{4.6}$$

From (4.3) we see that (4.6) is equivalent to

$$\alpha_{m_n-1} u_{0,n} = 0.$$

Since $d_{0,n} \neq 0$, we obtain

$$\alpha_{m_n-1} = 0.$$

Now we consider the equality (4.4) with $\alpha_{m_n-1} = 0$, i.e.,

$$\alpha_0 u_{0,n} + \cdots + \alpha_{m_n-2} u_{m_n-2,n} = 0.$$

Applying to both sides the operator $(L - \lambda_n I)^{m_n-2}$ and using (4.5) as in the previous step, we get

$$\alpha_{m_n-2} u_{0,n} = 0,$$

whence $\alpha_{m_n-2} = 0$.

Repeating these steps, we find that equality (4.4) is true only for $\alpha_k = 0, k = 0, \dots, m_n - 1$. This proves the linear independence of the system $\{u_{k,n}, k = 0, \dots, m_n - 1\}$.

Since $d_{0,n} \neq 0$, the system $u_{0,n}, u_{1,n}, \dots, u_{m_n-1,n}$ is linearly independent and, therefore, there is a basis of the root subspace

$$H_{\lambda_n} := \mathrm{Ker}(L - \lambda_n I)^{m_n}.$$

We expand the function $P_n f$ on $\mathrm{Ker}(L-\lambda_n I)^{m_n}$ by this basis

$$P_n f = C_{m_n-1,n}(f)u_{0,n} + C_{m_n-2,n}(f)u_{1,n} + \cdots + C_{0,n}(f)u_{m_n-1,n}.$$

Thus, for each element f of the space $L_2(0,b)$ we associate the element of the sequence space

$$\{(C_{0,n}(f),\ldots,C_{m_n-1,n}(f)),\lambda_n\in\sigma(L)\}\in\prod_{\lambda_n\in\sigma(L)}\mathbb{C}^{m_n},$$

i.e., we introduce the Fourier transform

$$\widehat{f}=\{(C_{0,n}(f),\ldots,C_{m_n-1,n}(f)),\lambda_n\in\sigma(L)\}.$$

In the space of sequences, we introduce the inner Cauchy convolution. Let ξ and η be any elements of the $X:=\prod_{\lambda_n\in\sigma(L)}\mathbb{C}^{m_n}$, then we will call their convolution the sequence

$$\mu:=\left\{\left(\xi_{0,n}\eta_{0,n};\xi_{0,n}\eta_{1,n}+\xi_{1,n}\eta_{0,n};\ldots,\sum_{k=0}^{m_n-1}\xi_{k,n}\eta_{m_n-1-k,n}\right),\lambda_n\in\sigma(L)\right\},$$

which we denote by $\xi *_X \eta$. Introduced convolutions $*_X$ and $*$ are associated between themselves by the Fourier transform. $\square$

Theorem 4.2. *For arbitrary functions f and g from the space $W_2^1[0,b]$ the equality*

$$(f*g)^{\frown}=\widehat{f}*_X\widehat{g} \tag{4.7}$$

holds.

Proof. Let $f=u_{m_n-1,n}$. Then the equality $(u_{m_n-1,n}*g)^{\frown}=\widehat{u}_{m_n-1,n}*_X\widehat{g}$ follows from the following chain of equalities

$$\begin{aligned}(u_{m_n-1,n}*g)^{\frown}&=(P_ng)^{\frown}=(C_{0,n}(g),C_{1,n}(g),\ldots,C_{m_n-1,n}(g))\\&=(1,0,\ldots,0)*_X(C_{0,n}(g),C_{1,n}(g),\ldots,C_{m_n-1,n}(g)).\end{aligned}$$

Let now $f=u_{s,n}$, then

$$\begin{aligned}(u_{s,n}*g)^{\frown}&=((L-\lambda_nI)^{m_n-1-s}u_{m_n-1,n}*g)^{\frown}=((L-\lambda_nI)^{m_n-1-s}(u_{m_n-1,n}*g))^{\frown}\\&=((L-\lambda_nI)^{m_n-1-s}P_ng))^{\frown}=(P_n(L-\lambda_nI)^{m_n-1-s}g))^{\frown}\\&=(C_{0,n}(L-\lambda_nI)^{m_n-1-s}g)),C_{1,n}(L-\lambda_nI)^{m_n-1-s}g)),\ldots\\&\quad\ldots,C_{m_n-1,n}(L-\lambda_nI)^{m_n-1-s}g)).\end{aligned}$$

On the other hand, the relations

$$\begin{gathered}C_{m_n-1,n}((L-\lambda_nI)g)=C_{m_n-2,n}(g),\\ \ldots,\\ C_{1,n}((L-\lambda_nI)g)=C_{0,n}(g)\end{gathered}$$

and

$$C_{0,n}((L-\lambda_nI)g)=0$$

are valid, i.e., the action of the operator $(L - \lambda_n I)$ is equivalent to a shift in the sequence space. Then action of the operator $(L - \lambda_n I)^{m_n - 1 - s}$ corresponds to a shift to the right on $m_n - 1 - s$ position. This implies the equality

$$(u_{s,n} * g)\widehat{\ } = \widehat{u_{s,n}} *_X \widehat{g}, \quad s = 0, \dots, m_n - 1.$$

From the fact that the theorem holds for all elements of the basis $\{u_{s,n}\}$ follows the required equality (4.7) follow in the whole space. □

5. Coefficient functionals and a boundary condition

Elements of the basis of the each root subspace have good properties.

Lemma 5.1. *Let λ_n be zeros of the entire function $\Delta(\lambda)$ with corresponding multiplicities m_n. Then the elements (4.1) and (4.3) of the root subspace H_{λ_n} have the following properties:*

$$u_{p,n} * u_{q,n} = \begin{cases} 0, & \text{for } p + q < m_n - 1, \\ u_{p+q-m_n+1}, & \text{for } p + q \geq m_n - 1, \end{cases} \tag{5.1}$$

$0 \leq p, q \leq m_n - 1$.

Proof. At first, we note that the element (4.1) is idempotent with respect to the convolution, i.e.,

$$u_{m_n-1,n} * u_{m_n-1,n} = u_{m_n-1,n}.$$

By acting on the element $u_{m_n-1,n}$ of the basis of root subspace of the operator P_n, we get

$$P_n(u_{m_n-1,n}) = u_{m_n-1,n}.$$

On the other hand, by replacing the function f by $u_{m_n-1,n}$ in the formula (4.2), we have

$$P_n(u_{m_n-1,n}) = u_{m_n-1,n} * u_{m_n-1,n},$$

which proves the property that the element $u_{m_n-1,n}$ is idempotent. Further, from (4.3) it is easy to see that each basis element can be presented by an idempotent element $u_{m_n-1,n}$ in the form

$$u_{q,n} = (L - \lambda_n I)^{m_n-1-q} u_{m_n-1,n}, q = 0, \dots, m_n - 2. \tag{5.2}$$

Then consider the convolution of two elements of the basis

$$\begin{aligned} u_{q,n} * u_{p,n} &= [(L - \lambda_n I)^{m_n-1-q} u_{m_n-1,n}] * [(L - \lambda_n I)^{m_n-1-p} u_{m_n-1,n}] \\ &\quad \text{(by using bilinear property of the convolution } *) \\ &= (L - \lambda_n I)^{2m_n-2-q-p} (u_{m_n-1,n} * u_{m_n-1,n}) \\ &\quad \text{(by using idempotence property of the element } u_{m_n-1,n}) \\ &= (L - \lambda_n I)^{m_n-1-(q+p-m_n+1} u_{m_n-1,n}. \end{aligned}$$

It is clear that if the inequality $q+p-m_n+1 \geq 0$ holds, then from (5.2) it follows that

$$u_{p,n} * u_{q,n} = u_{p+q-m_n+1,n}$$

and otherwise

$$u_{p,n} * u_{q,n} = 0. \qquad \square$$

In the following lemma we give the coefficient functionals in the boundary condition.

Lemma 5.2. *The coefficient functionals in the projector expansion*

$$P_n f = \sum_{k=0}^{m_n-1} C_{m_n-1-k,n}(f) u_{k,n}$$

in the root subspace $\operatorname{Ker}(L-\lambda_n I)^{m_n}$ *are of the form*

$$C_{k,n}(f) = -iU_\mu \left\{ \int_0^\mu f(\xi) \frac{(i(\mu-\xi))^k}{k!} \exp(i\lambda_n(\mu-\xi)) d\xi \right\}, 0 \leq k \leq m_n - 1.$$

Proof. By the definition we have $P_n f = f * u_{m_n-1,n}$, but on other hand $P_n f = C_{m_n-1,n}(f)u_{0,n} + C_{m_n-2,n}(f)u_{1,n} + \cdots + C_{0,n}(f)u_{m_n-1,n}$. Then we get the equality

$$f * u_{m_n-1,n} = \sum_{k=0}^{m_n-1} C_{m_n-1-k,n}(f) u_{k,n}. \tag{5.3}$$

As for $u_{0,n}$, we have

$$f * u_{m_n-1,n} * u_{0,n} = \sum_{k=0}^{m_n-1} C_{m_n-1-k,n}(f) u_{k,n} * u_{0,n},$$

then by using Lemma (4.1), we get

$$f * u_{0,n} = C_{0,n}(f) u_{0,n}. \tag{5.4}$$

Let us rewrite the left side of (5.4) by using formula (3.2)

$$\begin{aligned} f * u_{0,n} &= -iU_\mu \left\{ \int_\mu^x f(\xi) u_{0,n}(x+\mu-\xi) d\xi \right\} \\ &= -iU_\mu \left\{ \int_0^x f(\xi) u_{0,n}(x+\mu-\xi) d\xi \right\} - iU_\mu \left\{ \int_0^\mu f(\xi) u_{0,n}(x+\mu-\xi) d\xi \right\}. \end{aligned}$$

Since

$$\begin{aligned} &iU_\mu \left\{ \int_0^x f(\xi) d_{0,n} \exp(i\lambda_n(x+\mu-\xi)) d\xi \right\} \\ &\quad = d_{0,n} \exp(i\lambda x) i \int_0^x f(\xi) \exp(-i\lambda_n \xi) d\xi \left[1 - \lambda_n \int_0^b \exp(i\lambda_n \mu) \overline{\sigma(\mu)} d\mu \right] \\ &\quad = u_{0,n} i \Delta(\lambda_n) \int_0^x f(\xi) \exp(-i\lambda_n \xi) d\xi, \end{aligned}$$

the first term is equal to zero. And

$$\begin{aligned} f * u_{0,n} &= -iU_\mu\left\{\int_0^\mu f(\xi)u_{0,n}(x+\mu-\xi)d\xi\right\} \\ &= -iU_\mu\left\{\int_0^\mu f(\xi)d_{0,n}\exp(i\lambda_n(x+\mu-\xi))d\xi\right\} \\ &= -d_{n,0}\exp(-i\lambda_n\xi)iU_\mu\left\{\int_0^\mu f(\xi)\exp(i\lambda_n(\mu-\xi))d\xi\right\} \\ &= -iU_\mu\left\{\int_0^\mu f(\xi)\exp(i\lambda_n(\mu-\xi))d\xi\right\}u_{0,n}. \end{aligned}$$

By comparing the right- and the left-hand side of (5.4), we get

$$C_{0,n}(f) = -iU_\mu\left\{\int_0^\mu f(\xi)\exp(i\lambda_n(\mu-\xi))d\xi\right\}.$$

Now suppose that

$$C_{k,n}(f) = -iU_\mu\left\{\int_0^\mu f(\xi)\frac{(i(\mu-\xi))^k}{k!}\exp(i\lambda_n(\mu-\xi))d\xi\right\},$$

$k = 0 : s-1, 1 \le s \le m_n - 1$. Let us prove that

$$C_{s,n}(f) = -iU_\mu\left\{\int_0^\mu f(\xi)\frac{(i(\mu-\xi))^s}{s!}\exp(i\lambda_n(\mu-\xi))d\xi\right\}, 1 \le s \le m_n - 1.$$

By convolving the both sides of (5.3) with $u_{s,n}$, we have

$$f * u_{m_n-1,n} * u_{s,n} = \sum_{k=0}^{m_n-1} C_{m_n-1-k,n}(f)u_{k,n} * u_{0,n} \tag{5.5}$$

The equality (5.5) is changed to

$$f * u_{s,n} = \sum_{k=m_n-1-s}^{m_n-1} C_{m_n-1-k,n}(f)u_{k+s-m_n+1,n}.$$

Let us relate the index k by $l = m_n - 1 - k$, then we get

$$f * u_{s,n} = \sum_{l=0}^{s} C_{l,n}(f)u_{s-1,n}. \tag{5.6}$$

As in the previous case by using formula (3.2), we write

$$\begin{aligned} f * u_{s,n} &= iU_\mu\left\{\int_\mu^x f(\xi)u_{s,n}(x+\mu-\xi)d\xi\right\} \\ &= iU_\mu\left\{\int_0^x f(\xi)u_{s,n}(x+\mu-\xi)d\xi\right\} - iU_\mu\left\{\int_0^\mu f(\xi)u_{s,n}(x+\mu-\xi)d\xi\right\}. \end{aligned}$$

Since

$$iU_\mu\left\{\int_0^x f(\xi)\sum\nolimits_{j=0}^{s}\frac{d_{j,n}}{j!}\frac{(i(x+\mu-\xi))^{s-j}}{(s-j)!}\exp(i\lambda_n(x+\mu-\xi))d\xi\right\}$$
$$=\sum_{j=0}^{s}\frac{d_{j,n}}{j!}\exp(i\lambda_n x)\sum_{p=0}^{s-j}\frac{(ix)^{s-j-p}}{(s-j-p)!}$$
$$\times iU_\mu\left\{\int_0^x f(\xi)\frac{(i(\mu-\xi))^p}{p!}\exp(i\lambda_n(\mu-\xi))d\xi\right\}$$
$$=\sum_{j=0}^{s}\frac{\partial^j}{\partial\lambda^j}G(f;x,\lambda)|_{\lambda=\lambda_n}u_{s-j,n}$$

the first term is equal to zero, here

$$G(f;x,\lambda):=i\Delta(\lambda)\int_0^x f(\xi)\exp(-i\lambda\xi)d\xi.$$

It is obvious that if λ_n is zero of the function $\Delta(\lambda)$, then

$$\frac{\partial^j}{\partial\lambda^j}G(f;x,\lambda)|_{\lambda=\lambda_n}=0,\ \ j=0,\dots,m_n-1.$$

We have

$$f*u_{s,n}=-iU_\mu\left\{\int_0^\mu f(\xi)u_{s,n}(x+\mu-\xi)d\xi\right\}$$
$$=-iU_\mu\left\{\int_0^\mu f(\xi)\sum\nolimits_{j=0}^{s}\frac{d_{j,n}}{j!}\frac{(i(x+\mu-\xi))^{s-j}}{(s-j)!}\exp(i\lambda_n(x+\mu-\xi))d\xi\right\}$$
$$=-\sum_{j=0}^{s}\frac{d_{j,n}}{j!}\exp(i\lambda_n x)\sum_{p=0}^{s-j}\frac{(ix)^{s-j-p}}{(s-j-p)!}$$
$$\times iU_\mu\left\{\int_0^\mu f(\xi)\frac{(i(\mu-\xi))^p}{p!}\exp(i\lambda_n(\mu-\xi))d\xi\right\}$$
$$=\sum_{j=0}^{s}C_{j,n}(f)u_{s-j,n}-iU_\mu\left\{\int_0^\mu f(\xi)\frac{(i(\mu-\xi))^s}{s!}\exp(i\lambda_n(\mu-\xi))d\xi\right\}u_{0,n}.$$

By comparing the right- and the left-hand sides of the equality (5.6), we get

$$C_{s,n}(f)=-iU_\mu\left\{\int_0^\mu f(\xi)\frac{(i(\mu-\xi))^s}{s!}\exp(i\lambda_n(\mu-\xi))d\xi\right\}.\qquad\square$$

Hence we have explicitly constructed a biorthogonal system to the system $\{u_{k,n},\lambda_n\in\sigma(L)\}$.

Theorem 5.3. *The chosen basis system $\{u_{k,n},k=0,\dots,m_n-1,\lambda_n\in\sigma(L)\}$ is minimal in $L_2(0,b)$, i.e., there exists a biorthogonal system of the form*

$$h_{k,n}(\xi)=\int_\xi^b\overline{\sigma(\mu)}\frac{\partial}{\partial\mu}\Big(\frac{(i(\mu-\xi))^k}{k!}\exp(i\lambda_n(\mu-\xi))\Big)d\mu,k=0,\dots,m_n-1.$$

6. On the Sedletskiy formula for the remainder term

It is well known that the partial sum of the Fourier series is written in the form

$$S_R(f;x) = -\frac{1}{2\pi i}\oint_{|\lambda|=R}(L-\lambda I)^{-1} f d\lambda = \sum_{|\lambda_n|<R} P_{\lambda_n} f.$$

By using (3.3), we rewrite the partial sum as

$$S_R(f;x) = -\frac{1}{2\pi i}\oint_{|\lambda|=R}\frac{\exp(i\lambda x)}{\Delta(\lambda)} * f(x)d\lambda.$$

In this formula instead of f we put the function $\exp(i\mu x)$. Then we have

$$\begin{aligned}
S_R(\exp(i\mu x);x) &= -\frac{1}{2\pi i}\oint_{|\lambda|=R}\frac{\exp(i\lambda x)}{\Delta(\lambda)} * \exp(i\mu x)d\lambda \\
&\quad \text{(by applying Lemma 3.7)} \\
&= \frac{1}{2\pi i}\oint_{|\lambda|=R}\frac{\exp(i\lambda x)\Delta(\mu) - \exp(i\mu x)\Delta(\lambda)}{\Delta(\lambda)(\mu-\lambda)}d\lambda \\
&= \frac{\Delta(\mu)}{2\pi i}\oint_{|\lambda|=R}\frac{\exp(i\lambda x)}{\Delta(\lambda)}\frac{d\lambda}{\mu-\lambda} + \frac{1}{2\pi i}\oint_{|\lambda|=R}\frac{\exp(i\mu x)}{\lambda-\mu}d\lambda \\
&\quad \text{(by using the Cauchy formula to the second term)} \\
&= \exp(i\mu x) + \frac{\Delta(\mu)}{2\pi i}\oint_{|\lambda|=R}\frac{\exp(i\lambda x)}{\Delta(\lambda)}\frac{d\lambda}{\mu-\lambda}.
\end{aligned}$$

Hence it follows that the formula for the remainder term for $f(x) = \exp(i\mu x)$ is

$$\begin{aligned}
Q_R(\exp(i\mu x);x) &:= S_R(\exp(i\mu x);x) - \exp(i\mu x) \\
&= \frac{\Delta(\mu)}{2\pi i}\oint_{|\lambda|=R}\frac{\exp(i\lambda x)}{\Delta(\lambda)}\frac{d\lambda}{\mu-\lambda},
\end{aligned} \tag{6.1}$$

which is valid for any μ. Since an arbitrary element f of the space $L_2(0,b)$ is represented in the form

$$f(x) = \frac{1}{2\pi}\int_{-\infty}^{\infty}\hat{f}(-\mu)\exp(i\mu x)d\mu,$$

where $\widehat{f}(\mu)$ is the Fourier transform of the function $f(x)$, from (6.1), we obtain the integral form of the remainder term for an arbitrary function $f(x)$ in the space $L_2(0,b)$,

$$\begin{aligned}
Q_R(f;x) &:= S_R(f(x);x) - f(x) \\
&= \frac{1}{2\pi i}\oint_{|\lambda|=R}\frac{\exp(i\lambda x)}{\Delta(\lambda)}d\lambda\left(\frac{1}{2\pi}\int_{-\infty}^{\infty}\frac{\hat{f}(-\mu)\Delta(\mu)}{\mu-\lambda}d\mu\right).
\end{aligned}$$

A similar formula was proved in another way by A.M. Sedletskiy in the work [6]. Let us write the set of conditions

$$\text{(A)} \qquad \begin{cases} \max \operatorname{supp} \sigma(x) = b, \\ \min \operatorname{supp}(\sigma(x) + i) = 0, \end{cases}$$

$$\text{(B)} \qquad \sup |\mathrm{Im}\lambda_n| = M < \infty,$$

$$\text{(C)} \qquad \inf_{n \neq k} |\lambda_n - \lambda_k| > 0,$$

$$\text{(D)} \qquad \sup m_n = m < \infty.$$

Denote $\omega(\lambda) = |\Delta(\lambda)|^2$. Let us write the Muckenhoupt condition

$$\text{(E)} \qquad \sup_{l} \left(\frac{1}{|I|} \int_I \omega(\lambda) d\lambda \right) \left(\frac{1}{|I|} \int_I \omega^{-1}(\lambda) d\lambda \right) < \infty,$$

where I an arbitrary interval of the real axis. If the operator

$$S^+ : \sum_{\lambda_n \in \Lambda} P_n(x) \exp(i\lambda_n x) \mapsto \sum_{Re\lambda_n > 0} P_n(x) \exp(i\lambda_n x)$$

is bounded in the space $L_2(0,b)$, then we will say that the basis $\{u_{k,n}, k = 0, \dots, m_n - 1, \lambda_n \in \sigma(L)\}$ of the space $L_2(0,b)$ is Riesz basis.

The following theorems are true.

Theorem 6.1. *Let the conditions* (A), (B), (C), (D) *be valid and assume that the function* $\omega(\lambda) = |\Delta(\lambda)|^2$ *satisfies the Muckenhoupt condition* (E). *Then the system* $\{u_{k,n}, k = 0, \dots, m_n - 1, \lambda_n \in \sigma(L)\}$ *is a Riesz basis in the space* $L_2(0,b)$.

Theorem 6.2. *Assume the conditions of Theorem* 6.1, *except, perhaps, the condition* (D). *Then for every* $f \in L_2(0,b)$, *we have*

$$\|x^{\frac{1}{2}}(b-x)^{\frac{1}{2}} Q_r(f,x)\|_{C[0,b]} \to 0$$

continuously for $r \to \infty$.

Theorem 6.3. *Suppose the conditions of Theorem* 6.1. *Then for every* $f \in L_2(0,b)$ *the coefficient sequence satisfies*

$$\left\{ \{c_{k,n}\}_{k=0}^{\infty} \right\}_{n=0}^{\infty} \in l_2,$$

and $\|\{c_{k,n}\}\|_{l_2} \le C(L)\|f\|_{L_2}$.

References

[1] M. Ruzhansky, V. Turunen, *Pseudo-Differential Operators and Symmetries.* Birkhäuser, 2010.

[2] M. Ruzhansky, V. Turunen, *Quantization of Pseudo-differential Operators on the Torus.* J. Fourier Anal. Appl., **16** (2010), pp. 943–982.

[3] B.E. Kanguzhin and M.A. Sadybekov, *Differential operators on a segment. Distributions of the eigenvalues.* Gylym, Almaty, 1996 [in Russian].

[4] E.C. Titchmarsh, *The zeros of certain integral functions.* Proc. London Math. Soc. 1926, V. 25, No. 4, pp. 283–302.

[5] M.L. Cartwright, *The zeros of certain integral functions.* The Quarterly Journal of Math., Oxford series, 1930, V. 1, No. 1, pp. 38–59.

[6] A.M. Sedletskiy, *On biortogonal expansions of functions to exponent series on a real segment.* Usp. Mat. Nauk, 1982, V. 37, No. 5 (227), pp. 51–95 [in Russian].

Baltabek Kanguzhin
Department of Fundamental Mathematics
Faculty of Mechanics and Mathematics
Al-Farabi Kazakh National University
Al-Farabi ave., 71
500040, Almaty, Kazakhstan
e-mail: kanbalta@mail.ru

Niyaz Tokmagambetov
Department of Fundamental Mathematics
Faculty of Mechanics and Mathematics
Al-Farabi Kazakh National University
Al-Farabi ave., 71
500040, Almaty, Kazakhstan

and

Institute of Mathematics and Mathematical Modeling
Shevchenko str., 28
500010, Almaty, Kazakhstan
e-mail: tokmagam@list.ru

Fourier Analysis
Trends in Mathematics, 253–262

Global Existence for Quadratically Perturbed Massless Dirac Equations Under the Null Condition

Soichiro Katayama and Hideo Kubo

Dedicated to the memory of Professor Rentaro Agemi

Abstract. The aim of this note is to prove a global existence result for some nonlinear Dirac equations with small initial data, when the nonlinearity has a certain algebraic structure.

Mathematics Subject Classification (2010). Primary 35L03; Secondary 58J37.

Keywords. Nonlinear Dirac equation, null condition, global solution.

1. Introduction

In this paper we consider the Cauchy problem for nonlinear massless Dirac equation:

$$\mathcal{D}\psi \equiv (\gamma^0\partial_t + \gamma^j\partial_j)\psi = F(\psi), \qquad (t,x) \in (0,\infty)\times\mathbb{R}^3, \tag{1.1}$$

$$\psi(0,x) = \varepsilon\psi_0(x), \qquad x \in \mathbb{R}^3, \tag{1.2}$$

where $\psi(t,x)$ is a $\mathbb{C}^4$-valued unknown function, $\partial_0 = \partial_t = \partial/\partial t$, $\partial_j = \partial/\partial x_j$ $(j = 1,2,3)$, and γ^μ $(\mu = 0,1,2,3)$ are the Dirac matrices expressed as follows:

$$\gamma^0 = \begin{pmatrix} \sigma^0 & 0 \\ 0 & -\sigma^0 \end{pmatrix}, \quad \gamma^j = \begin{pmatrix} 0 & \sigma^j \\ -\sigma^j & 0 \end{pmatrix} \ (j = 1,2,3).$$

Here σ^μ $(\mu = 0,1,2,3)$ are the Pauli matrices:

$$\sigma^0 = \begin{pmatrix} 1 & 0 \\ 0 & 1 \end{pmatrix}, \ \sigma^1 = \begin{pmatrix} 0 & 1 \\ 1 & 0 \end{pmatrix}, \ \sigma^2 = \begin{pmatrix} 0 & -i \\ i & 0 \end{pmatrix}, \ \sigma^3 = \begin{pmatrix} 1 & 0 \\ 0 & -1 \end{pmatrix},$$

where $i = \sqrt{-1}$. Throughout this paper we use the standard summation convention of repeated upper and lower indices: For example, the expression $\gamma^j\partial_j$ means

$\sum_{j=1}^{3}\gamma^j\partial_j$. It is easy to check that for $\mu,\nu=0,1,2,3$ and $j=1,2,3$

$$\gamma^\mu\gamma^\nu=-\gamma^\nu\gamma^\mu\ (\mu\neq\nu),\ (\gamma^0)^2=I,\ (\gamma^0)^*=\gamma^0,\ \ (\gamma^j)^2=-I,\ (\gamma^j)^*=-\gamma^j$$

hold. Here A^* denotes the complex conjugate transpose of A. In addition, we assume that $\varepsilon>0$, $\psi_0\in\mathcal{S}(\mathbb{R}^3;\mathbb{C}^4)$, where $\mathcal{S}$ denotes the set of rapidly decreasing functions, and

$$F(\psi)=O(|\psi|^p),\quad p>1. \tag{1.3}$$

It was shown in Tzvetkov [6] that if $p>2$, then the problem (1.1)–(1.2) admits a unique global solution for sufficiently small ε. It was also conjectured that if $p=2$, then such a global existence result does not hold for some nonlinearity. This conjecture seems to remain open for smooth data, though it was shown in [6] that we can take some $\psi_0\in L^2$ such that the solution to the problem (1.1)–(1.2) with $F(\psi)=|\psi|\gamma^0\psi$ blows up in finite time, however small ε is. Nevertheless, the following interesting conjecture was given in the same paper: The problem (1.1)–(1.2) has a global solution for small ε, provided that the nonlinearity takes such special forms as $S_0(\psi,\psi)e$ or $S_1(\psi,\psi)e$ with an arbitrary constant vector $e\in\mathbb{C}^4$, where

$$S_0(\phi,\psi)=\langle\phi,\gamma^0\psi\rangle,\quad S_1(\phi,\psi)=\langle\phi,\gamma^0\gamma^5\psi\rangle, \tag{1.4}$$

$\langle\,\cdot\,,\,\cdot\,\rangle$ is the inner product on $\mathbb{C}^4$, and

$$\gamma^5=\begin{pmatrix}0&\sigma^0\\ \sigma^0&0\end{pmatrix}.$$

We note that $\gamma^\mu\gamma^5=-\gamma^5\gamma^\mu$ $(\mu=0,1,2,3)$.

The aim of this paper is to give an affirmative answer to the above conjecture in a slightly general form. For $\omega=(\omega_1,\omega_2,\omega_3)\in S^2$ we define

$$P(\omega)=\frac{1}{2}(I+\omega_j\gamma^0\gamma^j),$$

and introduce the null condition related to the Dirac operator:

$$F(P(\omega)X)=0\quad\text{for all }\omega\in S^2,\ X\in\mathbb{C}^4, \tag{1.5}$$

where S^2 denotes the unit sphere. We shall discuss the relation between the null condition for the wave equation due to Klainerman [3] and the above condition at the end of Section 4.

Now we are in a position to state our main result.

Theorem 1.1. *Let δ be a small positive number and let $\psi_0\in\mathcal{S}(\mathbb{R}^3;\mathbb{C}^4)$. Suppose that each component of $F(\psi)$ is a homogeneous polynomial of degree 2 in the components of ψ and $\overline{\psi}$. Assume that $F(\psi)$ satisfies* (1.5). *Then there exist positive numbers ε_0 and C such that for all $\varepsilon\in(0,\varepsilon_0]$ the problem* (1.1)–(1.2) *has a classical global solution ψ in $[0,\infty)\times\mathbb{R}^3$ enjoying*

$$|\psi(t,x)|\le C(1+|x|)^{-1}(1+|t-|x||)^{-1} \tag{1.6}$$

for $(t,x) \in [0,\infty) \times \mathbb{R}^3$, *together with*

$$|P(-|x|^{-1}x)\psi(t,x)| \le C(1+t+|x|)^{-2+\delta} \tag{1.7}$$

for $(t,x) \in [0,\infty) \times \mathbb{R}^3$ *satisfying* $0 \le t \le 2|x|$ *and* $|x| \ge 1$.

Moreover, there exists $\psi^+ \in L^2(\mathbb{R}^3;\mathbb{C}^4)$ *such that*

$$\lim_{t\to\infty} \|\psi(t) - U(t)\psi^+\|_{L^2(\mathbb{R}^3)} = 0, \tag{1.8}$$

where we put $U(t) = \exp(-t\gamma^0\gamma^j\partial_j)$, *which is the free Dirac propagator.*

Remark 1.2. One can check that if $F(\psi)$ satisfies the null condition (1.5) and each of its components is a homogeneous polynomial of degree 2 in the components of ψ and $\overline{\psi}$, then $F(\psi)$ can be written as $S_0(\psi,\psi)e_0+S_1(\psi,\psi)e_1$ with some $e_0, e_1 \in \mathbb{C}^4$, where S_0 and S_1 are given in (1.4). We shall prove this fact in the appendix.

Remark 1.3. Since $\psi = P(|x|^{-1}x)\psi+P(-|x|^{-1}x)\psi$, (1.6) and (1.7) suggest that the main part of ψ is $P(|x|^{-1}x)\psi$. This observation motivates the null condition (1.5).

Remark 1.4. One sees from (1.8) that the global solution obtained in Theorem 1.1 is asymptotically free.

2. Notations

For $x \in \mathbb{R}^3$, we write $r = |x|$, $\omega = x/r$. We set $\partial_r = (x/r)\cdot\nabla_x$. Denoting

$$\Omega_1 = (x_2\partial_3 - x_3\partial_2)I - \frac{1}{2}\gamma^2\gamma^3,$$

$$\Omega_2 = (x_3\partial_1 - x_1\partial_3)I - \frac{1}{2}\gamma^3\gamma^1,$$

$$\Omega_3 = (x_1\partial_2 - x_2\partial_1)I - \frac{1}{2}\gamma^1\gamma^2,$$

we define $\Gamma = (\Gamma_0, \Gamma_1, \dots, \Gamma_6) = (\partial_t, \partial_1, \partial_2, \partial_3, \Omega_1, \Omega_2, \Omega_3)$. For a multi-index $\alpha = (\alpha_0, \dots, \alpha_6)$, we put $\Gamma^\alpha = \Gamma_0^{\alpha_0}\cdots\Gamma_6^{\alpha_6}$. For a nonnegative integer m and a smooth function φ, we define

$$|\varphi(t,x)|_m = \sum_{|\alpha|\le m} |\Gamma^\alpha\varphi(t,x)|, \quad \|\varphi(t)\|_m = \sum_{|\alpha|\le m} \|\Gamma^\alpha\varphi(t)\|_{L^2(\mathbb{R}^3)}.$$

Note that the vector fields Γ are closed with respect to the commutator $[A,B] = AB - BA$. We also put $\widetilde{\Omega} = (\widetilde{\Omega}_1, \widetilde{\Omega}_2, \widetilde{\Omega}_3) = x \wedge \nabla_x$, and define $\widetilde{\Gamma}$ by replacing Ω_j with $\widetilde{\Omega}_j$ in the definition of Γ. The vector fields $\widetilde{\Gamma}$ are also closed with respect to the commutator. Moreover we see that $\sum_{|\alpha|\le m}\left|\widetilde{\Gamma}^\alpha\varphi(t,x)\right|$ and $\sum_{|\alpha|\le m}\left\|\widetilde{\Gamma}^\alpha\varphi(t)\right\|_{L^2(\mathbb{R}^3)}$ are equivalent to $|\varphi(t,x)|_m$ and $\|\varphi(t)\|_m$, respectively.

For two positive functions $f = f(y)$ and $g = g(y)$ on a suitable domain D in $\mathbb{R}^n$, we write $f(y) \lesssim g(y)$ if there exists $C > 0$ such that $f(y) \le Cg(y)$ for all $y \in D$.

3. Preliminaries

We consider the inhomogeneous Dirac equation:

$$\mathcal{D}\psi \equiv (\gamma^0\partial_t + \gamma^j\partial_j)\psi = f(t,x), \qquad (t,x) \in (0,\infty)\times\mathbb{R}^3, \tag{3.1}$$

$$\psi(0,x) = \psi_0(x), \qquad x \in \mathbb{R}^3. \tag{3.2}$$

Because of the commutator relation $[\Gamma_a, \mathcal{D}] = 0$ for $a = 0,\ldots,6$, the following L^2 estimates are valid.

Lemma 3.1. *Let m be a nonnegative integer. If ψ solves* (3.1), *then we have*

$$\|\psi(t)\|_m \lesssim \|\psi(0)\|_m + \int_0^t \|f(s)\|_m ds, \quad t \geq 0. \tag{3.3}$$

In order to derive pointwise estimates, we apply $\mathcal{D}$ to the equation (3.1) to obtain

$$(\partial_t^2 - \Delta)I\psi = \mathcal{D}f(t,x), \qquad (t,x) \in (0,\infty)\times\mathbb{R}^3, \tag{3.4}$$

$$\psi(0,x) = \psi_0(x), \quad (\partial_t\psi)(0,x) = -\gamma^0\gamma^j\partial_j\psi_0(x) + \gamma^0 f(0,x), \quad x \in \mathbb{R}^3. \tag{3.5}$$

Then we are able to adopt the pointwise estimates for the wave equation which are well studied, for instance, by Asakura [1], Yokoyama [7], and Katayama–Kubo [5].

Lemma 3.2. *Let μ, $\nu \geq 1$, m be a nonnegative integer, and ψ solve* (3.1)–(3.2). *If $\nu = 1$ or $\mu = 1$, then we have*

$$\begin{aligned}(1+r)(1+|t-r|)|\psi(t,x)|_m &\lesssim \sup_{y\in\mathbb{R}^3}(1+\rho)^3|\psi_0(y)|_{m+1} \\ &+ \log(2+t)\sup_{(\tau,y)\in[0,T)\times\mathbb{R}^3}\rho(1+\tau+\rho)^\nu(\min\{1+\rho, 1+|\tau-\rho|\})^\mu|f(\tau,y)|_{m+1}\end{aligned} \tag{3.6}$$

for $(t,x) \in [0,T)\times\mathbb{R}^3$. On the other hand, if $\nu > 1$ and $\mu > 1$, then we have

$$\begin{aligned}(1+r)(1+|t-r|)|\psi(t,x)|_m &\lesssim \sup_{y\in\mathbb{R}^3}(1+\rho)^3|\psi_0(y)|_{m+1} \\ &+ \sup_{(\tau,y)\in[0,T)\times\mathbb{R}^3}\rho(1+\tau+\rho)^\nu(\min\{1+\rho, 1+|\tau-\rho|\})^\mu|f(\tau,y)|_{m+1}\end{aligned} \tag{3.7}$$

for $(t,x) \in [0,T)\times\mathbb{R}^3$. Here we put $r = |x|$ and $\rho = |y|$.

In order to take advantage of the algebraic structure of the nonlinearity, we introduce the following crucial estimate for $P(-\omega)\psi(t,x)$.

Proposition 3.3. *We set $\Lambda = \{(t,x) \in [0,T]\times\mathbb{R}^3 \,|\, 0 \leq t \leq 2|x|,\ |x| \geq 1\}$. Let $\eta > 0$, $M > 0$, and m be a nonnegative integer. Suppose that F is sufficiently smooth and satisfies* (1.3) *for $p = 2$. If ψ solves* (1.1) *and satisfies*

$$\sup_{(t,x)\in\Lambda}|\psi(t,x)|_{m+1}(1+t)^{-\eta}(1+r)(1+|t-r|) \leq M, \tag{3.8}$$

then we have

$$\sum_{|\alpha|\leq m}|P(-\omega)\Gamma^\alpha\psi(t,x)| \lesssim (M+M^2)(1+t+r)^{-2+2\eta} \tag{3.9}$$

for all $(t,x) \in \Lambda$. Here we put $r = |x|$ and $\omega = x/r$.

Proof. By using the identity $\nabla_x = \omega\partial_r - r^{-1}O$ with $O = \omega \wedge (x \wedge \nabla_x)$, we get from (1.1)

$$(\gamma^0\partial_t + \omega_j\gamma^j\partial_r)\Gamma^\alpha\psi = r^{-1}O_j\gamma^j\Gamma^\alpha\psi + \Gamma^\alpha F(\psi).$$

Here O_j stands for the jth component of O. Multiplying the both sides of the equation by $P(-\omega)\gamma^0$, we get

$$(\partial_t - \partial_r)P(-\omega)\Gamma^\alpha\psi = P(-\omega)\gamma^0(r^{-1}O_j\gamma^j\Gamma^\alpha\psi + \Gamma^\alpha F(\psi)) =: G(\psi). \qquad (3.10)$$

Note that $r \geq C(1 + r + t)$ for $(t, x) \in \Lambda$, and that $|O\varphi| \lesssim |\widetilde{\Omega}\varphi|$ for a smooth function φ. Then, by (3.8) we get

$$|G(\psi(t,x))| \lesssim M(1+r+t)^{-2+\eta}(1+|t-r|)^{-1} + M^2(1+r+t)^{-2+2\eta}(1+|t-r|)^{-2}$$

for $(t, x) \in \Lambda$. Now, if we fix $(t_0, x_0) = (t_0, r_0\,\omega_0) \in \Lambda$ and integrate (3.10) along a ray $\{(s, (r_0 + (t_0 - s))\omega_0) \mid s \in [0, t_0]\}$, then we find (3.9) for $(t, x) = (t_0, x_0)$. This completes the proof. □

4. Proof of Theorem 1.1

Since the existence of local solutions of the problem (1.1)–(1.2) is well known, it suffices to derive a suitable *a priori* estimate. Let k be a positive integer, $0 < \eta < 1/4$, $M \geq 1$, $\varepsilon \in (0, \varepsilon_0]$, and $M\varepsilon \leq 1$. Suppose that

$$\sup_{(t,x)\in[0,T)\times\mathbb{R}^3} (1 + r)(1 + |t - r|)|\psi(t, x)|_k \leq M\varepsilon \qquad (4.1)$$

holds. In particular, we have $|\psi(t, x)|_k \lesssim M\varepsilon(1 + t)^{-1}$ for $(t, x) \in [0, T) \times \mathbb{R}^3$. It follows from Lemma 3.1 that

$$\|\psi(t)\|_{2k} \lesssim \varepsilon + \int_0^t \|F(\psi(s))\|_{2k} ds \lesssim M\varepsilon + \int_0^t M\varepsilon(1 + s)^{-1}\|\psi(s)\|_{2k} ds.$$

By the Gronwall inequality, we get

$$\|\psi(t)\|_{2k} \lesssim M\varepsilon(1 + t)^{CM\varepsilon}, \quad t \in [0, T) \qquad (4.2)$$

with some positive constant C. We are going to apply the Klainerman–Sobolev inequality

$$|x||v(x)| \lesssim \|v\|_2, \quad x \in \mathbb{R}^3,$$

where v is a smooth function in $\mathbb{R}^3$ vanishing sufficiently rapidly at infinity (see [4] for the proof of the inequality). Then we get

$$r\,|\psi(t, x)|_{2k-2} \lesssim \|\psi(t)\|_{2k} \lesssim M\varepsilon(1 + t)^{CM\varepsilon}, \quad (t, x) \in [0, T) \times \mathbb{R}^3.$$

On the other hand, we see from (4.1) that

$$(1 + t + r)(\min\{1 + r, 1 + |t - r|\})|\psi(t, x)|_k \lesssim M\varepsilon, \quad (t, x) \in [0, T) \times \mathbb{R}^3.$$

Hence, it follows from (3.6) with $\mu = \nu = 1$ that

$$\begin{aligned}(1 + r)(1 + |t - r|)|\psi(t, x)|_{2k-3} &\lesssim \varepsilon + M^2\varepsilon^2(1 + t)^{CM\varepsilon}\log(2 + t)\\ &\lesssim M\varepsilon(1 + t)^{\eta}\end{aligned} \qquad (4.3)$$

for $(t,x) \in [0,T) \times \mathbb{R}^3$ and $0 < \varepsilon \le \varepsilon_0$, provided that ε_0 satisfies $CM\varepsilon_0 \le \eta/2$. Applying Proposition 3.3, we get

$$\sum_{|\alpha| \le 2k-4} |P(-\omega)\,\Gamma^\alpha \psi(t,x)| \lesssim M\varepsilon(1+t+r)^{-2+2\eta}, \quad (t,x) \in \Lambda. \tag{4.4}$$

Now the null condition (1.5) comes into play. By (1.5), we have $F(\psi) = S_0(\psi,\psi)e_0 + S_1(\psi,\psi)e_1$ with $e_0, e_1 \in \mathbb{C}^4$ (see Remark 1.2). We define a quadratic form $N(\phi,\psi)$ associated with $F(\psi)$ by

$$N(\phi,\psi) = S_0(\phi,\psi)e_0 + S_1(\phi,\psi)e_1.$$

Then we see that (1.5) implies

$$N(P(\omega)\phi, P(\omega)\psi) = 0 \quad \text{for all } \omega \in S^2,\ \phi,\psi \in \mathbb{C}^4 \tag{4.5}$$

by the polarization identity.

Since a direct computation shows $N(\gamma^j\gamma^k\phi,\psi) + N(\phi,\gamma^j\gamma^k\psi) = 0$ for $j,k = 1,2,3$, we get

$$\Omega_j N(\phi,\psi) = N(\Omega_j\phi,\psi) + N(\phi,\Omega_j\psi) - \frac{1}{2}\gamma^k\gamma^l N(\phi,\psi)$$

for $(j,k,l) = (1,2,3), (2,3,1), (3,1,2)$. In a similar manner, it follows that

$$\Gamma^\alpha F(\psi) = \sum_{|\beta|+|\gamma| \le |\alpha|} A_{\alpha\beta\gamma} N(\Gamma^\beta\psi, \Gamma^\gamma\psi)$$

with appropriate constant matrices $A_{\alpha\beta\gamma}$. Writing $\Gamma^\beta\psi = P(\omega)\Gamma^\beta\psi + P(-\omega)\Gamma^\beta\psi$ and using (4.5), we get

$$\Gamma^\alpha F(\psi) = \sum_{|\beta|+|\gamma| \le |\alpha|} A_{\alpha\beta\gamma} \left(N(P(-\omega)\Gamma^\beta\psi, P(\omega)\Gamma^\gamma\psi) + N(P(\omega)\Gamma^\beta\psi, P(-\omega)\Gamma^\gamma\psi) \right).$$

Using (4.3) and (4.4) in Λ, and using (4.1) and (4.3) outside Λ, we obtain

$$\begin{aligned} &\sum_{|\alpha| \le 2k-4} |\Gamma^\alpha F(\psi(t,x))| \\ &\quad \lesssim M^2\varepsilon^2(1+r)^{-1}(1+t+r)^{-2+4\eta}(\min\{1+r, 1+|t-r|\})^{-1-\eta} \end{aligned} \tag{4.6}$$

for $(t,x) \in [0,T) \times \mathbb{R}^3$. Using (3.7) as $\nu = 2 - 4\eta$ (> 1) and $\mu = 1 + \eta$, we obtain

$$(1+r)(1+|t-r|)|\psi(t,x)|_{2k-5} \le C_0\varepsilon + C_1M^2\varepsilon^2, \quad (t,x) \in [0,T) \times \mathbb{R}^3.$$

Taking M with $M \ge 4C_0$, and choosing ε_0 in such a way that $C_1M\varepsilon_0 \le 1/4$, we get

$$(1+r)(1+|t-r|)|\psi(t,x)|_k \le \frac{1}{2}M\varepsilon, \quad (t,x) \in [0,T) \times \mathbb{R}^3,$$

provided that $k \ge 5$ and $0 < \varepsilon \le \varepsilon_0$. Thus, by the bootstrap argument, we find that (4.1) is true for small ε as long as the solution exists, and we get the global solution to the problem (1.1)–(1.2).

Moreover, (4.6) yields

$$\|F(\psi(t))\|_{L^2} \lesssim M^2\varepsilon^2(1+t)^{-2+4\eta}, \quad t \ge 0.$$

Denoting by $U(t)$ the Dirac propagator $\exp(-t\gamma^0\gamma^j\partial_j)$, we have

$$U(-t)\psi(t) = \varepsilon\psi_0 + \left(\int_0^\infty - \int_t^\infty\right) U(-s)\gamma^0 F(\psi(s))ds$$
$$=: \psi^+ - \int_t^\infty U(-s)\gamma^0 F(\psi(s))ds.$$

Therefore, we get $\psi^+ \in H^k(\mathbb{R}^3 : \mathbb{C}^4)$ and

$$\|\psi(t) - U(t)\psi^+\|_{L^2} \lesssim \int_t^\infty \|F(\psi(s))\|_{L^2} ds \lesssim M^2\varepsilon^2(1+t)^{-1+4\eta}, \quad t \geq 0.$$

This completes the proof of Theorem 1.1.

Remark 4.1. It seems natural to call (1.5) the null condition related to the Dirac operator, because it is closely connected to the null condition for systems of nonlinear wave equations (concerning the global existence theorem for nonlinear wave equations, we refer to [2], [3], [5], and the references cited therein). Moreover, instead of our direct approach, reduction to a system of nonlinear wave equations gives an alternative proof of Theorem 1.1. In fact, if we introduce the Dirac potential φ by $\mathcal{D}\varphi = \psi$ with $\varphi(0) = 0$, then (1.1) becomes

$$(\partial_t^2 - \Delta)\varphi = F(\mathcal{D}\varphi), \quad (t, x) \in (0, \infty) \times \mathbb{R}^3$$

with $\varphi(0) = 0$ and $(\partial_t\varphi)(0) = \varepsilon\gamma^0\psi_0$, because $\mathcal{D}^2 = (\partial_t^2 - \Delta)I$. The null condition for this system of wave equations is written as

$$F\left(\left((-1)\gamma^0 + \omega_j\gamma^j\right) X\right) = 0 \quad \text{for all } \omega = (\omega_1, \omega_2, \omega_3) \in S^2,\ X \in \mathbb{C}^4. \tag{4.7}$$

Since $(-\gamma^0 + \omega_j\gamma^j)X = (I + \omega_j\gamma^0\gamma^j)(-\gamma^0 X)$, we see that (4.7) is equivalent to (1.5). Therefore, one can obtain the existence part of Theorem 1.1 as a corollary to the classical result due to [2], [3]. We can also reduce the existence of ψ^+ satisfying (1.8) to that of $(\varphi_0^+, \varphi_1^+) \in H^1(\mathbb{R}^3) \times L^2(\mathbb{R}^3)$ satisfying $\lim_{t\to\infty} \|\partial\varphi(t) - \partial\varphi^+(t)\|_{L^2(\mathbb{R}^3)} = 0$ for the solution φ^+ to $(\partial_t^2 - \Delta)\varphi^+ = 0$ with $(\varphi^+, \partial_t\varphi^+) = (\varphi_0^+, \varphi_1^+)$ at $t = 0$, because we get $U(t)\psi^+ = \mathcal{D}\varphi^+$ by setting $\psi^+ = \gamma^0\varphi_1^+ + \gamma^j\partial_j\varphi_0^+$.

Appendix: On the null condition related to the Dirac operator

In this appendix we give a proof of the fact stated in Remark 1.2. To be more precise, we denote the jth component of $F(\psi)$ by $F_j(\psi)$ $(j = 1, \ldots, 4)$, and assume that each $F_j(\psi)$ can be written as

$$F_j(\psi) = {}^t\psi M_j^1\psi + {}^t\overline{\psi} M_j^2\psi + {}^t\overline{\psi} M_j^3\overline{\psi}, \tag{A.1}$$

where $M_j^k \in \mathcal{M}_4(\mathbb{C})$ $(k = 1, 2, 3)$ and M_j^1, M_j^3 are symmetric matrices. Here $\mathcal{M}_s(\mathbb{C})$ denotes the set of $s \times s$ complex matrices. We shall show that if

$$F_j(P(\omega)X) = 0 \quad \text{for all } \omega \in S^2,\ X \in \mathbb{C}^4, \tag{A.2}$$

then we have

$$M_j^1 = M_j^3 = 0, \quad M_j^2 = c_j^0\gamma^0 + c_j^1\gamma^0\gamma^5 \tag{A.3}$$

for some c_j^0, $c_j^1 \in \mathbb{C}$. This means that putting $e_k = {}^t(c_1^k, \ldots, c_4^k)$ for $k = 0, 1$, we get

$$F(\psi) = S_0(\psi, \psi)e_0 + S_1(\psi, \psi)e_1.$$

Our first step is to show that if (A.2) holds, then we have

$$M_j^1(\omega) := {}^tP(\omega)M_j^1P(\omega) = 0, \tag{A.4}$$

$$M_j^2(\omega) := {}^t\overline{P(\omega)}M_j^2P(\omega) = 0, \tag{A.5}$$

$$M_j^3(\omega) := {}^t\overline{P(\omega)}M_j^3\overline{P(\omega)} = 0 \tag{A.6}$$

for all $\omega \in S^2$. We see from (A.1) and (A.2) that

$${}^tXM_j^1(\omega)X + {}^t\overline{X}M_j^2(\omega)X + {}^t\overline{X}M_j^3(\omega)\overline{X} = 0 \quad \text{for all } \omega \in S^2,\ X \in \mathbb{C}^4. \tag{A.7}$$

We replace X by iX in (A.7) and combining the resulting identity with (A.7), we get

$${}^t\overline{X}M_j^2(\omega)X = 0 \qquad \text{for all } \omega \in S^2,\ X \in \mathbb{C}^4, \tag{A.8}$$

$${}^tXM_j^1(\omega)X + {}^t\overline{X}M_j^3(\omega)\overline{X} = 0 \qquad \text{for all } \omega \in S^2,\ X \in \mathbb{C}^4. \tag{A.9}$$

Since the equation in (A.8) is rewritten as $\langle X, M_j^2(\omega)X\rangle = 0$, we see that (A.5) holds. On the other hand, taking $X = X_0 \in \mathbb{R}^4$ in (A.9), we have

$${}^tX_0(M_j^1(\omega) + M_j^3(\omega))X_0 = 0 \quad \text{for all } \omega \in S^2,\ X_0 \in \mathbb{R}^4,$$

so that

$$M_j^1(\omega) + M_j^3(\omega) = 0 \quad \text{for all } \omega \in S^2, \tag{A.10}$$

because $M_j^1(\omega)$ and $M_j^3(\omega)$ are symmetric matrices. Moreover, by taking $X = (1+i)X_0$, (A.9) leads to

$${}^tX_0(M_j^1(\omega) - M_j^3(\omega))X_0 = 0 \quad \text{for all } \omega \in S^2,\ X_0 \in \mathbb{R}^4,$$

and hence $M_j^1(\omega) - M_j^3(\omega) = 0$ for all $\omega \in S^2$. Combining this with (A.10), we get (A.4) and (A.6).

Next we prove that (A.4) yields $M_j^1 = 0$. It follows from (A.4) that

$$(I + {}^t(\omega_k\gamma^k)\gamma^0)M_j^1(I + \gamma^0(\omega_k\gamma^k)) = 0 \quad \text{for all } \omega \in S^2. \tag{A.11}$$

We replace ω by $-\omega$ in (A.11) and subtract the resulting identity from (A.11). Then we get

$${}^t(\omega_k\gamma^k)\gamma^0M_j^1 + M_j^1\gamma^0(\omega_k\gamma^k) = 0 \quad \text{for all } \omega \in S^2,$$

which implies

$$\gamma^0\gamma^kM_j^1 + M_j^1\gamma^0\gamma^k = 0, \quad -\gamma^0\gamma^2M_j^1 + M_j^1\gamma^0\gamma^2 = 0, \tag{A.12}$$

for $k = 1, 3$, because ${}^t\gamma^1 = -\gamma^1$, ${}^t\gamma^2 = \gamma^2$, and ${}^t\gamma^3 = -\gamma^3$. Since M_j^1 is a symmetric matrix, we can express it as

$$M_j^1 = \begin{pmatrix} A_j & B_j \\ C_j & D_j \end{pmatrix},$$

where $A_j, B_j, C_j, D_j \in \mathcal{M}_2(\mathbb{C})$ satisfying ${}^tA_j = A_j$, ${}^tB_j = C_j$, and ${}^tD_j = D_j$. Then we have from (A.12)

$$A_j\sigma^k + \sigma^k D_j = 0,\ B_j\sigma^k + \sigma^k C_j = 0,\quad A_j\sigma^2 - \sigma^2 D_j = 0,\ B_j\sigma^2 - \sigma^2 C_j = 0$$

for $k = 1, 3$. These relations lead to $A_j = B_j = C_j = D_j = 0$ by a direct computation, and hence $M_j^1 = 0$.

Since (A.6) gives ${}^tP(\omega)\overline{M_j^3}P(\omega) = 0$, we see from the above argument that $\overline{M_j^3} = 0$, so that $M_j^3 = 0$.

It remains to consider the condition (A.5). Denoting

$$M_j^2 = \begin{pmatrix} P_j & Q_j \\ R_j & S_j \end{pmatrix},$$

with $P_j, Q_j, R_j, S_j \in \mathcal{M}_2(\mathbb{C})$, and proceeding as before, we obtain

$$P_j\sigma^k + \sigma^k S_j = 0,\ Q_j\sigma^k + \sigma^k R_j = 0$$

for $k = 1, 2, 3$, because $(\gamma^k)^* = -\gamma^k$. Therefore, by a direct calculation, we find that

$$P_j = -S_j = \begin{pmatrix} c_j^0 & 0 \\ 0 & c_j^0 \end{pmatrix},\quad Q_j = -R_j = \begin{pmatrix} c_j^1 & 0 \\ 0 & c_j^1 \end{pmatrix}$$

for some c_j^0, $c_j^1 \in \mathbb{C}$. Thus we have proved (A.3). This completes the proof.

Acknowledgment

The first author is partially supported by Grant-in-Aid for Scientific Research (C) (No. 23540241), JSPS. The second author is grateful to Professor Timothy Candy for valuable discussion as for Remark 4.1, and is partially supported by Grant-in-Aid for Scientific Research (B) (No. 24340024), JSPS.

References

[1] F. Asakura, *Existence of a global solution to a semi-linear wave equation with slowly decreasing initial data in three space dimensions*, Comm. Partial Differential Equations **11** (1986), 1459–1487.

[2] D. Christodoulou, *Global solutions of nonlinear hyperbolic equations for small initial data*, Comm. Pure Appl. Math. **39** (1986), 267–282.

[3] S. Klainerman, *The null condition and global existence to nonlinear wave equations*, in: Nonlinear Systems of Partial Differential Equations in Applied Mathematics, Part 1, Lectures in Appl. Math. **23**, Amer. Math. Soc., Providence, RI, 1986, pp. 293–326.

[4] S. Klainerman and T.C. Sideris, *On almost global existence for nonrelativistic wave equations in* 3D, Comm. Pure Appl. Math. **49** (1996), 307–321.

[5] S. Katayama and H. Kubo, *Decay estimates of a tangential derivative to the light cone for the wave equation and their application*, SIAM J. Math. Anal. **39** (2008), 1851–1862.

[6] N. Tzvetkov, *Existence of global solutions to nonlinear massless Dirac system and wave equation with small data*, Tsukuba J. Math. **22** (1988), 193–211.

[7] K. Yokoyama, *Global existence of classical solutions to systems of wave equations with critical nonlinearity in three space dimensions*, J. Math. Soc. Japan **52** (2000), 609–632.

Soichiro Katayama
Department of Mathematics
Wakayama University
930 Sakaedani
Wakayama 640-8510, Japan
Tel: +81-73-457-7343
e-mail: `katayama@center.wakayama-u.ac.jp`

Hideo Kubo
Department of Mathematics
Faculty of Sciences
Hokkaido University
Sapporo 060-0810, Japan
Tel: +81-11-706-2649
e-mail: `kubo@math.sci.hokudai.ac.jp`

Fourier Analysis
Trends in Mathematics, 263–284

Singular Partial Differential Operators and Pseudo-differential Boundary Conditions in Invisibility Cloaking

Matti Lassas and T. Zhou

Abstract. Transformation optics constructions have allowed the design of cloaking devices that steer electromagnetic, acoustic and quantum parameters waves around a region without penetrating it, so that this region is hidden from external observations. The material parameters used to describe these devices are anisotropic, and singular at the interface between the cloaked and uncloaked regions, making physical realization a challenge. These singular material parameters correspond to singular coefficient functions in the partial differential equations modeling these constructions and the presence of these singularities causes various mathematical problems and physical effects on the interface surface.

In this paper, we give a review on mathematical theory of cloaking. Moreover, we consider the two-dimensional cloaking, that is, cylindrical cloaking, for Maxwell's equations. For this case, we present results that generalizes earlier analogous results for the two-dimensional cloaking for the scalar equations. In particular, we consider nonsingular approximate invisibility cloaks based on the truncation of the singular transformations. Using such truncation we analyze the limit when the approximate cloaking approaches the ideal cloaking. We show that a non-local pseudo-differential boundary condition appears on the inner cloak interface. This effect in the two-dimensional (or cylindrical) invisibility cloaks, which seems to be caused by the infinite phase velocity near the interface between the cloaked and uncloaked regions, is very different to the behavior of the solutions in the three-dimensional cloaks.

Mathematics Subject Classification (2010). 35R30, 35Q60, 35Q61.

Keywords. Pseudodifferential boundary conditions, invisibility cloaking.

1. Introduction

Invisibility has been a subject of human fascination for millenia, from the Greek legend of Perseus versus Medusa to the more recent *The Invisible Man* and *Harry Potter*. Over the years, there have been occasional scientific prescriptions for invisibility in various settings, *e.g.*, [33, 7]. However, since 2005 there has been a wave of serious theoretical proposals [1, 47, 46, 43, 51] in the physics literature, and a widely reported experiment by Schurig et al. [53], for cloaking devices – structures that would not only render an object invisible but also undetectable to electromagnetic waves. The particular route to cloaking that has received the most attention is that of *transformation optics* [59], the designing of optical devices with customized effects on wave propagation, made possible by taking advantage of the transformation rules, in change of coordinates, for the material properties of optics: the index of refraction $n(x)$ for scalar optics, governed by the Helmholtz equation, and the electric permittivity $\varepsilon(x)$ and magnetic permeability $\mu(x)$ for vector optics, as described by Maxwell's equations. It is this approach to cloaking that we will examine in some detail. In particular, we discuss the appearance of pseudo-differential boundary conditions in cloaking constructions.

As it happens, two papers appeared in the same issue of *Science* with transformation optics-based proposals for cloaking. Leonhardt [43] gave a description, based on conformal mapping, of inhomogeneous indices of refraction $n(x)$ in two dimensions that would cause light rays to go around a region and emerge on the other side as if they had passed through empty space (for which $n \equiv 1$). (The region in question is then said to be *cloaked*.) On the other hand, Pendry, Schurig and Smith [51] gave a prescription for values of $\varepsilon(x)$ and $\mu(x)$ giving a cloaking device for electromagnetic waves, based on the fact that $\varepsilon(x)$ and $\mu(x)$ transform in the same way (2.5) as the conductivity tensor in electrostatics. In fact, they used exactly the same singular transformation (2.6), resulting in singular electromagnetic material parameters, as had already been used three years earlier to describe examples of nondetectability in the context of the Calderón Problem [28, 29]!

The papers [28, 29] considered the case of electrostatics, which can be considered as optics at frequency zero. In §2 we describe this case in more detail since it already contains the basic idea of transformation optics and also shows the importance of careful formulation and analysis of solutions in the setting of *singular* transformation optics. These articles give counterexamples to uniqueness in Calderón Problem, which is the inverse problem for electrostatics which lies at the heart of *Electrical Impedance Tomography.* This consists in determining the electrical conductivity of a medium filling a region Ω by making voltage and current measurements at the boundary $\partial\Omega$. The counterexamples were motivated by consideration of certain degenerating families of Riemannian metrics, which in the limit correspond to singular conductivities, *i.e.*, that are not bounded below or above, so that the corresponding PDE is no longer uniformly elliptic. A related example of a complete but noncompact two-dimensional Riemannian manifold with boundary having the same Dirichlet–Neumann map as a compact one was given in [39]. The techniques in [28, 29] are valid in dimensions three and higher, but the same construction has been shown to work in two dimensions [36]. We point out here that although we emphasize boundary observations using the Dirichlet–Neumann map or the set of Cauchy data, this is equivalent to scattering information [6]; see [56].

In §3 we consider cloaking for the Helmholtz equation. We place special emphasis on the singular parameters at the *cloaking surface* in both two-dimensional and three-dimensional spaces. For electromagnetism, the cloaking electric permittivity and magnetic permeability are obtained using the same transformation optics idea, hence are singular as well. The analysis of [43, 52] uses ray tracing which explains the behavior of the light rays but not the full electromagnetic waves. In [19], perfect cloaking was shown to indeed hold with respect to finite energy distribution solutions of the Helmholtz equation or Maxwell's equations, with passive objects (no internal current) being cloaked. In the case of cloaking active objects, for Helmholtz, such cloaking is always possible[1]. Moreover, in [19] it is shown that the finite energy solutions for the Helmholtz equation in the three-dimensional case satisfy a hidden boundary condition, namely waves inside the cloaked region satisfy the Neumann boundary conditions. However, for Maxwell's equations certain overdetermined boundary conditions emerge at the cloaking surface. While satisfied for passive cloaked objects, they cannot be satisfied for generic internal currents, *i.e.*, for active objects that are themselves radiating within the cloaked region.

The materials proposed for cloaking with electromagnetic waves are artificial materials referred to as *metamaterials*. The study of these materials has undergone an explosive growth in recent years. There is no universally accepted definition of metamaterials, which seem to be in the "know it when you see it" category. However, the label usually attaches to macroscopic material structures having a manmade one-, two- or three-dimensional cellular architecture, and producing combinations of material parameters not available in nature (or even in conventional composite materials), due to resonances induced by the geometry of the cells [58, 18]. Using metamaterial cells (or "atoms", as they are sometimes called), designed to resonate at the desired frequency, it is possible to specify the permittivity and permeability tensors fairly arbitrarily *at a given frequency*, so that they may have very large, very small or even negative eigenvalues. The use of resonance phenomenon also explains why the material properties of metamaterials strongly depend on the frequency, and broadband metamaterials may not be possible.

Another point of view in dealing with the singular anisotropic design for cloaking devices is to approximate the ideal cloaking parameters by nonsingular, or even nonsingular and isotropic, parameters [22, 23, 24, 35, 36, 45], which has its

[1] Since Helmholtz also governs acoustic waves, this allows the theoretical description of a 3D acoustic cloak, a spherically symmetric case of which was subsequently obtained in the physics literature [14, 17]; see [21].

advantages in practical fabrication. In the truncation based nonsingular approximate cloaking for three-dimensional Helmholtz equation [24], when it approaches the ideal cloaking, one can obtain above Neumann hidden boundary condition for the finite energy solution. Considering limits of the nonsingular and isotropic approximate cloaks in the three-dimensional case, one can obtain different types of Robin boundary conditions by varying slightly the way how the approximative cloak in constructed, see [22, 23, 24]. Similarly, for Maxwell's equations it has been studied how the approximate cloak behave in the limit when the approximate cloaks approach the ideal one [45]. We note that for Maxwell's equations there are various suggestions what kind of limiting cloaks are possible in three dimensions. These suggestions are based on constructions where additional layers (e.g., perfectly conducting layer) is attached inside the cloak [19] or where the ideal cloak corresponds to some of the possible self-adjoint extensions of Maxwell's equations [60, 61]. In the two-dimensional or cylindrical cloaking construction for Maxwell's equations, the eigenvalues of the permittivity and the permeability of the cloaking medium do not only contain eigenvalues approaching to zero (as in 3D) but also some of the eigenvalues approach infinity at the cloaking interface. Then, the electric flux density D and magnetic flux density B may blow up even when there are no sources inside the cloak and an incident plane wave is scattered from the cloak, see [20, 26]. However, if a soft-hard (SH)-surface is included inside the cloak, the solutions behave well. These above examples show the different behaviors the solutions may have, in different type of cloaking devices, near the interface between the cloaked and uncloaked regions.

In §4, we analyze the two-dimensional cloaking when there are sources present inside the cloaked region. We present results for Maxwell's equation on the similar phenomenon that was observed earlier for Helmholtz equation [41, 12], which can be interpreted as the invisible cloaking for TM-polarized electromagnetic waves in a waveguide. We start with the nonsingular approximate cloaking based on the truncation of the singular transformation. Taking the limit when the approximate cloaks approach the ideal cloak, we show that a non-local boundary condition appears on the inner cloak interface. This type of boundary behavior is very different from that the solutions have in three-dimensional case discussed in [24, 62]. The main result is formulated as Theorem 4.2. We note that in [49] a different type of formulation, based on a transmission problem, is given for the non-local boundary condition appearing in two-dimensional cloaking. In [49], also cloaking for more general second-order equations and quantitative convergence properties of the three- and two-dimensional approximative cloaks are analyzed. Physically speaking, the non-local boundary condition is possible due to the fact that the phase velocity of the waves in the invisibility cloak approaches infinity near the interface between the cloaked and uncloaked regions, even though the group velocity stays finite, see [13]. We note that as the most important experimental implementations of invisibility cloaks [53] have been based on cylindrical cloaks, the appearance of such boundary condition could also be studied in the present experimental configurations, at least on micro-wave frequencies. We note that the early experimental implementations of cloaking were actually for the so-called reduced parameter set, which does not have the same singular behavior as the material parameters studied here. However, the nonlocal boundary condition considered in this paper could be studied using numerical simulations or possibly even in experimental tests using the current metamaterials. We also study the eigenvalues, *i.e.*, resonances inside the ideal cloak corresponding to the non-local boundary condition. As these eigenvalues play an essential role in the study of almost trapped states [23] and in the development of the invisible sensors [2, 25], such resonances can be used to study analogous constructions in cylindrical geometry.

2. The case of electrostatics

2.1. Calderón's problem

Calderón's inverse problem, which forms the mathematical foundation of Electrical Impedance Tomography (EIT), is the question of whether an unknown conductivity distribution inside a domain in $\mathbb{R}^n$, modeling for example the Earth, a human thorax, or a manufactured part, can be determined from voltage and current measurements made on the boundary. A.P. Calderón's motivation to propose this problem [11] was geophysical prospection. One widely studied potential application of EIT is the early diagnosis of breast cancer [15]. The conductivity of a malignant breast tumor is typically 0.2 mho, significantly higher than normal tissue, which has been typically measured at 0.03 mho. See the book [31] and the special issue of *Physiological Measurement* [32] for applications of EIT to medical imaging and other fields.

For isotropic conductivities this problem can be mathematically formulated as follows. Let Ω be the measurement domain, and denote by $\sigma(x)$ the coefficient, bounded from above and below by positive constants, describing the electrical conductivity in Ω. In Ω the voltage potential u satisfies a divergence form equation,

$$\nabla \cdot \sigma \nabla u = 0. \tag{2.1}$$

To uniquely fix the solution u it is enough to give its value f on the boundary. In the idealized case, one measures, for all voltage distributions $u|_{\partial\Omega} = f$ on the boundary, the corresponding current fluxes $\nu\cdot\sigma\nabla u$ over the entire boundary, where ν is the exterior unit normal to $\partial\Omega$. Mathematically this amounts to the knowledge of the Dirichlet–Neumann (DN) map Λ_σ corresponding to σ, *i.e.*, the map taking the Dirichlet boundary values of the solution to (2.1) to the corresponding Neumann boundary values

$$\Lambda_\sigma : \quad u|_{\partial\Omega} \mapsto \nu\cdot\sigma\nabla u|_{\partial\Omega}.$$

Calderón's inverse problem is then to reconstruct σ from Λ_σ. We remark here that DN map is a pseudo-differential operator [42].

For what conductivities is there no cloaking? This is the question of uniqueness of determination of the conductivity from the DN map. For the isotropic case described above, Kohn and Vogelius showed that piecewise analytic conductivities are uniquely determined by the DN map [37]. Sylvester and Uhlmann proved that C^∞ smooth conductivities can be uniquely determined by the DN map in dimension $n \geq 3$ [55]. This was extended to conductivities having 3/2 derivatives [50, 9], and recently to those of C^1 or small Lipschitz [30], which is the best currently known result for scalar conductivities for $n \geq 3$. For conormal conductivities in $C^{1+\epsilon}$, uniqueness was shown in [27]. In the challenging two-dimensional case, unique identifiability of the conductivity from the DN map was shown for C^2 conductivities by Nachman [48], for Lipschitz conductivities by Brown and Uhlmann [10], and for the class of merely L^∞ conductivities by Astala and Päivärinta [3]. We are only briefly summarizing here the known uniqueness results for isotropic conductivities since, as will be seen below, these are not directly relevant to the subject of cloaking. For issues concerning stability, analytic and numerical reconstruction in EIT, we refer the reader to the surveys [8, 15, 57].

We now discuss the anisotropic case, that is when the conductivity depends on direction. Physically realistic models must incorporate anisotropy. In the human body, for example, muscle tissue is a highly anisotropic conductor, *e.g.*, cardiac muscle has a conductivity of 2.3 mho in the direction transversal to the fibers and 6.3 mho in the longitudinal direction.

An anisotropic conductivity on a domain $\Omega \subset \mathbb{R}^n$ is defined by a symmetric, positive semi-definite matrix-valued function, $\sigma = [\sigma^{ij}(y)]_{i,j=1}^n$ for $y \in \Omega$. In the absence of sources or sinks, an electrical potential $u(y)$ satisfies

$$\begin{aligned} \nabla\cdot \sigma\nabla u = \sum\nolimits_{j,k=1}^{n} \partial_j(\sigma^{jk}(y)\partial_k u) = 0 \text{ in } \Omega, \\ u|_{\partial\Omega} = f, \end{aligned} \tag{2.2}$$

where f is the prescribed voltage on the boundary. The resulting DN map (or voltage-to-current map) is then defined by

$$\Lambda_\sigma(f) = \sum\nolimits_{j,k=1}^{n} (\nu_j \sigma^{jk}\partial_k u)|_{\partial\Omega},$$

u being the solution of (2.2) and $\nu = (\nu_1, \dots, \nu_n)$ the unit normal vector of $\partial\Omega$.

Applying the divergence theorem, we have

$$Q_\sigma(f) =: \sum\nolimits_{j,k=1}^{n} \int_\Omega \sigma^{jk}(y)\frac{\partial u}{\partial y^j}\frac{\partial u}{\partial y^k}dy = \int_{\partial\Omega} \Lambda_\sigma(f) f dS, \tag{2.3}$$

where u solves (2.2) and dS denotes surface measure on $\partial\Omega$. $Q_\sigma(f)$ represents the power needed to maintain the potential f on $\partial\Omega$. By (2.3), knowing Q_σ is equivalent to knowing Λ_σ. If $G : \Omega \to \Omega$, $G = (G^1, G^2, \dots, G^n)$, is a diffeomorphism with $G|_{\partial\Omega} =$ Identity, then by making the change of variables $x = G(y)$ and setting $v = u \circ G^{-1}$ in the first integral in (2.3), we obtain

$$\Lambda_{G_*\sigma} = \Lambda_\sigma, \tag{2.4}$$

where

$$(G_*\sigma)^{jk}(x) = \frac{1}{\det\left[\frac{\partial G}{\partial y}(y)\right]} \sum\nolimits_{p,q=1}^{n} \frac{\partial G^j}{\partial y^p}(y)\,\frac{\partial G^k}{\partial y^q}(y)\sigma^{pq}(y)\bigg|_{y=G^{-1}(x)} \tag{2.5}$$

is the push-forward of the conductivity σ by G. Thus, there is a large (infinite-dimensional) class of conductivities which give rise to the same electrical measurements at the boundary. This was first observed in [38] following a remark by Luc Tartar. Calderón's inverse problem for anisotropic conductivities is then the question of whether two conductivities with the same DN operator must be push-forwards of each other. There are a number of positive results in this direction in two dimensions [4, 40, 39, 42, 54], but it was shown in [28, 29] in three dimensions and in [36] two dimensions that, if one allows singular maps, then in fact there are counterexamples, i.e., conductivities that are undetectable to electrostatic measurements at the boundary.

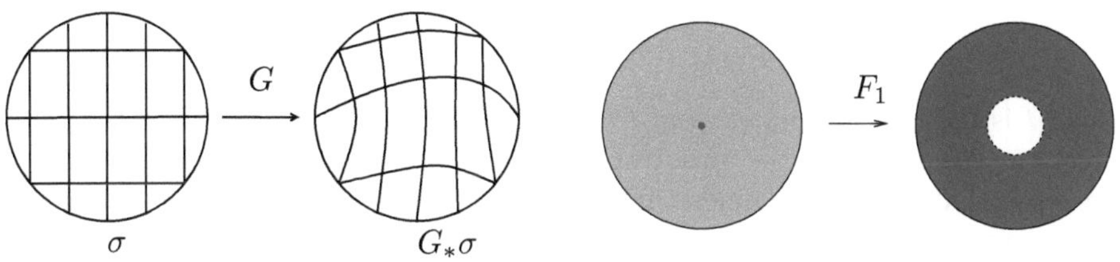

FIGURE 1. Left: The diffeomorphism $G : B_2 \to B_2$ satisfying $G|_{\partial B_2} = Id$ pushes the conductivity σ to a conductivity $G_*\sigma$ that appears the same as σ in all boundary measurements. Right: The blow-up map $F : B_2 \setminus \{0\} \to B_2 \setminus \overline{B}_1$ extends a hole of one point to a ball. The map F is used to define the cloaking construction.

2.2. Transformation optics for electrostatics

The fact that smooth diffeomorphisms that leave the boundary fixed give the same boundary information (2.4) can already be considered as a weak form of invisibility, with distinct conductivities being indistinguishable by external observations; however, nothing has been hidden yet. Using the invariance (2.4), examples of singular anisotropic conductivities in $\mathbb{R}^n, n \geq 3$, that are indistinguishable from a constant isotropic conductivity by the boundary Dirichlet-to-Neumann map, are given in [28, 29].

To give a precise description of the idea, let $B_R \subset \mathbb{R}^n$ $(n \geq 3)$ be an open ball with center at the origin O and with radius $R > 0$. Let $\sigma_0 = 1$ be the homogeneous conductivity corresponding to the vacuum Euclidean space. To build a spherical invisibility cloak, define a singular transformation $F : \mathbb{R}^n \backslash \{O\} \to \mathbb{R}^n \backslash B_1$, given by

$$x = F(y) := \begin{cases} y, & \text{for } |y| > 2, \\ \left(1 + \frac{|y|}{2}\right) \frac{y}{|y|}, & \text{for } 0 < |y| \leq 2. \end{cases} \tag{2.6}$$

Note that F blows up the origin to B_1 and $F|_{\partial B_2} =$ Identity. Then the cloaking conductivity σ on $B_2 \backslash B_1$ is the push-forward

$$\sigma^{jk}(x) := (F_* \sigma_0)^{jk}(x)$$

using (2.5). On the cloaked region B_1,

$$\sigma(x) = \sigma_a(x)$$

where σ_a is the conductivity of the arbitrary object to be cloaked.

In $\mathbb{R}^3$, denoting by $(r,\phi,\theta) \mapsto (r\sin\theta\cos\phi, r\sin\theta\sin\phi, r\cos\theta)$ the spherical coordinates, we have

$$\sigma = \begin{pmatrix} 2(r-1)^2\sin\theta & 0 & 0 \\ 0 & 2\sin\theta & 0 \\ 0 & 0 & 2(\sin\theta)^{-1} \end{pmatrix}, \quad 1 < |x| \leq 2.$$

This means that in the Cartesian coordinates the conductivity σ is given by

$$\sigma(x) = 2(I - P(x)) + 2|x|^{-2}(|x|-1)^2 P(x), \quad 1 < |x| < 2,$$

where I is the identity matrix and $P(x) = |x|^{-2}xx^t$ is the projection to the radial direction. We note that the anisotropic conductivity σ is singular on $\Sigma := \partial B_1$ in the sense that it is not bounded from below by any positive multiple of I. (See [36] for a similar calculation.)

Consider now the *Cauchy data* (an extension of Dirichlet-to-Neumann map) of all solutions in the Sobolev space $H^1(B_2)$ of the conductivity equation corresponding to σ, that is,

$$C_1(\sigma) = \big\{(u|_{\partial B_2}, \nu\cdot\sigma\nabla u|_{\partial B_2}) \;:\; u \in H^1(B_2)\cap L^\infty(B_2),\ \nabla\cdot\sigma\nabla u = 0\big\}.$$

Theorem 2.1 ([29]). *The Cauchy data of all H^1-solutions for the conductivities σ and σ_0 on B_2 coincide, that is, $C_1(\sigma) = C_1(\sigma_0)$.*

This means that all boundary measurements for the homogeneous conductivity $\sigma_0 = 1$ and the degenerated conductivity σ are the same.

Remark 2.2. We now make a simple but crucial observation: In order for the one-to-one correspondence between solutions of the conductivity equation for σ_0 and those for σ to hold, it is necessary to impose some regularity assumption on the electrical potentials u for σ. If, for example, we start with the Newtonian potential $K(x) = -\frac{1}{4\pi|x|}$, then this pushes forward to a (non-H^1) potential u_K for σ whose Cauchy data do *not* equal the Cauchy data of any potential for σ_0. Thus, it does not suffice to simply work with the transformation law in the exterior of the cloaked region. This comment is equally valid when one considers cloaking for the Helmholtz and Maxwell's equations.

The key to the proof of Theorem 2.1 and these generalized results is a removable singularity argument for the divergence form operator $\nabla\cdot\sigma_0\nabla$. It implies that solutions of the conductivity equation in the annulus pull back by a singular transformation to solutions of the conductivity equation in the whole ball.

In the two-dimensional case, the same basic construction works [36]. For a further study of the limits of visibility and invisibility in two dimensions, see [5].

3. Cloaking for the Helmholtz equation

Let us start with the cases of scalar optics or acoustics, governed by the Helmholtz equation

$$\lambda\nabla\cdot\sigma\nabla u+\omega^2 u=p(y)\qquad y\in\Omega\subset\mathbb{R}^n(n\geq 2)$$

in the medium described by (σ,λ), where, *e.g.*, σ can denote the inverse of the mass density, taking values to be positive semi-definite matrices $\sigma(y)=[\sigma^{ij}(y)]_{i,j=1}^n$ in anisotropic media; and $\lambda(y)$ is the scalar bulk modulus. The invisibility cloak will be constructed against the boundary Dirichlet-to-Neumann map, defined again by

$$\Lambda_{\sigma,\lambda}(u|_{\partial\Omega})=(\nu\cdot\sigma\nabla u)|_{\partial\Omega}.$$

The corresponding transformation law to the Helmholtz equation states that

$$\Lambda_{F_*\sigma,F_*\lambda}=\Lambda_{\sigma,\lambda}$$

if $F:\ \Omega\to\Omega$ is a diffeomorphism with $F|_{\partial\Omega}=$ Identity, where the push-forward of the density $F_*\sigma$ is defined as in the electrostatic case, *i.e.*, by (2.5), and $F_*\lambda$ is given by

$$F_*\lambda(x):=\det\left[\frac{\partial F}{\partial y}(y)\right]\lambda(y)\ \Big|_{y=F^{-1}(x)}.$$

Using the singular transformation (2.6), the cloaking medium in B_2 is given by

$$\sigma^{jk}=\begin{cases}\sigma_0^{jk} & \text{for } |x|>2\\ (F_*\sigma_0)^{jk} & \text{for } 1<|x|\leq 2\\ \sigma_a^{jk} & \text{for } |x|\leq 1\end{cases},\quad \lambda=\begin{cases}\lambda_0 & \text{for } |x|>2\\ F_*\lambda_0 & \text{for } 1<|x|<2\\ \lambda_a & \text{for } |x|\leq 1\end{cases}\tag{3.1}$$

where (σ_0,λ_0) corresponds to homogeneous background space and (σ_a,λ_a) are arbitrary smooth, nondegenerate medium in cloaked region B_1. If $(\sigma_0,\lambda_0)=(I,1)$, in $\mathbb{R}^3$, the degenerate singularity appears at the interface Σ as seen in the electrostatics, while in $\mathbb{R}^2$, the medium in the cloaking layer $B_2\backslash B_1$ is

$$\sigma(x)=\frac{|x|-1}{|x|}P(x)+\frac{|x|}{|x|-1}(I-P(x)),\quad \lambda(x)=\frac{|x|}{4(|x|-1)}$$

implying both degenerate and blowing-up singularity at the interface Σ.

4. Two-dimensional approximate cloaking for Maxwell's equations and nonlocal (pseudo-differential) boundary conditions

Let us consider time-harmonic Maxwell equations

$$\nabla\times\vec{H}(x)=-i\omega\vec{D}(x),\quad \nabla\times\vec{E}(x)=i\omega\vec{B}(x),\tag{4.1}$$

$$\vec{D}(x)=\varepsilon(x)\vec{E}(x),\qquad \vec{B}(x)=\mu(x)\vec{H}(x),\tag{4.2}$$

where $x=(x_1,x_2,x_3)\in\mathbb{R}^3$ and $\omega>0$ is the frequency.

Let us recall some basic facts in case when the material parameters $\varepsilon(x)$ and $\mu(x)$ are independent of the x_3-variable. In other words, we consider a waveguide-type geometry oriented along the x_3-axis. Below, we assume that the permittivity matrix $\varepsilon(x)=(\varepsilon_{jk}(x))_{j,k=1}^3$ and the permeability matrix $\mu(x)=(\mu_{jk}(x))_{j,k=1}^3$ are symmetric matrixes that are independent of the x_3 variable. Let $\widehat{e}_1 = (1,0,0)$, $\widehat{e}_2 = (0,1,0)$ and $\widehat{e}_3 = (0,0,1)$ denote the unit coordinate vectors. Let us consider an electromagnetic waves solving Maxwell's equations (4.1)–(4.2) that are the form

$$\vec{E}(x_1,x_2) := (\vec{E}_t(x_1,x_2) + E_3(x_1,x_2)\widehat{e}_3)e^{ikx_3}$$

where $\vec{E}_t := (E_1(x_1,x_2), E_2(x_1,x_2), 0)^t$. Similar notations apply to $\vec{H}$, $\vec{D}$ and $\vec{B}$. Then

$$\nabla \times \vec{E} = \nabla_t \times (E_3\widehat{e}_3) + ik\widehat{e}_3 \times \vec{E}_t + \nabla_t \times \vec{E}_t$$

where $\nabla_t = (\partial_1, \partial_2, 0)^t$, $\partial_j = \partial_{x_j}$ for $j = 1,2,3$. Here $\nabla_t \times (E_3\widehat{e}_3)$ is on the transverse direction while $\nabla_t \times \vec{E}_t$ is on the $\widehat{e}_3$ direction, that is,

$$\nabla_t \times (E_3\widehat{e}_3) + ik\widehat{e}_3 \times \vec{E}_t = i\omega\vec{B}_t, \qquad \nabla_t \times \vec{E}_t = i\omega B_3.$$

Similarly, we have

$$\nabla_t \times (H_3\widehat{e}_3) + ik\widehat{e}_3 \times \vec{H}_t = -i\omega\vec{D}_t, \qquad \nabla_t \times \vec{H}_t = -i\omega D_3.$$

Alternatively, we can write

$$J\begin{pmatrix} \partial_1 E_3 \\ \partial_2 E_3 \end{pmatrix} - ikJ\begin{pmatrix} E_1 \\ E_2 \end{pmatrix} = i\omega\begin{pmatrix} B_1 \\ B_2 \end{pmatrix} = i\omega\begin{pmatrix} (\mu\vec{H})_1 \\ (\mu\vec{H})_2 \end{pmatrix}$$
$$J\begin{pmatrix} \partial_1 H_3 \\ \partial_2 H_3 \end{pmatrix} - ikJ\begin{pmatrix} H_1 \\ H_2 \end{pmatrix} = -i\omega\begin{pmatrix} D_1 \\ D_2 \end{pmatrix} = -i\omega\begin{pmatrix} (\varepsilon\vec{E})_1 \\ (\varepsilon\vec{E})_2 \end{pmatrix}$$

and

$$(\nabla_t \times \vec{E}_t)\cdot\widehat{e}_3 = (\partial_1,\partial_2)J\begin{pmatrix} E_1 \\ E_2 \end{pmatrix} = i\omega(\mu\vec{H})_3$$
$$(\nabla_t \times \vec{H}_t)\cdot\widehat{e}_3 = (\partial_1,\partial_2)J\begin{pmatrix} H_1 \\ H_2 \end{pmatrix} = -i\omega(\varepsilon\vec{E})_3$$

where $J = \begin{pmatrix} 0 & 1 \\ -1 & 0 \end{pmatrix}$.

In the Transverse Magnetic (TM) mode of the electromagnetic field, the propagation direction is perpendicular to $\widehat{e}_3$ direction, and the magnetic field is parallel to $\widehat{e}_3$, that is, $k = 0$ and $\vec{H}_t = 0$. In this case we also have $E_3 = 0$. Analogously, we consider below the Transverse Electric (TE) mode, where one has $k = 0$, $\vec{E}_t = 0$ and $H_3 = 0$.

Let us first consider the TM mode of the electromagnetic field. Then we have

$$J\begin{pmatrix} \partial_1 H_3 \\ \partial_2 H_3 \end{pmatrix} = -i\omega\begin{pmatrix} D_1 \\ D_2 \end{pmatrix}, \qquad \begin{pmatrix} \partial_1 \\ \partial_2 \end{pmatrix}\cdot J\begin{pmatrix} E_1 \\ E_2 \end{pmatrix} = i\omega\mu_{33}H_3$$

Moreover, we assume that $\varepsilon_{13} = \varepsilon_{23} = 0$. Then $\vec{D}(x) = (D_1(x_1,x_2), D_2(x_1,x_2), 0)$ and

$$\begin{pmatrix} D_1 \\ D_2 \end{pmatrix} = \widetilde{\varepsilon}\begin{pmatrix} E_1 \\ E_2 \end{pmatrix} \quad \text{where } \widetilde{\varepsilon} := \begin{pmatrix} \varepsilon_{11} & \varepsilon_{12} \\ \varepsilon_{12} & \varepsilon_{22} \end{pmatrix}.$$

Then we obtain in $\mathbb{R}^2$ the two-dimensional Maxwell's equations for the vertical component $u(x_1,x_2) := H_3(x_1,x_2)$ of the magnetic field (of the TM-polarized electromagnetic waves),

$$\kappa(x_1,x_2)\,\nabla\cdot\varrho(x_1,x_2)\nabla u(x_1,x_2) + \omega^2 u(x_1,x_2) = 0, \quad (x_1,x_2)\in\mathbb{R}^2, \tag{4.3}$$

where

$$\varrho(x_1,x_2) := J\,\widetilde{\varepsilon}(x_1,x_2)^{-1}\,J^{-1}, \quad \kappa(x_1,x_2) = \mu_{33}(x_1,x_2)^{-1}.$$

Note that $J^2 = -I$.

To consider the TE mode of the electromagnetic field one has to assume that $\mu_{13} = \mu_{23} = 0$ and define $\widetilde{\mu} := \begin{pmatrix} \mu_{11} & \mu_{12} \\ \mu_{12} & \mu_{22} \end{pmatrix}$. Then, the vertical component $E_3(x_1,x_2)$ of the electric field satisfies the equation (4.3) with $\varrho(x_1,x_2)$ and $\kappa(x_1,x_2)$ being $J\widetilde{\mu}(x_1,x_2)^{-1}J^{-1}$ and $\varepsilon_{33}(x_1,x_2)^{-1}$, correspondingly.

Let us next consider the approximate cylindrical invisibility cloaks for Maxwell's equations for electromagnetic field for which, the propagation direction is perpendicular to $\widehat{e}_3$ direction, that is, $k=0$. In particular, we will consider what happens when such approximate cloaks approach to a perfect one and show that in the limit a pseudo-differential boundary condition at the inside of the cloak.

To start, let $1<R<2$, $\rho = 2(R-1)$ and introduce the coordinate transformation $F_R : \mathbb{R}^2\backslash B_\rho \to \mathbb{R}^2\backslash B_R$,

$$x := F_R(y) = \begin{cases} y, & \text{for } |y|>2, \\ \left(1+\frac{|y|}{2}\right)\frac{y}{|y|}, & \text{for } \rho<|y|\le 2. \end{cases}$$

Let us consider also the three-dimensional coordinate transformation $G_R : (\mathbb{R}^2\backslash B_\rho)\times\mathbb{R} \to (\mathbb{R}^2\backslash B_R)\times\mathbb{R}$ defined by $G_R(x_1,x_2,x_3) = (F_R(x_1,x_2),x_3)$. Then the permittivity and the permeability of the cylindrical cloak are defined by

$$\widetilde{\varepsilon}_R(x) = \begin{cases} ((G_R)_*I)(x) & \text{for } |x|>R, \\ \varepsilon_a(x_1,x_2)\delta^{jk} & \text{for } |x|\le R, \end{cases} \qquad \widetilde{\mu}_R(x) = \begin{cases} ((G_R)_*I)(x) & \text{for } |x|>R, \\ \mu_a(x_1,x_2)\delta^{jk} & \text{for } |x|\le R, \end{cases}$$

where $\varepsilon_a(x)$ and $\mu_a(x)$ are scalar functions modeling the cloaked object. As we are interested in the case when we have sources inside the cloak, for simplicity we assume below that $\varepsilon_a = 1$ and $\mu_a = 1$. Note that the obtained medium satisfies both $(\widetilde{\varepsilon}_R)_{13} = (\widetilde{\varepsilon}_R)_{23} = 0$ and $(\widetilde{\mu}_R)_{13} = (\widetilde{\mu}_R)_{23} = 0$ so that one can consider both the TM and TE modes of the electromagnetic waves.

Let us next consider an electromagnetic wave that is in the TM mode. We define the corresponding approximate cloaking medium (ϱ_R, κ_R) as

$$\varrho_R^{jk}(x) = \begin{cases} \varrho^{jk}(x) & \text{for } |x|>R, \\ \varrho_a\delta^{jk} & \text{for } |x|\le R. \end{cases} \qquad \kappa_R(x) = \begin{cases} \kappa(x) & \text{for } |x|>R, \\ \kappa_a & \text{for } |x|\le R, \end{cases} \tag{4.4}$$

where $\varrho^{jk} := F_*I$ and $\kappa := F_*1$.

Note that then $\varrho^{jk}(x) = ((F_R)_*\,I)^{jk}(x)$ and $\kappa(x) = ((F_R)_*\,1)(x)$ for $|x|>R$ and, as we consider the TM mode, $\varrho^{jk}(x) = (J\,\widetilde{\varepsilon}_R^{-1}J^{-1})^{jk}$ and $\kappa_R = (\mu_R)_{33}^{-1}$. Recall that we consider here the case when that $\varrho_a = 1$ and $\kappa_a = 1$ are inside B_1.

Observe that, for each $R > 1$, the medium is nonsingular, i.e., is bounded from above and below with, however, the lower bound going to 0, and the upper bound going to ∞ as $R \to 1^+$. Consider the solutions of

$$\begin{aligned}(\kappa_R \nabla\cdot \varrho_R \nabla + \omega^2) u_R &= p \quad \text{in } \Omega \\ u_R|_{\partial\Omega} &= f,\end{aligned} \tag{4.5}$$

where u_R can be considered as the vertical component of the magnetic field of an electromagnetic wave in the TM mode.

FIGURE 2. Left: The vertical component $H_3(x) = \operatorname{Re} e^{i\omega x_1}$ of magnetic field of a TM polarized plane wave. Right: The vertical component of magnetic field, $H_3(x) = u(x)$, satisfying (4.5) when R is close to one, $f = \operatorname{Re} e^{i\omega x_1}$, and the source term $p = 0$ vanishes. Physically, the solution corresponds to the situation when a TM polarized plane wave scatters from an approximative invisibility cloak. Observe that the solution is then close to zero inside the cloak.

As ϱ_R and κ_R are now non-singular everywhere on Ω, we have the standard transmission conditions on $\Sigma_R := \{x : \ |x| = R\}$,

$$\begin{aligned}u_R|_{\Sigma_R+} &= u_R|_{\Sigma_R-}, \\ e_r\cdot \varrho_R \nabla u_R|_{\Sigma_R+} &= e_r\cdot \varrho_R \nabla u_R|_{\Sigma_R-},\end{aligned} \tag{4.6}$$

where e_r is the radial unit vector and $\pm$ indicates when the trace on Σ_R is computed as the limit $r \to R^{\pm}$.

Let $\Omega = B_3$. Then u_R defines two functions $v_R^{\pm}$ such that

$$u_R(x) = \begin{cases} v_R^+(F_R^{-1}(x)), & \text{for } R < |x| < 3, \\ v_R^-(x), & \text{for } |x| \le R, \end{cases}$$

and $v_R^{\pm}$ satisfy

$$\begin{aligned}(\nabla^2 + \omega^2) v_R^+(y) &= p(F_R(y)) \quad \text{in } \rho < |y| < 3, \\ v_R^+|_{\partial B_3} &= f,\end{aligned}$$

and

$$(\nabla^2 + k_a^2 \omega^2) v_R^-(x) = k_a^2 p(x), \quad \text{in } |x| < R. \tag{4.7}$$

where $k_a^2 = (\varrho_a \kappa_a)^{-1}$ is a constant. In this paper we consider the case when $k_a = 1$ for simplicity. Moreover, if we assume ω^2 is not an eigenvalue of the transmission problem, then by the transformation law we have

$$e_r \cdot \varrho_R \nabla u_R\big|_{\partial\Omega} = e_r \cdot \nabla v_R^+\big|_{\partial\Omega}.$$

This implies that the DN-map $\Lambda_{\varrho_R,\kappa_R}$ at $\partial\Omega$, for the approximate cloaking medium (4.4), that physically corresponds to the impedance map that maps the boundary value of the magnetic field to the tangential boundary value of the electric flux, is the same as the DN-map at $\partial\Omega$, denoted by Λ_ρ, of a nearly vacuum domain with a small inclusion present in B_ρ.

Next, using polar coordinates (r,θ), $r = |y|$, and $(\widetilde{r},\theta)$, $\widetilde{r} = |x|$, the transmission conditions (4.6) on the surface Σ_R yield

$$\begin{aligned} v_R^+(\rho,\theta) &= v_R^-(R,\theta), \\ \rho\,\partial_r v_R^+(\rho,\theta) &= R\,\partial_{\widetilde{r}} v_R^-(R,\theta). \end{aligned} \tag{4.8}$$

We consider the source term $p(x)$ where $p(x) \in C^\infty(\mathbb{R}^2)$ with $\operatorname{supp} p \subset B_{R_0}$ $(0 < R_0 < 1)$. It generates a radiating wave $w(x) \in C^\infty(\mathbb{R}^2)$, namely the solution of

$$(\nabla^2 + \omega^2) w = p \qquad \text{in } \mathbb{R}^2 \tag{4.9}$$

satisfying the Sommerfeld radiation condition. Moreover, one can write w as

$$w(\widetilde{r},\theta) = \sum_{n=-\infty}^{\infty} p_n H_{|n|}^{(1)}(\omega\widetilde{r}) e^{in\theta}, \qquad \text{for } r > R_0 \tag{4.10}$$

where $H_{|n|}^{(1)}(z)$ and $J_{|n|}(z)$ denote the Hankel and Bessel functions, see [16].

In $B_R \backslash \overline{B_{R_0}}$ the function $v_R^-(x)$ differs from w by an entire solution to the homogeneous equation of (4.7), and thus we can write the following ansatz of spherical expansions, for $\widetilde{r} \in (R_0, R)$,

$$v_R^-(\widetilde{r},\theta) = \sum_{n=-\infty}^{\infty} (a_n J_{|n|}(\omega\widetilde{r}) + p_n H_{|n|}^{(1)}(\omega\widetilde{r})) e^{in\theta},$$

with yet undefined $a_n = a_n(\omega; R)$. Similarly, for $\rho < r < 3$,

$$v_R^+(r,\theta) = \sum_{n=-\infty}^{\infty} (c_n H_{|n|}^{(1)}(\omega r) + b_n J_{|n|}(\omega r)) e^{in\theta}, \tag{4.11}$$

with as yet unspecified $b_n = b_n(\omega; R)$ and $c_n = c_n(\omega; R)$.

In this analysis, we are interested in the case without a boundary incident wave, namely, taking $f \equiv 0$. we obtain, together with transmission conditions (4.8), the following explicit formulas of the coefficients in terms of Bessel and

Hankel functions

$$
\begin{aligned}
a_n &= \frac{R(H^{(1)}_{|n|})'(\omega R)l_1 - \rho H^{(1)}_{|n|}(\omega R)l_2}{D_n} p_n := \frac{A_n}{D_n} p_n \\
b_n &= \frac{R\{(H^{(1)}_{|n|})'(\omega R)J_{|n|}(\omega R) - J'_{|n|}(\omega R)H^{(1)}_{|n|}(\omega R)\}H^{(1)}_{|n|}(3\omega)}{D_n} p_n \\
c_n &= -\frac{R\{(H^{(1)}_{|n|})'(\omega R)J_{|n|}(\omega R) - J'_{|n|}(\omega R)H^{(1)}_{|n|}(\omega R)\}J_{|n|}(3\omega)}{D_n} p_n
\end{aligned}
\tag{4.12}
$$

where

$$D_n := \rho J_{|n|}(\omega R)l_2 - RJ'_{|n|}(\omega R)l_1$$

and

$$
\begin{aligned}
l_1 &= J_{|n|}(\omega\rho)H^{(1)}_{|n|}(3\omega) - H^{(1)}_{|n|}(\omega\rho)J_{|n|}(3\omega), \\
l_2 &= J'_{|n|}(\omega\rho)H^{(1)}_{|n|}(3\omega) - (H^{(1)}_{|n|})'(\omega\rho)J_{|n|}(3\omega).
\end{aligned}
\tag{4.13}
$$

For small arguments $0 < x \ll 1$,

$$
\begin{aligned}
J_{|n|}(x) &\sim \begin{cases} 1 & n = 0, \\ \frac{1}{n!}\left(\frac{x}{2}\right)^n & n \geq 1, \end{cases} &
J'_{|n|}(x) &\sim \begin{cases} -\frac{x}{2} & n = 0, \\ \frac{1}{2(n-1)!}\left(\frac{x}{2}\right)^{n-1} & n \geq 1, \end{cases} \\
H^{(1)}_{|n|}(x) &\sim \begin{cases} \frac{2i}{\pi}\ln\left(\frac{x}{2}\right) & n = 0, \\ -\frac{i(n-1)!}{\pi}\left(\frac{2}{x}\right)^n & n \geq 1, \end{cases} &
(H^{(1)}_{|n|})'(x) &\sim \begin{cases} 2ix^{-1}/\pi & n = 0, \\ in!2^n x^{-n-1}/\pi & n \geq 1, \end{cases}
\end{aligned}
\tag{4.14}
$$

where we denote $f \sim g$ if $f - g = o(g)$ as $x \to 0$. Thus, for $n \geq 1$, we have

$$D_n = \frac{-i2^n\omega^{-n-1}(n-1)!}{\pi} J_n(3\omega)\big[\omega RJ'_n(\omega R) + nJ_n(\omega R)\big]\rho^{-n} + O(\rho^{-n+1}), \tag{4.15}$$

as $\rho \to 0^+$ (or equivalently, as $R \to 1^+$).

Now we observe that, e.g., $|a_n| \to \infty$ as $R \to 1^+$, if the factor in P_n

$$\big[\omega R(J_{|n|})'(\omega R) + |n|J_{|n|}(\omega R)\big]\Big|_{R=1} = 0. \tag{4.16}$$

One can verify the numerator of a_n is not vanishing in this situation. This implies that if is outside a discrete set and if ω is such that (4.16) is satisfied by functions $J_{|n|}$ and $H^{(1)}_{|n|}$ for some n, then there are sources p for which the H^1-norm of the solution u_R goes to infinity in the cloaked region (i.e., when resonance happens) as $R \to 1^+$ (i.e., $\rho \to 0$).

Next we consider the nonresonant frequencies ω for which

$$
\begin{cases}
\big[\omega RJ'_{|n|}(\omega R) + |n|J_{|n|}(\omega R)\big]\Big|_{R=1} \neq 0, \\
J_{|n|}(3\omega) \neq 0,
\end{cases}
\quad \text{for any } n \in \mathbb{Z}. \tag{4.17}
$$

Next we consider the Fourier coefficients of the solution $v_R^- := u_R|_{B_R}$ in the cloaked region

$$v_{R,n}^-(\widetilde{r}) = \int_0^{2\pi} e^{-in\theta} v_R^-(\widetilde{r}, \theta) d\theta. \tag{4.18}$$

Next we consider the approximative cloaking material parameters in $B_2 \backslash \overline{B_R}$, $1 < R < 2$ and the solutions when $R \to 1^+$.

Lemma 4.1. *Assume that in $\Omega = B_3$ we have the material parameters (ϱ_R, κ_R). Moreover, suppose ω is such that (4.17) holds. When $R > 1$ is sufficiently close to 1, then for any source $p \in L^2(\Omega)$ supported compactly in B_1 and for $f \in H^{1/2}(\partial\Omega)$ the Helmholtz equation (4.5) has a unique solution u_R. Let $v_{R,n}^-(\widetilde{r})$ be the Fourier coefficients given in (4.18). Then the limits $\lim_{R\to 1+} v_{R,n}^-(\widetilde{r})$ exist and we have for all $n \in \mathbb{Z}$*

$$\lim_{R\to 1^+} (\widetilde{r}\partial_{\widetilde{r}} v_{R,n}^-(\widetilde{r}) + |n| v_{R,n}^-(\widetilde{r}))\big|_{\widetilde{r}=R} = 0, \tag{4.19}$$

$$\lim_{R\to 1^+} (\widetilde{r}\partial_{\widetilde{r}} v_{R,n}^-(\widetilde{r}) + |n| v_{R,n}^-(\widetilde{r}))\big|_{\widetilde{r}=1} = 0. \tag{4.20}$$

Proof. As we suppose that ω is such that (4.17) holds, we have

$$\begin{aligned} v_R^-(\widetilde{r}, \theta) &= \sum_{n=-\infty}^{\infty} (a_n J_{|n|}(\omega\widetilde{r}) + p_n H_{|n|}^{(1)}(\omega\widetilde{r}))e^{in\theta} \\ &= \sum_{n=-\infty}^{\infty} \left[\frac{A_n}{D_n} J_{|n|}(\omega\widetilde{r}) + H_{|n|}^{(1)}(\omega\widetilde{r})\right] p_n e^{in\theta} \end{aligned}$$

by (4.12). Denote

$$\Phi_n(\widetilde{r}) := \frac{A_n}{D_n} J_{|n|}(\omega\widetilde{r}) + H_{|n|}^{(1)}(\omega\widetilde{r}) \qquad \text{and} \qquad \widetilde{\Phi}(\widetilde{r}, \theta) := \Phi_n(\widetilde{r}) e^{in\theta}.$$

Apparently $\widetilde{\Phi}(\widetilde{r}, \theta)$ satisfies the Helmholtz equation

$$(\Delta + \omega^2)\widetilde{\Phi} = 0 \qquad \text{in } B_R.$$

To prove the existence of the limits $\lim_{R\to 1+} v_{R,n}^-(\widetilde{r})$ and (4.19), it is sufficient to show

$$\widetilde{r}\partial_{\widetilde{r}}\Phi_n(\widetilde{r}) + n\Phi_n(\widetilde{r})\big|_{\widetilde{r}=R} \to 0 \quad \text{as } R \to 1^+. \tag{4.21}$$

Indeed, by

$$\partial_{\widetilde{r}}\Phi_n(\widetilde{r}) = \frac{A_n}{D_n}\omega J'_{|n|}(\omega\widetilde{r}) + \omega(H_{|n|}^{(1)})'(\omega\widetilde{r}),$$

and $\lim_{R\to 1} D_n(R) \neq 0$ (corresponding to the non-resonant case), we have

$$\begin{aligned} &\widetilde{r}\partial_{\widetilde{r}}\Phi_n(\widetilde{r}) + n\Phi_n(\widetilde{r})\big|_{\widetilde{r}=R} \\ &= \frac{R}{D_n}(nl_1 + \omega\rho l_2)\left((H_{|n|}^{(1)})'(\omega R) J_{|n|}(\omega R) - J'_{|n|}(\omega R) H_{|n|}^{(1)}(\omega R)\right) \end{aligned} \tag{4.22}$$

where l_1 and l_2 are given by (4.13).

Therefore, for $n \geq 1$, as $\rho \to 0^+$,

$$nl_1+\omega\rho l_2 = \omega\rho\left(J_{|n|-1}(\omega\rho)H_{|n|}^{(1)}(3\omega) - H_{|n|-1}^{(1)}(\omega\rho)J_{|n|}(3\omega)\right) = \mathcal{O}(\rho^{-n+2}). \quad (4.23)$$

Combining (4.22), (4.23) and (4.15), one has

$$\widetilde{r}\partial_{\widetilde{r}}\Phi_n(\widetilde{r}) + |n|\Phi_n(\widetilde{r})\big|_{\widetilde{r}=R} = \mathcal{O}(\rho^2) \qquad \text{as } R \to 1^+, \text{ (i.e., } \rho \to 0^+),$$

which proves (4.21).

For $n = 0$, from (4.14),

$$\begin{aligned}\partial_{\widetilde{r}}\Phi_0(R) &= \frac{A_0}{D_0}\omega J_0'(\omega R) + \omega(H_0^{(1)})'(\omega R) \\ &\sim \frac{W_n(\omega R)}{RJ_0'(\omega R)\ln\left(\frac{\omega\rho}{2}\right)} \to 0 \quad \text{as } \rho \to 0^+\end{aligned} \quad (4.24)$$

where $W_n(x) = H_0^{(1)}(x)J_0'(x) - (H_0^{(1)})'(x)J_0(x)$, i.e., the boundary condition (4.21) is satisfied for $n = 0$ and moreover,

$$\widetilde{r}\partial_{\widetilde{r}}\Phi_0(\widetilde{r})\big|_{\widetilde{r}=R} = \mathcal{O}\left(\frac{1}{\ln\left(\frac{\omega\rho}{2}\right)}\right) \qquad \text{as } R \to 1^+.$$

This proves (4.19). The equation (4.20) follows similarly by evaluating (4.22) and (4.24) at $\widetilde{r} = 1$ instead of $\widetilde{r} = R$. □

Now we are ready to prove our main result of the limit of the waves u_R of the physical approximate cloaking medium as $R \to 1^+$. We recall that $\Sigma = \partial B_1$. We denote the nonlocal pseudo-differential operator on $\mathbb{S}^1$ of order 1

$$\mathcal{A} := (-\partial_\theta^2)^{1/2},$$

that is, $\mathcal{A}$ is the square root of the positive Laplace operator on the unit circle.

We are now ready to show that a non-local boundary condition appears inside the cloak when $R \to 1$.

Theorem 4.2. *Let ω be such that (4.17) is satisfied. Assume that u_R is the solution of (4.5) with $f = 0$ and $p \in C_0^\infty(B_{R_0})$ with $R_0 < 1$. Then as $R \to 1^+$, u_R converges uniformly in compact subsets of $B_3 \setminus \Sigma$ to the limit u_1 satisfying*

$$(\nabla^2 + \omega^2)u_1 = p \qquad \text{in } B_1, \quad (4.25)$$

$$\partial_{\widetilde{r}}u_1 + \mathcal{A}u_1\big|_{\partial B_1} = 0, \quad (4.26)$$

and

$$u_1\big|_{B_2\setminus\overline{B_1}} = 0. \quad (4.27)$$

Proof. Let $R_0 < R_1 < R$. Recall that solution $w \in C^\infty(\mathbb{R}^2)$ of (4.9) is the radiating solution produced by source p in $\mathbb{R}^2$ and it has the expansion (4.10) for $\widetilde{r} > R_1$. Consider its Fourier coefficients

$$w_n(\widetilde{r}) := \int_0^{2\pi} e^{-in\theta}w(\widetilde{r},\theta)d\theta$$

and denote $P_n(|x|) := -(\nabla^2+\omega^2-n^2/|x|^2)w_n(|x|)$. As $w \in C^\infty(\mathbb{R}^2)$, we see using integration by parts that

$$\|P_n(|x|)\|_{L^2(B_3)} \le C_M(1+|n|)^{-M} \qquad \text{for arbitrary } M>0. \tag{4.28}$$

The Fourier coefficients,

$$v^-_{R,n}(\widetilde{r}) = \int_0^{2\pi} e^{-in\theta} v^-_R(\widetilde{r},\theta)d\theta, \quad v^+_{R,n}(r) = \int_0^{2\pi} e^{-in\theta} v^+_R(r,\theta)d\theta,$$

satisfy the following problem

$$\begin{aligned} &\left(-\nabla^2-\omega^2+\frac{n^2}{|x|^2}\right) v^-_{R,n}(|x|) = P_n(|x|) \quad \text{for } 0\le |x| \le R,\\ &\left(-\nabla^2-\omega^2+\frac{n^2}{|y|^2}\right) v^+_{R,n}(|y|) = 0 \qquad \text{for } \rho \le |y| \le 2, \end{aligned} \tag{4.29}$$

$$\begin{aligned} &v^+_{R,n}|_{\partial B_2} = 0,\\ &v^-_{R,n}|_{\partial B^-_R} = v^+_{R,n}|_{\partial B^+_\rho}, \quad R(v^-_{R,n})'(|x|)|_{\partial B^-_R} = \rho(v^+_{R,n})'(|y|)|_{\partial B^+_\rho}. \end{aligned} \tag{4.30}$$

By the transmission conditions (4.30), we see for $V^\pm_{R,n}(x) := v^\pm_{R,n}(|x|)$

$$\int_{\partial B_R} \partial_{\widetilde{r}} V^-_{R,n}\overline{V^-_{R,n}}dS_x = \frac{R}{\rho}\int_{\partial B_\rho} \frac{\rho}{R}\partial_r V^+_{R,n}\overline{V^+_{R,n}}dS_y = \int_{\partial B_\rho} \frac{1}{-}\partial_r V^+_{R,n}\overline{V^+_{R,n}}dS_y.$$

Thus, using integration by parts, we obtain

$$\begin{aligned} \int_{B_R} P_n\overline{V_{R,n}}dy &= \int_{B_R}\left(|\nabla V^-_{R,n}|^2 + \left(-\omega^2+\frac{n^2}{|x|^2}\right)|V^-_{R,n}|^2\right)dx\\ &\quad + \frac{1}{-}\int_{B_2\setminus\overline{B_\rho}}\left(|\nabla V^+_{R,n}|^2 + \left(-\omega^2+\frac{n^2}{|y|^2}\right)|V^+_{R,n}|^2\right)dy\\ &\ge \int_{B_R}\left(-\omega^2+\frac{n^2}{R^2}\right)|V^-_{R,n}|^2dx + \frac{1}{-}\int_{B_2\setminus\overline{B_\rho}}\left(-\omega^2+\frac{n^2}{2^2}\right)|V^+_{R,n}|^2dy. \end{aligned}$$

For $|n| \ge N_0$ with $\frac{N_0^2}{2^2} \ge \max\{\omega^2,\omega^2\}$, as $R<2$, the above and that

$$\left|\int_{B_R} P_n\overline{V^-_{R,n}}dy\right| \le \|V^-_{R,n}\|_{L^2}\|P_n\|_{L^2}$$

imply first that

$$\left(\|V^-_{R,n}\|_{L^2(B_R)} + \|V^+_{R,n}\|_{L^2(B_2\setminus\overline{B_\rho})}\right) \le C_{N_0}\|P_n\|_{L^2(B_3)}$$

and second that

$$\left(\|V^-_{R,n}\|_{H^1(B_R)} + \|V^+_{R,n}\|_{H^1(B_2\setminus\overline{B_\rho})}\right) \le C'_{N_0}\|P_n\|_{L^2(B_3)} \tag{4.31}$$

where C_{N_0} and C'_{N_0} are independent of R and n.

By Lemma 4.1, for each $n\in\mathbb{Z}$, $\widetilde{r}\in[0,1)$ and $r\in(0,2)$ there exist limits

$$v^-_n(\widetilde{r}) = \lim_{R\to1^+} v^-_{R,n}(\widetilde{r}), \qquad v^+_n(r) = \lim_{R\to1^+} v^+_{R,n}(r) \tag{4.32}$$

and we denote $V_n^{\pm}(x) := v_n^{\pm}(|x|)$. Let now $0 < r_1 < R_1 < 1$. Then by (4.31) the restrictions $v_{R,n}^-|_{(r_1,R_1)}$, $R > R_1$ are uniformly bounded in $H^1(r_1, R_1)$. By Sobolev embedding theorem, the set $\{v_{R,n}^-|_{[r_1,R_1]};\ R_0 < 1 < R\}$ is relatively compact in $H^s(r_1, R_1) \subset C([r_1, R_1])$, $1/2 < s < 1$. Thus any sequence $(v_{R_j,n}^-|_{[r_1,R_1]})_{j=1}^{\infty}$ with $R_j \to 1$ has a subsequence converging in $C([r_1, R_1])$ which limit has to coincide with v_n^- by (4.32). Thus $v_{R,n}^-$ have to converge to v_n^- in $C([r_1, R_1])$ and hence $V_{R,n}^-$ have to converge to V_n^- in $C(\overline{B}_{R_1} \setminus B_{r_1})$ as $R \to 1$. Similarly, for all $\rho_1 > 0$ we see using (4.31) that $V_{R,n}^+$ have to converge to V_n^+ in $C(\overline{B}_2 \setminus B_{\rho_1})$ as $R \to 1$. Now, as the Sobolev norm $u \mapsto \|u\|_{H^1(B_{R_1}\setminus \overline{B}_{r_1})}$ is a lower semi-continuous function in $L^2(B_{R_1} \setminus \overline{B}_{r_1})$ and (4.31) holds, we see that

$$\|V_n^-\|_{H^1(B_{R_1}\setminus \overline{B}_{r_1})} \le C'_{N_0}\|P_n\|_{L^2(B_3)}$$

for all $0 < r_1 < R_1 < 1$. Hence, by (4.32) we see using, e.g., monotone convergence theorem and [34] that

$$\begin{aligned}\|V_n^-\|_{H^1(B_1)} = \|V_n^-\|_{H^1(B_1\setminus\{0\})} &= \lim_{r_1\to 0, R_1\to 1} \|V_n^-\|_{H^1(B_{R_1}\setminus \overline{B}_{r_1})} \\ &\le C'_{N_0}\|P_n\|_{L^2(B_3)}.\end{aligned} \tag{4.33}$$

By (4.28) we see that $u_1(\widetilde{r}, \theta) = \sum_{n\in\mathbb{Z}} v_n^-(\widetilde{r})e^{in\theta}$ is a well-defined function in $H^1(B_1)$ satisfying (4.25). By (4.31),

$$\begin{aligned}\|V_{R,n}^-\|_{C(\overline{B}_{R_1}\setminus B_{r_1})} = \|v_{R,n}^-\|_{C([r_1,R_1])} &\le C_{N_0,r_1,R_1}\|v_{R,n}^-\|_{H^1(r_1,R_1)} \\ &\le C'_{N_0,r_1,R_1}\|P_n\|_{L^2(B_3)},\end{aligned}$$

where C'_{N_0,r_1,R_1} does not depend on R or $n > N_0$. Thus, using (4.28) and the convergence of $V_{R,n}^-$ to V_n^- in $C(\overline{B}_{R_1} \setminus B_{r_1})$ we see that u_R converge to u_1 uniformly in compact subsets of $B_1 \setminus \{0\}$. Using equation (4.29), we see the uniform convergence in a neighborhood of zero, too. Similarly, we see the uniform convergence in compact subsets of $B_3 \setminus \overline{B}_1$. By (4.28) and (4.33), we see that

$$\begin{aligned}\|u_1\|_{H^1(B_1)}^2 &\le \sum_{n=-\infty}^{\infty} (1+n^2)\|V_n^-\|_{H^1(B_1)}^2 \\ &\le \sum_{|n|<-N_0} (1+n^2)\|V_n^-\|_{H^1(B_1)}^2 + \sum_{|n|\ge N_0} (1+n^2)(C'_{N_0}\|P_n\|_{L^2(B_3)})^2 < \infty.\end{aligned}$$

Hence, $u_1 \in H^1(B_1)$ and we have $u_1|_{\partial B_1} \in H^{1/2}(\partial B_1)$. Moreover, by (4.25) we have $\nabla^2 u_1 = -\omega^2 u_1 + p \in L^2(B_1)$. For $\psi \in H^{1/2}(\partial B_1)$ there is an extension $E_\psi \in H^1(B_1)$ such that $E_\psi|_{\partial B_1} = \psi$ and $\|E_\psi\|_{H^1} \le C'\|\psi\|_{H^{1/2}}$ with some $C' > 0$. Thus, analogously to [44], we see that the Neumann boundary value $\partial_r u_1|_{\partial B_1} \in H^{-1/2}(\partial B_1)$ is well defined by the identity

$$\langle \partial_r u_1|_{\partial B_1}, \psi\rangle_{H^{-1/2}(\partial B_1)\times H^{1/2}(\partial B_1)} = \int_{B_1} ((-\omega^2 u_1 + p)E_\psi - \nabla u_1 \cdot \nabla E_\psi)dx,$$

for any $\psi \in H^{1/2}(\partial B_1)$ and

$$\|\partial_r u_1|_{\partial B_1}\|_{H^{-1/2}(\partial B_1)} \leq C'(\| - \omega^2 u_1 + p\|_{L^2(B_1)} + \|u_1\|_{H^1(B_1)}).$$

Using this for (4.29) we obtain

$$\|\partial_r V_n^-\|_{H^{-1/2}(\partial B_1)} \leq C_2(1+n^2)\|V_n^-\|_{H^1(B_1 \setminus \overline{B}_{R_0})}$$

where $C_2 > 0$ is independent of n. Thus using (4.20), (4.28), and (4.33) we see that the boundary value $\partial_r u_1 + (-\partial_\theta^2)^{1/2} u_1$, vanishes. Hence, the boundary condition (4.26) is satisfied.

Equation (4.27) for $u_1|_{B_3 \setminus \overline{B_1}}$ follows using (4.28), (4.31) and the fact that for any fixed $n \in \mathbb{Z}$ we have in (4.11)

$$c_n(\rho) = \mathcal{O}(\rho^{|n|}), \quad b_n(\rho) = \mathcal{O}(\rho^{|n|}) \qquad \text{as } \rho \to 0^+, \quad |n| \geq 1,$$
$$c_0(\rho) = \mathcal{O}((\log \rho)^{-1}), \quad b_0(\rho) = \mathcal{O}((\log \rho)^{-1}) \qquad \text{as } \rho \to 0^+. \qquad \square$$

Above, we studied an electromagnetic wave in the TM mode in a cylindrical cloak. As noted above, the same considerations hold for an electromagnetic wave in the TE mode when we change the notations so that ϱ_R denotes $J\widetilde{\mu}_R^{-1}J^{-1}$ and $\kappa_r(x_1, x_2)$ denotes $(\varepsilon_R)_{33}^{-1}$, and finally, u_R is the vertical component E_3 of the electric field. Then, we see that as $R \to 1$, the vertical component u_R of the electric field of tends to a function u_1 that satisfies the boundary condition (4.26) at the inside of the cloak.

Summarizing, the above theorem yields that when the approximate cylindrical invisibility cloaks corresponding to material parameters $\widetilde{\varepsilon}_R$ and $\widetilde{\mu}_R$ approach to a perfect one and we have a source inside the cloak, in the limit when $R \to 1$ the solutions of two-dimensional Maxwell's equation approach to a field that satisfies the non-local, pseudo-differential boundary condition (4.26) at the inside of the cloak.

Acknowledgment

M.L. has been supported by Academy of Finland projects 260522 and CoE-250215.

References

[1] A. Alu and N. Engheta, *Achieving transparency with plasmonic and metamaterial coatings*, Phys. Rev. E, **72** (2005), 016623.

[2] A. Alu and N. Engheta, *Cloaking a sensor*, Phys. Rev. Lett. **102** (2009), 233901.

[3] K. Astala and L. Päivärinta, *Calderón's inverse conductivity problem in the plane*, Ann. of Math., **163** (2006), 265–299.

[4] K. Astala, M. Lassas and L. Päivärinta, *Calderón's inverse problem for anisotropic conductivity in the plane*, Comm. PDE, **30** (2005), 207–224.

[5] K. Astala, M. Lassas and L. Päivärinta, *The borderlines of invisibility and visibility of Calderón's inverse problem*, preprint, *Arxiv*:1109. 2749 (2011).

[6] Y. Berezanskii, *The uniqueness theorem in the inverse problem of spectral analysis for the Schrödinger equation*, (Russian) Trudy Moskov. Mat. Obsch., **7** (1958), 1–62.

[7] Y. Benveniste and T. Miloh, *Neutral inhomogeneities in conduction phenomeno*, Jour. Mech. Phys. Solids., **47** (1999), 1873.

[8] L. Borcea, *Electrical impedance tomography*, Inverse Problems, **18** (2002), R99–R136.

[9] R. Brown and R. Torres, *Uniqueness in the inverse conductivity problem for conductivities with* 3/2 *derivatives in* $L^p, p > 2n$, J. Fourier Analysis Appl., **9** (2003), 1049–1056.

[10] R. Brown and G. Uhlmann, *Uniqueness in the inverse conductivity problem with less regular conductivities in two dimensions,* Comm. PDE, **22** (1997), 10009–10027.

[11] A.P. Calderón, *On an inverse boundary value problem,* Seminar on Numerical Analysis and its Applications to Continuum Physics (Rio de Janeiro, 1980), pp. 65–73, Soc. Brasil. Mat., Rio de Janeiro, 1980.

[12] X. Chen, *Implicit boundary conditions in transformation-optics cloaking for electromagnetic waves*, Progress in Electromagnetics Research, **121** (2011), 521–534.

[13] H. Chen and C.T. Chan, *Time delays and energy transport velocities in three dimensional ideal cloaking*, J. Appl. Phys. **104** (2008), 033113.

[14] H. Chen and C.T. Chan, *Acoustic cloaking in three dimensions using acoustic metamaterials*, Appl. Phys. Lett. **91** (2007), 183518.

[15] M. Cheney, D. Isaacson and J. Newell, *Electrical impedance tomography*, SIAM Rev., **41** (1999), 85–101.

[16] D. Colton and R. Kress, *Inverse acoustic and electromagnetic scattering theory*, Springer-Verlag, Berlin (1992).

[17] S. Cummer, et al., *Scattering theory derivation of a* 3D *acoustic cloaking shell*, Phys. Rev. Lett., **100** (2008), 024301.

[18] G. Eleftheriades and K. Balmain, eds., *Negative-Refraction Metamaterials*, IEEE/ Wiley, Hoboken (2005).

[19] A. Greenleaf, Y. Kurylev, M. Lassas and G. Uhlmann, *Full-wave invisibility of active devices at all frequencies*, Comm. Math. Phys. **275** (2007), 749–789.

[20] A. Greenleaf, Y. Kurylev, M. Lassas and G. Uhlmann, *Improvement of cylindrical cloaking with the SHS lining*, Optics Express **15** (2007), 12717–12734.

[21] A. Greenleaf, Y. Kurylev, M. Lassas and G. Uhlmann, *Comment on "Scattering theory derivation of a* 3D *acoustic cloaking shell"*, *Arxiv*:0801.3279 (2008).

[22] A. Greenleaf, Y. Kurylev, M. Lassas, and G. Uhlmann, *Approximate quantum and acoustic cloaking.* J. Spectral Theory. **1** (2011), 27–80.

[23] A. Greenleaf, Y. Kurylev, M. Lassas, and G. Uhlmann, *Approximate quantum cloaking and almost trapped states*, Phys. Rev. Lett. **101** (2008) 220404.

[24] A. Greenleaf, Y. Kurylev, M. Lassas, and G. Uhlmann, *Isotropic transformation optics: approximate acoustic and quantum cloaking*, New J. Phys. **10** (2008) 115024.

[25] A. Greenleaf, Y. Kurylev, M. Lassas, and G. Uhlmann, *Cloaking a sensor via transformation optics*, Phys. Rev. E, **83** (2011), 016603.

[26] A. Greenleaf, Y. Kurylev, M. Lassas, U. Leonhardt and G. Uhlmann, *Cloaked electromagnetic, acoustic, and quantum amplifiers via transformation optics*, Proceedings of the National Academy of Sciences (PNAS), **109** (2012), no. 26, 10169–10174.

[27] A. Greenleaf, M. Lassas and G. Uhlmann, *The Calderón problem for conormal potentials, I: Global uniqueness and reconstruction*, Comm. Pure Appl. Math **56** (2003), no. 3, 328–352.

[28] A. Greenleaf, M. Lassas and G. Uhlmann, *Anisotropic conductivities that cannot detected in EIT*, Physiolog. Meas. (special issue on Impedance Tomography), **24** (2003), 413–420.

[29] A. Greenleaf, M. Lassas and G. Uhlmann, *On nonuniqueness for Calderón's inverse problem*, Math. Res. Lett. **10** (2003), no. 5-6, 685–693.

[30] B. Haberman, D. Tataru, *Uniqueness in Calderón's problem with Lipschitz conductivities*, Duke Math. J., in press; *arXiv*:1108.6068v1.

[31] A. Hoffman, et al., *Negative refraction in semiconductor metamaterials*, Nature Materials, doi:10.1038/nmat2033 (14 Oct. 2007).

[32] D. Isaacson, J. Mueller and S. Siltanen, Special issue on electrical impedance tomography of *Physiological Measurement*, 24, 2003

[33] M. Kerker, *Invisible bodies*, J. Opt. Soc. Am., **65** (1975), 376.

[34] T. Kilpeläinen, J. Kinnunen and O. Martio, *Sobolev spaces with zero boundary values on metric spaces*, Potential Anal., **12** (2000), 233–247.

[35] R. Kohn, D. Onofrei, M. Vogelius, and M. Weinstein, *Cloaking via change of variables for the Helmholtz equation*, Comm. Pure Appl. Math. **63** (2010), 0973–1016.

[36] R. Kohn, H. Shen, M. Vogelius and M. Weinstein, *Cloaking via change of variables in electrical impedance tomography*, Inverse Problems, **24** (2008), 015016.

[37] R. Kohn, M. Vogelius, *Determining conductivity by boundary measurements II, Interior results*, Comm. Pure Appl. Math., **38** (1985), 643–667.

[38] R. Kohn, M. Vogelius, *Identification of an unknown conductivity by means of measurements at the boundary*, in Inverse Problems, SIAM-AMS Proc., **14** (1984).

[39] M. Lassas, M. Taylor and G. Uhlmann, *The Dirichlet-to-Neumann map for complete Riemannian manifolds with boundary*, Comm. Geom. Anal., **11** (2003), 207–222.

[40] M. Lassas and G. Uhlmann, *Determining Riemannian manifold from boundary measurements*, Ann. Sci. École Norm. Sup. **34** (2001), 771–787.

[41] M. Lassas and T. Zhou, *Two-dimensional invisibility cloaking for Helmholtz equation and non-local boundary conditions*, Math. Res. Lett., **18** (2011), 10001–10015.

[42] J. Lee and G. Uhlmann, *Determining anisotropic real-analytic conductivities by boundary measurements*, Comm. Pure Appl. Math., **42** (1989), 1097–1112.

[43] U. Leonhardt, *Optical Conformal Mapping*, Science **312** (23 June, 2006), 1777–1780.

[44] J.-L. Lions and E. Magenes, *Non-homogeneous boundary value problems and applications*, Vol. 1, Springer-Verlag (1972).

[45] H.Y. Liu and T. Zhou, *On approximate electromagnetic cloaking by transformation media*, SIAM J. Appl. Math. **71** (2011), no. 1, 218–241.

[46] D. Miller, *On perfect cloaking*, Opt. Exp., **14** (2006), 12457–12466.

[47] G. Milton and N.-A. Nicorovici, *On the cloaking effects associated with anomalous localized resonance*, Proc. Royal Soc. A, **462** (2006), 3027–3059.

[48] A. Nachman, *Global uniqueness for a two-dimensional inverse boundary value problem*, Ann. of Math., **143** (1996), 71–96.

[49] H.-M. Nguyen, *Approximate cloaking for the Helmholtz equation via transformation optics and consequences for perfect cloaking*, Comm. Pure Appl. Math., **65** (2012), 155–188.

[50] L. Päivärinta, A. Panchenko and G. Uhlmann, *Complex geometrical optics for Lipschitz conductivities*, Rev. Mat. Iberoam., **19** (2003), 57–72.

[51] J.B. Pendry, D. Schurig and D.R. Smith, *Controlling electromagnetic fields*, Science **312** (2006), 1780–1782.

[52] J.B. Pendry, D. Schurig and D.R. Smith, *Calculation of material properties and ray tracing in transformation media*, Opt. Exp. **14** (2006), 9794.

[53] D. Schurig, J. Mock, B. Justice, S. Cummer, J. Pendry, A. Starr and D. Smith, *Metamaterial electromagnetic cloak at microwave frequencies*, Science, **314** (2006), no. 5801, pp. 977–980.

[54] J. Sylvester, *An anisotropic inverse boundary value problem*, Comm. Pure Appl. Math., **43** (1990), 201–232.

[55] J. Sylvester and G. Uhlmann, *A global uniqueness theorem for an inverse boundary value problem*, Ann. of Math., **125** (1987), 153–169.

[56] G. Uhlmann, *Scattering by a metric*, Chap. 6.1.5, in Encyclopedia on Scattering, R. Pike and P. Sabatier, eds., Academic Pr. (2002), 1668–1677.

[57] G. Uhlmann, *Developments in inverse problems since Calderón's foundational paper*, Chapter 19 in Harmonic Analysis and Partial Differential Equations, M. Christ, C. Kenig and C. Sadosky, eds., University of Chicago Press (1999), 295–345.

[58] R. Walser, in: W.S. Weiglhofer and A. Lakhtakia (eds.), *Introduction to Complex Mediums for Electromagnetics and Optics*, SPIE Press, Bellingham, WA, USA, 2003

[59] A. Ward and J. Pendry, *Refraction and geometry in Maxwell's equations*, Jour. Modern Opt., **43** (1996), 773–793.

[60] R. Weder, *A rigorous analysis of high order electromagnetic invisibility cloaks*, Jour. Phys. A: Math. Theor., **41** (2008), 065207.

[61] R. Weder, *The boundary conditions for electromagnetic invisibility cloaks*, J. Phys. A: Math. Theor. **41** (2008) 415401.

[62] B. Zhang, H. Chen, B.-I. Wu, and J.A. Kong, *Extraordinary surface voltage effect in the invisibility cloak with an active device inside*, Phys. Rev. Lett. **100** (2008), 063904.

Matti Lassas
Institute of Mathematics
Helsinki University of Technology
FIN-02015, Finland
e-mail: `matti.lassas@helsinki.fi`

T. Zhou
Department of Mathematics
Massachusetts Institute of Technology
Cambridge MA 02139, USA
e-mail: `tzhou@math.mit.edu`

Fourier Analysis
Trends in Mathematics, 285–309

Perturbed Besov Spaces by a Short-range Type Potential in an Exterior Domain

Tokio Matsuyama

Abstract. The aim of this work is to provide the equivalence relation between perturbed Besov spaces by a short-range potential in an exterior domain outside a convex obstacle, and the free ones.

Mathematics Subject Classification (2010). Primary 58C07; Secondary 58C40.

Keywords. Besov spaces, resolvent estimates, functional calculus.

1. Introduction

Let Ω be an exterior domain of $\mathbb{R}^n$ with a smooth boundary $\partial\Omega$ such that $\mathbb{R}^n \setminus \Omega$ is compact and $0 \notin \overline{\Omega}$. Throughout this paper, we will assume that $n \geq 3$ and $\mathbb{R}^n \setminus \Omega$ is convex. The convexity assumption on the obstacle will be needed for the construction of identification operators in §3. We consider the Schrödinger operator $-\Delta + V(x)$ on Ω, where the potential V is assumed to be the following:

(V) V is a real-valued measurable function on Ω satisfying

$$-c_0|x|^{-\delta_0} \leq V(x) \leq c_1|x|^{-\delta_0} \tag{1.1}$$

for some constants $0 < c_0 < (n-2)^2/4$, $c_1 > 0$ and $\delta_0 > (n+2)/2$. When $n = 3$, we assume $\delta_0 > 2$.

This paper will reveal the equivalence relation between perturbed Besov spaces $\dot{B}^s_{p,q,V}(\Omega)$ generated by the operator $-\Delta + V$ on $L^2(\Omega)$, and the free ones $\dot{B}^s_{p,q}(\Omega)$. When one makes Strichartz estimates for the propagator $e^{it\sqrt{-\Delta+V}}$, these estimates would be obtained by TT^* argument after establishing dispersive estimates. The norms of data in these decay estimates include the perturbed Besov spaces. However, it is not clear whether the equivalence relation

$$\dot{B}^s_{p,q,V}(\Omega) \cong \dot{B}^s_{p,q}(\Omega)$$

This paper was supported by Grant-in-Aid for Scientific Research (C) (No. 21540198), Japan Society for the Promotion of Science.

holds or not. Hence they would prevent us from using TT^* argument. Then a problem of equivalence relation between these spaces naturally arises. Thus, we need to go back to functional calculus for operators $\varphi(\theta(-\Delta+V))$ $(\theta>0)$ with a function $\varphi(\lambda)\in C_0^\infty((-\infty,\infty))$. For instance, when Ω is the whole space $\mathbb{R}^3$, Georgiev & Visciglia developed the functional calculus to discuss the equivalence relation, and as a result, they succeeded to obtain Strichartz estimates for the propagator $e^{it\sqrt{-\Delta+V}}$ (see [6]). After them, D'Ancona & Pierfelice further developed the argument of [6], and treated a more general class of potentials, say, Kato's class (see [4]). In exterior version of this problem, the present author developed the functional calculus and proved the equivalence relation mentioned above, and obtained the Strichartz estimates for $e^{it\sqrt{-\Delta+V}}$ provided that $\mathbb{R}^3\setminus\Omega$ is compact and convex (see [10]). However, the proofs of some delicate arguments are skipped there, and throughout all the arguments, the spatial dimension is restricted to $n=3$. The aim of the present paper is to provide an almost complete proof of the argument in [10], and prove L^∞-L^∞ resolvent estimates in general spatial dimensions (see Proposition 3.3). By using this result the exterior problem version of extended Hardy inequality will be proved in Lemma 4.1, and as a result, the isomorphic relation between perturbed Sobolev spaces and the free ones will be proved for higher spatial dimensions (see Theorem 2.2). Though the result on Besov spaces (see Theorems 2.3) are proved in [10], they will be given for the completeness.

In this paper we denote by $\mathscr{B}(X,Y)$ the space of all bounded linear operators from a Banach space X to another one Y. When $X=Y$, we denote $\mathscr{B}(X)=\mathscr{B}(X,X)$. $\mathscr{R}(T)$ stands for the range of an operator T. We often use the weighted Sobolev spaces. For a non-negative integer m, real number κ and a domain G of $\mathbb{R}^n$, we define the weighted Sobolev spaces to be

$$H^m_\kappa(G)=\{f:\langle x\rangle^\kappa\partial_x^\alpha f\in L^2(G),\ |\alpha|\le m\}\quad\text{and}\quad L^2_\kappa(G)=H^0_\kappa(G).$$

The rest of this paper is organized as follows. In §2 the main results will be stated. In §3 representation formulae for perturbed resolvent via the free one will be given, and several properties of generalized Fourier transforms will be discussed. §4 is devoted to the proof of the equivalence relation among the perturbed Sobolev spaces and the free ones. Functional calculus will be developed in §5. In §6 the equivalence relation with the perturbed Besov spaces and the free ones will be proved. In §7 it will be proved that potentials V satisfying assumption (V) are of generic type; zero is neither an eigenvalue nor a resonance of operator $-\Delta+V$.

2. Results

Let us introduce some operators and function spaces. We denote by A_V a self-adjoint realization in $L^2(\Omega)$ of the Schrödinger operator $-\Delta+V$ with domain $H^2(\Omega)\cap H^1_0(\Omega)$. Then A_V is non-negative on $L^2(\Omega)$ on account of (1.1). Later on, a class of potentials of generic type will be introduced, and we will see that zero is neither an eigenvalue nor a resonance of A_V. Also, it is known that, under assumption (V), A_V has no embedded strictly positive eigenvalues nor strictly positive resonances. Hence the continuous spectrum of A_V coincides with the interval $[0,\infty)$. We denote by H_V a square root of A_V:

$$H_V = A_V^{1/2}.$$

Let us define the perturbed Besov spaces $\dot{B}^s_{p,q,V}(\Omega)$ generated by $-\Delta + V$. As usual, let $\{\Phi_\ell(\lambda)\}_{\ell=-\infty}^{\infty}$ be the Littlewood–Paley partition of unity: $\Phi(\lambda)$ is a non-negative function having its support in $\{\lambda : 1/2 \le \lambda \le 2\}$ such that

$$\sum_{\ell=-\infty}^{+\infty} \Phi(2^{-\ell}\lambda) = 1 \quad (\lambda \neq 0),$$

where we put $\Phi_\ell(\lambda) = \Phi(2^{-\ell}\lambda)$, $\ell \in \mathbb{Z}$. For any $p, q \ge 1$ and for any $s \in \mathbb{R}$ we define the perturbed Besov spaces $\dot{B}^s_{p,q,V}(\Omega)$ as the completion of $C_0^\infty(\Omega)$ with respect to the following quasi-norm:

$$\|u\|_{\dot{B}^s_{p,q,V}(\Omega)} = \left\{ \sum_{\ell=-\infty}^{\infty} \left(2^{\ell s}\|\Phi_\ell(H_V)u\|_{L^p(\Omega)}\right)^q \right\}^{1/q}.$$

The quasi-norm $\|u\|_{\dot{B}^s_{p,q,V}(\Omega)}$ is independent of the choice of Φ. We shall use also the perturbed Sobolev spaces over Ω:

$$\dot{H}^{\sigma}_V(\Omega) = \dot{B}^{\sigma}_{2,2,V}(\Omega).$$

In particular case $V = 0$, denoting $H = (-\Delta)^{1/2}$ and replacing $\Phi_\ell(H_V)$ by $\Phi_\ell(H)$ in the definition, we write

$$\dot{B}^s_{p,q}(\Omega) = \dot{B}^s_{p,q,0}(\Omega) \quad \text{and} \quad \dot{H}^s(\Omega) = \dot{B}^s_{2,2,0}(\Omega).$$

As a related study, Georgiev & Visciglia investigated the perturbed Besov and Sobolev spaces over the whole space $\mathbb{R}^3$ (see [6], and also D'Ancona & Pierfelice [4]).

To state the results, we introduce a class of potentials of generic type in $L^2_{-s}(\Omega)$. Let $R(\lambda^2 \pm i0)$ be the resolvent operators:

$$R(\lambda^2 \pm i0) = \mathrm{s} - \lim_{\varepsilon \searrow 0} (A - (\lambda^2 \pm i\varepsilon))^{-1} \quad \text{in } \mathscr{B}(L^2_s(\Omega), H^2_{-s}(\Omega))$$

for some $s > 1/2$ and for any $\lambda > 0$, where A is the self-adjoint realization in $L^2(\Omega)$ of Dirichlet Laplacian $-\Delta$ with domain $H^2(\Omega) \cap H^1_0(\Omega)$. These limits are called the limiting absorption principle (see, e.g., Mochizuki [13]). It is proved in Proposition 2.5 of [12] that $R(0) = R(0 \pm i0)$ exist in $\mathscr{B}(L^2_s(\Omega), H^2_{-s}(\Omega))$ for some $s > 1/2$, where the assumption $s > n/2$ is required in [12], but it can be relaxed to $s > 1/2$ by a trivial modifications. Thus, referring to Yajima [19], we define the null space of $I + R(0)V$:

$$\mathcal{M} = \left\{ u \in L^2_{-s}(\Omega) \cap L^\infty(\Omega) : u + R(0)Vu = 0 \text{ in } \Omega \text{ and } u = 0 \text{ on } \partial\Omega \right\}$$

for some $1/2 < s \le \delta_0/2$. When $n = 3$, we can remove L^∞ assumption on u in the definition of $\mathcal{M}$ (see §7). Now, any $u \in \mathcal{M}$ satisfies the boundary value problem for the stationary Schrödinger equation:

$$\begin{cases} \{-\Delta + V(x)\}u(x) = 0 & \text{in } \Omega, \\ u(x) = 0 & \text{on } \partial\Omega. \end{cases} \tag{2.1}$$

Conversely, any function $u \in L^2_{-s}(\Omega)$ for some $1/2 < s \le \delta_0/2$ satisfying (2.1) belongs to $\mathcal{M}$, since Vu belongs to $L^2_s(\Omega)$ for such an s. Hence the eigenspace, denoted by $\mathcal{E}$, of $-\Delta + V$ with eigenvalue 0 is a subspace of $\mathcal{M}$. Elements in $\mathcal{M} \setminus \mathcal{E}$ are called resonances of $-\Delta + V$. Then we define a class of potentials as follows:

Definition 2.1. We say that V is of generic type if $\mathcal{M} = \{0\}$.

In the final section we will prove that the potential V satisfying assumption (V) is of generic type. This meas that zero is neither an eigenvalue nor a resonance of operator $-\Delta + V$.

We shall prove here the following:

Theorem 2.2. *Let $n \ge 3$. Assume that the measurable potential V is non-negative and satisfies assumption* (V). *Then for any real s satisfying $|s| \le 3/2$ there exist two constants $c = c(s) > 0$ and $C = C(s) > 0$ such that*

$$c\|f\|_{\dot{H}^s(\Omega)} \le \|f\|_{\dot{H}^s_V(\Omega)} \le C\|f\|_{\dot{H}^s(\Omega)} \tag{2.2}$$

for any $f \in C_0^\infty(\Omega)$. When $n = 3$, $s = \pm 3/2$ are excluded.

Let us give a few remarks on Theorem 2.2. When $n = 3$, the isomorphism (2.2) can be found in the previous paper [10] without proof. Hence Theorem 2.2 generalizes [10] to higher spatial dimensions. The proof is based on the extended Hardy inequality (see Lemma 4.1), which would not be still known in the previous literatures. In particular case $\Omega = \mathbb{R}^3$, we refer to Theorem VII.4 of Georgiev & Visciglia [6]. It seems that when $n \ge 4$, we can expect the range of the regularity number s to be $|s| < n/2$. However, it is possible only when the potential $V(x)$ is of $C^{[n/2]+1}(\overline{\Omega})$.

As to the isomorphic relation of Besov spaces, the Hölder continuity of the potential V is required. More precisely we have:

Theorem 2.3. *Let $n = 3$. In addition to the assumptions on V in Theorem 2.2, suppose that V belongs to $C^{0,\alpha}_{\mathrm{loc}}(\overline{\Omega})$. Then there exists a real $\tilde{p} > 2$ such that the following assertion holds: For any real $p \ge \tilde{p}$ there exist constants $\delta_p > 0$, $c > 0$ and $C > 0$ such that*

$$c\|f\|_{\dot{B}^s_{p,q}(\Omega)} \le \|f\|_{\dot{B}^s_{p,q,V}(\Omega)} \le C\|f\|_{\dot{B}^s_{p,q}(\Omega)} \tag{2.3}$$

for any $|s| < \delta_p$ and for any $f \in C_0^\infty(\Omega)$.

The restriction $n = 3$ comes from Proposition 3.2 in §3. Finally, interpolating between Theorems 2.2–2.3, we have:

Theorem 2.4. *Let $n = 3$ and let V be as in Theorem 2.3. Then for $1 < p < \infty$, we have*

$$\dot{B}^0_{p,2,V}(\Omega) \cong \dot{B}^0_{p,2}(\Omega).$$

3. Resolvent estimates and the generalized Fourier transform

In this section we shall review a spectral analysis related with several properties of the perturbed resolvent $R_V(z) = (A_V - z)^{-1}$. We shall also define the generalized Fourier transform.

It is well known that if the measurable potential V satisfies assumption (V), then the limiting absorption principle

$$R_V(\lambda^2 \pm i0) = \text{s-} \lim_{\varepsilon \searrow 0} (A_V - (\lambda^2 \pm i\varepsilon))^{-1} \text{ in } \mathscr{B}(L^2_s(\Omega), H^2_{-s}(\Omega)) \tag{3.1}$$

holds for any $s > 1/2$ and $\lambda > 0$ (see Mochizuki [13]). In particular case $V = 0$ and $\Omega = \mathbb{R}^n$, we have also (3.1). Namely, let us denote by A_0 the self-adjoint realization in $L^2(\mathbb{R}^n)$ of $-\Delta$ with domain $H^2(\mathbb{R}^n)$, and by $R_0(z) = (A_0 - z)^{-1}$ the free resolvent operator in $\mathbb{R}^n$. Then we have

$$R_0(\lambda^2 \pm i0) = \text{s-} \lim_{\varepsilon \searrow 0} (A_0 - (\lambda^2 \pm i\varepsilon))^{-1} \quad \text{in } \mathscr{B}(L^2_s(\mathbb{R}^n), H^2_{-s}(\mathbb{R}^n))$$

for any $s > 1/2$ and $\lambda > 0$ (see, e.g., Agmon [1]). Also, we have the uniform resolvent estimates

$$\|R_0(\lambda^2 \pm i0) f\|_{L^2_{-s}(\mathbb{R}^n)} \leq C\lambda^{-1} \|f\|_{L^2_s(\mathbb{R}^n)}, \quad \forall \lambda > 0 \tag{3.2}$$

for any $s > 1/2$ (see Ben-Artzi & Klainerman [2]). We should also refer to the result of the limiting absorption principle in the critical case $s = 1/2$, where Ruzhansky & Sugimoto found (see [15]). For small $\lambda \geq 0$, the continuity of $R_0(\lambda^2 \pm i0)$ with respect to $\lambda \geq 0$ and the existence of $R_0(0 \pm i0)$ in $\mathscr{B}(L^2_s(\mathbb{R}^n), H^2_{-s}(\mathbb{R}^n))$ for $s > 1/2$ imply that $R_0(\lambda^2 \pm i0)$ are bounded in $\mathscr{B}(L^2_s(\mathbb{R}^n), H^2_{-s}(\mathbb{R}^n))$ for any $s > 1/2$ and $0 \leq \lambda \leq 1$. Thus, combining (3.2) with this fact, we get the uniform resolvent estimates:

$$\|R_0(\lambda^2 \pm i0) f\|_{L^2_{-s}(\mathbb{R}^n)} \leq C(1 + \lambda)^{-1} \|f\|_{L^2_s(\mathbb{R}^n)} \tag{3.3}$$

for any $s > 1/2$ and $\lambda \geq 0$.

3.1. Resolvent estimates

In this subsection we will investigate the relation among the resolvent operators $R_V(\lambda^2 \pm i0)$, $R(\lambda^2 \pm i0)$ and $R_0(\lambda^2 \pm i0)$. For this purpose, we need identify the space $L^2(\mathbb{R}^n)$ that $-\Delta$ acts on, with $L^2(\Omega)$ that $-\Delta + V$ acts on. Following Mochizuki [13] (see also Kuroda [9]), we define *identification operators*

$$J \in \mathscr{B}(L^2(\mathbb{R}^n), L^2(\Omega)), \quad J_* \in \mathscr{B}(L^2(\Omega), L^2(\mathbb{R}^n)) \tag{3.4}$$

as follows. Let $j(x) = j(r\omega) \in C^\infty(\mathbb{R}^n)$ ($r = |x|, \omega = x/|x| \in \mathbb{S}^{n-1}$) be a non-decreasing function in r for each ω such that $j(x) = 0$ for $x \in \mathbb{R}^n \setminus \Omega$, and equal to 1 for $x \in \widetilde{\Omega} := \{x : \text{dist}(x, \mathbb{R}^n \setminus \Omega) > 1\}$, which is possible if we assume that $\mathbb{R}^n \setminus \Omega$ is convex. In fact, for each $\omega \in \mathbb{S}^{n-1}$ taking $0 < r_1(\omega) < r_2(\omega)$ such that $r_1(\omega)\omega \in \partial(\mathbb{R}^n \setminus \Omega)$ and $r_2(\omega)\omega \in \partial\widetilde{\Omega}$, we let $j(r\omega) = 0$ for $r \leq r_1(\omega)$, $j(r\omega) = 1$

for $r \geq r_2(\omega)$, and $j(r\omega)$ is non-decreasing in $[r_1(\omega), r_2(\omega)]$. For $f \in L^2_{\mathrm{loc}}(\mathbb{R}^n)$ we define the operator J to be

$$(Jf)(x) = j(x)f(x), \quad x \in \Omega,$$

and for $g \in L^2_{\mathrm{loc}}(\overline{\Omega})$ define

$$(J_* g)(x) = \begin{cases} j(x)g(x), & x \in \Omega, \\ 0, & x \in \mathbb{R}^n \setminus \Omega. \end{cases}$$

Then $J \in \mathscr{B}(L^2_s(\mathbb{R}^n), L^2_s(\Omega))$ and $J_* \in \mathscr{B}(L^2_s(\Omega), L^2_s(\mathbb{R}^n))$ for any $s \in \mathbb{R}$, and hence, (3.4) holds. It is readily checked that J_* is the adjoint operator of J: For any $f \in L^2(\mathbb{R}^n)$ and $g \in L^2(\Omega)$,

$$(Jf, g)_{L^2(\Omega)} = (f, J_* g)_{L^2(\mathbb{R}^n)}.$$

Put

$$W_* = J_*(-\Delta + V(x)) + \Delta J_*.$$

By elementary calculus, we can write

$$W_* g = j(x)V(x)g - (\Delta j)g - 2\nabla j \cdot \nabla g.$$

Then we have the "resolvent equations":

$$J_* R_V(\lambda^2 \pm i0) = R_0(\lambda^2 \pm i0)\left\{J_* - W_* R_V(\lambda^2 \pm i0)\right\},$$

which implies that

$$\left\{J_* + R_0(\lambda^2 \pm i0)W_*\right\} R_V(\lambda^2 \pm i0) = R_0(\lambda^2 \pm i0)J_*. \tag{3.5}$$

As a byproduct of resolvent equations, if $\mathscr{R}(R_V(\lambda^2 \pm i0))$ is contained in the domain of $J_* + R_0(\lambda^2 \pm i0)W_*$, we have

$$\mathscr{R}(R_0(\lambda^2 \pm i0)J_*) \subset \mathscr{R}(J_* + R_0(\lambda^2 \pm i0)W_*) \subset L^2_{-s}(\mathbb{R}^n), \quad \forall \lambda \geq 0 \tag{3.6}$$

for any $s > 1/2$ (see Lemma 3.1 in [10]), where $\mathscr{R}(T)$ is the range of an operator T. We notice that the second inclusion in (3.6) is an immediate consequence of the uniform resolvent estimates (3.3).

We often use the following Banach spaces; we define

$$L^2_{\nabla,-s}(\Omega) = \left\{u \in L^2_{-s}(\Omega) : \nabla j \cdot \nabla u \in L^2(\Omega) \text{ and } u = 0 \text{ on } \partial\Omega\right\}$$

with norm

$$\|u\|_{L^2_{\nabla,-s}(\Omega)} = \|u\|_{L^2_{-s}(\Omega)} + \|\nabla j \cdot \nabla u\|_{L^2(\Omega)}.$$

An L^p-version $(1 \leq p \leq \infty)$ is defined by

$$L^p_\nabla(\Omega) = \{u \in L^p(\Omega) : \nabla j \cdot \nabla u \in L^p(\Omega) \text{ and } u = 0 \text{ on } \partial\Omega\}$$

with norm

$$\|u\|_{L^p_\nabla(\Omega)} = \|u\|_{L^p(\Omega)} + \|\nabla j \cdot \nabla u\|_{L^p(\Omega)}.$$

Referring to (3.6), we have:

Proposition 3.1. *Assume that the measurable potential V satisfies assumption* (V). *Let $1/2 < s \le \delta_0/2$. If the domain of $J_* + R_0(\lambda^2 \pm i0)W_*$ is $L^2_{\nabla,-s}(\Omega)$ for $\lambda \ge 0$, there exists a family of bounded operators $S^\pm_\lambda$ from $\mathscr{R}(J_* + R_0(\lambda^2 \pm i0)W_*)$ to $L^2_{\nabla,-s}(\Omega)$ such that*

$$S^\pm_\lambda \circ \{J_* + R_0(\lambda^2 \pm i0)W_*\} = I, \tag{3.7}$$

$$R_V(\lambda^2 \pm i0) = S^\pm_\lambda R_0(\lambda^2 \pm i0)J_* \tag{3.8}$$

for any $\lambda \ge 0$. Furthermore, there exists a constant $C > 0$ such that

$$\left\|S^\pm_\lambda R_0(\lambda^2 \pm i0)J_* f\right\|_{L^2_{-s}(\Omega)} \le C\left\|R_0(\lambda^2 \pm i0)J_* f\right\|_{L^2_{-s}(\mathbb{R}^n)} \tag{3.9}$$

for any $\lambda \ge 0$ and $f \in L^2_s(\Omega)$.

The proof of Proposition 3.1 is identical to the one in Proposition 3.2 from [10] that treats three-dimensional case. Hence we omit it.

The L^p-version of Proposition 3.1 is important to develop the functional calculus.

Proposition 3.2. *Let $n = 3$. Assume that the measurable potential V satisfies assumption* (V). *Then there exists a real $\tilde{p} > 2$ such that the following assertion holds: For any real p with $\tilde{p} \le p < \infty$, if the domain of $J_* + R_0(\lambda^2 \pm i0)W_*$ is $L^p_\nabla(\Omega)$ for $\lambda \ge 0$, there exists a family of bounded operators $S^\pm_\lambda$ from $\mathscr{R}(J_* + R_0(\lambda^2 \pm i0)W_*)$ to $L^p_\nabla(\Omega)$ such that equations* (3.7)–(3.8) *hold for any $\lambda \ge 0$. Furthermore, there exists a constant $C > 0$ such that*

$$\left\|S^\pm_\lambda R_0(\lambda^2 \pm i0)J_* f\right\|_{L^p(\Omega)} \le C\|R_0(\lambda^2 \pm i0)J_* f\|_{L^p(\mathbb{R}^n)}, \quad \forall\lambda \ge 0 \tag{3.10}$$

for $f \in L^p(\Omega)$.

The proof of Proposition 3.2 is given in [10, Proposition 6.2]. As to the estimates for $n \ge 4$, it is open problem, since we need to develop the analysis of free resolvent from another point of view.

We need L^∞-estimates for the free resolvent $R(\lambda^2 \pm i0) = (A - (\lambda^2 \pm i0))^{-1}$ to obtain L^∞-estimate for the generalized Fourier transform. Referring to equation (3.5), let us define an operator

$$V_* = [-J_*, \Delta] = -J_*\Delta + \Delta J_*.$$

Then

$$V_* g = -(\Delta j)g - 2\nabla j \cdot \nabla g,$$

and we have the resolvent equations:

$$\{J_* + R_0(\lambda^2 \pm i0)V_*\}\, R(\lambda^2 \pm i0) = R_0(\lambda^2 \pm i0)J_*. \tag{3.11}$$

We notice that

$$L^\infty_\nabla(\Omega) \subset L^2_{\nabla,-s}(\Omega) \quad \text{provided } s > n/2. \tag{3.12}$$

Then we have:

Proposition 3.3. *Let $n \geq 3$. If the domain of $J_* + R_0(\lambda^2 \pm i0)V_*$ is $L^\infty_\nabla(\Omega)$ for any $\lambda \geq 0$, there exists a family of bounded operators $S^\pm_{0,\lambda}$ from $\mathscr{R}(J_* + R_0(\lambda^2 \pm i0)V_*)$ to $L^\infty_\nabla(\Omega)$ such that*

$$S^\pm_{0,\lambda} \circ \{J_* + R_0(\lambda^2 \pm i0)V_*\} = I, \tag{3.13}$$

$$R(\lambda^2 \pm i0) = S^\pm_{0,\lambda} R_0(\lambda^2 \pm i0) J_* \tag{3.14}$$

for any $\lambda \geq 0$. Furthermore, there exists a constant $C > 0$ such that

$$\left\| R(\lambda^2 \pm i0) f \right\|_{L^\infty(\Omega)} \leq C \left\| R_0(\lambda^2 \pm i0) J_* f \right\|_{L^\infty(\mathbb{R}^n)}, \quad \forall \lambda \geq 0 \tag{3.15}$$

for $f \in C_0^\infty(\Omega)$.

Proof. First, we prove that there exists a constant $C > 0$ such that

$$\left\| R_0(\lambda^2 \pm i0) V_* g \right\|_{L^2_{-s}(\mathbb{R}^n)} \leq C(1+\lambda)^{-1} \left(\|g\|_{L^\infty(\Omega)} + \|\nabla j \cdot \nabla g\|_{L^\infty(\Omega)} \right) \tag{3.16}$$

for any $\lambda \geq 0$ and $s > 1/2$. In fact, by using the uniform resolvent estimates (3.3) we can write

$$\|R_0(\lambda^2 \pm i0) V_* g\|_{L^2_{-s}(\mathbb{R}^n)} \leq C(1+\lambda)^{-1} \left\{ \|(\Delta j) g\|_{L^2_s(\Omega)} + \|\nabla j \cdot \nabla g\|_{L^2_s(\Omega)} \right\}$$

for any $\lambda \geq 0$ and $s > 1/2$. Here the right members are estimated as

$$\|(\Delta j) g\|_{L^2_s(\Omega)} + \|\nabla j \cdot \nabla g\|_{L^2_s(\Omega)} \leq C \left(\|g\|_{L^\infty(\Omega)} + \|\nabla j \cdot \nabla g\|_{L^\infty(\Omega)} \right)$$

for any $s \in \mathbb{R}$. Hence we get the estimates (3.16). Now, the estimates (3.16) and $J_* \in \mathscr{B}(L^\infty_\nabla(\Omega), L^2_{-s}(\mathbb{R}^n))$ for $s > n/2$ imply that $J_* + R_0(\lambda^2 \pm i0)V_*$ map from $L^\infty_\nabla(\Omega)$ to $\mathscr{R}(J_* + R_0(\lambda^2 \pm i0)V_*)$ for any $\lambda \geq 0$. On the other hand, thanks to Proposition 3.1, $J_* + R_0(\lambda^2 \pm i0)V_*$ are injective from $L^2_{\nabla,-s}(\Omega)$ onto $\mathscr{R}(J_* + R_0(\lambda^2 \pm i0)V_*)$ for $\lambda \geq 0$, and thanks to the embedding (3.12), the operators $J_* + R_0(\lambda^2 \pm i0)V_*$ are also injective from $L^\infty_\nabla(\Omega)$ to $\mathscr{R}(J_* + R_0(\lambda^2 \pm i0)V_*)$. Hence, the inverses $S^\pm_{0,\lambda}$ of $J_* + R_0(\lambda^2 \pm i0)V_*$ exist and map from $\mathscr{R}(J_* + R_0(\lambda^2 \pm i0)V_*)$ to $L^\infty_\nabla(\Omega)$ for any $\lambda \geq 0$, which proves (3.13). Also, equations (3.14) follow from (3.11) and (3.13).

Next, we prove the boundedness of $S^\pm_{0,\lambda}$ from $\mathscr{R}(J_* + R_0(\lambda^2 \pm i0)V_*)$ to $L^\infty_\nabla(\Omega)$, where $\mathscr{R}(J_* + R_0(\lambda^2 \pm i0)V_*)$ are endowed with the induced topology of $L^\infty(\mathbb{R}^n)$. We note that the operators $R_0(\lambda^2 \pm i0)V_*$ map from $L^\infty_\nabla(\Omega)$ to $L^\infty(\mathbb{R}^n)$ for each $\lambda \geq 0$. Indeed, we write

$$\left[R_0(\lambda^2 \pm i0) V_* g \right](x) = \int_{\mathbb{R}^n} K_\lambda(|x-y|)(V_* g)(y)\, dy = \left[K_\lambda(|x|) * (V_* g) \right](x) \tag{3.17}$$

$$\text{with} \quad K_\lambda(|x|) = -\frac{i\lambda^{(n/2)-1}}{2^{(n/2)+1}\pi^{(n/2)-1}|x|^{(n/2)-1}} H^{(1)}_{(n/2)-1}(\lambda|x|), \tag{3.18}$$

where $H^{(1)}_{(n/2)-1}$ is the Hankel function of the first kind. By using the property of Hankel functions we have

$$|K_\lambda(|x|)| \leq \begin{cases} c_1 |x|^{-(n-2)}, & \lambda|x| \leq 1, \quad n \geq 3, \\ c_2 (1+\lambda)^{(n-3)/2} |x|^{-(n-1)/2}, & \lambda|x| \geq 1, \quad n \geq 1 \end{cases} \tag{3.19}$$

for each $\lambda \geq 0$ (see, e.g., (13.7.2) from [17]). Then we can estimate the right member as

$$\begin{aligned} |[K_\lambda(|x|) * (V_* g)](x)| &\leq c(\lambda, n, L)(\|g\|_{L^\infty(\Omega)} + \|\nabla j \cdot \nabla g\|_{L^\infty(\Omega)}) \\ &\quad \times \left(\int_{\mathbb{R}^n} \frac{dy}{|x-y|^{n-2}\langle y\rangle^L} + \int_{\mathbb{R}^n} \frac{dy}{|x-y|^{(n-1)/2}\langle y\rangle^L} \right) \\ &\leq c(\lambda, n, L)\langle x\rangle^{-(n-1)/2}(\|g\|_{L^\infty(\Omega)} + \|\nabla j \cdot \nabla g\|_{L^\infty(\Omega)}) \end{aligned} \tag{3.20}$$

for any $L > (n+1)/2$, by explicit computations

$$\int_{\mathbb{R}^n} \frac{dy}{|x-y|^{n-2}\langle y\rangle^L} + \int_{\mathbb{R}^n} \frac{dy}{|x-y|^{(n-1)/2}\langle y\rangle^L} \leq c(n, L)\langle x\rangle^{-(n-1)/2}.$$

Hence $R_0(\lambda^2 \pm i0)V_*$ map from $L^\infty_\nabla(\Omega)$ to $L^\infty(\mathbb{R}^n)$ for each $\lambda \geq 0$. Now we claim that $\mathscr{R}(J_* + R_0(\lambda^2 \pm i0)V_*)$ are closed subspaces in $L^\infty(\mathbb{R}^n)$. In fact, let $\{f_{\pm,k}\}_{k=1}^\infty$ be sequences in $L^\infty_\nabla(\Omega)$ such that

$$\{J_* + R_0(\lambda^2 \pm i0)V_*\}f_{\pm,k} \to g_{\lambda,\pm} \quad \text{in } L^\infty(\mathbb{R}^n) \text{ as } k \to \infty,$$

which implies that

$$\big\|\{J_* + R_0(\lambda^2 \pm i0)V_*\}(f_{\pm,k} - f_{\pm,\ell})\big\|_{L^\infty(\mathbb{R}^n)} \to 0 \quad \text{as } k, \ell \to \infty. \tag{3.21}$$

In particular, $\{J_* + R_0(\lambda^2 \pm i0)W_*\}(f_{\pm,k} - f_{\pm,\ell})\}_{k,\ell}$ are uniformly bounded in $L^\infty(\mathbb{R}^n)$ with respect to k, ℓ. Then the family $\{f_{\pm,k} - f_{\pm,\ell}\}_{k,\ell}$ can be identified with a family of operators from the Banach space $\mathscr{B}(L^\infty_\nabla(\Omega), L^\infty(\mathbb{R}^n))$ to the another one $L^\infty(\mathbb{R}^n)$, and hence, by using principle of uniform boundedness we get

$$\sup_{k,\ell} \|f_{\pm,k} - f_{\pm,\ell}\|_{L^\infty_\nabla(\Omega)} < \infty,$$

which implies that there exist $\tilde{f}_\pm \in L^\infty_\nabla(\Omega)$ and subsequences $\{f_{\pm,k_m} - f_{\pm,\ell_m}\}_{m=1}^\infty$ such that

$$f_{\pm,k_m} - f_{\pm,\ell_m} \to \tilde{f}_\pm \quad \text{weakly}^* \text{ in } L^\infty_\nabla(\Omega) \text{ as } m \to \infty. \tag{3.22}$$

Next, we claim that

$$J_*(f_{\pm,k} - f_{\pm,\ell}) \to 0 \quad \text{in } L^\infty(\mathbb{R}^n) \text{ as } k, \ell \to \infty. \tag{3.23}$$

If (3.23) does not hold, there exist a constant $\varepsilon_0 > 0$ and subsequences $\{f_{\pm,k_m} - f_{\pm,\ell_m}\}_{m=1}^\infty$ of $\{f_{\pm,k} - f_{\pm,\ell}\}$ such that

$$\|J_*(f_{\pm,k_m} - f_{\pm,\ell_m})\|_{L^\infty(\mathbb{R}^n)} \geq \varepsilon_0 \quad (m = 1, 2, \ldots).$$

Since $\left[R_0(\lambda^2 \pm i0)V_*(f_{\pm,k_m} - f_{\pm,\ell_m})\right](x)$ decay to 0 as $|x| \to \infty$ by (3.20), it follows that

$$\left|\left[\{J_* + R_0(\lambda^2 \pm i0)W_*\}(f_{\pm,k_m} - f_{\pm,\ell_m})\right](x)\right| \geq \varepsilon_0/2, \quad |x| \gg 1.$$

This contradicts (3.21). Thus (3.23) is true. Now the limits (3.22)–(3.23) imply that $\tilde{f}_\pm = 0$ in Ω and

$$f_{\pm,k_m} - f_{\pm,\ell_m} \to 0 \quad \text{weakly}^* \text{ in } L^\infty(\Omega) \text{ as } m \to \infty;$$

hence, we find the limits $f_\pm \in L^\infty(\Omega)$ satisfying $f_\pm = 0$ on $\partial\Omega$ such that

$$f_{\pm,k_m} \to f_\pm \quad \text{weakly}^* \text{ in } L^\infty(\Omega) \text{ as } m \to \infty. \tag{3.24}$$

On the other hand, we see from (3.21)–(3.23) that

$$\psi^\lambda_{\pm,m} := R_0(\lambda^2 \pm i0)V_*(f_{\pm,k_m} - f_{\pm,\ell_m}) \to 0 \quad \text{in } L^\infty(\mathbb{R}^n) \text{ as } m \to \infty,$$

and $V_*(f_{\pm,k_m} - f_{\pm,\ell_m})$ satisfy

$$V_*(f_{\pm,k_m} - f_{\pm,\ell_m}) = (-\Delta - \lambda^2)\psi^\lambda_{\pm,m}.$$

Hence

$$\langle V_*(f_{\pm,k_m} - f_{\pm,\ell_m}), \varphi\rangle = \langle \psi^\lambda_{\pm,m}, (-\Delta - \lambda^2)\varphi\rangle \to 0$$

for any $\varphi \in C_0^\infty(\mathbb{R}^n)$ as $m \to \infty$. Here both members are the duality pair of $L^\infty(\mathbb{R}^n)$ and $L^1(\mathbb{R}^n)$. Therefore, observing the definition of V_* and (3.24), we have

$$\nabla j \cdot \nabla(f_{\pm,k_m} - f_{\pm,\ell_m}) \to 0 \quad \text{weakly}^* \text{ in } L^\infty(\Omega) \text{ as } m \to \infty.$$

Hence there exist the weak* limits

$$f_{\pm,k_m} \to f_\pm \quad \text{weakly}^* \text{ in } L^\infty_{\nabla,-s}(\Omega) \text{ as } m \to \infty;$$

thus we conclude $g_{\lambda,\pm} = \{J_* + R_0(\lambda^2 \pm i0)V_*\}f_\pm \in \mathscr{R}(J_* + R_0(\lambda^2 \pm i0)V_*)$. This proves that $\mathscr{R}(J_* + R_0(\lambda^2 \pm i0)V_*)$ are closed in $L^\infty(\mathbb{R}^n)$, as desired. Then, by using the closed graph theorem, the inverses $S^\pm_{0,\lambda}$ belong to $\mathscr{B}(\mathscr{R}(J_* + R_0(\lambda^2 \pm i0)V_*), L^\infty_\nabla(\Omega))$ for each $\lambda \geq 0$.

Finally, we prove the uniform estimates (3.15). As a first step, we prove that

$$\left\|S^\pm_\lambda u_\lambda\right\|_{L^\infty(\Omega)} \leq C\|u_\lambda\|_{L^\infty(\mathbb{R}^n)}, \quad \forall\lambda \geq 0 \tag{3.25}$$

for $u_\lambda \in \mathscr{R}(J_* + R_0(\lambda^2 \pm i0)V_*)$. We notice from (3.12) that

$$\|J_* + R_0(\lambda^2 \pm i0)V_*\|_{\mathscr{B}(L^\infty_\nabla(\Omega),L^\infty(\mathbb{R}^n))} \geq C\|J_* + R_0(\lambda^2 \pm i0)V_*\|_{\mathscr{B}(L^\infty_\nabla(\Omega),L^2_{-s}(\mathbb{R}^n))}$$

for $s > n/2$. Hence it is sufficient to prove (3.25) that there exists a constant $C > 0$ such that

$$\|J_* + R_0(\lambda^2 \pm i0)V_*\|_{\mathscr{B}(L^\infty_\nabla(\Omega),L^2_{-s}(\mathbb{R}^n))} \geq C, \quad \forall\lambda \geq 0. \tag{3.26}$$

To begin with, we prove the estimates (3.26) for large λ. Indeed, we find from the estimates (3.16) that there exists $\lambda_0 > 1$ such that

$$\left\|R_0(\lambda^2 \pm i0)V_*\right\|_{\mathscr{B}(L^\infty_\nabla(\Omega),L^2_{-s}(\mathbb{R}^n))} \leq \frac{1}{2}\|J_*\|_{\mathscr{B}(L^\infty_\nabla(\Omega),L^2_{-s}(\mathbb{R}^n))}$$

for any $\lambda \geq \lambda_0$, and hence, we get

$$\left\|J_* + R_0(\lambda^2 \pm i0)V_*\right\|_{\mathscr{B}(L^\infty_\nabla(\Omega),L^2_{-s}(\mathbb{R}^n))} \geq \frac{1}{2}\|J_*\|_{\mathscr{B}(L^\infty_\nabla(\Omega),L^2_{-s}(\mathbb{R}^n))}$$

for any $\lambda \geq \lambda_0$. This proves (3.26) for $\lambda \geq \lambda_0$. For $\lambda \in [0, \lambda_0]$, we suppose that

$$\inf_{\lambda\in[0,\lambda_0]} \left\|J_* + R_0(\lambda^2 \pm i0)V_*\right\|_{\mathscr{B}(L^\infty_\nabla(\Omega),L^2_{-s}(\mathbb{R}^n))} = 0,$$

and lead to a contradiction. By the compactness of $[0, \lambda_0]$ there exists a sequence $\{\lambda_j\}_{j=1}^{\infty} \subset [0, \lambda_0]$ which converges to some $\lambda_* \in [0, \lambda_0]$, but

$$\left\| J_* + R_0(\lambda_j^2 \pm i0)V_* \right\|_{\mathscr{B}(L^\infty_\nabla(\Omega), L^2_{-s}(\mathbb{R}^n))} \to 0 \quad (j \to \infty).$$

Hence

$$\left\| J_* + R_0(\lambda_*^2 \pm i0)V_* \right\|_{\mathscr{B}(L^\infty_\nabla(\Omega), L^2_{-s}(\mathbb{R}^n))} = 0.$$

This contradicts the injection property of $J_* + R_0(\lambda_*^2 \pm i0)V_*$. Thus we conclude (3.26) for $\lambda \in [0, \lambda_0]$. In conclusion, we get (3.26) for any $\lambda \geq 0$.

We now turn to the proof of (3.15). By using the regularity theory for elliptic equations, we obtain the fact that if $f \in C_0^\infty(\Omega)$, then $R(\lambda^2 \pm i0)f \in L^\infty_\nabla(\Omega)$ for each $\lambda \geq 0$. Then, if the domain of $R(\lambda^2 \pm i0)$ is $C_0^\infty(\Omega)$, then $\mathscr{R}(R(\lambda^2 \pm i0))$ is contained in $\mathscr{R}(J_* + R(\lambda^2 \pm i0)V_*)$. Hence observing equation (3.5), we have

$$\mathscr{R}(R_0(\lambda^2 \pm i0)J_*) \subset \mathscr{R}(J_* + R(\lambda^2 \pm i0)V_*).$$

Thus, this inclusion and (3.25) imply the required estimate (3.15). The proof of Proposition 3.3 is finished. □

We conclude this subsection by proving L^p-estimates for $R_V(\lambda^2 \pm i0)V$. Observing the proof of the following lemma, we can understand the exact value of $\tilde{p}$ appearing in Proposition 3.2.

Lemma 3.4. *Let $n = 3$, and let V and $\tilde{p}$ be as in Proposition 3.2. Then there exists a constant $C > 0$ such that for any real $p \geq \tilde{p}$,*

$$\left\| R_V(\lambda^2 \pm i0)Vf \right\|_{L^p(\Omega)} \leq C(1+\lambda)^{-2/p}\|f\|_{L^p(\Omega)}, \quad \forall \lambda \geq 0. \tag{3.27}$$

Proof. By using (3.10) from Proposition 3.2, we have:

$$\left\| R_V(\lambda^2 \pm i0)Vf \right\|_{L^p(\Omega)} \leq C \left\| R_0(\lambda^2 \pm i0)J_*Vf \right\|_{L^p(\mathbb{R}^3)}. \tag{3.28}$$

It is sufficient to prove that there exists a number $\tilde{p} > 2$ such that for any $p > \tilde{p}$,

$$\left\| R_0(\lambda^2 \pm i0)J_*Vf \right\|_{L^p(\Omega)} \leq C(1+\lambda)^{-2/p}\|f\|_{L^p(\Omega)}. \tag{3.29}$$

By using the uniform resolvent estimates (3.3) we have

$$\begin{aligned} \left\| R_0(\lambda^2 \pm i0)J_*Vf \right\|_{L^2_{-s}(\mathbb{R}^3)} &\leq C(1+\lambda)^{-1}\|\langle\cdot\rangle^{s-\delta_0} f\|_{L^2(\Omega)} \\ &\leq C(1+\lambda)^{-1}\|f\|_{L^2_{-s}(\Omega)}, \quad \forall \lambda \geq 0 \end{aligned} \tag{3.30}$$

for any $1/2 < s \leq \delta_0/2$. Next, we prove that

$$\left\| \langle\cdot\rangle^\mu R_0(\lambda^2 \pm i0)J_*Vf \right\|_{L^\infty(\mathbb{R}^3)} \leq C\|f\|_{L^\infty(\Omega)} \tag{3.31}$$

for any $0 \leq \mu \leq \min\{\delta_0 - 2, 1\}$. By using the well-known formula:

$$\left[R_0(\lambda^2 \pm i0)f\right](x) = \frac{1}{4\pi}\int_{\mathbb{R}^3} \frac{e^{\pm i\lambda|x-y|}}{|x-y|} f(y)\, dy.$$

we have

$$\left|\langle x\rangle^{\mu}\left[R_0(\lambda^2\pm i0)J_*Vf\right](x)\right|\le\frac{1}{4\pi}\langle x\rangle^{\mu}\int_{\mathbb{R}^n}\frac{|J_*Vf(y)|}{|x-y|}\,dy,$$

and the right member can be estimated as

$$C\langle x\rangle^{\mu}\left(\int_{\mathbb{R}^n}\frac{dy}{|x-y|\langle y\rangle^{\delta_0}}\right)\|f\|_{L^\infty(\Omega)}\le C\langle x\rangle^{\mu-\min\{\delta_0-2,1\}}\|f\|_{L^\infty(\Omega)},$$

on account of $\delta_0>2$, which implies (3.31). Interpolating between (3.30) and (3.31), we obtain

$$\left\|\langle\cdot\rangle^{\mu_p}R_0(\lambda^2\pm i0)J_*Vf\right\|_{L^p(\mathbb{R}^3)}\le C(1+\lambda)^{-2/p}\|\langle\cdot\rangle^{-s_p}f\|_{L^p(\Omega)}$$

for all $2\le p\le\infty$, where

$$\mu_p=-\frac{2s}{p}+\mu\left(1-\frac{2}{p}\right),\quad s_p=\frac{2s}{p}.$$

If $\tilde{p}=2+2s/\mu$, then we have $\mu_p\ge 0$ for any $p\ge\tilde{p}$, and hence, (3.29) holds. This proves the lemma. □

3.2. Generalized Fourier transforms

Following Wilcox [18], let us define the generalized Fourier transforms in an arbitrary exterior domain. We recall the resolvent operator

$$R(|\xi|^2\pm i0)=\lim_{\varepsilon\to+0}(A-(|\xi|^2\pm i\varepsilon))^{-1},$$

where A is the self-adjoint realization of $-\Delta$ on $L^2(\Omega)$. Introducing a function $j(x)\in C^\infty(\mathbb{R}^n)$ vanishing in a neighbourhood of $\mathbb{R}^n\setminus\Omega$ and equal to one for large $|x|$, let us define the generalized Fourier transform as follows:

$$(\mathscr{F}_\pm f)(\xi)=\lim_{R\to\infty}(2\pi)^{-n/2}\int_{\Omega_R}\overline{\psi_\pm(x,\xi)}f(x)\,dx\quad\text{in}\quad L^2(\mathbb{R}^n),$$

where we put

$$\begin{gathered}\psi_\pm(x,\xi)=j(x)e^{ix\cdot\xi}+[R(|\xi|^2\pm i0)M_\xi(\cdot)](x)\\ \text{with}\quad M_\xi(x)=-(A-|\xi|^2)(j(x)e^{ix\cdot\xi}).\end{gathered}$$

Notice that we can write formally

$$M_\xi(x)=\{\Delta j(x)+2i\xi\cdot\nabla j(x)\}e^{ix\cdot\xi},\tag{3.32}$$

hence, supp $M_\xi(\cdot)$ is compact for any fixed $\xi\in\mathbb{R}^n$. The kernels $\psi_\pm(x,\xi)$ are called eigenfunction of the operator A with eigenvalue $|\xi|^2$ in the sense that, formally, $(A-|\xi|^2)\psi_\pm(x,\xi)=0$, but $\psi_\pm(x,\xi)\notin L^2(\Omega)$. Similarly, the inverse transform is defined by

$$(\mathscr{F}_\pm^* g)(x)=\lim_{R\to\infty}(2\pi)^{-n/2}\int_{B_R(0)}\psi_\pm(x,\xi)g(\xi)\,d\xi\quad\text{in } L^2(\Omega).$$

We treat $\mathscr{F}_+f$ only and drop the subscript +, since $\mathcal{F}_-f$ can be dealt with by essentially the same method. The transform $\mathscr{F}f$ thus defined obeys the following properties (see, e.g., Shenk II [16, Theorem 1 and Corollary 5.1]):

(i) $\mathscr{F}$ is the unitary mapping

$$\mathscr{F} : L^2(\Omega) \to L^2(\mathbb{R}^n).$$

Hence

$$\mathscr{F}\mathscr{F}^* = I.$$

(ii) $\mathscr{F}$ is fulfilled with the generalized Parseval identity:

$$(\mathscr{F}f, \mathscr{F}g)_{L^2(\mathbb{R}^n)} = (f,g)_{L^2(\Omega)}, \quad f,g \in L^2(\Omega). \tag{3.33}$$

(iii) $\mathscr{F}$ diagonalizes the operator A in the sense that

$$\mathscr{F}(\varphi(A)f)(\xi) = \varphi(|\xi|^2)(\mathscr{F}f)(\xi),$$

where $\varphi(A)$ is the operator defined by the spectral representation theorem for self-adjoint operators.

The next lemma is useful in the subsequent argument.

Lemma 3.5. *Let $n \geq 3$, and let $\mathbb{R}^n \setminus \Omega$ be a compact and convex domain with a smooth boundary $\partial\Omega$. Then there exists a constant $C > 0$, independent of ξ, such that*

$$\left\| \left[R(|\xi|^2 \pm i0)M_\xi\right](\cdot)\right\|_{L^\infty(\Omega)} \leq C(1+|\xi|)^{(n-1)/2}, \quad \forall \xi \in \mathbb{R}^n, \tag{3.34}$$

where $M_\xi(x)$ is the function defined in (3.32).

Proof. Applying Proposition 3.3, we have

$$\| \left[R(|\xi|^2 \pm i0)M_\xi\right](\cdot)\|_{L^\infty(\Omega)} \leq C \left\| \left[R_0(|\xi|^2 \pm i0)M_\xi\right](\cdot)\right\|_{L^\infty(\mathbb{R}^n)}, \quad \forall \xi \in \mathbb{R}^n.$$

We write

$$\left[R_0(|\xi|^2 \pm i0)M_\xi(\cdot)\right](x) = \int_{\mathbb{R}^n} K_{|\xi|}(|x-y|)M_\xi(y)\,dy,$$

where $K_{|\xi|}(|x|)$ is defined in (3.18). We notice from (3.32) that $M_\xi(x)$ has a compact support in x and satisfies an estimate

$$\|M_\xi(\cdot)\|_{L^\infty(\mathbb{R}^n)} \leq C(1+|\xi|);$$

hence, by using asymptotic behaviour (3.19) of $K_{|\xi|}(|x|)$,

$$\begin{aligned} &\left|\int_{\mathbb{R}^n} K_{|\xi|}(|x-y|)M_\xi(y)\,dy\right| \\ &\leq C(1+|\xi|)^{(n-1)/2} \int_{\mathbb{R}^n} \left(\frac{1}{|x-y|^{n-2}\langle y\rangle^L} + \frac{1}{|x-y|^{(n-1)/2}\langle y\rangle^L}\right) dy \\ &\leq C(L)(1+|\xi|)^{(n-1)/2} \end{aligned}$$

for any $L > (n+1)/2$. Thus we obtain (3.34), as desired. □

4. Proof of Theorem 2.2

In this section we prove the equivalence relation between $\dot{H}^s_V(\Omega)$ and $\dot{H}^s(\Omega)$. We will need the extended Hardy inequality in exterior domains. When $\Omega = \mathbb{R}^n$, the inequality (4.1) below is a special version of a result of Muckenhoupt (see Theorem 1 in [14], and also Theorem 9.2 in D'Ancona, Georgiev and Kubo [3]).

Lemma 4.1 (Extended Hardy inequality). *Let $n \geq 3$, and let $\mathbb{R}^n \setminus \Omega$ be a compact and convex domain with a smooth boundary $\partial\Omega$. Denote by $(\mathscr{F} f)(\xi)$, $\xi \in \mathbb{R}^n$, the generalized Fourier transform of $f \in C_0^\infty(\Omega)$. Then for any real $a \in [0, n/2)$, we have*

$$\left\| \frac{(\mathscr{F} f)(\xi)}{|\xi|^a} \right\|_{L^2(\mathbb{R}^n)} \leq C \| |x|^a f \|_{L^2(\Omega)} \tag{4.1}$$

with a constant $C = C(a, n, \Omega)$ independent of f.

Proof. Thanks to Lemma 3.5, the proof is done by the similar argument of [3]. For the completeness we give here a proof. We must prove that

$$\int_{\mathbb{R}^n} \left| \int_\Omega \overline{\psi_\pm(x,\xi)} f(x)\, dx \right|^2 |\xi|^{-2a}\, d\xi \leq C \int_\Omega |f(x)|^2 |x|^{2a}\, dx,$$

where

$$\psi_\pm(x,\xi) = j(x) e^{ix\cdot\xi} + \left[R(|\xi|^2 \pm i0) M_\xi(\cdot) \right](x).$$

Split the first integral as $\mathrm{I} + \mathrm{II}$, with

$$\mathrm{I} = \sum_{j\in\mathbb{Z}} \int_{2^j < |\xi|^{-a} \leq 2^{j+1}} \left| \int_{\Omega \cap \{|x|^a > 2^j\}} \overline{\psi_\pm(x,\xi)} f(x)\, dx \right|^2 |\xi|^{-2a}\, d\xi$$

and

$$\mathrm{II} = \sum_{j\in\mathbb{Z}} \int_{2^j < |\xi|^{-a} \leq 2^{j+1}} \left| \int_{\Omega \cap \{|x|^a \leq 2^j\}} \overline{\psi_\pm(x,\xi)} f(x)\, dx \right|^2 |\xi|^{-2a}\, d\xi.$$

We can estimate

$$\mathrm{I} \leq \sum_{j\in\mathbb{Z}} 2^{2j+2} \left\| \mathscr{F}(f \cdot \chi_{\{|x|^a > 2^j\}}) \right\|^2_{L^2(\mathbb{R}^n)},$$

where χ_A is the characteristic function of the set A. Thus, by the generalized Parseval identity (3.33), we obtain

$$\mathrm{I} \leq \sum_{j\in\mathbb{Z}} 2^{2j+2} \left\| f \cdot \chi_{\{|x|^a > 2^j\}} \right\|^2_{L^2(\Omega)} \leq \| f \cdot h \|^2_{L^2(\Omega)},$$

where the function $h(x)$ is defined by

$$h(x) = \sum_{j\in\mathbb{Z}} \chi_{\{|x|^a > 2^j\}} 2^{j+1},$$

and hence satisfies

$$h(x) \leq 2|x|^a.$$

This concludes the estimate for I.

To estimate II, we begin by noticing that

$$\mathrm{II} \leq C \int_{\mathbb{R}^n} \left(\int_{|x| \leq 1/|\xi|} \chi_\Omega(x) |f(x)|\, dx \right)^2 \{ |\xi|^{-2a} + |\xi|^{-2(a-\frac{n-1}{2})} \}\, d\xi,$$

where we used Lemma 3.5 and $(1+|\xi|)^{(n-1)/2} \leq c(n)\{1+|\xi|^{(n-1)/2}\}$. Hence all we have to do is to prove that

$$\mathrm{II}' := \int_{\mathbb{R}^n} \left(\int_{|x| \leq 1/|\xi|} \chi_\Omega(x) |f(x)|\, dx \right)^2 |\xi|^{-2a}\, d\xi \leq C \| |x|^a f \|^2_{L^2(\Omega)} \tag{4.2}$$

for $a \in [0, n/2)$. For, the following estimate is an immediate consequence of (4.2):

$$\mathrm{II}'' := \int_{\mathbb{R}^n} \left(\int_{|x| \leq 1/|\xi|} \chi_\Omega(x) |f(x)|\, dx \right)^2 |\xi|^{-2(a-\frac{n-1}{2})}\, d\xi \leq C \| |x|^{a-\frac{n-1}{2}} f \|^2_{L^2(\Omega)}$$

for $a \in [(n-1)/2, n/2)$. Clearly, the right member of the previous inequality is bounded by $L^2(\Omega)$-norm of $|x|^a f$ on account of $0 \notin \overline{\Omega}$. Hence estimates for II′ and II″ imply the estimate for II with $a \in [(n-1)/2, n/2)$. Thus, this estimate together with the one for I we obtain (4.1) for $a \in [(n-1)/2, n/2)$. The case $a \in [0, (n-1)/2]$ follows from interpolation between (4.1) for $a = (n-1)/2$ and the trivial one $\|\mathscr{F} g\|_{L^2(\mathbb{R}^n)} = \|g\|_{L^2(\Omega)}$.

Consider the lowest integer $J \in \mathbb{Z}$ such that $2^J \geq \|f\|_{L^1(\Omega)}$; then define $r_J = \infty$ and, for $j < J$, choose any non-decreasing sequence of positive numbers r_j such that

$$\int_{|x|^a \leq r_j} \chi_\Omega(x) |f(x)|\, dx = 2^j\,;$$

finally, define the sets for $-\infty < j \leq J$

$$A_j = \{ x : r_{j-2} < |x|^a \leq r_{j-1} \}, \quad B_j = \{ \xi : r_{j-1} < |\xi|^{-a} \leq r_j \}.$$

We notice the following property:

$$\int_{|x|^a \leq r_j} \chi_\Omega(x)|f(x)|\, dx = 4 \int_{|x|^a \leq r_{j-2}} \chi_\Omega(x)|f(x)|\, dx = 4 \int_{A_j} \chi_\Omega(x)|f(x)|\, dx.$$

Hence, for $\xi \in B_j$,

$$\begin{aligned}
\left(\int_{|x| \leq 1/|\xi|} \chi_\Omega(x)|f(x)|\, dx \right)^2 &\leq \left(\int_{|x|^a \leq r_j} \chi_\Omega(x)|f(x)|\, dx \right)^2 \\
= \left(4 \int_{A_j} \chi_\Omega(x)|f(x)|\, dx \right)^2 &\leq 16 \int_{A_j} \chi_\Omega(x)|f(x)|^2 |x|^{2a}\, dx \cdot \int_{A_j} \chi_\Omega(x)|x|^{-2a}\, dx \\
&\leq c r_j^{-2+(n/a)} \int_{A_j} \chi_\Omega(x)|f(x)|^2|x|^{2a}\, dx
\end{aligned}$$

by Cauchy–Schwarz inequality and the explicit computations (valid for $a < n/2$)

$$\int_{A_j} \chi_\Omega(x)|x|^{-2a}\, dx \le c r_{j-1}^{-2+(n/a)}.$$

Thus we can write

$$\begin{aligned} \mathrm{II}' &= \sum_{j=-\infty}^{J} \int_{B_j} \left(\int_{|x|\le 1/|\xi|} \chi_\Omega(x)|f(x)|\, dx \right)^2 |\xi|^{-2a}\, d\xi \\ &\le c \sum_{j=-\infty}^{J} r_{j-1}^{-2+(n/a)} \left(\int_{A_j} \chi_\Omega(x)|f(x)|^2|x|^{2a}\, dx \right) \left(\int_{B_j} |\xi|^{-2a}\, d\xi \right) \end{aligned}$$

whence

$$\mathrm{II}' \le c \sum_{j=-\infty}^{J} \int_{A_j} \chi_\Omega(x)|f(x)|^2|x|^{2a}\, dx \le c \int_\Omega |x|^{2a}|f(x)|^2\, dx$$

by the explicit computation

$$\int_{B_j} |\xi|^{-2a}\, d\xi \le c_n r_{j-1}^{2-(n/a)}.$$

Since the A_j are disjoint sets, the proof is complete. □

We now turn to the proof of Theorem 2.2. Once the extended Hardy inequality is established, the proof is identical to that of Theorem VII.4 of [6]. In the present case, since we treat the higher-dimensional case, we give the proof for the completeness.

Since $V \ge 0$ on Ω, the operator $-\Delta+V$ satisfies the strong maximum principle (see Theorem 3.5 in [7]), and as a result, we have

$$\left\|(-\Delta+V)^{-k}g\right\|_{L^2(\Omega)} \le \left\|(-\Delta)^{-k}g\right\|_{L^2(\Omega)}, \quad \forall k \in \mathbb{N}. \tag{4.3}$$

Estimates (4.3) and the interpolation argument imply:

$$\|f\|_{\dot H_V^{-s}(\Omega)} \le \|f\|_{\dot H^{-s}(\Omega)}, \quad \forall s \in \mathbb{R}^+, \tag{4.4}$$

that by the duality argument are equivalent to:

$$\|f\|_{\dot H^{s}(\Omega)} \le \|f\|_{\dot H_V^{s}(\Omega)}, \quad \forall s \in \mathbb{R}^+.$$

Hence it is sufficient to prove that

$$\|f\|_{\dot H_V^{s}(\Omega)} \le C\|f\|_{\dot H^{s}(\Omega)}, \quad 0 \le \forall s \le 3/2,$$

which, in turn, are equivalent to:

$$\left\|(-\Delta+V)^{(3/4)-\varepsilon}f\right\|_{L^2(\Omega)} \le C\left\|(-\Delta)^{(3/4)-\varepsilon}f\right\|_{L^2(\Omega)}, \quad 0 \le \forall \varepsilon \le 3/4. \tag{4.5}$$

For the proof of (4.5), we write the left-hand side as

$$\begin{aligned}\left\|(-\Delta+V)^{(3/4)-\varepsilon}f\right\|_{L^2(\Omega)} &= \left\|(-\Delta+V)^{-(1/4)-\varepsilon}(-\Delta+V)f\right\|_{L^2(\Omega)}\\ &\le \left\|(-\Delta+V)^{-(1/4)-\varepsilon}(-\Delta)f\right\|_{L^2(\Omega)}+\left\|(-\Delta+V)^{-(1/4)-\varepsilon}Vf\right\|_{L^2(\Omega)}\\ &= \mathrm{I}+\mathrm{II}.\end{aligned}$$

As a preliminary stage, by using (4.4) with $s=1$ and the (classical) Hardy inequality, we have

$$\left\|(-\Delta+V)^{-1/2}g\right\|_{L^2(\Omega)} \le \left\|(-\Delta)^{-1/2}g\right\|_{L^2(\Omega)} \le C\,\||x|g\|_{L^2(\Omega)}\,.$$

Interpolating between this inequality and the trivial one: $\|g\|_{L^2(\Omega)}=\|g\|_{L^2(\Omega)}$, we get

$$\left\|(-\Delta+V)^{-(1/4)-\varepsilon}g\right\|_{L^2(\Omega)} \le C(\varepsilon)\left\||x|^{(1/2)+2\varepsilon}g\right\|_{L^2(\Omega)},\quad 0\le\forall\varepsilon\le 1/4.$$

Then, by taking $g=Vf$, we obtain

$$\begin{aligned}\mathrm{II}=\left\|(-\Delta+V)^{-(1/4)-\varepsilon}Vf\right\|_{L^2(\Omega)} &\le C(\varepsilon)\left\||x|^{(1/2)+2\varepsilon}Vf\right\|_{L^2(\Omega)}\\ &\le C(\varepsilon)\left\||x|^{-(3/2)+2\varepsilon}f\right\|_{L^2(\Omega)}\end{aligned}\tag{4.6}$$

for $0\le\forall\varepsilon\le 1/4$, since

$$|x|^{(1/2)+2\varepsilon}V(x)\le c_1|x|^{(1/2)+2\varepsilon}|x|^{-\delta_0}\le c_1|x|^{-(3/2)+2\varepsilon}$$

which is possible on account of the assumption (1.1). We notice that the inequality

$$0\le\frac{3}{2}-2\varepsilon<\frac{n}{2}$$

holds if $0\le\varepsilon\le 1/4$. When $n=3$, $\varepsilon=0$ should be excluded. In the sequel, we treat only the case $n\ge 4$. Applying Lemma 4.1 to the right-hand side of (4.6), we obtain

$$\mathrm{II}\le C(\varepsilon)\left\|(-\Delta)^{(3/4)-\varepsilon}f\right\|_{L^2(\Omega)},\quad 0\le\forall\varepsilon\le 1/4\tag{4.7}$$

As to the estimate for I, interpolating between the inequality (4.3) with $k=1$, and the trivial one $\|g\|_{L^2(\Omega)}=\|g\|_{L^2(\Omega)}$, we get

$$\left\|(-\Delta+V)^{-(1/4)-\varepsilon}g\right\|_{L^2(\Omega)}\le\left\|(-\Delta)^{-(1/4)-\varepsilon}g\right\|_{L^2(\Omega)},\quad -1/4\le\forall\varepsilon\le 3/4.$$

If we choose $g=-\Delta f$ in the previous inequality, we get

$$\mathrm{I}=\left\|(-\Delta+V)^{-(1/4)-\varepsilon}(-\Delta)f\right\|_{L^2(\Omega)}\le\left\|(-\Delta)^{(3/4)-\varepsilon}f\right\|_{L^2(\Omega)},\quad 0\le\forall\varepsilon\le 3/4.\tag{4.8}$$

Thus, combining (4.7) and (4.8), we conclude (4.5) for $0\le\forall\varepsilon\le 1/4$. Since inequality (4.5) for $\varepsilon=3/4$ is trivial, the required inequality (4.5) for $0\le\forall\varepsilon\le 3/4$ can be obtained by interpolation between $\varepsilon=1/4$ and $\varepsilon=3/4$. In conclusion, (4.5) is proved for $0\le\forall\varepsilon\le 3/4$. This completes the proof of Theorem 2.2.

5. Functional calculus

In this section we will always assume that $n = 3$, and proceed with L^p-boundedness of the operators $\Phi_\ell(H_V)$ and $\Phi_\ell(H)$ in order to prove the equivalence of the perturbed Besov spaces and the free ones, where we recall that

$$H = (-\Delta)^{1/2} \text{ on } L^2(\Omega).$$

We begin by adopting the results of [10].

Lemma 5.1. *Assume that the measurable potential V is non-negative on Ω, satisfies assumption* (V) *and belongs to $C^{0,\alpha}_{\mathrm{loc}}(\overline{\Omega})$. Then for any $y \in \Omega$ there exists a measurable function $K(t,x,y)$ such that $K \in C([0,\infty); \mathscr{D}'(\Omega))$ and K satisfies the following parabolic initial-boundary value problem:*

$$\begin{cases} \partial_t K(t,x,y) - \Delta K(t,x,y) + V(x)K(t,x,y) = 0, & t>0, \quad x \in \Omega, \\ K(0,x,y) = \delta_y, & \qquad\quad x \in \Omega, \\ K(t,x,y) = 0, & t>0, \quad x \in \partial\Omega. \end{cases} \tag{5.1}$$

Moreover, K satisfies the following pointwise estimates:

(i) *When x, y are in any unbounded connected component of Ω, then there exist positive constants c_1, c_2 depending on Ω such that for all $t > 0$,*

$$0 \le K(t,x,y) \le \left(\frac{\rho(x)}{\sqrt{t}\wedge 1} \wedge 1\right)\left(\frac{\rho(y)}{\sqrt{t}\wedge 1} \wedge 1\right) \frac{e^{-\frac{|x-y|^2}{c_2 t}}}{c_1 B(x,\sqrt{t})}, \tag{5.2}$$

where $\rho(x) = \mathrm{dist}(x, \partial\Omega)$ and $B(x,\sqrt{t}) = \mathrm{Vol}(\{z : |x-z| \le \sqrt{t}\})$.

(ii) *When x, y are in any bounded connected component of Ω, then there exist positive constants c_1, c_2, μ depending on Ω such that for all $t > 0$,*

$$0 \le K(t,x,y) \le \left(\frac{\rho(x)}{\sqrt{t}\wedge 1} \wedge 1\right)\left(\frac{\rho(y)}{\sqrt{t}\wedge 1} \wedge 1\right) \frac{e^{-\mu t-\frac{|x-y|^2}{c_2 t}}}{c_1 B(x,\sqrt{t})}. \tag{5.3}$$

Outline of proof. The regularity assumption $V \in C^{0,\alpha}_{\mathrm{loc}}(\overline{\Omega})$ assures the existence of the fundamental solution $K(t,x,y)$ to (5.1) (see, e.g., Friedman [5]). The fact that $K(t,x,y) \ge 0$ follows from the maximum principle, since $V \ge 0$. Furthermore, consider the solution $\Gamma(t,x,y)$ to the following initial-boundary value problem:

$$\begin{cases} \partial_t \Gamma(t,x,y) - \Delta \Gamma(t,x,y) = 0, & t>0, \quad x \in \Omega, \\ \Gamma(0,x,y) = \delta_y, & \qquad\quad x \in \Omega, \\ \Gamma(t,x,y) = 0, & t>0, \quad x \in \partial\Omega. \end{cases}$$

Since $V \ge 0$, we use again the maximum principle to deduce that $K(t,x,y) \le \Gamma(t,x,y)$. Since the pointwise estimates for $\Gamma(t,x,y)$ are given by the right-hand side of (5.2)–(5.3) (see [20]), we obtain (5.2)–(5.3). The proof is complete. □

Among other things, we shall prove here the following:

Proposition 5.2. *Suppose that V is non-negative on Ω, satisfies assumption* (V) *and belongs to $C^{0,\alpha}_{\rm loc}(\overline{\Omega})$ $(0<\alpha<1)$. Then for any $\varphi \in C_0^\infty(\mathbb{R})$ and any $1 \le p \le \infty$, there exists a constant $C = C(\varphi, p) > 0$ such that*

$$\|\varphi(\theta A_V)\|_{\mathscr{B}(L^p(\Omega))} \le C, \quad \forall \theta > 0. \tag{5.4}$$

Proof. Consider the rescaled potential

$$V_\theta(x) = \theta V(\sqrt{\theta} x),$$

and the operator

$$A_\theta = -\Delta + V_\theta.$$

The proof of Proposition 5.2 can be done along that of Theorem 2.1 in Jensen & Nakamura [8]. (5.4) is a consequence of

$$\|\varphi(A_\theta)\|_{\mathscr{B}(L^p(\Omega))} \le C$$

uniformly in $\theta > 0$. The main points are to obtain the three estimates uniformly in θ:

(i) a pointwise estimate for the kernel $K_\theta(t,x,y)$ of e^{-tA_θ};
(ii) an L^2-L^2 estimate for the operator $(A_\theta + 1)^{-1/2}$;
(iii) an L^2-L^2 estimate for the operator $\partial_x (A_\theta + 1)^{-1/2}$.

Step (i) directly follows from Lemma 5.1:

(a) When x, y are in any unbounded connected component of Ω, then there exist positive constants c_1, c_2 depending on Ω such that for all $t > 0$,

$$0 \le K_\theta(t,x,y) \le \left(\frac{\rho(x)}{\sqrt{t}\wedge 1} \wedge 1\right)\left(\frac{\rho(y)}{\sqrt{t}\wedge 1} \wedge 1\right) \frac{e^{-\frac{|x-y|^2}{c_2 t}}}{c_1 B(x,\sqrt{t})},$$

where $\rho(x) = \mathrm{dist}(x, \partial\Omega)$ and $B(x,\sqrt{t}) = \mathrm{Vol}(\{z : |x-z| \le \sqrt{t}\})$.

(b) When x, y are in any bounded connected component of Ω, then there exist positive constants c_1, c_2, μ depending on Ω such that for all $t > 0$,

$$0 \le K_\theta(t,x,y) \le \left(\frac{\rho(x)}{\sqrt{t}\wedge 1} \wedge 1\right)\left(\frac{\rho(y)}{\sqrt{t}\wedge 1} \wedge 1\right) \frac{e^{-\mu t-\frac{|x-y|^2}{c_2 t}}}{c_1 B(x,\sqrt{t})}.$$

Step (ii) is obvious, and step (iii) can be proved by the standard energy method:

$$\left\|(A_\theta + 1)^{-1/2}\right\|_{\mathscr{B}(L^2(\Omega))} + \left\|\partial_x (A_\theta + 1)^{-1/2}\right\|_{\mathscr{B}(L^2(\Omega))} \le C$$

uniformly in $\theta > 0$ (see Lemma VI.2 from [6]). □

In what follows, we denote by $\{\Phi_j(\lambda)\}_{j=-\infty}^{\infty}$ the Littlewood–Paley partition of unity. As an immediate consequence of Proposition 5.2, we have:

Lemma 5.3. *For any $j \in \mathbb{Z}$,*

$$\|\Phi_j(H)\|_{\mathscr{B}(L^p(\Omega))} \le C, \qquad \left\|\Phi_j(H)(-\Delta)^{-1}\right\|_{\mathscr{B}(L^p(\Omega))} \le C2^{-2j}, \tag{5.5}$$

$$\|\Phi_j(H)\Delta\|_{\mathscr{B}(L^p(\Omega))} \le C2^{2j}, \quad \left\|\Phi_j(H)(-\Delta)^{-1/2}\right\|_{\mathscr{B}(L^p(\Omega))} \le C2^{j}. \tag{5.6}$$

The following lemma provides L^p boundedness of the operators $|x|^{-\delta}\Phi_j(H)$, where $H = (-\Delta)^{1/2}$.

Lemma 5.4. *For any $p > 2$ and $0 \le \delta \le 2/p$, there exists a constant $C > 0$ such that*

$$\left\||x|^{-\delta}\Phi_j(H)\right\|_{\mathscr{B}(L^p(\Omega))} \le C2^{j\delta}, \quad \forall j \in \mathbb{Z}. \tag{5.7}$$

The proof of Lemma 5.4 is identical to the one in Lemma VI.5 from [6]. We may omit it.

Let us recall that A_0 is the self-adjoint realization in $L^2(\mathbb{R}^3)$ of $-\Delta$ with domain $H^2(\mathbb{R}^3)$, and we put $H_0 = A_0^{1/2}$. The following lemma is useful to the proof of Lemma 5.6 below.

Lemma 5.5. *Let V be as in Proposition 5.2. Then*

$$\left\|\Phi_j(H_0)A_0^{-1}J_*V\right\|_{\mathscr{B}(L^p(\Omega),L^p(\mathbb{R}^n))} \le C2^{-2j} \tag{5.8}$$

for any $1 \le p \le \infty$. Furthermore, there is a real $\tilde{p} > 2$ such that for any real $p \ge \tilde{p}$

$$\left\|\Phi_j(H_V)A_V^{-1}V\right\|_{\mathscr{B}(L^p(\Omega))} \le C\min(2^{-2j}, 2^{-2j/p}) \tag{5.9}$$

with a certain constant $C > 0$.

Proof. The functional calculus implies that

$$\Phi_j(H_0)A_0^{-1}J_*V = K_j * (J_*V),$$

where the kernel of the operator has the form:

$$K_j(x) = \int_{\mathbb{R}^n} e^{ix\cdot\xi}\Phi(2^{-j}|\xi|)|\xi|^{-2}\,d\xi,$$

and by a scaling argument we can write

$$K_j(x) = 2^j K_0(2^j x). \tag{5.10}$$

By using the Young inequality we have

$$\|K_j * (J_*Vg)\|_{L^p(\mathbb{R}^3)} \le \|K_j\|_{L^1(\mathbb{R}^3)}\|J_*Vg\|_{L^p(\mathbb{R}^3)}.$$

It follows from (5.10) and the properties of the Fourier transforms $\mathscr{F}$ and $\mathscr{F}^{-1}$ on $\mathbb{R}^3$ that

$$\|K_j\|_{L^1(\mathbb{R}^n)} = 2^{-2j}\left\|\mathscr{F}^{-1}\left[\Phi(\xi)|\xi|^{-2}\right]\right\|_{L^1(\mathbb{R}^n)} \le C2^{-2j},$$

and by assumption (1.1) on V, we have

$$\|J_*Vg\|_{L^p(\mathbb{R}^n)} \le C\left\||\cdot|^{-\delta_0}g\right\|_{L^p(\Omega)} \le C\|g\|_{L^p(\Omega)}.$$

Summarizing all the estimates obtained now, we get

$$\|K_j * (J_* V g)\|_{L^p(\mathbb{R}^n)} \le C 2^{-2j} \|g\|_{L^p(\Omega)},$$

which implies (5.8).

Let us turn to the proof of (5.9). We write

$$\Phi_j(H_V) A_V^{-1} V = \frac{1}{\pi i} \int_0^\infty \Phi_j(\lambda) \lambda^{-1} \left[R_V(\lambda^2 + i0) - R_V(\lambda^2 - i0) \right] V \, d\lambda.$$

We divide the estimate into two cases $j > 0$ and $j < 0$. Let $\tilde{p} > 2$ be as in Lemma 3.4. When $j > 0$, by using (3.27) from Lemma 3.4, we have

$$\left\| \Phi_j(H_V) A_V^{-1} V \right\|_{\mathscr{B}(L^p(\Omega))} \le C \int_0^\infty \Phi_j(\lambda) \lambda^{-1-2/p} \, d\lambda \le C 2^{-2j/p}$$

for any $p \ge \tilde{p}$. When $j < 0$, by using (3.27) from Lemma 3.4 again, we have

$$\left\| \Phi_j(H_V) A_V^{-1} V \right\|_{\mathscr{B}(L^p(\Omega))} \le C \int_0^\infty \Phi_j(\lambda) \lambda^{-1} \, d\lambda \le C$$

for any $p \ge \tilde{p}$. Thus these estimates imply (5.9). The proof of Lemma 5.5 is complete. □

The goal of this section is the following lemma.

Lemma 5.6. *Let V and $\tilde{p}$ be as in Lemma 5.5. Then for any real $p > \tilde{p}$ there exist two constants $C > 0$ and $\delta_p > 0$ such that for any $j, k \in \mathbb{Z}$,*

$$\|\Phi_j(H_V)\Phi_k(H)\|_{\mathscr{B}(L^p(\Omega))} \le C 2^{-\delta_p |j-k|}, \tag{5.11}$$

$$\|\Phi_j(H)\Phi_k(H_V)\|_{\mathscr{B}(L^p(\Omega))} \le C 2^{-\delta_p |j-k|}. \tag{5.12}$$

Proof. We can restrict our attention to prove (5.11), since the proof of (5.12) can be done along the similar argument of (5.11) by some trivial changes. We first note two consequences of (5.4): For any $j \in \mathbb{Z}$,

$$\|\Phi_j(H_V) A_V\|_{\mathscr{B}(L^p(\Omega))} \le C 2^{2j}, \qquad \|\Phi_j(H)\Delta\|_{\mathscr{B}(L^p(\Omega))} \le C 2^{2j}, \tag{5.13}$$

$$\left\|\Phi_j(H_V) A_V^{-1}\right\|_{\mathscr{B}(L^p(\Omega))} \le C 2^{-2j}, \quad \left\|\Phi_j(H)\Delta^{-1}\right\|_{\mathscr{B}(L^p(\Omega))} \le C 2^{-2j}. \tag{5.14}$$

We divide the proof into two cases $j > k$ and $j < k$.

The case $j > k$: We can write

$$\begin{aligned} \Phi_j(H_V)\Phi_k(H) &= \Phi_j(H_V) A_V^{-1} A_V \Phi_k(H) \\ &= \Phi_j(H_V) A_V^{-1} (-\Delta) \Phi_k(H) + \Phi_j(H_V) A_V^{-1} V \Phi_k(H) \\ &= \mathrm{I} + \mathrm{II}. \end{aligned} \tag{5.15}$$

By using (5.13)–(5.14) the first term can be estimated as

$$\|\mathrm{I}\|_{\mathscr{B}(L^p(\Omega))} \le C 2^{-2|j-k|}.$$

To estimate the second term we use the following identity:

$$\Phi_j(H_V) A_V^{-1} V \Phi_k(H) = \Phi_j(H_V) A_V^{-1} V_\delta |x|^{-\delta} \Phi_k(H),$$

where we put $V_\delta = V|x|^\delta$. Then we have

$$\|\mathrm{II}\|_{\mathscr{B}(L^p(\Omega))} \le \left\|\Phi_j(H_V)A_V^{-1}V_\delta\right\|_{\mathscr{B}(L^p(\Omega))} \left\||\cdot|^{-\delta}\Phi_k(H)\right\|_{\mathscr{B}(L^p(\Omega))}.$$

It is readily seen that if $0 < \delta < \delta_0$, then the potential V_δ satisfies the same assumption of the potential V for a suitable choice of the constants that appear in (1.1). Hence, by using Lemma 5.4 and (5.9) from Lemma 5.5 we conclude that

$$\|\mathrm{II}\|_{\mathscr{B}(L^p)(\Omega)} \le C2^{k\delta}\min(2^{-2j}, 2^{-2j/p}).$$

If δ is so small that $\delta_p = \delta < 2/p$, it follows that

$$2^{k\delta_p}\min(2^{-2j}, 2^{-2j/p}) \le 2^{-\delta_p|j-k|},$$

and hence, we conclude that

$$\|\mathrm{II}\|_{\mathscr{B}(L^p)(\Omega)} \le C2^{-\delta_p|j-k|}.$$

The case $j < k$: Going back to (5.15), we have, by using (5.14),

$$\|\Phi_j(H_V)\Phi_k(H)\|_{\mathscr{B}(L^p(\Omega))} \le C2^{-2j}\left(\|(-\Delta)\Phi_k(H)\|_{\mathscr{B}(L^p(\Omega))} + \|V\Phi_k(H)\|_{\mathscr{B}(L^p(\Omega))}\right).$$

Applying (5.13) to the first term, and applying the decay assumption (1.1) on V and Lemma 5.4 to the second term, we get

$$\|\Phi_j(H_V)\Phi_k(H)\|_{\mathscr{B}(L^p(\Omega))} \le C\left(2^{-2|j-k|} + 2^{-2j+k\delta_p}\right).$$

Here if either $j < k < 0$ or $j < 0 < k$, then $2^{-2j+k\delta_p} \le 2^{-\delta_p|j-k|}$. Finally, if $0 < j < k$, then $2^{-2j+k\delta_p} \le 2^{-2|j-k|}$. In any case we conclude (5.11), as desired. The proof of Lemma 5.6 is complete. □

6. Proof of Theorem 2.3

The proof can be done by the same manner as Theorem VII.3 of [6]. Since

$$\sum_{k\in\mathbb{Z}}\Phi_k(H) = I,$$

it follows that

$$\sum_{j\in\mathbb{Z}} 2^{jsq}\|\Phi_j(H_V)f\|^q_{L^p(\Omega)} = \sum_{j\in\mathbb{Z}} 2^{jsq}\left\|\sum_{k\in\mathbb{Z}}\Phi_j(H_V)\Phi_k(H)f\right\|^q_{L^p(\Omega)}.$$

We write $\Phi_k = \Phi_k(\Phi_{k-1} + \Phi_k + \Phi_{k+1}) = \Phi_k\tilde{\Phi}_k$ and we have

$$\sum_{j\in\mathbb{Z}} 2^{jsq}\left\|\sum_{k\in\mathbb{Z}}\Phi_j(H_V)\Phi_k(H)f\right\|^q_{L^p(\Omega)} = \sum_{j\in\mathbb{Z}} 2^{jsq}\left\|\sum_{k\in\mathbb{Z}}\Phi_j(H_V)\Phi_k(H)\tilde{\Phi}_k(H)f\right\|^q_{L^p(\Omega)};$$

now applying (5.11) from Lemma 5.6 to the right member, we obtain

$$\sum_{j\in\mathbb{Z}} 2^{jsq} \|\Phi_j(H_V)f\|^q_{L^p(\Omega)} \le \sum_{j\in\mathbb{Z}} 2^{jsq} \left(\sum_{k\in\mathbb{Z}} 2^{-\delta_p|j-k|} \left\|\tilde{\Phi}_k(H)f\right\|_{L^p(\Omega)} \right)^q.$$

provided that $p > \tilde{p}$. Here we apply the following fact (see, e.g., Proposition VIII.3 from [6]) to the right-hand side: For any $1 \le q \le \infty$, $s \in \mathbb{R}$ and for any couple of sequences $\{a_k\}$, $\{b_k\}$ the following inequality holds:

$$\left\| \left\{ 2^{js} \left(\sum_k a_k b_{j-k} \right) \right\} \right\|_{\ell^q} \le \|\{2^{js} a_j\}\|_{\ell^q} \|\{2^{js} b_j\}\|_{\ell^1}.$$

Then we have

$$\sum_{j\in\mathbb{Z}} 2^{jsq} \|\Phi_j(H_V)f\|^q_{L^p(\Omega)} \le \left(\sum_{j\in\mathbb{Z}} 2^{js} 2^{-\delta_p|j|} \right)^q \sum_{j\in\mathbb{Z}} 2^{jsq} \left\|\tilde{\Phi}_j(H)f\right\|^q_{L^p(\Omega)}.$$

We note that if $|s| < \delta_p$, then there exists a constant $C = C(s) > 0$ such that

$$\sum_{j\in\mathbb{Z}} 2^{js} 2^{-\delta_p|j|} \le C,$$

which implies that

$$\|f\|_{\dot{B}^s_{p,q,V}(\Omega)} \le C\|f\|_{\dot{B}^s_{p,q}(\Omega)}.$$

The other inequality is obtained by interchanging H_V and H and using (5.12) instead of (5.11). The proof of Theorem 2.3 is complete.

7. Potentials of generic type

In this section we briefly check that assumption (V) assures that zero is neither an eigenvalue nor a resonance of $-\Delta + V$, i.e., $\mathcal{M} = \{0\}$. Let $u \in L^2_{-s}(\Omega) \cap L^\infty(\Omega)$ for some $1/2 < s \le \delta_0/2$ be a solution to the integral equation of the form:

$$u + R(0)Vu = 0 \text{ in } \Omega \quad \text{and} \quad u = 0 \text{ on } \partial\Omega. \tag{7.1}$$

We must prove that $u(x) = 0$ in Ω. When $n = 3$, assuming $\delta_0 > 2$, we have proved the triviality of solution in Lemma 2.1 from [10]. In this case, we may only assume that $u \in L^2_{-s}(\Omega)$. Hence we assume $n \ge 4$. The solution u to (7.1) satisfies the boundary value problem for the stationary Schrödinger equation:

$$-\Delta u + V(x)u = 0 \quad \text{in } \Omega \tag{7.2}$$

with the homogeneous Dirichlet boundary condition. Now let us consider the extension $\tilde{u}$ of u to $\mathbb{R}^n$. More precisely, we define $\tilde{u}$ to be $\tilde{u}(x) = \psi(x)u(x)$, where $\psi(x) \in C^\infty(\mathbb{R}^n)$ equals zero in a bounded open neighbourhood $\mathcal{O}$ of $\mathbb{R}^n \setminus \Omega$ and one in $\mathbb{R}^n \setminus \widetilde{\mathcal{O}}$ such that the closure of $\widetilde{\mathcal{O}}$ is compact and $\mathcal{O} \Subset \widetilde{\mathcal{O}}$. This $\tilde{u}$ satisfies the equation $-\Delta\tilde{u} = f$ in $\mathbb{R}^n$, where

$$f = -\psi V u - 2\nabla\psi \cdot \nabla u - (\Delta\psi)u \in L^2_s(\mathbb{R}^n). \tag{7.3}$$

Then we have proved in Lemma 2.2 from [11] that

$$\lim_{R\to\infty}\frac{1}{R^n}\int_{R<|x|<2R}|\tilde{u}(x)|^2\,dx=0.$$

Thanks to this asymptotic behaviour, $\tilde{u}(x)$ has no harmonic polynomials in its representation, and as a result, we can write

$$\tilde{u}(x)=\frac{\Gamma(n/2)}{4(n-2)\pi^{n/2}}\int_{\mathbb{R}^n}\frac{f(y)}{|x-y|^{n-2}}\,dy.$$

Thus, when $n\ge 4$, by the argument in Lemma 2.1 of [11] we obtain for $|\alpha|=0,1$:

$$\partial_x^\alpha u(x)=\begin{cases}O\left(|x|^{-(\delta_0-2+|\alpha|)}\right), & \text{if } 2<\delta_0<n,\\ O\left(|x|^{-(n-2+|\alpha|)}\log|x|\right), & \text{if } \delta_0=n,\\ O\left(|x|^{-(n-2+|\alpha|)}\right), & \text{if } \delta_0>n,\end{cases}\qquad \text{as } |x|\to\infty. \tag{7.4}$$

Finally, multiplying equation (7.2) by u and integrating by parts, we get

$$\int_{\Omega\cap\{|x|<R\}}\left\{|\nabla u(x)|^2+V(x)|u(x)|^2\right\}dx=\int_{|x|=R}u_r(x)\overline{u(x)}\,dS_R, \tag{7.5}$$

where $u_r=\partial u/\partial r$ $(r=|x|)$ and dS_R is the $(n-1)$-dimensional surface element. Hence, by using the asymptotic behaviour (7.4) and (classical) Hardy inequality, we conclude from (7.5) that $u=0$ in Ω, provided $\delta_0>(n+2)/2$. In conclusion, integral equation (7.1) has only a trivial solution.

Acknowledgment

The author would like to thank Professor Shigetoshi Kuroda for indicating him a class of potentials of generic type introduced by the paper of Yajima [19]. He would like to thank also Professor Michael Ruzhansky and Doctor Tsukasa Iwabuchi for giving him some useful comments on Proposition 3.3.

References

[1] S. Agmon, *Spectral properties of Schrödinger operators and scattering theory,* Ann. Scuola Norm. Sup. Pisa, Ser. IV, **2** (1975), 151–218.

[2] M. Ben-Artzi and S. Klainerman, *Decay and regularity for the Schrödinger equation,* J. Analyse Math. **58** (1992), 25–37.

[3] P. D'Ancona, V. Georgiev, and H. Kubo, *Weight decay estimates for the wave equation,* J. Differential Equations. **177** (2001), 146–208.

[4] P. D'Ancona and V. Pierfelice, *On the wave equation with a large rough potential,* J. Functional Analysis **227** (2005), 30–77.

[5] A. Friedman, *Partial Differential Equations of Parabolic Type,* Englewood Cliff. NJ; Prentice Hall Inc. 1964.

[6] V. Georgiev and N. Visciglia, *Decay estimates for the wave equation with potential,* Comm. Partial Differential Equations. **29** (2003), 101–153.

[7] D. Gilberg and N.S. Trudinger, *Elliptic Partial Differential Equations of 2nd Order,* Springer-Verlag, Berlin, Heidelberg, New York, Tokyo, 1983.

[8] A. Jensen and S. Nakamura, *Mapping properties of functions of Schrödinger operators between L^p-spaces and Besov spaces,* Spectral and Scattering Theory and Applications, Vol. 23, Tokyo: Adv. Stud. Pure Math., Mathematical Society of Japan, 187–209 (1994).

[9] S.T. Kuroda, *Scattering theory for differential operators, III; exterior problems,* Spectral Theory and Differential Equations, Proceedings of the Symposium held at Dundee, Scotland, pp. 227–241, (1974).

[10] T. Matsuyama, *Dispersion for* 3D *wave equation with a potential outside a convex obstacle,* submitted.

[11] T. Matsuyama, *Representation formula of the resolvent for wave equation with a potential outside a convex obstacle,* Progress in Mathematics **301** (2012), 213–225, Birkhäuser.

[12] T. Matsuyama, *Global well-posedness for the exterior initial-boundary value problem to the Kirchhoff equation,* J. Math. Soc. Japan **62** (2010), 1167–1204.

[13] K. Mochizuki, *Spectral and Scattering Theory for Second Order Elliptic Differential Operators in an Exterior Domain,* Lecture Notes Univ. Utah, Winter and Spring 1972.

[14] B. Muckenhoupt, *Weighted norm inequalities for the Fourier transform,* Trans. Amer. Math. Soc. **276** (1983), 729–742.

[15] M. Ruzhansky and M. Sugimoto, *A smoothing property of Schrödinger equations in the critical case*, Math. Ann. **335** (2006), 645–673.

[16] N.A. Shenk II, *Eigenfunction expansions and scattering theory for the wave equation in an exterior region*, Arch. Rational Mech. Anal. **21** (1966), 120–150.

[17] E.C. Titchmarsh, *Eigenfunction Expansions Associated with Second-Order Differential Equations,* part II, Oxford University Press, 1958.

[18] C.H. Wilcox, *Scattering Theory for the D'Alembert Equation in Exterior Domains,* Lecture Notes Math. **442** (1975).

[19] K. Yajima, *Dispersive estimates for Schrödinger equations with threshold resonance and eigenvalue*, Comm. Math. Phys. **259** (2005), 475–509.

[20] Qi S. Zhang, *The global behavior of heat kernels in exterior domains,* J. Functional Analysis **200** (2003), 160–176.

Tokio Matsuyama
Department of Mathematics
Faculty of Science and Engineering
Chuo University
1-13-27, Kasuga, Bunkyo-ku
Tokyo 112-8551, Japan
e-mail: `tokio@math.chuo-u.ac.jp`

Fourier Analysis
Trends in Mathematics, 311–334

On the Cauchy Problem for Hyperbolic Operators with Double Characteristics, a Transition Case

Tatsuo Nishitani

Abstract. We discuss the well-posedness of the Cauchy problem for hyperbolic operators with double characteristics when the operator is effectively hyperbolic outside a submanifold of codimension 1 of the double characteristic manifold. We assume that there is no null bicharacteristic tangent to the double characteristic manifold. We derive microlocal weighted energy estimates which proves the well-posedness of the Cauchy problem in all Gevrey classes assuming that the ratio of the imaginary part of the subprincipal symbol to the real eigenvalue of the Hamilton map is bounded on the double characteristic manifold and that the strict Ivrii-Petkov-Hörmander condition holds on the transition manifold.

Mathematics Subject Classification (2010). Primary 35L15; Secondary 35B30.

Keywords. Cauchy problem, well-posedness, effectively hyperbolic, transition case, null bicharacteristics.

1. Introduction

Let

$$P(x,D) = -D_0^2 + \sum_{|\alpha|\le 2, \alpha_0<2} a_\alpha(x)D^\alpha = p(x,D) + P_1 + P_0$$

be a second-order differential operator defined in a neighborhood of the origin of $\mathbb{R}^{n+1}$ where $x = (x_0, x') = (x_0, x_1, \dots, x_n)$. We assume that the principal symbol $p(x,\xi)$ vanishes exactly of order 2 on a C^∞ manifold Σ and

$$\text{rank}\Big(\sum_{j=0}^{n} d\xi_j \wedge dx_j\Big|_\Sigma\Big) = \text{constant}. \tag{1.1}$$

If the Hamilton map F_p (the linearization of the Hamilton flow H_p on Σ, see [9], [7]) has non-zero real eigenvalues on Σ then the Cauchy problem is C^∞ well posed for any lower-order terms (see [14], [11], [12]). On the other hand if the Hamilton map F_p has no non-zero real eigenvalues on Σ, assuming that the spectral structure of the Hamilton map F_p is stable on Σ, that is either $\mathrm{Ker}\, F_p^2 \cap \mathrm{Im}\, F_p^2 = \{0\}$ or $\mathrm{Ker}\, F_p^2 \cap \mathrm{Im}\, F_p^2 \neq \{0\}$ throughout Σ, the well-posedness of the Cauchy problem are discussed in [7], [10], [3], [4], [5].

In this note we restrict ourselves to the case $\mathrm{codim}\, \Sigma = 3$ and the spectral structure of F_p changes across a codimension 1 submanifold of Σ. Our aim is, studying this simplest transition case, to elucidate the effects of the spectral transition of F_p on Σ to the well-posedness of the Cauchy problem. By assumption, near any point $\rho \in \Sigma$, one can write

$$p(x,\xi) = -\xi_0^2 + \phi_1(x,\xi')^2 + \phi_2(x,\xi')^2$$

where $d\phi_1$ and $d\phi_2$ are linearly independent at ρ and $\Sigma = \{\xi_0 = 0, \phi_1 = 0, \phi_2 = 0\}$. We assume that

$$\left\{ \begin{array}{l} \text{the spectral structure of } F_p \text{ changes across} \\ \text{a submanifold } S \text{ of codimension 1 of } \Sigma. \end{array} \right. \tag{1.2}$$

Let $\rho \in \Sigma$. Then the Taylor expansion of $p(x,\xi)$ around ρ starts with $p_\rho(X)$, a second-order homogeneous polynomial in $X = (x,\xi)$, called the localization of p at ρ which is a hyperbolic polynomial in the direction $(0,\ldots,0,1,0,\ldots,0)$ (see [9]). Let $Q(X,Y)$ be the quadratic form associated to p_ρ then the Hamilton map at ρ is given by

$$\sigma(X, F_pY) = Q(X,Y), \quad X, Y \in \mathbb{R}^{2(n+1)}.$$

Lemma 1.1. *Assume* (1.1) *and* (1.2). *Then we have either* $\{\xi_0,\phi_1\} \neq 0$ *or* $\{\xi_0,\phi_2\} \neq 0$ *on* S.

Proof. Suppose that $\{\xi_0, \phi_j\} = 0$ at $\rho \in S$ for $j = 1, 2$. Then in a suitable symplectic basis in $T_\rho\Sigma$ we can write (see [7])

$$p_\rho = -\xi_0^2 + \xi_1^2 + \xi_2^2, \tag{1.3}$$

$$p_\rho = -\xi_0^2 + \mu(x_1^2 + \xi_1^2), \quad \mu > 0 \tag{1.4}$$

according to $\{\phi_1, \phi_2\}(\rho) = 0$ or $\{\phi_1, \phi_2\}(\rho) \neq 0$. If the first case occurs then from (1.1) we have $\mathrm{rank}\,(d\xi \wedge dx|_\Sigma) = 0$ and hence p_ρ takes the form (1.3) everywhere Σ in a suitable symplectic basis. In the second case we note that $\pm i\mu$ are the eigenvalues of F_p. From the continuity of $F_p(\rho)$ with respect to $\rho \in \Sigma$ it follows that F_p has still non-zero pure imaginary eigenvalues near ρ on Σ and hence p_ρ takes the form (1.4) in a suitable symplectic basis. Therefore, in both cases, the spectral structure of F_p does not change near ρ. This proves the assertion. □

We now assume that $\{\xi_0, \phi_j\}(\rho) \neq 0$ with some j. Considering

$$\tilde{\phi}_i = \sum_{j=1}^{2} O_{ij}\phi_j$$

with a smooth orthogonal matrix (O_{ij}) we may assume without restrictions that

$$\{\xi_0, \phi_2\} > 0, \quad \{\xi_0, \phi_1\} = O(|\phi|) \tag{1.5}$$

near ρ. Here and in what follows $f = O(|\phi|)$, $\phi = (\phi_1, \phi_2)$ means that f is a linear combination of ϕ_1 and ϕ_2 near the reference point. We look for non-zero eigenvalues μ of F_p; $F_pX = \mu X$. Since $\mu \neq 0$ it suffices to consider F_p on the image of F_p, that is on the vector space $\langle H_{\xi_0}, H_{\phi_1}, H_{\phi_2}\rangle$ spanned by H_{ξ_0}, H_{ϕ_1}, H_{ϕ_2}. Let $X = \alpha H_{\xi_0} + \beta H_{\phi_1} + \gamma H_{\phi_2}$ and consider $F_pX = \mu X$. Since

$$F_pX = -\gamma\{\phi_2, \xi_0\}H_{\xi_0} + \gamma\{\phi_2, \phi_1\}H_{\phi_1} + (\alpha\{\xi_0, \phi_2\} + \beta\{\phi_1, \phi_2\})H_{\phi_2}$$

we have

$$-\gamma\{\phi_2, \xi_0\} = \mu\alpha,\ \gamma\{\phi_2, \phi_1\} = \mu\beta,\ \alpha\{\xi_0, \phi_2\} + \beta\{\phi_1, \phi_2\} = \mu\gamma$$

so that

$$\begin{pmatrix} 0 & 0 & \{\xi_0, \phi_2\} \\ 0 & 0 & -\{\phi_1, \phi_2\} \\ \{\xi_0, \phi_2\} & \{\phi_1, \phi_2\} & 0 \end{pmatrix} \begin{pmatrix} \alpha \\ \beta \\ \gamma \end{pmatrix} = \mu \begin{pmatrix} \alpha \\ \beta \\ \gamma \end{pmatrix}.$$

Thus the characteristic equation is

$$\mu\big(\mu^2 - (\{\xi_0, \phi_2\}^2 - \{\phi_1, \phi_2\}^2)\big) = 0. \tag{1.6}$$

Lemma 1.2. *If the spectral structure of F_p changes across S then we have*

$$\{\xi_0, \phi_2\}^2 - \{\phi_1, \phi_2\}^2 = 0 \text{ on } S.$$

Therefore we have one of the following cases;

(i) $\{\xi_0, \phi_2\}^2 - \{\phi_1, \phi_2\}^2 < 0$ *in* $\Sigma \setminus S$, *that is* p *is noneffectively hyperbolic in* $\Sigma \setminus S$ *with* $\operatorname{Ker} F_p^2 \cap \operatorname{Im} F_p^2 = \{0\}$ *and noneffectively hyperbolic on* S *with* $\operatorname{Ker} F_p^2 \cap \operatorname{Im} F_p^2 \neq \{0\}$,

(ii) $\{\xi_0, \phi_2\}^2 - \{\phi_1, \phi_2\}^2 > 0$ *in* $\Sigma \setminus S$, *that is* p *is effectively hyperbolic in* $\Sigma \setminus S$ *and noneffectively hyperbolic on* S *with* $\operatorname{Ker} F_p^2 \cap \operatorname{Im} F_p^2 \neq \{0\}$,

(iii) $\{\xi_0, \phi_2\}^2 - \{\phi_1, \phi_2\}^2$ *changes the sign across* S, *that is* p *is effectively hyperbolic in the one side of* $\Sigma \setminus S$, *noneffectively hyperbolic in the other side with* $\operatorname{Ker} F_p^2 \cap \operatorname{Im} F_p^2 = \{0\}$ *and noneffectively hyperbolic on* S *with* $\operatorname{Ker} F_p^2 \cap \operatorname{Im} F_p^2 \neq \{0\}$.

Proof. Note that if $\{\xi_0, \phi_2\}^2 - \{\phi_1, \phi_2\}^2 \neq 0$ at $\rho \in S$ then $F_p(\rho)$ has non-zero real eigenvalues or non-zero pure imaginary eigenvalues according to $\{\xi_0, \phi_2\}^2 - \{\phi_1, \phi_2\}^2 > 0$ or $\{\xi_0, \phi_2\}^2 - \{\phi_1, \phi_2\}^2 < 0$. By the continuity of F_p with respect to ρ, $F_p(\rho)$ has still non-zero real eigenvalues or pure imaginary eigenvalues near ρ on Σ and then the spectral structure of F_p does not change near ρ on Σ. This proves the first assertion. Since

$$F_p^2 H_{\phi_2} = (\{\xi_0, \phi_2\}^2 - \{\phi_1, \phi_2\}^2)H_{\phi_2}$$

it is clear that $0 \neq H_{\phi_2} \in \operatorname{Ker} F_p^2 \cap \operatorname{Im} F_p^2$ if $\{\xi_0, \phi_2\}^2 - \{\phi_1, \phi_2\}^2 = 0$ and this proves the assertion. □

As for the case (i) more general geometrical settings of transitions are studied in [6]. They proved microlocal a priori estimate assuming that there is no null bicharacteristic tangent to the double characteristic manifold, while in [16] the well-posedness is discussed without this restriction but the codimension of Σ is restricted to 3. Some cases of transition from effectively hyperbolic to noneffectively hyperbolic are studied in [2].

2. Case (ii)

In what follows we study the case (ii) in Lemma 1.2. Let $e(\rho) > 0$, $\rho \in \Sigma \setminus S$ be the real eigenvalue of $F_p(\rho)$. Then from (1.6) one has

$$e(\rho)^2 = \{\xi_0, \phi_2\}^2 - \{\phi_1, \phi_2\}^2$$

and hence $e(\rho)^2$ extends smoothly on Σ. We assume that $|e(\rho)|$ vanishes "simply" on S, that is with some constant $C > 0$ we assume

$$C^{-1}\text{dist}(\rho, S) \le |e(\rho)| \le C\text{dist}(\rho, S) \tag{2.1}$$

on Σ where $\text{dist}(\rho, S)$ denotes the distance from ρ to S on Σ. From (1.6), (2.1) and Lemma 1.2 one can write

$$\{\xi_0, \phi_2\}^2 - \{\phi_1, \phi_2\}^2 = \theta^2 + c_1\phi_1 + c_2\phi_2$$

in a neighborhood of ρ where S is defined by $\{\theta = 0\} \cap \Sigma$ and $d\theta \neq 0$ on S. Since

$$\{\xi_0, \phi_2\}^2 - \{\phi_1, \phi_2\}^2 = \{\xi_0 + \phi_1, \phi_2\}\{\xi_0 - \phi_1, \phi_2\}$$

and hence we have either $\{\xi_0 + \phi_1, \phi_2\} = 0$ or $\{\xi_0 - \phi_1, \phi_2\} = 0$ on S. Since the arguments are completely parallel we may assume that

$$\{\xi_0 - \phi_1, \phi_2\} = 0 \quad \text{on} \quad S$$

and hence $\{\xi_0, \phi_2\} = \{\phi_1, \phi_2\} > 0$ on S. Thus we can write

$$\{\xi_0 - \phi_1, \phi_2\} = \tilde{\theta}^2 + c_1'\phi_1 + c_2'\phi_2 \tag{2.2}$$

near ρ where $\tilde{\theta} = c\theta$ with a non-zero factor c. Since we have $\{\xi_0 + \phi_1, \phi_2\}\tilde{\theta}^2 = e^2$ on Σ and

$$\frac{\{\xi_0 + \phi_1, \phi_2\}}{2\{\phi_1, \phi_2\}} = 1$$

on S then for any $\epsilon > 0$ there is a neighborhood of ρ where we have

$$(1 - \epsilon)e(\rho) \le \sqrt{2\{\phi_1, \phi_2\}}|\tilde{\theta}| \le (1 + \epsilon)e(\rho). \tag{2.3}$$

Define

$$\hat{\theta} = \tilde{\theta} - \frac{\{\phi_1, \tilde{\theta}\}}{\{\phi_1, \phi_2\}}\phi_2 + \frac{\{\phi_2, \tilde{\theta}\}}{\{\phi_1, \phi_2\}}\phi_1 \tag{2.4}$$

so that

$$S = \{\hat{\theta} = 0\} \cap \Sigma, \ \{\hat{\theta}, \phi_j\} = O(|\phi|), \ j = 1, 2.$$

Note that (2.2) and (2.3) still holds with $\hat{\theta}$ instead of $\tilde{\theta}$. Here we note

Proposition 2.1. *Assume that there is no null bicharacteristic tangent to* Σ. *Then we have*

$$\{\xi_0, \hat{\theta}\} = 0, \quad \{\{\xi_0 - \phi_1, \phi_2\}, \phi_2\} = 0$$

on S.

We will give a sketch of the proof in Section 4.

Lemma 2.2. *Assume that* $\{\{\xi_0 - \phi_1, \phi_2\}, \phi_2\} = 0$ *on* S. *Then we can write*

$$\{\xi_0 - \phi_1, \phi_2\} = \hat{\theta}^2 + c_0\hat{\theta}\phi_1 + c_1\phi_1^2 + c_2\phi_2.$$

Proof. Recall $\{\xi_0 - \phi_1, \phi_2\} = \hat{\theta}^2 + c_1\phi_1 + c_2\phi_2$. Thus

$$\{\{\xi_0 - \phi_1, \phi_2\}, \phi_2\} = O(\hat{\theta}|\phi|) + c_1\{\phi_1, \phi_2\} + O(|\phi|) = O(|(\hat{\theta}, \phi)|)$$

which implies that $c_1 = O(|(\hat{\theta}, \phi)|)$ and hence the result. □

Lemma 2.3. *Assume that* $\{\xi_0, \hat{\theta}\} = 0$, $\{\{\xi_0 - \phi_1, \phi_2\}, \phi_2\} = 0$ *on* S. *Then we have*

$$\{\xi_0 - \phi_1, \hat{\theta}\} = c\hat{\theta} + c_0\hat{\theta}\phi_1 + c_1\phi_1^2 + c_2\phi_2.$$

Proof. Note that $\{\xi_0 - \phi_1, \hat{\theta}\} = \alpha\hat{\theta} + \beta\phi_1 + \gamma\phi_2$. On the other hand we see

$$\{\hat{\theta}, \{\xi_0 - \phi_1, \phi_2\}\} = O(|\phi|), \quad \{\xi_0 - \phi_1, \{\hat{\theta}, \phi_2\}\} = O(|(\hat{\theta}, \phi)|).$$

Then from the Jacobi identity it follows that $\beta = O(|(\hat{\theta}, \phi)|)$ and hence

$$\{\xi_0 - \phi_1, \hat{\theta}\} = c\hat{\theta} + c_0\hat{\theta}\phi_1 + c_1\phi_1^2 + c_2\phi_2$$

which is the assertion. □

We make a dilation of the coordinate x_0; $x_0 \to \mu x_0$ with small $\mu > 0$ so that we have

$$\begin{aligned} P(x, \xi, \mu) &= \mu^2 P(\mu x_0, x', \mu^{-1}\xi_0, \xi') \\ &= p(\mu x_0, x', \xi_0, \mu\xi') + \mu P_1(\mu x_0, x', \xi_0, \xi') + \mu^2 P_0(\mu x_0, x') \\ &= p(x, \xi, \mu) + P_1(x, \xi, \mu) + P_0(x, \mu). \end{aligned}$$

In what follows we often write such symbols dropping μ. Let

$$g_0 = \mu^2 dx_0^2 + |dx'|^2 + \langle\xi'\rangle_\mu^{-2}|d\xi'|^2, \quad \langle\xi'\rangle_\mu^2 = \mu^{-2} + |\xi'|^2$$

then it is clear that $a(\mu x_0, x', \mu\xi') = a(x, \xi', \mu) \in S(\langle\mu\xi'\rangle^m, g_0)$ if $a(x, \xi') \in S^m_{1,0}$. Here we note that

$$\langle\mu\xi'\rangle\langle\xi'\rangle_\mu^{-1} = \mu.$$

We assume that our assumptions are satisfied globally and hence one can assume

$$\begin{cases} p(x,\xi) = -\xi_0^2 + \phi_1(x,\xi')^2 + \phi_2(x,\xi')^2, \quad \phi_j \in S(\langle \mu\xi' \rangle, g_0), \\ \{\xi_0 - \phi_1, \phi_1\} = c_1\phi_1 + c_2\phi_2, \quad c_j \in \mu S(1, g_0), \\ \{\xi_0 - \phi_1, \phi_2\} = \mu\hat{\theta}^2\langle \mu\xi' \rangle + c_0\hat{\theta}\phi_1 + c_1\phi_1^2\langle \mu\xi' \rangle^{-1} + c_2\phi_2, \\ \{\xi_0 - \phi_1, \hat{\theta}\} = c'\hat{\theta} + c_0'\hat{\theta}\phi_1\langle \mu\xi' \rangle^{-1} + c_1'\phi_1^2\langle \mu\xi' \rangle^{-2} + c_2'\phi_2\langle \mu\xi' \rangle^{-1} \\ \{\phi_1, \phi_2\} \geq c\mu\langle \mu\xi' \rangle, \quad c > 0, \\ e(\rho)(1-\epsilon)\langle \mu\xi' \rangle \leq \langle \xi' \rangle_\mu^{1/2}\sqrt{2\{\phi_1,\phi_2\}}|\hat{\theta}| \leq e(\rho)(1+\epsilon)\langle \mu\xi' \rangle \end{cases}$$

where c', c_j, $c_j' \in \mu S(1, g_0)$ and $\hat{\theta} \in S(1, g_0)$ is defined by (2.4) so that

$$\{\hat{\theta}, \phi_j\} = c_{j1}\phi_1 + c_{j2}\phi_2, \quad c_{jk} \in \mu S(\langle \mu\xi' \rangle^{-1}, g)$$

and $\sup|\hat{\theta}|$, $\sup|\phi_j\langle \mu\xi' \rangle^{-1}|$ can be assumed to be enough small working in a small neighborhood of the reference point.

Recall that

$$P(x,D) = (p + P_{\rm sub})^w + R, \quad R \in S(1, g_0)$$

where $p = -\xi_0^2 + \phi_1^2 + \phi_2^2$. We assume

$$|\mathsf{Im}P_{\rm sub}(\rho)| \leq Ce(\rho) \text{ on } \Sigma, \quad \mathrm{Tr}^+F_p(\rho) + \mathsf{Re}P_{\rm sub}(\rho) \geq c > 0 \text{ on } S \tag{2.5}$$

with some $c > 0$ and $C > 0$. Then the global version of (2.5) is

$$P_{\rm sub} = \hat{K} + K, \quad \hat{K} \geq c\mu\langle \mu\xi' \rangle, \quad K = c_0\hat{\theta}\langle \mu\xi' \rangle + c_1\phi_1 + c_2\phi_2 \tag{2.6}$$

with $c_j \in \mu S(1, g_0)$ since $\mathrm{Tr}^+F_p = 0$ on Σ.

Theorem 2.4. *Assume the case* (ii) *and that there is no null bicharacteristic tangent to* Σ. *Then under the conditions* (2.1) *and* (2.5) *the Cauchy problem is well posed in any Gevrey class* $\gamma^{(s)}$, $s \geq 1$.

Remark 2.5. With $X^\pm = \{\xi_0, \phi_2\}H_{\xi_0} - \{\phi_1, \phi_2\}H_{\phi_1} \pm e(\rho)H_{\phi_2}$ it is easy to see

$$F_p(\rho)X^\pm = \pm e(\rho)X^\pm$$

on Σ and this implies that for any point $\rho \in \Sigma \setminus S$, there exist exactly two null bicharacteristics passing ρ transversally to Σ with tangents $X^\pm$ (see [13]). Note

$$d\xi_0(X^\pm) = \pm e(\rho)\{\xi_0, \phi_2\}, \quad d\phi_1(X^\pm) = \pm e(\rho)\{\phi_1, \phi_2\},$$
$$d\phi_2(X^\pm) = -e(\rho)^2, \quad d\hat{\theta}(X^\pm) = -\{\xi_0, \phi_2\}\{\xi_0, \hat{\theta}\}$$

and $X^\pm$ converges to $\{\phi_1, \phi_2\}H_{\xi_0 - \phi_1} \in T\Sigma$ when ρ approaches to S. Thus we see

$$\{\phi_1, \phi_2\}H_{\xi_0 - \phi_1} \in TS, \quad \{\phi_1, \phi_2\}H_{\xi_0 - \phi_1} \in T\Sigma \setminus TS$$

according to $\{\xi_0, \hat{\theta}\} = 0$ and $\{\xi_0, \hat{\theta}\} \neq 0$ respectively.

Remark 2.6. It is clear from the proof that the Cauchy problem is well posed in much wider class which consists of all C_0^∞ functions such that its Fourier transform is bounded by constant times

$$\exp\big(- C\log^N \langle\xi'\rangle\big)$$

with some $C > 0$ where N is some fixed large positive number.

3. Idea of the proof of Theorem 2.4

Our strategy for proving Theorem 2.4 is to derive a weighted energy estimate. We first discuss the weights. To simplify notations let us set

$$\hat{\phi}_i = \phi_i\langle\mu\xi'\rangle^{-1}, \quad \theta = \hat{\theta}.$$

Without restrictions we may assume (see [4])

$$\hat{\phi}_2 = \hat{\phi}_2(x)$$

which is independent of ξ'. Let us introduce χ, $\chi_2 \in C^\infty(\mathbb{R})$

$$\chi(s) = \begin{cases} 1, & s \le d_1, \\ 0, & s \ge d_2 \end{cases} \qquad \chi_2(s) = \begin{cases} 1, & s \ge d_3, \\ 0, & s \le d_2 \end{cases}$$

where $0 < d_1 < d_2 < d_3$ such that $\chi\chi_2 = 0$ and χ is non increasing. Define

$$w = \sqrt{\hat{\phi}_2^2 + \langle\xi'\rangle_\mu^{-1}}$$

and put

$$\chi(x,\xi') = \chi(\hat{\phi}_1^2 w^{-1}), \;\; \chi_2(x,\xi') = \chi_2(\hat{\phi}_1^2 w^{-1}).$$

We simply write χ, χ_2 for $\chi(x,\xi')$ and $\chi_2(x,\xi')$ if there is no confusion. It is easy to check that $w \in S(w, w^{-2}|dx|^2 + \langle\xi'\rangle_\mu^{-2}|d\xi'|^2)$ and we see easily

$$\chi, \; \chi_2 \in S(1,g), \quad g = w^{-2}|dx|^2 + w^{-1}\langle\xi'\rangle_\mu^{-2}|d\xi'|^2.$$

We also set

$$\psi = \nu\theta^2\phi_1 + \chi_2\phi_1^3\langle\mu\xi'\rangle^{-2} \tag{3.1}$$

where $0 < \nu \ll 1$ is a positive parameter. Let us rewrite p as

$$\begin{aligned} p &= -(\xi_0 + \phi_1)(\xi_0 - \phi_1) + \phi_2^2 \\ &= -(\xi_0 + \phi_1 - \psi)(\xi_0 - \phi_1 + \psi) + 2\psi\phi_1 - \psi^2 + \phi_2^2 \\ &= -(\xi_0 + \lambda)(\xi_0 - \lambda) + \phi_2^2 + 2\nu a^2\theta^2\phi_1^2 + 2\chi_2 a^2\phi_1^4\langle\mu\xi'\rangle^{-2} \\ &= -(\xi_0 + \lambda)(\xi_0 - \lambda) + q \end{aligned} \tag{3.2}$$

where

$$\begin{cases} \lambda = \phi_1 - \psi, \\ q = \phi_2^2 + 2\nu a^2\theta^2\phi_1^2 + 2\chi_2 a^2\phi_1^4\langle\mu\xi'\rangle^{-2}, \\ a = \sqrt{1 - \nu\theta^2/2 - \chi_2\phi_1^2\langle\mu\xi'\rangle^{-2}/2}. \end{cases}$$

As in [10] we move $c_1\phi_1$ in (2.6) into the principal part so that

$$\begin{aligned}&-(\xi_0+\lambda)(\xi_0-\lambda)+c_1\phi_1\\&\quad=-(\xi_0+\lambda+c_1)(\xi_0-\lambda-c_1)-2c_1\psi-c_1^2.\end{aligned}$$

Then K turns to be $c_0\theta\langle\mu\xi'\rangle+c_2\phi_2+d_0\theta\phi_1+d_1\phi_1^2\langle\mu\xi'\rangle^{-1}$. We denote $P-K$ by P again and hence we are led to consider $P=-M\Lambda+Q$ with $M=\xi_0+\lambda+c_1$, $\Lambda=\xi_0-\lambda-c_1$ and $Q=q+\hat{K}$.

Let us define

$$\begin{aligned}&\phi=\langle\xi'\rangle_\mu^{1/2}(\hat{\phi}_2+w),\\&\Phi=i\{\log(\hat{\phi}_2+i\omega)-\log(\hat{\phi}_2-i\omega)\}=-2\arg(\hat{\phi}_2+i\omega)\end{aligned}$$

where

$$\omega=\sqrt{\hat{\phi}_1^4+\langle\xi'\rangle_\mu^{-1}}.$$

It is easy to see $\omega\in S(\omega,\omega^{-1}(|dx'|^2+\langle\xi'\rangle_\mu^{-2}|d\xi'|^2))$. We put

$$\rho^2=\hat{\phi}_2^2+\omega^2=\hat{\phi}_2^2+\hat{\phi}_1^4+\langle\xi'\rangle_\mu^{-1}$$

then $\rho\in S(\rho,\rho^{-2}|dx|^2+\rho^{-1}\langle\xi'\rangle_\mu^{-2}|d\xi'|^2)$. Let us introduce the weight symbol

$$T=\exp\big(\ell(\chi^2\log\phi+\Phi)\big)$$

where ℓ is a real parameter. It is easy to check

Lemma 3.1. *We have*

$$\begin{aligned}&\{f,\log\phi\}=\frac{\{f,\hat{\phi}_2\}}{w}+\langle\xi'\rangle_\mu^{-1/2}\{f,\langle\xi'\rangle_\mu^{1/2}\}+\frac{\{f,\langle\xi'\rangle_\mu^{-1}\}}{2w(\hat{\phi}_2+w)},\\&\{f,\Phi\}=2\frac{\omega\{f,\hat{\phi}_2\}-\hat{\phi}_2\{f,\omega\}}{\rho^2}.\end{aligned}$$

Lemma 3.2. *We have*

$$\phi\in S(\phi,g),\quad \Phi\in S(1,g_1)$$

with $g_1=(\rho^{-1}+\omega^{-1/2})^2|dx|^2+\omega^{-1}\langle\xi'\rangle_\mu^{-2}|d\xi'|^2$.

Proof. With $\tilde{\phi}=\hat{\phi}_2+w$ it is enough to show $\tilde{\phi}\in S(\tilde{\phi},g)$. Note that

$$\partial_{x,\xi'}^\alpha\tilde{\phi}=\frac{\partial_{x,\xi'}^\alpha\hat{\phi}_2}{w}\tilde{\phi}+\frac{\partial_{x,\xi'}^\alpha\langle\xi'\rangle_\mu^{-1}}{2w}$$

for $|\alpha|=1$ and $\langle\xi'\rangle_\mu^{-1}\le 2w\tilde{\phi}$. Since $\phi_2\langle\mu\xi'\rangle^{-1}\in S(1,g_0)\subset S(1,g)$ and $w^{-1}\in S(w^{-1},g)$ we get the assertion by induction on $|\alpha|$. The proof of the second assertion is similar. □

Here we note that

$$g,\ g_1\le\bar{g}=\langle\xi'\rangle_\mu|dx|^2+\langle\xi'\rangle_\mu^{-3/2}|d\xi'|^2.$$

Lemma 3.3. *We have*

$$e^{\chi^2 \log\phi} \in S(e^{\chi^2 \log\phi}, \log^2 \langle\xi'\rangle_\mu g), \quad e^{\Phi} \in S(1, g_1)$$

and $T \in S(T, \log^2 \langle\xi'\rangle_\mu \bar{g})$.

Proof. Note that $\partial_{x,\xi'}^\alpha \log\phi = \phi^{-1}\partial_{x,\xi'}^\alpha\phi$ and $\phi^{-1} \in S(\phi^{-1}, g)$ for $|\alpha| = 1$. Since $|\log\phi| \le C\log\langle\xi'\rangle_\mu$ the assertion is clear. The proof of the assertion $e^{\Phi} \in S(1, g_1)$ is easy. □

We consider

$$PT = T\tilde{P}, \quad \tilde{P} = -\tilde{M}\tilde{\Lambda} + \tilde{Q} \tag{3.3}$$

and try to get an energy estimate for $\tilde{P}$, which is a weighted energy estimate with weight T. Let us write $\tilde{M} = D_0 - \tilde{m}(x, D')$, $\tilde{\Lambda} = D_0 - \tilde{\lambda}(x, D')$ and fix an arbitrary small $\varepsilon > 0$ and apply

Proposition 3.4. *In* (3.3) *replacing* $\tilde{\Lambda}$, $\tilde{M}$ *by* $\tilde{\Lambda} - i\gamma\langle\mu D'\rangle^{2\varepsilon}$, $\tilde{M} - i\gamma\langle\mu D'\rangle^{2\varepsilon}$ *we have*

$$\begin{aligned}
2\mathsf{Im}(\tilde{P}u, \tilde{\Lambda}u) \ge{}& \frac{d}{dx_0}(\|\tilde{\Lambda}u\|^2 + ((\mathsf{Re}\,\tilde{Q})u, u) + \gamma^2\|\langle\mu D'\rangle^{2\varepsilon}u\|^2)\\
&+ \gamma\|\langle\mu D'\rangle^{\varepsilon}\tilde{\Lambda}u\|^2 + 2\gamma\mathsf{Re}(\langle\mu D'\rangle^{2\varepsilon}\tilde{Q}u, u) + 2((\mathsf{Im}\,\tilde{m})\tilde{\Lambda}u, \tilde{\Lambda}u)\\
&+ 2\mathsf{Re}(\tilde{\Lambda}u, (\mathsf{Im}\,\tilde{Q})u) + \mathsf{Im}([D_0 - \mathsf{Re}\tilde{\lambda}, \mathsf{Re}\,\tilde{Q}]u, u)\\
&+ 2\mathsf{Re}((\mathsf{Re}\,\tilde{Q})u, (\mathsf{Im}\tilde{\lambda})u) + \frac{\gamma^3}{2}\|\langle\mu D'\rangle^{3\varepsilon}u\|^2 + 2\gamma^2(\langle\mu\xi'\rangle^{4\varepsilon}(\mathsf{Im}\tilde{\lambda})u, u).
\end{aligned}$$

Here $\langle\mu D'\rangle^{2\varepsilon}$ is introduced to control error terms such that $O(\|\log^N\langle D'\rangle u\|)$ and hence we can choose $\varepsilon > 0$ as small as we please, which determines the well-posed Gevrey class $\gamma^{(1/2\varepsilon)}$.

Let $Op(\phi)$ be the Weyl quantized pseudodifferential operator with symbol ϕ and we denote $Op(\phi)Op(\psi) = Op(\phi\#\psi)$. We often use the same letter to denote a symbol and the operator with its symbol if there is no confusion. Thus we denote

$$Op(\phi\psi)u = \phi\psi u, \quad Op(\phi)Op(\psi)u = \phi(\psi u).$$

We note that

Lemma 3.5. *Let* $a \in S(\langle\mu\xi'\rangle^2, g_0)$. *Then we have*

$$a\#T = T\#a - i\ell\{a, \chi^2\log\phi + \Phi\}T + T\#R$$

with some $R \in S(\langle\mu\xi'\rangle^{1/2}\log^3\langle\xi'\rangle_\mu, \bar{g})$.

The cut-off symbols χ, χ_2 are used efficiently when we estimate the term $\mathsf{Re}(\tilde{\Lambda}u, (\mathsf{Im}\,\tilde{Q})u)$. We will come back to this point later. On the other hand because of this χ some troubles occur in the terms $\mathsf{Im}\,\tilde{m}$ and $\mathsf{Im}\,\tilde{\lambda}$. The main parts of $\mathsf{Im}\,\tilde{m}$

and $\mathrm{Im}\,\tilde{\lambda}$ are $\{\xi_0\pm\phi_1,\chi^2\log\phi+\Phi\}$. Note $\{\xi_0\pm\phi_1,\chi^2\log\phi\}=\{\xi_0\pm\phi_1,\chi^2\}\log\phi+\chi^2\{\xi_0\pm\phi_1,\log\phi\}$ and it is expected that $\chi^2\{\xi_0\pm\phi_1,\log\phi\}$ is dominant. Note

$$\begin{aligned}\{\xi_0\pm\phi_1,\chi^2\}\log\phi&=4\chi\chi'\hat{\phi}_1w^{-1}\{\xi_0\pm\phi_1,\hat{\phi}_1\}\log\phi\\&\quad-2\chi\chi'\hat{\phi}_1^2w^{-3}\{\xi_0\pm\phi_1,\hat{\phi}_2\}\hat{\phi}_2\log\phi\\&\quad-\chi\chi'\hat{\phi}_1^2w^{-3}\{\xi_0\pm\phi_1,\langle\xi'\rangle_\mu^{-1}\}\log\phi.\end{aligned}$$

Since $\{\xi_0\pm\phi_1,\hat{\phi}_1\}$ is a linear combination of $\hat{\phi}_j$ the main term is

$$-2\chi\chi'\hat{\phi}_1^2w^{-2}\hat{\phi}_2w^{-1}\{\xi_0\pm\phi_1,\hat{\phi}_2\}\log\phi$$

and this can not be controlled, in general, by $\{\xi_0\pm\phi_1,\hat{\phi}_2\}w^{-1}$ which is the main term in $\{\xi_0\pm\phi_1,\log\phi\}$ because $|\chi\chi'\hat{\phi}_1^2w^{-2}\hat{\phi}_2|\geq c>0$ if $\chi'\neq 0$ and $|\hat{\phi}_2|\sim\langle\xi'\rangle_\mu^{-1/2}$. The key observation to manage this inconvenience is

Lemma 3.6. *We have*

$$\hat{\phi}_2\log\phi\geq 0$$

and hence

$$\delta=-2\chi\chi'\hat{\phi}_1^2w^{-3}\hat{\phi}_2\log\phi\geq 0.$$

Proof. Note that $\phi=\langle\xi'\rangle_\mu^{1/2}(\hat{\phi}_2+w)\geq\langle\xi'\rangle_\mu^{1/2}\langle\xi'\rangle_\mu^{-1/2}=1$ if $\hat{\phi}_2\geq 0$ and

$$\phi=\langle\xi'\rangle_\mu^{1/2}(\hat{\phi}_2+w)=\frac{\langle\xi'\rangle_\mu^{-1/2}}{|\hat{\phi}_2|+w}\leq 1$$

for $\hat{\phi}_2\leq 0$. Then the first assertion is clear. The second assertion is now obvious because $\chi'\leq 0$. □

Lemma 3.7. *Assume that $f\in S(\langle\mu\xi'\rangle,g_0)$ then we have*

$$\{f,\chi^2\log\phi\}=\chi^2w^{-1}\{f,\hat{\phi}_2\}+\{f,\chi^2\}\log\phi+R,\quad R\in S(1,g).$$

Moreover if $\{f,\hat{\phi}_1\}$ is a linear combination of $\hat{\phi}_j$ then we have

$$\{f,\chi^2\log\phi\}=\{f,\hat{\phi}_2\}(\chi^2w^{-1}+\delta)+R$$

with $R\in S(\log\langle\xi'\rangle_\mu,g)$ where $0\leq\delta\in S(w^{-1}\log\langle\xi'\rangle_\mu,g)$.

Proof. Note that with $\tilde{\phi}=\hat{\phi}_2+w$

$$\begin{aligned}\{f,\chi^2\log\phi\}&=\{f,\chi^2\}\log\phi+\chi^2\{f,\tilde{\phi}\}\tilde{\phi}^{-1}+\chi^2\{f,\langle\xi'\rangle_\mu^{1/2}\}\langle\xi'\rangle_\mu^{-1/2}\\&=\{f,\chi^2\}\log\phi+\frac{\chi^2\{f,\hat{\phi}_2\}}{w}+\frac{\chi^2\{f,\langle\xi'\rangle_\mu^{-1}\}}{w\tilde{\phi}}+\chi^2\{f,\langle\xi'\rangle_\mu^{1/2}\}\langle\xi'\rangle_\mu^{-1/2}\\&=\{f,\chi^2\}\log\phi+\frac{\chi^2\{f,\hat{\phi}_2\}}{w}+R\end{aligned}$$

with $R \in S(1,g)$ because $2w\tilde{\phi} \geq \langle\xi'\rangle_\mu^{-1}$. This proves the first assertion. Note

$$\{f,\chi^2\}\log\phi = 2\chi\chi'\{f,\hat{\phi}_1^2 w^{-1}\}\log\phi = 4\chi\chi'\hat{\phi}_1 w^{-1}\{f,\hat{\phi}_1\}\log\phi$$
$$- 2\chi\chi'\hat{\phi}_1^2 w^{-2}w^{-1}\{f,\hat{\phi}_2\}\hat{\phi}_2\log\phi - \chi\chi'\hat{\phi}_1^2 w^{-3}\{f,\langle\xi'\rangle_\mu^{-1}\}\log\phi.$$

Let $\{f,\hat{\phi}_1\}$ be a linear combination of $\hat{\phi}_j$ then it is clear that $\chi'\hat{\phi}_1 w^{-1}\{f,\hat{\phi}_1\} \in S(1,g)$ because $w^{-1}\hat{\phi}_2 \in S(1,g)$. It is also clear that $w^{-2}\{f,\langle\xi'\rangle_\mu^{-1}\} \in S(1,g)$ for $w^{-2} \in S(\langle\xi'\rangle_\mu, g)$ and hence the second assertion. □

Corollary 3.8. *We have*

$$\{\xi_0 \pm \phi_1, \chi^2\log\phi\} = \{\xi_0 \pm \phi_1, \hat{\phi}_2\}(\chi^2 w^{-1} + \delta) + R$$

with $R \in S(\log\langle\xi'\rangle_\mu, g)$.

Lemma 3.9. *We have*

$$\operatorname{Im}\tilde{\lambda} = \mu\ell\big(\theta^2 e_1(\chi^2 w^{-1} + \delta) + a_1\chi\theta w^{-1/2} + \theta^2 e_1\omega\rho^{-2} + a_2\theta\omega^{1/2}\rho^{-1}\big) + R,$$
$$\operatorname{Re}\tilde{\lambda} = b_1\theta^2 w^{-1/2} + b_2\theta^2\langle\mu\xi'\rangle^{-1/4}\omega\rho^{-2} + R$$

where $e_1 = 1 + \nu\mu^{-1}\{\phi_1,\hat{\phi}_2\}$, $a_i \in S(1,\bar{g})$, $b_i \in S(\log\langle\xi'\rangle_\mu, \bar{g})$, $R \in S(\log^2\langle\xi'\rangle_\mu, \bar{g})$. *We have also*

$$\operatorname{Im}\tilde{m} = \mu\ell\big(e_2(\chi^2 w^{-1} + \delta) + \omega\rho^{-2}\big) + R$$

with $e_2 = \mu^{-1}(\{\xi_0 + \phi_1, \hat{\phi}_2\} - \nu\theta^2\{\phi_1,\hat{\phi}_2\})$ *and* $R \in S(\log^2\langle\xi'\rangle_\mu, \bar{g})$.

Lemma 3.10. *We have*

$$C(\operatorname{Im}\tilde{\lambda}\, u, u) \geq \mu\ell\{(\chi^2\theta^2 w^{-1}u, u) + (\theta^2\omega\rho^{-2}u, u)\} - C'\|\langle\mu D'\rangle^\varepsilon u\|^2,$$
$$C(\operatorname{Im}\tilde{m}\, u, u) \geq \mu\ell\{(\chi^2 w^{-1}u, u) + (\omega\rho^{-2}u, u)\} - C'\|\langle\mu D'\rangle^\varepsilon u\|^2$$

with some $C > 0$ *independent of* μ.

Proof. Recall $\operatorname{Im}\tilde{\lambda} = \mu\ell\big(e_1\theta^2(\chi^2 w^{-1} + \delta) + a_1\chi\theta w^{-1/2} + e_1\theta^2\omega\rho^{-2} + a_2\theta\omega^{1/2}\rho^{-1}\big) + R$. With some $C_1 > 0$, $C > 0$ we have

$$C_1 + Ce_1\theta^2(\chi^2 w^{-1} + \delta) + a_1\chi\theta w^{-1/2} - \theta^2\chi^2 w^{-1} \geq 0.$$

Since $C_1 + Ce_1\theta^2(\chi^2 w^{-1} + \delta) + a_1\chi\theta w^{-1/2} - \theta^2\chi^2 w^{-1} \in S(w^{-1}\log\langle\xi'\rangle_\mu, g) \subset S(\langle\xi'\rangle_\mu^{1/2+\varepsilon}, \bar{g})$ from the Fefferman–Phong inequality (see Theorem 18.6.8 in [8]) it follows that

$$C\big((e_1\theta^2(\chi^2 w^{-1} + \delta) + a_1\chi\theta w^{-1/2})u, u\big) \geq (\theta^2\chi^2 w^{-1}u, u) - C_2\|\langle\mu D'\rangle^\varepsilon u\|^2.$$

Since $|(Ru, u)| \leq C\|\langle\mu D'\rangle^\varepsilon u\|^2$ the first assertion holds. Repeating the same arguments we get the second assertion. □

The main term in $\operatorname{Re}(\langle\mu D'\rangle^{2\varepsilon}\tilde{Q}u, u)$ is $(\langle\mu\xi'\rangle^{2\varepsilon}(q + \mu\langle\mu\xi'\rangle)u, u)$ and from which we get a positive contribution.

Proposition 3.11. *We have with some $c > 0$*

$$(\langle\mu\xi'\rangle^{2\varepsilon}(q+\mu\langle\mu\xi'\rangle)u,u) \geq c(\|\langle\mu\xi'\rangle^{\varepsilon}\phi_2 u\|^2 + \nu\|\langle\mu\xi'\rangle^{\varepsilon}\theta\phi_1 u\|^2$$
$$+ \|\langle\mu\xi'\rangle^{\varepsilon}\phi_1^2\langle\mu\xi'\rangle^{-1}u\|^2) - C\mu\|\langle\mu D'\rangle^{1/2+\varepsilon}u\|^2.$$

Proof. We first study $\phi_2^2 + \chi_2 a^2\phi_1^4\langle\mu\xi'\rangle^{-2} + \mu\langle\mu\xi'\rangle$. Note that if $\chi_2 \neq 1$ then we have $\hat{\phi}_1^4 \leq d_3^2(\hat{\phi}_2^2 + \langle\xi'\rangle_\mu^{-1})$ so that $\phi_2^2 + \chi_2 a^2\phi_1^4\langle\mu\xi'\rangle^{-2} + \mu\langle\mu\xi'\rangle \geq c\phi_1^4\langle\mu\xi'\rangle^{-2}$. Thus we have

$$\phi_2^2 + \chi_2 a^2\phi_1^4\langle\mu\xi'\rangle^{-2} + \mu\langle\mu\xi'\rangle \geq c\phi_1^4\langle\mu\xi'\rangle^{-2} \tag{3.4}$$

with some $c > 0$. Let us study

$$2w^2\langle\mu\xi'\rangle^2 + \chi_2 a^2\phi_1^4\langle\mu\xi'\rangle^{-2} - c\phi_1^4\langle\mu\xi'\rangle^{-2} = F^2,$$
$$F = w\langle\mu\xi'\rangle\sqrt{2 + \chi_2 a^2\hat{\phi}_1^4 w^{-2} - c\hat{\phi}_1^4 w^{-2}}$$

where $F \in S(w\langle\mu\xi'\rangle, g)$. Write

$$F^2 = F\#F + R$$

with $R \in \mu^2 S(w^{-2}, g)$ and recalling $w^2\langle\mu\xi'\rangle^2 = \phi_2^2 + \mu\langle\mu\xi'\rangle$ we conclude that

$$((\phi_2^2 + \mu\langle\mu\xi'\rangle)u,u) + (\chi_2 a^2\phi_1^4 u,u) \geq c\|\phi_1^2\langle\mu\xi'\rangle^{-1}u\|^2 - C\mu\|\langle\mu D'\rangle^{1/2}u\|^2.$$

The rest of the proof is clear. □

Applying Lemma 3.9 we derive a positive contribution

$$\begin{aligned}\mu\ell\Big\{&\nu((\chi^2 w^{-1}+\delta)(\theta^2\phi_1)u, \theta^2\phi_1 u) + ((\chi^2 w^{-1}+\delta)\tilde{\Lambda}u, \tilde{\Lambda}u)\\ &+ (\theta^2\omega\langle\mu\xi'\rangle^2 u,u) + (\chi^2\theta^2 w\langle\mu\xi'\rangle^2 u,u)\\ &+ \|\rho^{-1}\omega^{1/2}\tilde{\Lambda}u\|^2 + \|\chi w^{-1/2}\tilde{\Lambda}u\|^2\Big\}\end{aligned} \tag{3.5}$$

from the term

$$2((\mathsf{Im}\,\tilde{m})\tilde{\Lambda}u, \tilde{\Lambda}u) + 2\mathsf{Re}((\mathsf{Re}\,\tilde{Q})u, (\mathsf{Im}\tilde{\lambda})u).$$

As an example we show how to derive the positive contributions $(\chi^2\theta^2 w\langle\mu\xi'\rangle^2 u,u)$ and $(\theta^2\omega\langle\mu\xi'\rangle^2 u,u)$. Let us consider

$$\mathsf{Re}\,((\phi_2^2 + \mu\langle\mu\xi'\rangle)u, \theta^2 e_1(\chi^2 w^{-1}+\delta)u).$$

Recall that $\phi_2^2 + \mu\langle\mu\xi'\rangle = w^2\langle\mu\xi'\rangle^2$ and put $r = (\chi^2 w^{-1}+\delta)$. Note that one has

$$\mathsf{Re}\,(\theta^2 e_1 r\# w^2\langle\mu\xi'\rangle^2) = \theta^2 w^2\langle\mu\xi'\rangle^2 e_1 r + R$$

with $R \in \mu^2(w^{-2}\log\langle\xi'\rangle_\mu, g)$ and hence we see

$$\mathsf{Re}\,(w^2\langle\mu\xi'\rangle^2 u, \theta^2 e_1 r u) \geq (\theta^2 w^2\langle\mu\xi'\rangle^2 e_1 r u,u) - C\mu\|\langle\mu\xi'\rangle^{1/2+\varepsilon}u\|^2.$$

Taking a small $c > 0$ such that $e_1 r w - c\chi^2 \geq 0$ we consider

$$\begin{aligned}&\theta^2 w^2\langle\mu\xi'\rangle^2 e_1 r - c\chi^2\theta^2 w\langle\mu\xi'\rangle^2\\ &= \theta w^{1/2}\langle\mu\xi'\rangle\#(e_1 r w - c\chi^2)\#\theta w^{1/2}\langle\mu\xi'\rangle + R\end{aligned}$$

where $R \in \mu^2 S(w^{-2} \log \langle \xi' \rangle_\mu, g)$. From the Fefferman–Phong inequality it follows that

$$(\theta^2 w^2 \langle \mu\xi' \rangle^2 e_1 r u, u) - c(\chi^2 \theta^2 w \langle \mu\xi' \rangle^2 u, u)$$
$$\geq -C\mu \| w^{1/2} \langle \mu\xi' \rangle^{1/2+\varepsilon} u\|^2 - C\mu \| \langle \mu D' \rangle^{1/2+\varepsilon} u\|^2.$$

Thus we get

$$\mathsf{Re}\,((\phi_2^2 + \mu\langle \mu\xi' \rangle)u, \theta^2 e_1 r u)$$
$$\geq c(\chi^2 \theta^2 w \langle \mu\xi' \rangle^2 u, u) - C\mu \| \langle \mu\xi' \rangle^{1/2+\varepsilon} u\|^2.$$

We next show how to derive the positive term $(\theta^2 \omega^2 \langle \mu\xi' \rangle^2 u, u)$. Consider

$$\mathsf{Re}\,((\phi_2^2 + \chi_2 a^2 \phi_1^4 \langle \mu\xi' \rangle^{-2} + \mu\langle \mu\xi' \rangle)u, \theta^2 e_1 \omega \rho^{-2} u).$$

We first note that $\chi_2 \in S(1, \omega^{-2}|dx|^2 + \omega^{-1}\langle \xi' \rangle_\mu^{-2} |d\xi'|^2)$ because on the support of χ_2' we have

$$C^{-1} w \leq \omega \leq C w \tag{3.6}$$

with some $C > 0$. Taking this into account we can write

$$\mathsf{Re}\,\{(\phi_2^2 + \chi_2 a^2 \phi_1^4 \langle \mu\xi' \rangle^{-2} + \mu\langle \mu\xi' \rangle) \# \theta^2 e_1 \omega \rho^{-2}\}$$
$$= (\phi_2^2 + \chi_2 a^2 \phi_1^4 \langle \mu\xi' \rangle^{-2} + \mu\langle \mu\xi' \rangle)\theta^2 e_1 \omega \rho^{-2} + R$$

with $R \in \mu^2 S(\omega^{-2}, \bar{g})$. Thanks to (3.4) one can find $c > 0$ such that

$$e_1(\phi_2^2 + \chi_2 a^2 \phi_1^4 \langle \mu\xi' \rangle^{-2} + \mu\langle \mu\xi' \rangle) \geq 2c\rho^2 \langle \mu\xi' \rangle^2.$$

Let us write

$$(\phi_2^2 + \chi_2 a^2 \phi_1^4 \langle \mu\xi' \rangle^{-2} + \mu\langle \mu\xi' \rangle)\theta^2 e_1 \omega \rho^{-2} - c\theta^2 \omega \langle \mu\xi' \rangle^2$$
$$= \theta^2 \omega \langle \mu\xi' \rangle^2 \{e_1(\phi_2^2 + \chi_2 a^2 \phi_1^4 \langle \mu\xi' \rangle^{-2} + \mu\langle \mu\xi' \rangle)\rho^{-2}\langle \mu\xi' \rangle^{-2} - c\} = F^2.$$

Since $F^2 = F \# F + R$ with $R \in \mu^2 S(\omega^{-2}, \bar{g})$ we conclude

$$\mathsf{Re}\,((\phi_2^2 + \chi_2 a^2 \phi_1^4 \langle \mu\xi' \rangle^{-2} + \mu\langle \mu\xi' \rangle)u, \theta^2 e_1 \omega \rho^{-2} u)$$
$$\geq c(\theta^2 \omega \langle \mu\xi' \rangle^2 u, u) - C\mu \| \langle \mu D' \rangle^{1/2} u\|^2.$$

Lemma 3.12. *We have with some $C > 0$ independent of μ that*

$$C(\theta^2 \omega \langle \mu\xi' \rangle^2 u, u) \geq \mu^{1/2}(\theta^2 \langle \mu\xi' \rangle^{3/2} u, u) - C'\mu \| \langle \mu D' \rangle^{1/2} u\|^2.$$

Proof. Consider

$$M\theta^2 \omega \langle \mu\xi' \rangle^2 - \mu^{1/2}\theta^2 \langle \mu\xi' \rangle^{3/2} = \theta^2 \langle \mu\xi' \rangle^2 \omega(M - \mu^{1/2}\omega^{-1}\langle \mu\xi' \rangle^{-1/2})$$

which can be written as

$$\theta\langle \mu\xi' \rangle \# \omega(M - \mu^{1/2}\omega^{-1}\langle \mu\xi' \rangle^{-1/2}) \# \theta\langle \mu\xi' \rangle + R$$

with $R \in \mu^2 S(\omega^{-2}, \bar{g})$. Since

$$0 \leq \omega(M - \mu^{1/2}\omega^{-1}\langle \mu\xi' \rangle^{-1/2}) \in S(1, \langle \mu\xi' \rangle^{1/2}|dx|^2 + \langle \mu\xi' \rangle^{-3/2}|d\xi'|^2)$$

then from the Fefferman–Phong inequality one has the assertion. □

Lemma 3.13. *We have*

$$\mu^{4/3}\|\langle\mu D'\rangle^{1/3+\varepsilon}u\|^2$$
$$\le C\big(\|\langle\mu\xi'\rangle^{\varepsilon}\phi_2 u\|^2+\|\langle\mu\xi'\rangle^{\varepsilon}\phi_1^2\langle\mu\xi'\rangle^{-1}u\|^2\big)+C\mu^2\|\langle\mu D'\rangle^{\varepsilon}u\|^2$$

with some $C>0$.

Proof. Note that

$$\{\phi_1\langle\mu\xi'\rangle^{-1/3},\phi_2\}=\{\phi_1,\phi_2\}\langle\mu\xi'\rangle^{-1/3}+\phi_1\langle\mu\xi'\rangle^{-1/3}A,\quad A\in\mu S(1,g_0)$$

which proves

$$\mathsf{Re}\,(i[\phi_1\langle\mu\xi'\rangle^{-1/3},\phi_2]u,u)$$
$$\ge(\{\phi_1,\phi_2\}\langle\mu\xi'\rangle^{-1/3}u,u)-C\|\phi_1\langle\mu\xi'\rangle^{-1/3}u\|^2-C\mu^2\|u\|^2.$$

Since $\{\phi_1,\phi_2\}\langle\mu\xi'\rangle^{-1/3}\ge c\mu\langle\mu\xi'\rangle^{2/3}$ this shows

$$\mu\|\langle\mu D'\rangle^{1/3}u\|^2\le C\big(\mu^{-1/3}\|\phi_2 u\|^2+\mu^{1/3}\|\phi_1\langle\mu\xi'\rangle^{-1/3}u\|^2\big)+C\mu^{5/3}\|u\|^2.$$

Since we have $\phi_1\langle\mu\xi'\rangle^{-1/3}\#\phi_1\langle\mu\xi'\rangle^{-1/3}=\mathsf{Re}\,(\phi_1^2\langle\mu\xi'\rangle^{-1}\#\langle\mu\xi'\rangle^{1/3})+R$ where $R\in\mu^2 S(1,g_0)$ we get

$$C\mu^{1/3}\|\phi_1\langle\mu\xi'\rangle^{-1/3}u\|^2\le C_1\mu^{-1/3}\|\phi_1^2\langle\mu\xi'\rangle^{-1}u\|^2$$
$$+2^{-1}\mu\|\langle\mu D'\rangle^{1/3}u\|^2+C_2\mu^2\|u\|^2$$

and hence the assertion. □

The main term of $\mathsf{Re}(\tilde\Lambda u,(\mathsf{Im}\,\tilde Q)u)$ consists of

$$\ell(\tilde\Lambda u,\{q,\chi^2\log\phi+\Phi\}u).$$

Since $\chi\chi_2=0$ instead of $\{q,\chi^2\log\phi\}$ it suffices to consider $\{\phi_2^2+2\nu a^2\theta^2\phi_1^2,\chi^2\log\phi\}$. Let us study $\nu\{\theta^2\phi_1^2,\chi^2\log\phi\}=2\nu\theta\phi_1\{\theta\phi_1,\chi^2\log\phi\}$ which is the most delicate term. Since $\{\theta,\hat\phi_1\}$ is a linear combination of $\hat\phi_i$ it follows from Lemma 3.7 that

$$2\nu\theta\phi_1\{\theta\phi_1,\chi^2\log\phi\}=2\nu\theta\phi_1\Big[\{\theta\phi_1,\hat\phi_2\}(\chi^2w^{-1}+\delta)+R\Big]$$

with $R\in S(\log\langle\xi'\rangle_\mu,g)$. Since $\{\theta,\hat\phi_2\}$ is a linear combination of $\hat\phi_i$ we get

$$\nu\{\theta^2\phi_1^2,\chi^2\log\phi\}=2\nu\theta\phi_1\Big[\{\phi_1,\hat\phi_2\}\theta\big(\chi^2w^{-1}+\delta\big)+R'\Big]$$
$$=2\nu\theta^2\phi_1\{\phi_1,\hat\phi_2\}(\chi^2w^{-1}+\delta)+c\theta\phi_1$$

with $c\in S(\log\langle\xi'\rangle_\mu,g)$. Recall that $0\le\chi^2w^{-1}+\delta\in S(w^{-1}\log\langle\xi'\rangle_\mu,g)$ which is contained in $S(\langle\xi'\rangle_\mu^{1/2}\log\langle\xi'\rangle_\mu,\bar g)$. From the Fefferman–Phong inequality it follows that

$$((\chi^2w^{-1}+\delta)u,u)\ge-C_1\|\log^{1/2}\langle\xi'\rangle_\mu u\|^2\ge-C_2\|\langle\mu D'\rangle^{\varepsilon}u\|^2$$

with some $C_i>0$. Thus with $L=\chi^2w^{-1}+\delta+C_2\langle\mu\xi'\rangle^{2\varepsilon}$ we have $(Lu,u)\ge0$ so that

$$|\mathsf{Re}(Lu,v)|\le(Lu,u)+(Lv,v).$$

Then we have with $r = \chi^2 w^{-2} + \delta$

$$\mathsf{Re}(rv, w) \le (rv, v) + (rw, w) + C(\|\langle \mu D'\rangle^{\varepsilon} v\|^2 + \|\langle \mu D'\rangle^{\varepsilon} w\|^2).$$

With $v = \epsilon^{-1/2}\nu\theta^2\phi_1 u$ and $w = \epsilon^{1/2}\tilde{\Lambda}u$ we get

$$\begin{aligned}&\nu\mathsf{Re}((\chi^2 w^{-1} + \delta)(\theta^2\phi_1)u, \tilde{\Lambda}u)\\ &\quad\le \epsilon^{-1}\nu^2((\chi^2 w^{-1} + \delta)(\theta^2\phi_1)u, \theta^2\phi_1 u) + \epsilon((\chi^2 w^{-1} + \delta)\tilde{\Lambda}u, \tilde{\Lambda}u)\\ &\qquad + \epsilon C\|\langle \mu D'\rangle^{\varepsilon}\tilde{\Lambda}u\|^2 + \epsilon^{-1}\nu^2 C\|\langle \mu D'\rangle^{\varepsilon}\theta^2\phi_1 u\|^2.\end{aligned}$$

Choosing $0 < \epsilon \ll 1$ and $0 < \nu$ small enough so that $\epsilon^{-1}\nu^2 \ll \nu$ one can control the term $\mu\ell\nu\mathsf{Re}\,((\chi^2 w^{-1} + \delta)(\theta^2\phi_1)u, \tilde{\Lambda}u)$ by (3.5).

Consider $\{q, \Phi\}$ where Φ commutes against q much better than ϕ. To see this we consider, for example

$$\{\theta^2 a^2\phi_1^2, \Phi\} = 2\theta a\phi_1\{\theta a\phi_1, \Phi\} = 2\theta a\phi_1\left[\frac{\omega\{\theta a\phi_1, \hat{\phi}_2\}}{\rho^2} - \frac{\hat{\phi}_2\{\theta a\phi_1, \omega\}}{\rho^2}\right].$$

Since

$$\begin{aligned}&\frac{\omega\{\theta a\phi_1, \hat{\phi}_2\}}{\rho^2} = c_1\frac{\omega\theta}{\rho^2} + c_2\frac{\theta\omega\hat{\phi}_1}{\rho^2} + c_3\frac{\omega\hat{\phi}_1 O(|\hat{\phi}|)}{\rho^2}, \qquad c_i \in S(1, g_0),\\ &\frac{\hat{\phi}_2\{\theta a\phi_1, \omega\}}{\rho^2} - c_1'\frac{\langle\xi'\rangle_\mu^{-1}}{\rho\omega} + c_2'\frac{\theta\hat{\phi}_2\hat{\phi}_1^4}{\rho^2\omega} + c_3'\frac{\hat{\phi}_2\hat{\phi}_1^4 O(|\hat{\phi}|)}{\rho^2\omega} \quad \in S(1, g_1)\end{aligned}$$

and hence we see

$$\begin{aligned}&\theta\phi_1\frac{\omega\{\theta a\phi_1, \hat{\phi}_2\}}{\rho^2} = c\frac{\omega\theta^2\phi_1}{\rho^2} + c'\theta\phi_1, \quad c, c' \in S(1, g_0),\\ &\theta\phi_1\frac{\hat{\phi}_2\{\theta a\phi_1, \omega\}}{\rho^2} = c\theta\phi_1, \quad c \in S(1, g_1).\end{aligned}$$

Since one can write

$$c\rho^{-2}\omega\theta^2\phi_1 = \mathsf{Re}\,(c\rho^{-1}\omega^{1/2}\#\rho^{-1}\omega^{1/2}\theta^2\phi_1) + c_1\theta^2\phi_1 + c_2\mu^{1/4}\theta\langle\mu\xi'\rangle^{3/4} + R$$

with $c_i \in S(1, \bar{g})$ and $R \in \mu S(\rho^{-1}, \bar{g})$ so that we have

Lemma 3.14. *We have*

$$\begin{aligned}&\nu|(c\rho^{-2}\omega\theta^2\phi_1 u, \tilde{\Lambda}u)|\\ &\quad\le C\epsilon^{-1}\nu^2\|\rho^{-1}\omega^{1/2}\theta^2\phi_1 u\|^2 + \epsilon\|\rho^{-1}\omega^{1/2}\tilde{\Lambda}u\|^2\\ &\qquad + C'(\|\theta\phi_1 u\|^2 + \mu^{1/2}\|\theta\langle\mu\xi'\rangle^{3/4}u\|^2 + \mu\|\langle\mu D'\rangle^{1/2}u\|^2 + \|\tilde{\Lambda}u\|^2).\end{aligned}$$

Here taking ϵ, ν small enough so that $0 < \epsilon \ll 1$ and $\epsilon^{-1}\nu^2 \ll \nu$ again one can control the term $\mu\nu\ell|(c\rho^{-2}\omega\theta^2\phi_1 u, \tilde{\Lambda}u)|$ by (3.5).

We turn to consider the commutator $([D_0 - \mathsf{Re}\tilde{\lambda}, \mathsf{Re}\,\tilde{Q}]u, u)$. We study

$$\{\xi_0 - \phi_1, \phi_2^2\} = 2\mu\phi_2(\theta^2\langle\mu\xi'\rangle + c_0\theta\phi_1 + c_1\phi_1^2\langle\mu\xi'\rangle^{-1} + c_2\phi_2)$$

as one of the main terms where $c_j \in S(1, g_0)$. Since the terms $c_0\theta\phi_1\phi_2$, $\phi_2\phi_1^2\langle\mu\xi'\rangle^{-1}$ and $c_2\phi_2^2$ are easily managed we are concerned with the term $\phi_2\theta^2\langle\mu\xi'\rangle$. Write

$$\theta^2\phi_2\langle\mu\xi'\rangle = (1-\chi^2)\theta^2\phi_2\langle\mu\xi'\rangle + \chi^2\theta^2\phi_2\langle\mu\xi'\rangle.$$

Let us consider $M\theta^2\phi_1^2 - (1-\chi^2)\theta^2\phi_2\langle\mu\xi'\rangle$ with a large positive constant M. Write

$$\begin{aligned} &M^2\theta^2\phi_1^2 - (1-\chi^2)\theta^2\phi_2\langle\mu\xi'\rangle \\ &\quad = M^2\theta^2\phi_1^2(1-(1-\chi^2)\hat{\phi}_2\hat{\phi}_1^{-2}/M^2) = (M\theta\phi_1)^2 F \end{aligned}$$

where $0 < c \le F = 1-(1-\chi^2)\hat{\phi}_2\hat{\phi}_1^{-2}/M^2 \in S(1,g) \subset S(1,\bar{g})$. Note

$$(M\theta\phi_1)^2 F = M\theta\phi_1 \# F \# M\theta\phi_1 + R$$

with $R \in \mu^2 S(w^{-2}, g)$. Then from the Fefferman–Phong inequality it follows that

$$\begin{aligned} M^2(\theta^2\phi_1^2 u, u) &\ge ((1-\chi^2)\theta^2\phi_2\langle\mu\xi'\rangle u, u) \\ &\quad - C\mu^{1/2}\|\langle\mu\xi'\rangle^{-1/2}\theta\phi_1 u\|^2 - C\mu^2\|w^{-1}u\|^2 \end{aligned}$$

from which we obtain

$$M(\theta^2\phi_1^2 u, u) \ge ((1-\chi^2)\theta^2\phi_2\langle\mu\xi'\rangle u, u) - C\mu^{1/2}\|\langle\mu D'\rangle^{1/2}u\|^2.$$

We next consider

$$2w\chi^2\theta^2\langle\mu\xi'\rangle^2 - \chi^2\theta^2\phi_2\langle\mu\xi'\rangle = \chi^2\theta^2\langle\mu\xi'\rangle^2 w(2-\hat{\phi}_2 w^{-1}) = (w^{1/2}\theta\langle\mu\xi'\rangle)^2 F$$

where $0 \le F = \chi^2(2-\hat{\phi}_2 w^{-1}) \in S(1,g) \subset S(1,\bar{g})$. Since $w^{1/2} \in S(w^{1/2}, w^{-2}|dx'|^2 + \langle\xi'\rangle_\mu^{-2}|d\xi'|^2)$ then one has

$$(w^{1/2}\theta\langle\mu\xi'\rangle)^2 F = w^{1/2}\theta\langle\mu\xi'\rangle \# F \# w^{1/2}\theta\langle\mu\xi'\rangle + R, \quad R \in \mu^2 S(w^{-2}, g)$$

and then from the Fefferman–Phong inequality we have

$$\begin{aligned} &2(w\chi^2\theta^2\langle\mu\xi'\rangle^2 u, u) \\ &\quad \ge (\chi^2\theta^2\phi_2\langle\mu\xi'\rangle u, u) - C\mu^{1/2}\|w^{1/2}\theta\langle\mu\xi'\rangle^{1/2}u\|^2 - C\mu^2\|w^{-1}u\|^2 \end{aligned}$$

which proves

$$2(w\chi^2\theta^2\langle\mu\xi'\rangle^2 u, u) \ge (\chi^2\theta^2\phi_2\langle\mu\xi'\rangle u, u) - C\mu^{1/2}\|\langle\mu D'\rangle^{1/2}u\|^2.$$

Combining these two inequalities we get

$$\begin{aligned} &|\mu(\theta^2\phi_2\langle\mu\xi'\rangle u, u)| \\ &\quad \le C\mu\big((\theta^2\phi_1^2 u, u) + (w\chi^2\theta^2\langle\mu\xi'\rangle^2 u, u)\big) + C\mu^{3/2}\|\langle\mu D'\rangle^{1/2}u\|^2. \end{aligned}$$

Thus we have

Lemma 3.15. *We have*

$$\begin{aligned} &|(\{\xi_0 - \phi_1, \phi_2^2\}u, u)| \\ &\quad \le C\mu(\chi^2\theta^2 w\langle\mu\xi'\rangle^2 u, u) + C\mu((q+\mu\langle\mu\xi'\rangle)u, u) + C\mu^{3/2}\|\langle\mu D'\rangle^{1/2}u\|^2. \end{aligned}$$

Recall that our original operator is $P+K$ and hence we must estimate the term $(Ku, \tilde{\Lambda}u)$ where

$$K = c_0\theta\langle\mu\xi'\rangle + c_2\phi_2 + d_0\theta\phi_1 + d_1\phi_1^2\langle\mu\xi'\rangle^{-1}.$$

Here we show how the parameter ℓ works in estimating this term and how the weight $e^{\ell\Phi}$ is used in order to control this term in the region $\hat{\phi}_1^2 \geq d_1 w$. Let us consider $(b\theta\langle\mu\xi'\rangle u, \tilde{\Lambda}u)$ with $b \in \mu S(1, g_0)$ since other terms are easily controlled. Let us write

$$(b\theta\langle\mu\xi'\rangle u, \tilde{\Lambda}u) = (b\chi^2\theta\langle\mu\xi'\rangle u, \tilde{\Lambda}u) + (b(1-\chi^2)\theta\langle\mu\xi'\rangle u, \tilde{\Lambda}u).$$

We first consider $(b\chi^2\theta\langle\mu\xi'\rangle u, \tilde{\Lambda}u)$. Note that one can write

$$b\chi^2\theta\langle\mu\xi'\rangle = \mu\chi w^{-1/2}\#\tilde{b}\chi\theta w^{1/2}\langle\mu\xi'\rangle + c\langle\mu\xi'\rangle^{3/4}\theta + R$$

where $c \in \mu^{5/4}S(1, \bar{g})$, $R \in \mu^{3/2}S(\langle\mu\xi'\rangle^{1/2}, \bar{g})$ and $\tilde{b} = \mu^{-1}b \in S(1, g_0)$. This proves that with any small $\delta > 0$ we have

$$\begin{aligned}|(b\chi^2\theta\langle\mu\xi'\rangle u, \tilde{\Lambda}u)| &\leq \delta\mu\ell\|\chi w^{-1/2}\tilde{\Lambda}u\|^2 + C\delta^{-1}\mu\ell^{-1}\|\tilde{b}\chi\theta w^{1/2}\langle\mu\xi'\rangle u\|^2 \\ &\quad + C\mu^{5/2}\|\theta\langle\mu\xi'\rangle^{3/4}u\|^2 + C\mu^3\|\langle\mu D'\rangle^{1/2}u\|^2 + C\|\tilde{\Lambda}u\|^2.\end{aligned}$$

Here we remark that $\tilde{b}\chi\theta w^{1/2}\langle\mu\xi'\rangle\#\tilde{b}\chi\theta w^{1/2}\langle\mu\xi'\rangle = \tilde{b}^2\chi^2\theta^2 w\langle\mu\xi'\rangle^2 + R$ with $R \in \mu S(\langle\mu\xi'\rangle, \bar{g})$ and hence

$$C\mu\ell^{-1}\|\tilde{b}\chi\theta w^{1/2}\langle\mu\xi'\rangle u\|^2 \leq C'\mu\ell^{-1}(\chi^2\theta^2 w\langle\mu\xi'\rangle^2 u, u) + C'\mu^2\|\langle\mu D'\rangle^{1/2}u\|^2.$$

From these inequalities we get

$$\begin{aligned}|(b\chi^2\theta\langle\mu\xi'\rangle u, \tilde{\Lambda}u)| &\leq \delta\mu\ell\|\chi w^{-1/2}\tilde{\Lambda}u\|^2 + C\delta^{-1}\ell^{-1}\mu(\chi^2\theta^2 w\langle\mu\xi'\rangle^2 u, u) \\ &\quad + C\mu^{5/2}\|\theta\langle\mu\xi'\rangle^{3/4}u\|^2 + C\mu^2\|\langle\mu D'\rangle^{1/2}u\|^2 + C\|\tilde{\Lambda}u\|^2.\end{aligned}$$

We turn to estimate $(b(1-\chi^2)\theta\langle\mu\xi'\rangle u, \tilde{\Lambda}u)$. If $1-\chi^2 \neq 0$ then we have $\hat{\phi}_1^2 \geq d_1 w$ and hence $C\omega^2 \geq w^2 \geq \hat{\phi}_2^2$. Thus we have

$$\omega^2 \leq \hat{\phi}_2^2 + \omega^2 = \rho^2 \leq (C+1)\omega^2$$

so that $\omega^{1/2} \leq \rho w^{-1/2} \leq (1+C')\omega^{1/2}$. On the other hand since (3.6) is valid on the support of χ' we see

$$\rho w^{-1/2}(1-\chi^2) \in S(\omega^{1/2}, \omega^{-2}|dx|^2 + \omega^{-1}\langle\xi'\rangle_\mu^{-2}|d\xi'|^2).$$

Then from similar arguments it follows that

$$(1-\chi^2)b\theta\langle\mu\xi'\rangle = \mu\rho^{-1}\omega^{1/2}\#\rho w^{-1/2}(1-\chi^2)\tilde{b}\theta\langle\mu\xi'\rangle + c\langle\mu\xi'\rangle^{3/4}\theta + R$$

where $c \in \mu^{5/4}S(1, g_0)$, $R \in \mu^{3/2}S(\langle\mu\xi'\rangle^{1/2}, \bar{g})$ and $\tilde{b} = \mu^{-1}b \in S(1, g_0)$. Repeating similar arguments we conclude that

$$\begin{aligned}&|((1-\chi^2)b\theta^2\langle\mu\xi'\rangle u, \tilde{\Lambda}u)| \\ &\quad \leq \delta\mu\ell\|\rho^{-1}\omega^{1/2}\tilde{\Lambda}u\|^2 + C\delta^{-1}\ell^{-1}\mu(\omega\theta^2\langle\mu\xi'\rangle^2 u, u) + C\mu^{5/2}\|\theta\langle\mu\xi'\rangle^{3/4}u\|^2 \\ &\qquad + C\mu^2\|\langle\mu D'\rangle^{1/2}u\|^2 + C\|\tilde{\Lambda}u\|^2.\end{aligned}$$

Combining these inequalities we get

Lemma 3.16. *We have*

$$|(b\theta\langle\mu\xi'\rangle u, \tilde{\Lambda}u)| \le \delta\mu\ell\{\|\rho^{-1}\omega^{1/2}\tilde{\Lambda}u\|^2 + \|\chi w^{-1/2}\tilde{\Lambda}u\|^2\}$$
$$+ C\delta^{-1}\ell^{-1}\mu\{(\theta^2\omega\langle\mu\xi'\rangle^2 u, u) + (\chi^2\theta^2 w\langle\mu\xi'\rangle^2 u, u)\}$$
$$+ C'(\|\tilde{\Lambda}u\|^2 + \mu^2\|\langle\mu D'\rangle^{1/2}u\|^2 + \mu^{5/2}\|\theta\langle\mu\xi'\rangle^{3/4}u\|^2).$$

Thanks to Lemma 3.16 the term $|(b\theta\langle\mu\xi'\rangle u, \tilde{\Lambda}u)|$ can be controlled by (3.5) taking $\delta > 0$ small and then ℓ large.

4. Bicharacteristic geometry

We follow the arguments given in [15]. To simplify notations let us set $\Xi_0 = \xi_0 - \phi_1$, $X_0 = x_0$ and extend (X_0, Ξ_0) to a full symplectic coordinates (X, Ξ). Switching the notation from (X, Ξ) to (x, ξ) and writing θ for $\hat{\theta}$ one can write

$$p = -\xi_0(\xi_0 + 2\phi_1) + \phi_2^2$$

where we have

$$\{\xi_0, \phi_1\} = O(|\phi|), \quad \{\xi_0, \phi_2\} = \theta^2 + O(|\phi|), \quad \{\theta, \phi_j\} = O(|\phi|)$$

with $\phi = (\phi_1, \phi_2)$. We assume

$$\{\xi_0, \theta\}(\rho) = \mu \neq 0$$

at some $\rho \in S$. Let us take

$$\xi_0, x_0, \phi_1, \phi_2, \theta, \psi_1, \ldots, \psi_r, \quad r + 5 = 2(n+1)$$

to be a system of local coordinates around ρ. Note that we can assume that ψ_j are independent of x_0 taking $\psi_j(0, x', \xi')$ as new ψ_j. Moreover we can assume that

$$\{\psi_j, \phi_k\} = O(|\phi|), \quad k = 1, 2$$

taking $\psi_j - \{\psi_j, \phi_1\}\phi_2/\{\phi_2, \phi_1\} - \{\psi_j, \phi_2\}\phi_1/\{\phi_1, \phi_2\}$ as new ψ_j. Our Hamilton system is

$$\begin{cases} \dot{x} = \dfrac{\partial}{\partial \xi} p(x, \xi), \\ \dot{\xi} = -\dfrac{\partial}{\partial x} p(x, \xi). \end{cases}$$

Let $\gamma(s) = (x(s), \xi(s))$ be a solution to the Hamilton system and we consider $\xi_0(s)$, $x_0(s)$, $\phi_j(\gamma(s))$, $\theta(\gamma(s))$, $\psi_j(\gamma(s))$ and recall that

$$\frac{d}{ds} f(\gamma(s)) = \{p, f\}(\gamma(s)).$$

Let us change the parameter from s to $t = 1/s$ so that we have

$$\frac{d}{ds} = -tD, \quad D = t\frac{d}{dt}$$

and hence $tD(t^pF) = t^{p+1}(DF + pF)$ for $p \in \mathbb{N}$. Let us introduce new unknowns;

$$\begin{cases} \xi_0(s) = t^4\Xi_0(t),\ x_0(s) = tX_0(t),\ \phi_1(\gamma(s)) = t^2\Phi_1(t),\ \phi_2(\gamma(s)) = t^3\Phi_2(t), \\ \theta(\gamma(s)) = t\Theta(t),\ \psi_j(\gamma(s)) = t^2\Psi_j(t). \end{cases}$$

Recall

$$\{\xi_0, \phi_2\} = \theta^2 + \kappa\phi_1 + C\phi_2, \quad \delta = \{\phi_1, \phi_2\}(\rho) > 0.$$

Let us set

$$V = (X_0, \Phi_1, \Xi_0, \Phi_2, \Theta, \Psi), \quad \Psi = (\Psi_1, \ldots, \Psi_r)$$

then it is not difficult to see

$$\begin{cases} DX_0 = -X_0 + 2\Phi_1 + tG(t, V), \\ D\Phi_1 = -2\Phi_1 + 2\delta\Phi_2 + tG(t, V), \\ D\Xi_0 = -4\Xi_0 + 2\kappa\Phi_1\Phi_2 + 2\Phi_2\Theta^2 + tG(t, V), \\ D\Phi_2 = -3\Phi_2 + 2\kappa\Phi_1^2 + 2\delta\Xi_0 + 2\Phi_1\Theta^2 + tG(t, V), \\ D\Theta = -\Theta + 2\mu\Phi_1 + tG(t, V), \\ D\Psi_j = -2\Psi_j + tG(t, V) \end{cases} \tag{4.1}$$

where $G(t, V)$ denotes a smooth function in (t, V) such that $G(t, 0) = 0$.

Let us define the class of formal series in t and $\log 1/t$

$$\mathcal{E} = \Big\{ \sum_{0 \le j \le i} t^i (\log 1/t)^j V_{ij} \mid V_{ij} \in \mathbb{C}^N \Big\}$$

in which we look for our formal solutions to the reduced Hamilton system (4.1).

Proposition 4.1. *Assume that* $\{\xi_0, \theta\}(\rho) \neq 0$. *Then there exists a formal solution* $V \in \mathcal{E}$ *satisfying* (4.1) *with* $\Phi_1(0) \neq 0$, $X_0(0) \neq 0$.

Proof. Let us set

$$\begin{cases} X_0 = \sum_{0 \le j \le i} t^i (\log 1/t)^j \beta_{ij}^{(0)}, \quad \Xi_0 = \sum_{0 \le j \le i} t^i (\log 1/t)^j \alpha_{ij}^{(0)} \\ \Phi_1 = \sum_{0 \le j \le i} t^i (\log 1/t)^j \beta_{ij}^{(1)}, \quad \Phi_2 = \sum_{0 \le j \le i} t^i (\log 1/t)^j \alpha_{ij}^{(1)} \\ \Theta = \sum_{0 \le j \le i} t^i (\log 1/t)^j \theta_{ij}, \quad \Psi_k = \sum_{0 \le j \le i} t^i (\log 1/t)^j \gamma_{ij}^{(k)}. \end{cases} \tag{4.2}$$

Equating the constant terms of both sides of (4.1) one has

$$-\beta_{00}^{(0)} + 2\beta_{00}^{(1)} = 0, \quad -2\beta_{00}^{(1)} + 2\delta\alpha_{00}^{(1)} = 0,$$
$$-4\alpha_{00}^{(0)} + 2\kappa\beta_{00}^{(1)}\alpha_{00}^{(1)} + 2\alpha_{00}^{(1)}\theta_{00}^2 = 0,$$
$$-3\alpha_{00}^{(1)} + 2\kappa(\beta_{00}^{(1)})^2 + 2\delta\alpha_{00}^{(0)} + 2\beta_{00}^{(1)}\theta_{00}^2 = 0,$$
$$-\theta_{00} + 2\mu\beta_{00}^{(1)} = 0, \quad -2\gamma_{00}^{(k)} = 0.$$

With $b = \beta_{00}^{(1)}$ we see $\beta_{00}^{(0)} = 2b$, $\alpha_{00}^{(1)} = \delta^{-1}b$, $\theta_{00} = 2\mu b$. It follows from the third equation that

$$2\alpha_{00}^{(0)} = \kappa\delta^{-1}b^2 + 4\delta^{-1}\mu^2 b^3 = \delta^{-1}b(\kappa b + 4\mu^2 b^2).$$

Inserting these into the fourth equation we have

$$-3\delta^{-1}b + 3\kappa b^2 + 12\mu^2 b^3 = 3b\Big(-\frac{1}{\delta} + \kappa b + 4\mu^2 b^2\Big) = 0.$$

Let us study

$$-\frac{1}{\delta} + \kappa b + 4\mu^2 b^2 = 0.$$

Since $\delta > 0$ it is clear that this equation has non-zero real roots $b = b(\kappa, \mu, \delta)$, one is positive and the other one is negative.

Let us choose one of such b. Then

$$\bar{V} = \Big(\beta_{00}^{(0)}, \beta_{00}^{(1)}, \alpha_{00}^{(0)}, \alpha_{00}^{(1)}, \theta_{00}, \gamma_{00}^{(k)}\Big) = (2b, b, b\delta^{-2}/2, \delta^{-1}b, 2\mu b, 0)$$

is uniquely determined. We look for a formal solution to (4.1) in the form $\bar{V} + V$, $V \in \mathcal{E}^{\#}$ where

$$\mathcal{E}^{\#} = \Big\{ \sum_{1\le i, 0\le j\le i} t^i (\log 1/t)^j V_{ij} \mid V_{ij} \in \mathbb{C}^N \Big\}.$$

Let us denote

$$V^I = {}^t(X_0, \Phi_1, \Xi_0, \Phi_2, \Theta), \;\; V^{II} = \Psi.$$

Then (4.1) becomes

$$\begin{cases} DV^I = A_I V^I + F_I t + G_I(t, V), \\ DV^{II} = -2V^{II} + F_{II} t + G_{II}(t, V) \end{cases} \tag{4.3}$$

where

$$G_J(t, V) = \sum_{2\le i, 0\le j\le i} G_{J,ij} t^i (\log 1/t)^j, \;\; G_{J,ij} = G_{J,ij}(V_{pq} \mid p \le i-1)$$

and F_J are constant vectors. Making more precise looks on A_I we see

$$A_I = \begin{bmatrix} -1 & 2 & 0 & 0 & 0 \\ 0 & -2 & 0 & 2\delta & 0 \\ 0 & 2\kappa\delta^{-1}b & -4 & 2\delta^{-1} & 8\mu\delta^{-1}b^2 \\ 0 & 2(\kappa b + \delta^{-1}) & 2\delta & -3 & 8\mu b^2 \\ 0 & 2\mu & 0 & 0 & -1 \end{bmatrix}.$$

Then we have

Lemma 4.2. *A_I has real eigenvalues 1, −6, −1. Other real eigenvalues of A_I are non positive.*

Proof. We have

$$|\lambda - A_I| = \begin{vmatrix} \lambda+1 & -2 & 0 & 0 & 0 \\ 0 & \lambda+2 & 0 & -2\delta & 0 \\ 0 & -2\kappa\delta^{-1}b & \lambda+4 & -2\delta^{-1} & -8\mu\delta^{-1}b^2 \\ 0 & -2\delta^{-1} & -\lambda\delta-6\delta & \lambda+5 & 0 \\ 0 & -2\mu & 0 & 0 & \lambda+1 \end{vmatrix}$$

$$= (\lambda-1)(\lambda+1)(\lambda+6)\big(\lambda^2+5\lambda+8-4\kappa\delta b\big)$$

where we have used $-4\mu^2b^2\delta = \kappa b\delta - 1$. Let us put $\nu = \kappa b\delta$ so that

$$|\lambda - A_I| = (\lambda+1)(\lambda+6)(\lambda-1)(\lambda^2+5\lambda+8-4\nu).$$

Recall that ν verifies

$$1-\nu = 4\mu^2b^2\delta \geq 0$$

then it is clear that a real root of $\lambda^2+5\lambda+8-4\nu = 0$, if exists, is less than or equal to -1. □

Proof of Proposition 4.1. Note that (4.3) implies that with $A = A_I \oplus (-2I)$

$$(iV_{ij} - (j+1)V_{ij+1}) = AV_{ij} + \delta_{i1}\delta_{j0}F + G_{ij}$$

where $G_{ij} = 0$ for $i = 0, 1$. Then we have

$$\begin{cases} (I-A)V_{11} = 0, \\ (I-A)V_{10} = V_{11} + F. \end{cases}$$

Choose $V_{11} \in \operatorname{Ker}(I-A)$ so that

$$F + V_{11} \in \operatorname{Im}(I-A).$$

Then we can take $V_{10} \neq 0$ so that

$$(I-A)V_{10} = F + V_{11}$$

since $\operatorname{Ker}(I-A) \neq \{0\}$ by Lemma 4.2. We turn to the case $i \geq 2$;

$$(iI-A)V_{ij} = (j+1)V_{ij+1} + G_{ij}. \tag{4.4}$$

With $j = i$, (4.4) turns to

$$(iI-A)V_{ii} = G_{ii}(V_{pq} \mid p \leq i-1).$$

Since $iI - A$ is non singular for $i \geq 2$ by Lemma 4.2 one has

$$V_{ii} = (iI-A)^{-1}G_{ii}(V_{pq} \mid p \leq i-1).$$

Recurrently one can solve V_{ij} by

$$V_{ij} = (iI-A)^{-1}\big\{(j+1)V_{ij+1} + G_{ij}(V_{pq} \mid p \leq i-1)\big\}$$

for $j = i-1$, $i-2$,...,0. This proves the assertion. □

We next assume $\{\xi_0, \theta\} = 0$ on S and $\{\{\xi_0, \phi_2\}, \phi_2\}(\rho) \neq 0$ so that

$$\{\xi_0, \phi_1\} = O(|\phi|), \quad \{\xi_0, \phi_2\} = \theta^2 + O(|\phi|),$$
$$\{\theta, \phi_j\} = O(|\phi|), \quad \{\xi_0, \theta\} = O(|(\phi, \theta)|)$$

with $\phi = (\phi_1, \phi_2)$. Let us introduce new unknowns;

$$\begin{cases} \xi_0(s) = t^4\Xi_0(t), \ x_0(s) = tX_0(t), \ \phi_1(\gamma(s)) = t^2\Phi_1(t), \ \phi_2(\gamma(s)) = t^3\Phi_2(t), \\ \theta(\gamma(s)) = t^2\Theta(t), \ \psi_j(\gamma(s)) = t^2\Psi_j(t). \end{cases}$$

Recall

$$\{\xi_0, \phi_2\} = \theta^2 + \kappa\phi_1 + C\phi_2$$

where $\kappa = \delta^{-1}\{\{\xi_0, \phi_2\}, \phi_2\}(\rho) \neq 0$ by assumption. Let us set

$$V = (X_0, \Phi_1, \Xi_0, \Phi_2, \Theta, \Psi), \quad \Psi = (\Psi_1, \ldots, \Psi_r)$$

then we see

$$\begin{cases} DX_0 = -X_0 + 2\Phi_1 + tG(t, V), \\ D\Xi_0 = -4\Xi_0 + 2\kappa\Phi_1\Phi_2 + tG(t, V), \\ D\Phi_1 = -2\Phi_1 + 2\delta\Phi_2 + tG(t, V), \\ D\Phi_2 = -3\Phi_2 + 2\kappa\Phi_1^2 + 2\delta\Xi_0 + tG(t, V), \\ D\Theta = -2\Theta + tG(t, V), \\ D\Psi_j = -2\Psi_j + tG(t, V) \end{cases} \tag{4.5}$$

where $G(t, V)$ denotes a smooth function in (t, V) such that $G(t, 0) = 0$ as before.

Proposition 4.3. *Assume that* $\{\xi_0, \theta\} = 0$ *on* S *and* $\{\{\xi_0, \phi_2\}, \phi_2\}(\rho) \neq 0$. *Then there is a formal solution* $V \in \mathcal{E}$ *to the reduced Hamilton system* (4.5) *such that* $\Phi_1(0) \neq 0$, $X_0(0) \neq 0$.

Proof. Let us put $(X_0, \Phi_1, \Xi_0, \Phi_2, \Psi)$ as in (4.2) and insert this into (4.5). Equating the constant terms of both sides of (4.5) one has

$$-4\alpha_{00}^{(0)} + 2\kappa\beta_{00}^{(1)}\alpha_{00}^{(1)} = 0, \quad -\beta_{00}^{(0)} + 2\beta_{00}^{(1)} = 0,$$
$$-2\beta_{00}^{(1)} + 2\delta\alpha_{00}^{(1)} = 0, \quad -3\alpha_{00}^{(1)} + 2\kappa(\beta_{00}^{(1)})^2 + 2\delta\alpha_{00}^{(0)} = 0,$$
$$-2\theta_{00} = 0, \quad -2\gamma_{00}^{(k)} = 0.$$

With $b = \beta_{00}^{(1)}$ we obtain $\beta_{00}^{(0)} = 2b$, $\alpha_{00}^{(1)} = \delta^{-1}b$. It follows from the first equation that

$$2\alpha_{00}^{(0)} = \kappa\delta^{-1}b^2.$$

Inserting these into the fourth equation we have

$$-3\delta^{-1}b + 2\kappa b^2 + \kappa b^2 = 3b\Big(-\frac{1}{\delta} + \kappa b\Big) = 0.$$

We choose a root $b = (\kappa\delta)^{-1} \neq 0$ then

$$\bar{V} = \Big(\beta_{00}^{(0)}, b, \alpha_{00}^{(0)}, \alpha_{00}^{(1)}, \theta_{00}, \gamma_{00}^{(k)}\Big) = (\kappa\delta)^{-1}(2, 1, \delta^{-2}/2, \delta^{-1}, 0, 0)$$

is uniquely determined. We substitute $\bar{V}+V$ for V in (4.5) to get the equation for V. Let us denote

$$V^I = {}^t(X_0, \Phi_1, \Xi_0, \Phi_2), \quad V^{II} = (\Theta, \Psi).$$

Then (4.5) becomes (4.3) with

$$A_I = \begin{bmatrix} -1 & 2 & 0 & 0 \\ 0 & -2 & 0 & 2\delta \\ 0 & 2\delta^{-2} & -4 & 2\delta^{-1} \\ 0 & 4\delta^{-1} & 2\delta & -3 \end{bmatrix}.$$

Lemma 4.4. *The eigenvalues of A_I are -1, -4, -6 and* 1.

Proof. We see

$$|\lambda - A_I| = (\lambda+1)(\lambda+4)(\lambda+6)(\lambda-1)$$

where we have used $\kappa\delta b = 1$ which proves the assertion. □

The rest of the proof of Proposition 4.3 is just a repetition of that of Proposition 4.1. □

Proof of Proposition 2.1. Repeating a similar but much simpler arguments as in [15] we can conclude that there is a solution to (4.1) and (4.5) which is asymptotic to the formal solution given in Propositions 4.1 and 4.3 respectively. Thus we get a solution $(x(s), \xi(s))$ to the Hamilton system which is tangent to Σ.

References

[1] R. Beals, *Characterization of pseudodifferential operators and applications.* Duke Math. J. **44** (1977), 45–57

[2] E. Bernardi and A. Bove, *Geometric transition for a class of hyperbolic operators with double characteristics.* Japan J. Math. **23** (1997), 1–87.

[3] E. Bernardi and T. Nishitani, *On the Cauchy problem for non-effectively hyperbolic operators, the Gevrey* 5 *well-posedness.* J. Anal. Math. **116** (2008), 197–240.

[4] E. Bernardi and T. Nishitani, *On the Cauchy problem for noneffectively hyperbolic operators: The Gevrey* 4 *well-posedness.* Kyoto J. Math. **51** (2011), 767–810.

[5] E. Bernardi and T. Nishitani, *On the Cauchy problem for noneffectively hyperbolic operators: The Gevrey* 3 *well-posedness.* J. Hyperbolic Differ. Equ. **8** (2011), 615–650.

[6] E. Bernardi, C. Parenti and A. Parmeggiani, *The Cauchy problem for hyperbolic operators with double characteristics in presence of transition.* Comm. Partial Differential Equations **37** (2012), 1315–1356.

[7] L. Hörmander, *The Cauchy problem for differential equations with double characteristics.* J. Anal. Math. **32** (1977), 118–196.

[8] L. Hörmander, *The Analysis of Linear Partial Differential Operators, III.* Springer, Berlin-Heidelberg-New York-Tokyo, 1985.

[9] V.Ja. Ivrii and V.M. Petkov, *Necessary conditions for the Cauchy problem for non strictly hypebolic equations to be well posed.* Uspehi Mat. Nauk. **29** (1974), 3–70.

[10] V.Ja. Ivrii, *The well posedness of the Cauchy problem for non strictly hyperbolic operators III: The energy integral.* Trans. Moscow Math. Soc. **34** (1978), 149–168.

[11] N. Iwasaki, *The Cauchy problem for effectively hyperbolic equations (general case).* J. Math. Kyoto Univ. **25** (1985), 727–743.

[12] K. Kajitani and T. Nishitani, *The Hyperbolic Cauchy Problem.* Lecture Notes in Math. **1505**, Springer, 1991.

[13] G. Komatsu and T. Nishitani, *Continuation of bicharacteristics for effectively hyperbolic operators.* Publ. RIMS, Kyoto Univ. **28** (1992), 885–911.

[14] T. Nishitani, *Local energy integrals for effectively hyperbolic operators I, II.* J. Math. Kyoto Univ. **24** (1984), 625–658 and 659–666.

[15] T. Nishitani, *Non effectively hyperbolic operators, Hamilton map and bicharacteristics.* J. Math. Kyoto Univ. **44** (2004), 55–98.

[16] T. Nishitani, *On the Cauchy problem for noneffectively hyperbolic operators, a transition casse.* Studies in Phase Space Analysis with Applications to PDEs, pp. 259–290, Birkhäuser, 2013.

Tatsuo Nishitani
Department of Mathematics, Osaka University
Machikaneyama 1-1, Toyonaka, 560-0043,
Osaka Japan
e-mail: nishitani@math.sci.osaka-u.ac.jp

Fourier Analysis
Trends in Mathematics, 335–347

Differentiation Functor and Its Application in the Optimization Control Theory

Simon Ya. Serovajsky

Abstract. The differentiation of operators is defined as the functor from the category of Banach spaces with fixed points and germs of operators as morphisms to the category of Banach spaces with linear continuous operators. It is used for obtaining necessary conditions of optimality for abstract optimization control problems. Classical and extended operator derivatives are considered. An optimization control problem for a nonlinear elliptic equation is analyzed as an example.

Mathematics Subject Classification (2010). 47H99; 49K27; 18A05 .

Keywords. Differentiation, optimization, functors.

1. Introduction

Optimization control problems are analyzed as a rule using necessary conditions of optimality. These conditions characterize usually a local extremum because the properties of functions or functionals are similar in the neighborhood of the local and the absolute extremum. First-order necessary conditions of optimality (for example, the stationary condition, Euler equation, the variational inequality, the maximum principle, etc.) are used frequently. It includes the first derivative of the state functional [1]. So the optimal control theory has serious relations with the differentiation theory.

The differentiation is an operation of the local linearization [2]. It supposes that the nonlinear phenomenon has become weakly apparent in a small enough set. Then the regular enough nonlinear object can be approximated by a linear one. For example, the smooth curve can be approximated in a neighbourhood of a point by its tangent in this point. Note that the derivative is used for the definition of the tangent. If the local structure of the nonlinear object is analyzed by means of its linear approximation, we use the differentiation. It is true for the optimization control theory too. The differentiation technique is applied here for obtaining necessary conditions of optimality and gradient methods.

The differentiation relates with the local structure of the object. Two functions (functionals, operators) that are equal in a neighbourhood of a point have the same derivative in this point. So the derivative characterizes in reality the local structure of the class of objects, but not a concrete object. These objects are equivalent in some way. This equivalence class is the germ of functions (functionals, operators) in this point [3]. Then the differentiation relates with the germs theory naturally enough.

Thus the differentiation transforms the germ of operators to a linear operator, which is its derivative in the given point. This map can be interpreted as a functor. It transforms the category, which has germs of operators as morphisms, to the category, which has linear operators as morphisms. So our problem allows an interpretation in the categories theory [4]. The differentiation functor was defined in [5] without the germs theory. These results were used for the analysis of unconditional extremum problems there. We exact these results using germs and apply it for conditional extremum problems. We will consider classic and extended operator derivatives [6] here.

2. Differentiation functor

We consider the set of pairs (X, x), where X is a Banach space, and x is a fixed point of X. So we have the set of Banach spaces with fixed points. For all pairs (X, x) and (Y, y) determine an operator $L : X \to Y$, that is Fréchet differentiable at the point x such that $Lx = y$. Let us determine the equivalence on the set of such operators. Two operators are equivalent if they coincide at a neighbourhood of the point x. The relevant equivalence class, namely the germ of the operator L at the point x, is denoted by $[L]_x$.

Consider pairs of Banach spaces with fixed points (Y, y) and (Z, z) and an operator $M : Y \to Z$ that is Fréchet differentiable at the point y such that $My = z$. Then the superposition ML maps the set X to Z, besides $MLx = z$. By the Composite Function Theorem (see [7], p. 637) the operator ML is Fréchet differentiable at the point x, besides

$$(ML)'(x) = M'(y)L'(x).$$

Then we determine the composition of the germs $[L]_x$ and $[M]_x$ by the equality

$$[L]_x \circ [M]_y = [ML]_x.$$

Thus we get the category Γ with Banach spaces with fixed points as the objects and germs of operators as the morphisms.

We now consider a morphism $[L]_x$ of the category Γ such that the relevant operator L is continuously differentiable at a neighbourhood O of the point x, besides its derivative $L'(x)$ is reversible. Then the inverse operator L^{-1} exists by the

Inverse Function Theorem [8]. It is defined and differentiable at a neighbourhood O' of the point $y = Lx$, besides the equality

$$(L^{-1})'(\xi) = \Big\{L'\big[L^{-1}(\xi)\big]\Big\}^{-1}$$

is true for all $\xi \in O'$. Then the inverse operator determines the morphism $[L^{-1}]_y$ of Γ, which is the inverse of $[L]_x$, namely

$$[L^{-1}]_y = [L]_x^{-1}.$$

Therefore the set of Banach spaces with fixed points and these morphisms is the subcategory Σ of the category Γ for which every morphism is an isomorphism.

We now define a map D from Γ to the category B of Banach spaces with linear continuous operators. For all object (X, x) and the morphism $[L]_x$ with the beginning (X, x) and the end (Y, y) of the category Γ we determine

$$D(X, x) = X, \ D[L]_x = L'(x).$$

Let $[M]_y$ be a morphism of this category with the beginning (Y, y). Using the Composite Function Theorem we get the equality

$$D\big([L]_x \circ [M]_y\big) = (ML)'(x) = D[L]_x \circ D[M]_y.$$

Then D is the functor from Γ to the category B. Let the germ $[L]_x$ be the morphism of the category Σ with the beginning (X, x) and the end (Y, y). Then from the equality $(L^{-1})'(y) = [L'(x)]^{-1}$ we have

$$D[L^{-1}]_y = \big(D[L]_x\big)^{-1}.$$

Definition 1. The functor D is called **the differentiation**, and its value $D\psi$ at the germ ψ is called **the derivative of the morphism** of the category Γ.

The similar definition without using germs theory was given in [5].

Let us now consider the operator $L : X \to Y$ that are Fréchet differentiable at the point x. The relevant germs set (the morphisms of the category Γ) can be endowed with the linear space structure. For all operators L, M and numbers a, b the linear combination $a[L]_x + b[M]_x$ of the germs $[L]_x$ and $[M]_x$ is the germ at the point x of the operators that are equivalent to $aL + bM$. Thus the set of germs has the structure of the factor-space with operations

$$a[L]_x + b[M]_x = [aL + bM]_x.$$

Using the linearity of the derivative we get

$$D\Big(a[L]_x + b[M]_x\Big) = aD[L]_x + bD[M]_x.$$

Denote by Ξ the transformation, which maps the arbitrary morphism of Γ to the value of the relevant germ of operators at this point. Particularly if (X, x) and (Y, y) are the beginning and the end of a morphism ψ, then an operator $L : X \to Y$ exists; it is Fréchet differentiable at the point x and satisfies the equality $Lx = y$, besides $\psi = [L]_x$. Then we get $\Xi\psi = Lx$. Thus the derivative of the morphism is characterized by the equality

$$\Xi[L]_{x+h} = \Xi[L]_x + D[L]_x h + \eta(h), \tag{2.1}$$

where $\|\eta(h)\| = o(\|h\|)$.

These notions are the suitable means for describing optimization control problems.

3. Applications to the optimization control theory

We consider a state equation of the control system

$$Ay = v, \tag{3.1}$$

where $A : Y \to V$ is a state operator, Y, V are Banach spaces, v is a control, y is the state function. Suppose the equation (3.1) has a unique solution $y = y(v)$ from the space Y for all $v \in V$. Determine the cost functional $I : V \to R$ by the equality

$$I(v) = J(v) + K[y(v)],$$

where $J : V \to R$, $K : Y \to R$ are some functionals. It is given a closed convex subset U of the set V. We have the following optimization control problem.

Problem 1. *Minimize the functional I on the set U.*

Theorem 1. *If u is a point of the local minimum for the functional I on the set U, $[J]_u$ and $[K]_y$ are morphisms of the category Γ, and $[A]_y$ is a morphism of the category Σ, where $y = y(u)$, then the variational inequality*

$$D[J]_u(v-u) + \Big[H\big(D[A]_y\big)\Big]^{-1} D[K]_y(v-u) \geq 0 \;\; \forall v \in U \tag{3.2}$$

is true, where H is the general cofunctor from B to the sets category, define by the object $\mathbb{R}$.

Proof. Let u be a point of the local minimum for the functional I on the set U. Then we get

$$I(w) - I(u) \geq 0 \;\; \forall w \in O,$$

where the subset O of U is a neighbourhood of the point u. So for all fixed $v \in U$ the positive number σ can be chosen so small, that $u + \sigma(v-u) \in O$. Therefore we have

$$I\big(u + \sigma(v-u)\big) - I(u) \geq 0.$$

If $[I]_u$ is the morphism of the category Γ then we have the equality

$$\Xi[I]_{u+\sigma(v-u)} = \Xi[I]_u + \sigma D[I]_u(v-u) + \eta\big(\sigma(v-u)\big)$$

because of (2.1). So the previous inequality can be transformed to

$$\sigma D[I]_u(v-u) + \eta\big(\sigma(v-u)\big) \geq 0.$$

After division by σ and passing to the limit as $\sigma \to 0$ we get

$$D[I]_u(v-u) \geq 0 \ \forall v \in U. \tag{3.3}$$

Let us now prove that the germ of functionals $[I]_u$ is in reality the morphism of the category Γ, and the variational inequality (3.3) can be transformed to (3.2). Indeed $[A]_y$ is the morphism of the category Σ; so it is the isomorphism. Then the inverse morphism $[A^{-1}]_u$ of the category Γ exists. By the Composite Function Theorem and the definition of the sum of germs the germ of functionals $[I]_u$ is defined by the formula

$$[I]_u = [J]_u + [A^{-1}]_u \circ [K]_y.$$

So it is the morphism of the category Γ. Using the linearity of the derivative we find

$$D[I]_u = D[J]_u + D\Big([A^{-1}]_u \circ [K]_y\Big) = D[J]_u + D[A^{-1}]_u \circ D[K]_y.$$

By the definition of the functor of the inverse morphism we get

$$D[I]_u = D[J]_u + \Big(D[A]_y\Big)^{-1} \circ D[K]_y.$$

The functor H maps a Banach space X and the linear continuous operator L to the adjoint space X^* and the adjoint operator L^* [4]. Using the relation between adjoint and inverse operators (see [7]) we obtain

$$D[I]_u = D[J]_u + H\Big(D[A]_y\Big)^{-1} D[K]_y = D[J]_u + \Big(HD[A]_y\Big)^{-1} D[K]_y.$$

Therefore the inequality (3.3) can be transformed to (3.2). □

The partial case of this result for unconditional extremum problems is given in [5].

The standard necessary condition of optimality can be obtained as a corollary of Theorem 1.

Corollary 1. *Under assumptions of Theorem* 1 *the optimal control satisfies the variational inequality*

$$\langle J'(v) + p, v-u \rangle \geq 0 \ \forall v \in U, \tag{3.4}$$

where $\langle \lambda, \varphi \rangle$ *is the value of the linear continuous functional* λ *at the point* φ, *and* p *is the solution of the adjoint equation*

$$[A'(y)]^* p = K'(y). \tag{3.5}$$

Proof. Using the definition of the functor of the differentiation and the enclosure of the functional derivative to the relevant adjoint space we transform the inequality (3.2) to the form

$$\Big\langle J'(v) + \big\{[A'(y)]^*\big\}^{-1} K'(y), v-u \Big\rangle \geq 0 \ \forall v \in U.$$

Take into account here the reversibility of the derivative $A'(y)$ and its adjoint operator because $[A]_y$ is the morphism of the category Σ. Determine the function p by the equation (3.5). So we transform the previous inequality to (3.4). □

Thus we have the state equation (3.1) for the optimal control, the adjoint equation (3.5) and the variational inequality (3.4) for finding three unknown functions u, y, p. Let us now consider an example.

4. Example

We consider the equation

$$-\Delta y + |y|^\rho y = v \tag{4.1}$$

in an open bounded n-dimensional set Ω with homogeneous boundary conditions (Dirichlet Problem), where $\rho > 0$. Determine the spaces

$$Y = H_0^1(\Omega) \cap L_q(\Omega),\ V = Y^* = H^{-1}(\Omega) + L_{q'}(\Omega),$$

where $q = \rho + 2$, $1/q + 1/q' = 1$. By monotone operators theory (see [9]) for all $v \in V$ the equation (4.1) has a unique solution $y = y(v)$ from the space Y, besides the map $v \to y(v)$ is weakly continuous. We determine the functional

$$I(v) = \frac{\chi}{2}\|v\|^2_{L_2(\Omega)} + \frac{1}{2}\Big\|y(v) - y_d\Big\|^2_{H_0^1(\Omega)},$$

where $\chi > 0$, $y_d \in H_0^1(\Omega)$ is a given function. Determine the set

$$U = \Big\{v \in V \Big|\ a \le v(x) \le b\Big\}.$$

We have following optimization control problem.

Problem 2. *Minimize the functional I on the set U.*

By a standard method (see, for example, [10], Chapter 1, Theorem 1.1) we prove the solvability of this problem. Determine the operator $A : Y \to V$ such that Ay is the left side of the equality (4.1). Then the boundary problem is transformed to the operator equation (3.1). Let J and K be the first and the second terms of the last equality. Then Problem 2 is transformed to Problem 1. The operator A and the functionals J and K are differentiable. So Theorem 1 is applicable if $[A]_y$ is the morphism of the category Σ.

Lemma 1. *The germ $[A]_y$ is a morphism of the category Σ if $n = 2$ or $\rho \le 4n/(n-2)$ for $n \ge 3$.*

Proof. The lemma will be proven if the derivative $A'(y)$ is reversible. Prove that the homogeneous Dirichlet problem for the linearized equation

$$-\Delta z + (\rho + 1)|y|^\rho z = h \tag{4.2}$$

has a unique solution in the space Y for all $h \in V$. Multiply the equality (4.2) by the function z and integrate it in x. Using the Green Formula we get

$$\int\limits_{\Omega} \sum_{i=1}^{n} \left(\frac{\partial z}{\partial x_i} \right)^2 dx + (\rho + 1) \int\limits_{\Omega} |y|^{\rho} z^2 dx = \int\limits_{\Omega} hz dx. \tag{4.3}$$

By the Sobolev embedding theorem we have $H_0^1(\Omega) \subset L_q(\Omega)$; so $Y = H_0^1(\Omega)$, $V = H^{-1}(\Omega)$. Using equality (4.3) we get an a priori estimate for the solution of the equation (4.2) in the space Y for all $h \in V$. The necessary result follows now from the classic theory of linear elliptic equations (see, for example, [10], Chapter 2, Theorem 1.1). □

Using Corollary 1 and Lemma 1 we obtain the following result.

Corollary 2. *Under the assumptions of Lemma 1 the solution of Problem 2 satisfies the variational inequality*

$$\int\limits_{\Omega} (\chi u + p)(v - u) dx \geq 0 \;\; \forall v \in U, \tag{4.4}$$

where p is the solution of the homogeneous Dirichlet problem for the equation

$$-\Delta p + (\rho + 1)|y|^{\rho} p = \Delta y_d - \Delta y. \tag{4.5}$$

Thus the necessary conditions of the optimality includes the state equation (4.1) for $v = u$, adjoint equation (4.2), and variational inequality (4.3). By the inequality (4.4) we find the control

$$u = F(p), \tag{4.6}$$

where $F(p) = a$ if $p > -\chi a$, $F(p) = -\chi^{-1} p$ if $-\chi b \leq p \leq \chi a$, and $F(p) = b$ if $p < -\chi b$. Then we get the homogeneous Dirichlet problem for the equation (4.5) and

$$-\Delta y + |y|^{\rho} y = -\chi^{-1} p.$$

If we find its solution, we can determine the control by the formula (4.6).

Corollary 2 is come to agreement with known results of the optimization control problems for systems described by nonlinear elliptic equations (see, for example, [11]). The additional constraints for the dimension of the set and the nonlinearity parameter are typical for these problems (see [11], [12]). We now note that Problem 2 is solvable without these assumptions. So it will be interesting enough to analyze it without any suppositions. However for the large enough values of the nonlinearity parameter ρ and the set dimension n without the embedding $H_0^1(\Omega) \subset L_q(\Omega)$ the equality (4.3) does not have a priori estimates of the solution of the equation (4.2) in the space Y for all $v \in V$. Moreover we have the following result.

Lemma 2. *The germ $[A]_y$ is not a morphism of the category Σ if the embedding $H_0^1(\Omega) \subset L_q(\Omega)$ does not hold.*

Proof. It is sufficient to prove the absence of the differentiability for the inverse operator A^{-1}. Let the function y be continuous. Then the term in the left side of the equality (4.2) belongs to the subspace $V_* = H^{-1}(\Omega)$ of V for all $z \in Y$. Suppose the operator A^{-1} is Gateaux differentiable at the point $v = Ay$. Then we have the convergence $[y(v+\sigma h) - y]/\sigma \to y'(v)h$ in Y as $\sigma \to 0$ for all $h \in V$. By the equality (4.2) for the controls $v + \sigma h$ and v after the division by σ and passing to the limit we get

$$-\Delta y'(v)h + (\rho+1)|y|^\rho y'(v)h = h.$$

So the linearized equation (4.2) has the solution $z = y'(v)h$ from Y for all $h \in V$. However we proved that it is impossible if the condition $h \in V_*$ is false. This contradiction completes the proof. □

Thus the germ $[A]_y$ for the equation (4.1) is the isomorphism for small enough values of the nonlinearity parameter and the dimension of the set. But it is false for large enough values of these parameters. We will determine a category such that $[A]_y$ become its isomorphism without any constraints, besides we will determine the differentiation as a functor for this case too. So we will extend Theorem 1. Using this result we will prove that the assertion of Corollary 2 is true without any constraints.

5. Extended differentiation

We consider again the linearized equation (4.2). Determine the spaces

$$Y(v) = H_0^1(\Omega) \cap \{z \,|\, |y(v)|^{\rho/2} z \in L_2(\Omega)\}, \quad V(v) = Y(v)^*.$$

The space $Y(v)$ is Hilbert one with the scalar product

$$(y,z) = \int_\Omega \nabla y \nabla z dx + (\rho+1) \int_\Omega |y(v)|^\rho yzdx.$$

Using (4.3) we get

$$\|z\|_{Y(v)} \le \|h\|_{V(v)},$$

where $V(v)$ has the norm of the adjoint space. So for all h from the space $V(v)$, which is narrower than V, the equation (4.2) has a unique solution in the space $Y(v)$, which is larger than Y. This property can be interpreted as a weaker form of the reversibility of the derivative $A'(v)$. We suppose now the operator A^{-1} can have some positive property, which is weaker than standard differentiability. We now consider an operator $L: V \to Y$, where V, Y are Banach spaces.

Definition 2 [6]. The operator L is called **extended Gateaux differentiable** (more exact, $(V_0, Y_0; V_*, Y_*)$-extended Gateaux differentiable) at the point $v \in V$ in the case of the existence of linear normalized spaces V_0, Y_0, V_*, Y_*, which satisfies continuous embeddings $V_* \subset V_0 \subset V$, $Y \subset Y_0 \subset Y_*$, and a linear continuous operator $L'(v): V_0 \to Y_0$ such that $[L(v+\sigma h) - Lv]/\sigma \to L'(v)h$ in Y_* for all $h \in V_*$ as $\sigma \to 0$. The operator L is called **extended Fréchet differentiable** at the point v, if $L(v+h) = Lv + L'(v)h + \eta(h)$, where $\|\eta(h)\|_{Y_*} = o(\|h\|_{V_*})$.

The operator is called **extended continuously differentiable** at this point, if it has $(V_u, Y_u; V_*, Y_*)$-extended Gateaux derivative $L'(u)$ at the arbitrary point u from a V_*-neighbourhood of the point v, besides $L'(u)h \to L'(v)h$ in Y_* for all $h \in V_*$ as $(u - v) \to 0$ in V_*.

It is obvious that $(V, Y; V, Y)$-extended derivatives coincide with its classic analogues. The relations between different types of the extended derivatives with the same set of spaces and its classical analogues are identical.

Lemma 3 [6]. *The operator* A^{-1} *for the equation* (4.1) *has a* $(V(v), Y(v); V_*, Y_*)$*-extended Gateaux derivative* D *at the arbitrary point* $v \in V$ *such that*

$$\int_\Omega \mu D h dx = \int_\Omega h p_\mu(v) dx \;\; \forall h \in V(v), \mu \in Y(v)^*, \tag{5.1}$$

where $p_\mu(v)$ *is the solution of the homogeneous Dirichlet problem*

$$-\Delta p_\mu(v) + (\rho + 1)|y(v)|^\rho p_\mu(v) = \mu, \tag{5.2}$$

$Y_* = H_0^1(\Omega)$, *and other spaces were determined before.*

The classical differentiability of the solution of our equation with respect to the absolute term is true for the small enough values of the nonlinearity parameter and the dimension of the set by Lemma 1; but this property does not hold for large enough values of these parameter by Lemma 2. So the properties of the boundary problem change with a jump. However this dependence is always extended differentiable because of Lemma 3. But the difference between the domain and the codomain of the operator and its derivative depends from the parameters of the equation. So the extended differentiability theory describes more exact properties of the dependence of the state function from the absolute term of the equation. We use these results for the analysis of Problem 2.

Corollary 3. *The solution of Problem* 2 *is determined always by the formula* (4.6).

Proof. Let u be a solution of Problem 2. Then we get

$$I\big[u + \sigma(v - u)\big] - I(u) \geq 0 \;\; \forall v \in U_*, \sigma \in [0, 1].$$

Divide by σ and pass to the limit as $\sigma \to 0$. Using Lemma 3 and the Green formula we obtain

$$\int_\Omega \Big[\chi u(v - u) + (\Delta y_d - \Delta y) D(v - u)\Big] dx \geq 0 \;\; \forall v \in U_*.$$

Determine $\mu = \Delta y_d - \Delta y$ in the equality (5.2); using (5.1) we have equation (4.5) and the variational inequality

$$\int_\Omega (\chi u + p)(v - u) dx \geq 0 \;\; \forall v \in U \cap (v + V_*).$$

Its solution is determined by formula (4.6). □

We have now the question if the extended differentiation can be interpreted as a functor. It requires analogues of the Inverse Function Theorem and the Composite Function Theorem for extended derivatives. Let us now consider the Banach spaces Y, V, an operator $A : Y \to V$, and points $y_0 \in Y$, $v_0 = Ay_0$. Let V_* be a Banach subspace of V with neighbourhood O_* of zero. Then the set $O = v_0 + O_*$ is the neighbourhood of the point u. Suppose the following assumption.

Property 1. *The operator A is reversible at the set O.*

Then we have the equality

$$Ay(v_\sigma) - Ay(v_0) = \sigma h$$

for all $v \in V_1$ and small enough number σ, where $v_\sigma = v_0 + \sigma h$. Suppose the following condition is true.

Property 2. *The operator A is Gateaux differentiable at a neighbourhood of the point y, besides its derivative is so regular that the Mean Integral Theorem*

$$A(y+h) - Ay = \left[\int_0^1 A'(y+\theta h)d\theta\right]h \ \forall h \in O_*$$

is true.

For all $v \in O$ determine the linear continuous operator $G(v) : Y \to V$ by formula

$$G(v)h = \left\{\int_0^1 A'\big[y + \theta y(v) - y)\big]d\theta\right\}h.$$

Then we get

$$G(v_\sigma)\big[y(v_\sigma) - y(v_0)\big] = \sigma h,$$

so

$$\big\langle \lambda, G(v_\sigma)\big[y(v_\sigma) - y(v_0)\big]\big\rangle = \sigma\langle\lambda, h\rangle \ \forall \lambda \in V^*.$$

For all $v \in O$ consider the Banach spaces Y_*, $V(v)$ and $Y(v)$ such that the embeddings

$$Y \subset Y(v) \subset Y_*, \ V_* \subset V(v) \subset V$$

are continuous and dense. Suppose the following assumption.

Property 3. *For all $v \in O$ a continuous extension $\overline{G}(v)$ of the operator $G(v)$ on the set $Y(v)$ exists such that its codomain is a subset of $V(v)$.*

By the previous equality we have

$$\Big\langle \overline{G}(v_\sigma)^*\lambda, \big[y(v_\sigma) - y(v_0)\big]/\sigma\Big\rangle = \langle\lambda, h\rangle \ \forall \lambda \in V^*, \ h \in V_*.$$

Determine the linear operator equation

$$\overline{G}(v_\sigma)^* p_\mu(v_\sigma) = \mu. \tag{5.3}$$

It transforms to

$$\overline{A}'(y_0)^* p_\mu(v_0) = \mu$$

for $v = v_0$, where $\overline{A}'(y_0) = \overline{G}(v_0)$ is the extension of the operator $A'(y_0) = G(v_0)$ to the set $Y(v_0)$.

Property 4. *For all $v \in O$, $\mu \in Y(v)^*$ equation (5.3) has a unique solution $p_\mu(v) \in V(v)^*$, besides $p_\mu(v_\sigma) \to p_\mu(v_0)$ *-weakly in V_*^* uniformly with respect to μ from the closed unit ball of the space Y_*^* with zero as the center for all $h \in V_*$ as $\sigma \to 0$.*

The extended differentiability of the inverse operator is guaranteed by the following result.

Lemma 4 [13]. *Under properties 1–4 the operator A^{-1} has $(V(v_0), Y(v_0); V_*, Y_*)$-extended Gateaux derivative D at the point u is definite by the equality*

$$\langle \mu, Dh\rangle = \langle p_\mu(v_0), h\rangle \; \forall \mu \in Y(v_0)^*,\; h \in V(v_0).$$

It has a $(V(v_0), Y(v_0); W, Y_*)$-extended Fréchet derivative, if the embedding of Banach space W to V_* is compact and dense.

Note that these assumptions are true for the operator A of the equation (7). So the assertions of Lemma 3 are the corollary of Lemma 4. Determine here $W = L_2(\Omega)$ for obtaining extended Fréchet differentiability.

Determine now the extended analogue of the Composite Function Theorem. Consider the Banach spaces X_*, X_0, X; Y_1, Y^0, Y_*, Y_0, Y; Z_1, Z^0, Z_*, with continuous embeddings

$$X_* \subset X_0 \subset X,\; Y_1 \subset Y^0 \subset Y_* \subset Y_0 \subset Y,\; Z_1 \subset Z^0 \subset Z_*.$$

We determine the operators $L : X \to Y_1$ and $M : Y \to Z_1$, and the points $x \in X$, $y = Lx$.

Lemma 5 [5]. *If the operator L has $(X_0, Y^0; X_*, Y_*)$-extended Gateaux (respectively, Fréchet) derivative D_L at the point x, and the operator M has a $(Y_0, Z^0; Y_*, Z_*)$-extended Fréchet derivative at the point y, then the composition ML has a $(X_0, Z^0;$ $X_*, Z_*)$-extended Gateaux (consequently, Fréchet) derivative D_{ML} at the point x such that $D_{ML} = D_M D_L$.*

Let us use these results to determine the functor of the extended differentiation.

6. Functor of the extended differentiation

We now determine a category Γ_E. Let triples (X, x, X_*) be its objects, where X, X_* are Banach spaces with continuous embedding $X \subset X_*$ and x is a fixed point of X. Consider two threes (X, x, X_*) and (Y, y, Y_*). Suppose existence of Banach spaces X_0, Y^0, Y_1 with continuous embeddings

$$X_* \subset X_0 \subset X,\; Y_1 \subset Y^0 \subset Y_* \subset Y,$$

and an operator $L : X \to Y_1$ which is $(X_0, Y^0; X_*, Y_*)$-extended Fréchet differentiable at the point x such that $Lx = y$. Determine

$$\{L\}_x = \big[\iota(Y_1, Y)L\big]_x,$$

where the term in the right side is the corresponding morphism of the category Γ, and $\iota(Y_1, Y)$ is the canonic embedding of the space Y_1 to Y. We choose it as the morphism of the category Γ_E for the relevant beginning and end objects.

We now consider a morphism $\{M\}_y$ with the beginning (Y, y, Y_*) and the end (Z, z, Z_*). Then we have the existence of spaces Y_0, Z^0, Z_1 with continuous embeddings

$$Y_* \subset Y_0 \subset Y,\ Z_1 \subset Z^0 \subset Z_* \subset Z,$$

and an operator $M : Y \to Z_1$, that is $(Y_0, Z^0; Y_*, Z_*)$-extended Fréchet differentiable at the point y such that $My = z$, besides

$$\{M\}_y = \big[\iota(Z_1, Z)L\big]_y.$$

It is obvious that the composition ML maps the space X to Z_1, besides $MLx = z$. By Lemma 5 it is $(X_0, Z^0; X_*, Z_*)$-extended Fréchet differentiable at the point x. So

$$(ML)_z = \iota(Z_1, Z)ML$$

is the morphism of the category Γ_E with the beginning (X, x, X_*) and the end (Z, z, Z_*). It can be interpreted as the composition $\{L\}_x \circ \{M\}_y$ of the morphisms $\{L\}_x$ and $\{M\}_y$. Thus Γ_E is the category in reality. Lemma 4 characterizes its isomorphisms.

Determine the map $D_E : \Gamma_E \to B$. For all object (X, x, X_*) we determine

$$D_E(X, x, X_*) \ = \ X_*.$$

For all morphism $\{L\}_x$ with the beginning (X, x, X_*) and the end (Y, y, Y_*) we choose as $D_E\{L\}_x$ the extended Fréchet derivative D_L of the operator L, which determine the morphism $\{L\}_x$, at the point x. Let $\{M\}_y$ be a morphism of the category Γ_E with the beginning (Y, y, Y_*) and the end (Z, z, Z_*). Then

$$D_E\{M\}_y = D_M,$$

where D_M is the extended Fréchet derivative of the operator M, which determine the morphism $\{M\}_y$, at the point y. By Lemma 5 we have $D_{ML} = D_M D_L$, where D_{ML} is the extended Fréchet derivative of the composition, which determine the morphism

$$\{ML\}_z \ = \ \{L\}_x \circ \{M\}_y.$$

So we have

$$D_E\Big(\{L\}_x \circ \{M\}_y\Big) \ = \ D_E\{L\}_x \circ D_E\{M\}_y.$$

Thus the map D_E is the functor in reality.

Definition 3. The functor D_E is called **the extended differentiation**, and its value $D_E\psi$ is called **the extended derivative** of the morphism ψ of the category Γ_E.

These results can be used for the analysis of Problem 1. Suppose the functionals J and K for Problem 1 are Fréchet differentiable on the spaces W and Y_*. Replace the classic derivatives by extended ones for Theorem 1. Using Lemmas 1 and 2 we get the following result.

Theorem 2. *Let the operator A satisfy the assumptions of Lemma 4. Then the point u of the local minimum of the functional I on the set U satisfies the variational* *inequality*

$$D[J]_u(v-u) + [H\,(D_E\{A\}_y)]\,D[K]_y(v-u) \geq 0 \ \ \forall v \in U.$$

Theorem 2 uses weaker assumptions than Theorem 1. So it is applicable for the larger class of problems. Particularly the control-state mapping for the equation (7) is extended differentiable for the general case, but it does not have Gateaux derivative. The functionals J and K for Problem 2 are Fréchet differentiable on the space W and Y_* of the definition of the extended derivative. Then Corollary 2 follows from Corollary 3.

These results can be extended to the problem of minimization of the functional $I = I[v, y(v)]$, where $y(v)$ is a solution of the equation $A[v, y(v)]$. The Inverse Function Theorem is replaced here by the Implicit Function Theorem.

References

[1] A.D. Ioffe, V.M. Tihomirov, *Theory of extremal problem.* Nauka, Moscow, 1974.

[2] J. Dieudonné, *Foundation of Modern Analysis.* Academic Press, 1960.

[3] N. Bourbaki, *Topologie generale.* Hermann, Paris, 1940, 1942.

[4] I. Bucur, A. Deleanu, *Introduction to the theory of categories and functors.* John Wiley and Sons LTD, London, New York, Sydney, 1968.

[5] S. Serovajsky, *Differentiation of operators and extremum conditions category interpretation* Izvestiya vuzov. Mathematics (2010), 2, 66–76.

[6] S. Serovajsky, *Calculation of functional gradients and extended differentiation of operators* Journal of Inverse and Ill-Posed Problems **13** (2005), 4, 383–396.

[7] L.V. Kantorovich, G.P. Akilov, *Functional Analysis.* Nauka, Moscow, 1977.

[8] J.P. Aubin, I. Ekcland, *Applied nonlinear analysis.* John Wiley and Sons, New York, etc., 1984.

[9] J.L. Lions *Quelques Méthodes de Résolution des Problèmes aux Limites non Linéaires.* Dunod, Gauthier-Villars, Paris, 1969.

[10] J.L. Lions *Contrôlc Optimal de Systèmes Gouvernés par des Equations aux Dérivées Partielles.* Dunod, Gauthier-Villars, Paris, 1968.

[11] A.V. Fursikov, *Optimal control of distributed systems. Theory and applications.* Amer. Math. Soc., Providence, 1999.

[12] J.L. Lions, *Contrôle de Systèmes Distribués Singuliers.* Gauthier-Villars, Paris, 1983.

[13] Serovajsky, S. *Differentiability of inverse operators.* Progress in partial differential equations: Asymptotic profiles, regularity and well-posedness. Springer Proceedings in Mathematics & Statistics 44, DOI 10.1007/978-3-319-00125-8_14, Springer Int. Publish. Switzerland, 2013, pp. 303–320.

Simon Ya. Serovajsky
al-Farabi Kazakh National University
71 al Farabi av.
050078 Almaty, Kazazkhstan
e-mail: `serovajskys@mail.ru`

Fourier Analysis
Trends in Mathematics, 349–355

The Solution of the Initial Mixed Boundary Value Problem for Hyperbolic Equations by Monte Carlo and Probability Difference Methods

Kanat Shakenov

Abstract. The initial mixed boundary value problem for equations of hyperbolic type is considered. It is solved by algorithms "random walk on spheres", "random walk on balls" and "random walk on lattices" of Monte Carlo methods and by probability difference methods.

Mathematics Subject Classification (2010). 65C05.

Keywords. Hyperbolic equation, random walk, Monte Carlo, probability, difference, method, approximation, expectation.

1. Introduction

We consider in a bounded closed domain $\Omega \in \mathbb{R}^n$, $(n = 2, 3)$ with boundary $\partial\Omega$ and for $t \in (0, T)$ the initial mixed boundary value problem

$$\partial_t^2 u(t,x) - \Delta_x u(t,x) + \gamma^2 u(t,x) = f(t,x), \quad (t,x) \in \Omega \times (0,T), \tag{1.1}$$

$$u(0,x) = \varphi(x), \qquad x \in \Omega, \tag{1.2}$$

$$\partial_t u(0,x) = \psi(x), \qquad x \in \Omega, \tag{1.3}$$

$$\alpha(t,x)u(t,x) + \beta(t,x)\frac{\partial u(t,x)}{\partial \mathbf{n}} = g(t,x), \quad (t,x) \in \partial\Omega \times (0,T), \tag{1.4}$$

where γ is a parameter, $f(t,x)$, $\varphi(x)$, $\psi(x)$, $\alpha(t,x)$, $\beta(t,x)$, $g(t,x)$ are given functions, and $\mathbf{n}$ is the normal to $\partial\Omega$. Problems of this type in different settings have been considered in [1], [2], [3]. In monograph [1] (especially pages 7–10) the initial value problem for equation (1.1) is considered. In the AIP Conference Preceedings [2] the mixed problem for elliptic equation by Monte Carlo and by probability difference methods is solved. In article [3] the initial Neumann boundary value problem for parabolic type equation by the algorithm "random walk" of Monte Carlo methods is solved.

The initial mixed boundary value problem (1.1)–(1.4) takes the following form after discretization with respect to time variable t and approximation error $O(\tau^2)$ and for $n = 3$:

$$\frac{u^{i+1}(x) - 2u^i(x) + u^{i-1}(x)}{\tau^2} - \Delta u^{i+1}(x) + \gamma^2 u^{i+1}(x) = f^i(x), \quad i = 1, 2, \ldots, M-1, \quad \tau = T/M, \quad x \in \Omega, \tag{1.5}$$

$$u^0(x) = \varphi(x), \quad x \in \Omega, \tag{1.6}$$

$$\frac{u^1(x) - u^0(x)}{\tau} = \psi(x), \quad x \in \Omega, \tag{1.7}$$

$$\alpha^{i+1}(x)u^{i+1}(x) + \beta^{i+1}(x)\frac{\partial u^{i+1}(x)}{\partial \mathbf{n}} = g^{i+1}(x), \quad i = 1, 2, \ldots, M-1, \quad x \in \Omega. \tag{1.8}$$

We can also write

$$\mathbf{L}u^{i+1}(x) = F^i\big(f^i(x), u^{i-1}(x), u^i(x)\big), \quad i = 1, 2, \ldots, M-1, \quad x \in \Omega, \tag{1.9}$$

$$\alpha^{i+1}(x)u^{i+1}(x) + \beta^{i+1}(x)\frac{\partial u^{i+1}(x)}{\partial \mathbf{n}} = g^{i+1}(x), \quad i = 1, 2, \ldots, M-1, \quad x \in \Omega, \tag{1.10}$$

where $\mathbf{L} \equiv \big(\Delta - \tau^2\gamma^2\big)$ is the elliptic (Helmholtz) operator.

2. Monte Carlo methods

The main idea of the Monte Carlo methods: we construct the probability value or the probability process in such a way that the mean value is the solution of the given problem. Then, as a rule, the variance is the precision of the solution. From (1.9) and (1.10) we obtain an integral equation of the second type to be solved:

$$u(x) = \int_\Omega k(y,x)u(y)dy + v(x), \quad x, y \in \Omega, \tag{2.1}$$

where $v(x)$ is a given function and $k(y,x)$ is the kernel and a given function. The integral equation (2.1) can be solved by Monte Carlo methods if the integral operator K of this equation satisfies the condition

$$\|K\|_{L^1(\Omega)} < 1. \tag{2.2}$$

Ifcondition (1.10) holds then the integral equation (2.1) can be solved by "random walk on spheres" and "random walk on balls" algorithms of the Monte Carlo methods. It is also possible to construct the ε-displaced estimations for $u(x)$, cf. [4], [2], [5], [6], [9], [10].

It can also be solved by algorithms "random walk on spheres" and "random walk on lattices", cf. [2], [5], [6], of the Monte Carlo and probability difference methods, cf. [11].

Let $\partial\Omega$ be a Lyapunov surface, and let the surface Ω be convex. Then the norm of the integral operator acting in $C(\overline{\Omega})$ is less than 1. Hence, it is possible to apply the Neumann–Ulam scheme to equation (1.9), cf. [9].

The integral equation (2.1) can also be solved by "random walk on spheres" and by "random walk on balls" algorithms of the Monte Carlo methods, cf. [10]. By reaching the ε-boundary the Markov chain is reflected with probability

$$p_{\partial\Omega_\varepsilon} = \frac{|\beta^{i+1}(x)|}{|\alpha^{i+1}(x)| + |\beta^{i+1}(x)|}$$

and adsorbed with probability

$$q_{\partial\Omega_\varepsilon} = \frac{|\alpha^{i+1}(x)|}{|\alpha^{i+1}(x)| + |\beta^{i+1}(x)|}.$$

At transition from one condition to the following condition, the "weight" of node, which is defined by the recurrence relation

$$Q_0 = 1, \quad Q_{i+1} = Q_i\,\frac{k(x_i, x_{i+1})}{p_\Omega(x_i, x_{i+1})}, \quad i = 0, 1, \ldots,$$

is taken into account. On a border, the "weight" of border, proportional to

$$Q_{\partial\Omega} = \frac{g^{i+1}(x)}{|\alpha^{i+1}(x)| + |\beta^{i+1}(x)|},$$

is taken into account.

Let us denote by h a step of the difference scheme in each coordinate direction and by e_i the coordinate unit vector in the ith coordinate direction. We approximate the domain Ω and the operator $\mathbf{L}$ by the finite difference method, $p^h(x, x \pm e_i h)$, $p^h(x, x + e_i h + e_j h)$, $p^h(x, x - e_i h \pm e_j h)$ and $p^h(x, y) = 0$, for the others $x, y \in \Omega \in R^n_h$. Function $p^h(x, y)$ is nonnegative, the sum in y is equal to 1 for each x. This means that $p^h(x, y)$ is the probability of transitions of some Markov chain ξ^h_n and also will be the coefficients in finite difference approximation.

3. Probability difference method

We divide a discrete border into the reflecting part $\partial\Omega^h_R$ and the absorbing part $\partial\Omega^h_A$. Then it is possible to construct the ε-displaced approximation of the unique decision in the point x. For example, it is possible to construct a Markov chain by "random walk on lattices" and to define $\{\xi^h_n\}$ along this chain.

Let the set $\partial\Omega^h_R$ approximate $\partial\Omega$ "from within". That is, either $x \in \overline{\Omega} \bigcap R^3_h$ or $x \in \partial\Omega$ or a straight line connecting x with one of the nearest nodes $x_i \pm e_i h$, $x_i \pm e_i h \pm e_j h$ or $x_i \pm e_i h \mp e_j h$ touches $\partial\Omega$. The set is determined in $\overline{\Omega} \bigcap R^3_h$. Let us define a digitization $\Omega_h = \Omega \bigcap R^3_h - \partial\Omega^h_R$ of interior Ω and a digitization of a stopping set $\partial\Omega^h_A = R^3_h - \Omega_h - \partial\Omega^h_R$. Then p^h gives the transition probabilities of the approximating chain $\{\xi^h_i\}$ in Ω_h. The chain breaks at the first contact with $\partial\Omega^h_A$. We notice that $\mathbf{E}_x\left\{\xi^h_{n+1} - \xi^h_n \,|\, \xi^h_n = y_i \in \partial\Omega^h_R\right\} = v(y)\,h/|v(y)|$. It is coordinated so that the reflection from the point $\partial\Omega_R^h$ happens along direction $v(y)$. Here $v(y)$ is the direction of hit into the interior node, see [2], [5], [6], [11].

After an approximation of domains Ω, $\partial\Omega$ and (1.9), (1.10), we obtain the problem by the finite difference method

$$\mathbf{L}^h u_h^{i+1} = F_h^i(f_h^i, u_h^{i-1}, u_h^i), \tag{3.1}$$

$$\alpha_h^{i+1} u_h^{i+1} + \beta_h^{i+1} \delta_h(u_h^{i+1}) = g_h^{i+1}, \tag{3.2}$$

where $\mathbf{L}^h$ is a finite difference approximation of the operator $\mathbf{L}$, δ_h is the finite difference approximation of the operator $\frac{\partial}{\partial \mathbf{n}}$. Then we have the following

Theorem 1. *The Neumann–Ulam scheme is applicable to the finite difference mixed problem* (3.1)–(3.2).

Proof. Let us consider the difference problem (1.5)–(1.7) and approximate it by x. We prove the theorem for the case $\Omega \in \mathbb{R}^1 \equiv [0,1]$. Then we divide $[0,1]$ in N parts with step $h = 1/N$ by x. Then we derive the following finite-difference problem for $k = 1, \ldots, N-1$,

$$h^2 u_k^{i+1} - 2h^2 u_k^{i+1} + h^2 u_k^{i-1} - \tau^2 u_{k+1}^{i+1} + 2\tau^2 u_k^{i+1} - \tau^2 u_{k-1}^{i+1} + h^2\tau^2\gamma^2 u_k^{i+1} = h^2\tau^2 f_k^i. \tag{3.3}$$

The first-order of approximation of $\partial_t u(x,0)$ is $O(\tau^2)$. We use the following obvious equalities: $\partial_\tau u(x,0) = du(x,0)/dt = \psi(x)$ at $t=0$, $u(x,0) = u_0(x) = \varphi(x)$ and

$$\partial_\tau u(x,0) = \frac{du(x,0)}{dt} = \frac{u(x,\tau) - u(x,0)}{\tau} = \partial_t u(x,0) + \frac{\tau}{2}\partial_t^2 u(x,0) + O(\tau^2). \tag{3.4}$$

From (1.1) when $t = 0$ and using the first initial condition we get:

$$\partial_t^2 u(x,0) = \partial_x^2 u(x,0) - \gamma^2 u(x,0) + f(x,0) = \frac{d^2\varphi(x)}{dx^2} - \gamma^2\varphi(x) + f(x,0). \tag{3.5}$$

Now from (3.5) we derive

$$\frac{\tau}{2}\partial_t^2 u(x,0) = \frac{\tau}{2}\Big(\frac{d^2\varphi(x)}{dx^2} - \gamma^2\varphi(x) + f(x,0)\Big). \tag{3.6}$$

We put into (3.4) the expression $\tau\partial_t^2 u(x,0)/2$ from (3.6), and we get

$$\partial_\tau u(x,0) - \frac{\tau}{2}\Big(\frac{d^2\varphi(x)}{dx^2} - \gamma^2\varphi(x) + f(x,0)\Big) = \partial_\tau u(x,0) + O(\tau^2).$$

Hence

$$\partial_\tau u(x,0) \approx \frac{u_k^1 - u_k^0}{\tau} = \psi_k + \frac{\tau}{2}\Big(\frac{\varphi_{k+1} - 2\varphi_k + \varphi_{k-1}}{h^2} - \gamma^2\varphi_k + f_k\Big) + O(\tau^2)$$

or

$$u_k^1 = \varphi_k + \tau\psi_k + \frac{\tau}{2}\Big(\frac{\varphi_{k+1} - 2\varphi_k + \varphi_{k-1}}{h^2} - \gamma^2\varphi_k + f_k\Big) + O(\tau^2). \tag{3.7}$$

Thus, we show that $\partial_t u(x,0)$ (second initial condition) is also approximated with accuracy of $O(\tau^2)$. The finite-difference equation (3.7) with boundary condition

(1.8) on every time layer $(i+1)$ is solved by the sweep method. Indeed, from (3.3) for $i = 1, 2, \dots, M-1, \;\; k = 1, 2, \dots, N-1$ we get

$$-\tau^2 u_{k+1}^{i+1} + (2\tau^2 + h^2 + h^2\tau^2\gamma^2) u_k^{i+1} - \tau^2 u_{k-1}^{i+1} = 2h^2 u_k^i - h^2 u_k^{i-1} + h^2\tau^2 f_k^i. \quad (3.8)$$

Sufficiency condition of convergence for the sweep method of system (3.8) is satisfied because the inequality is true: $|2\tau^2 + h^2 + h^2\tau^2\gamma^2| > |-\tau^2| + |-\tau^2|$, where τ is the step by time t, $\tau > 0$, h is the step by space variable x, $h > 0$, and parameter $\gamma^2 > 0$. Thus, the solution of system (3.8) exists and it can be solved numerically by the sweep method. Now, (3.8) is written for each $(i+1)$ in the matrix form

$$\mathbf{A}\mathbf{u} = \mathbf{F}, \quad (3.9)$$

where matrix $\mathbf{A}$ has 3 diagonals: diagonal elements are $2\tau^2 + h^2 + h^2\tau^2\gamma^2$, upper and lower diagonal elements are $-\tau^2$, the rest of the elements are zeros. The column vector $\mathbf{F}$ is known because of the right side of the equation, initial and boundary conditions $\mathbf{F} = \mathbf{F}(u_k^i, u_k^{i-1}, f_k^i)$, and the column vector $\mathbf{u}$ is an unknown vector to be found. The solution of system (3.9) can be written in the form

$$\mathbf{u} = \mathbf{A}^{-1}\mathbf{F}. \quad (3.10)$$

Referring to [7] we can write out all eigenvalues of the matrix $\mathbf{A}$:

$$\lambda_k(\mathbf{A}) = 2\tau^2 + h^2 + h^2\tau^2\gamma^2 - 2\sqrt{\tau^2\tau^2}\cos(k\vartheta), \;\; k = 1, 2, \dots, N, \;\; \vartheta = \frac{\pi}{N+1}.$$

If $N \gg 1$, then $\lambda_k(\mathbf{A}) \geq h^2 + h^2\tau^2\gamma^2 > 0$. From the condition $|\lambda_k(\mathbf{A})| < 1$ we get a relation between the steps τ and h:

$$\frac{-\sqrt{1-h^2}}{h\gamma} < \tau < \frac{\sqrt{1-h^2}}{h\gamma}.$$

When it is considered that $\tau > 0$ and for $\gamma > 0$ we get the condition (Courant type condition)

$$\tau < \frac{\sqrt{1-h^2}}{h\gamma}. \quad (3.11)$$

An analogous approach was carried out in the [8]. If condition (3.11) is true, then the iterational process for the solution of system (3.9) (ex., Jacobi method) converges. The existence of a solution of system (3.9) (the same for system (3.8)), the convergence of the numerical method of sweep (inversion of matrix $\mathbf{A}$) for that system, as well as the discrete solution are constructed on absorbing Markov's chain, that terminates on ε-border, since for (3.9) the iterational process converges and (1.8) is true. The mixed boundary condition (1.8) consists of two parts: Dirichlet's and Neumann's; due to Dirichlet's condition Markov's chain is absorbed on the ε-border, that is, it terminates. This proves the usability of the Neumann–Ulam scheme for the solution of (3.8). This system can be solved by "random walk on lattices", that is, by the probability difference method. The theorem is proved.

Now we prove the applicability of the Neumann–Ulam scheme to (3.8) differently. We write (3.8) on time layers $(i+1)$ in the form

$$\mathbf{u_r} = \mathbf{K}\mathbf{u_l} + \mathbf{\Phi}, \quad (3.12)$$

where $\mathbf{K}$ is a two-diagonal matrix (operator): its upper and lower diagonal elements are $\tau^2/(2\tau^2+h^2+h^2\tau^2\gamma^2)$; the rest of the elements are zero, $\mathbf{u_r}=\mathbf{u_r}(\mathbf{u}_k^{i+1})$, $\mathbf{u_l}=\mathbf{u_l}(\mathbf{u}_{k+1}^{i+1},\mathbf{u}_{k-1}^{i+1})$, $\mathbf{\Phi}=\mathbf{\Phi}\big(\mathbf{u}_k^i,\mathbf{u}_k^{i-1},\mathbf{f}_k^i\big)$. Now we show that the operator (matrix) $\mathbf{K}$ is a compressive operator. We calculate the eigenvalues $\lambda_k(\mathbf{K})$ of the matrix $\mathbf{K}$, cf. [7]:

$$\lambda_k(\mathbf{K})=\frac{-2\tau^2}{2\tau^2+h^2+h^2\tau^2\gamma^2}\cos(k\vartheta),\quad k=1,2,\dots,N,\quad \vartheta=\frac{\pi}{N+1}.$$

It is easy to see, that for $\tau>0$, $h>0$, $\gamma^2>0$, the condition for the spectral radius of the matrix $\mathbf{K}$,

$$\rho(\mathbf{K})<1 \tag{3.13}$$

is true for any $\tau>0$ and $h>0$. In that case the iterational process (by k, where $k=0$ and for $k=N$ the boundary conditions are true) converges for (3.12) also; condition (3.13) is now the necessary and sufficient condition of convergence of the iterational process for (3.12). It means that we can use the Neumann–Ulam scheme for the system (3.12), that is, that the system (and (3.8) also) can be solved by the probability difference methods: by constructing the Markov chains converging to the ε-border, and all its trajectories terminate on the ε-border because of Dirichlet's conditions present on the borders. Thus, the theorem is proved again. □

Analogous algorithms can be found in [8], [9], [10].

The algorithm of constructing the Markov chains from [11] allows us to construct Markov chains, and the estimate discrete solution of the system (3.8),

$$u^h=\mathbf{E}_x\left[\sum_{i=0}^{N_h-1}C_i^h f(\xi_i^h)\Delta t_i^h I_{\Omega^h}(\xi_i^h)+C_{N_h-1}^h r(\xi_{N_h}^h)+\sum_{i=0}^{N_h-1}C_i^h\Big(-g(\xi_i^h)\Big)d\mu_i^h\right]$$

is the unique discrete solution of the problem (the same discrete solution of problem (3.1), (3.2)), where $\mathbf{E}_x$ is an expectation.

Remark 2. The common error of the probability difference method depends on the following parameters:

1) τ step by time t, $\tau>0$, as $O(\tau^2)$;
2) h step by space variable x, $h>0$, as $O(h^2)$;
3) parameter $\gamma^2>0$, and
4) it depends on ε-border linearly.

References

[1] Michael Ruzhansky and James Smith, *Dispersive and Strichartz estimates for hyperbolic equations with constant cofficents.* MSJ Memoirs, Vol. 22, (2010).

[2] Kanat K. Shakenov. *Solution of Mixed Problem for Elliptic Equation by Monte Carlo and Probability-Difference Methods.* CP1076. 7th International Summer School and Conference. Melville, New York. American Institute of Physics (AIP). Conference Proceedings, Volume 1076, (2008) 213–218.

[3] A. Haji-Sheikh and E.M. Sparrow, *The floating random walk and its application to Monte Carlo solutions of heat equations.* SIAM. Journal on Applied Mathematics, Vol. 14, **2**, Printed in U.S.A. (1966), 370–389.

[4] K. Shakenov, *Solution of one problem of linear relaxational filtration by Monte Carlo methods.* International Conference on Computational Mathematics (ICCM 2002), Part I. ICM & MG Publisher. Novosibirsk, Russia, (2002), 276–280.

[5] K. Shakenov, *Solution of one mixed problem for equation of a relaxational filtration by Monte Carlo methods.* Notes on Numerical Fluid Mechanics and Multidisciplinary Design, Springer Berlin – Heidelberg. Volume 93/2006. Book. Advances in High Performance Computing and Computational Sciences. Chapter 71. (2006), 205–210.

[6] K.K. Shakenov, *Solution of problem for one model of relaxational filtration by probabilitly-difference and Monte Carlo methods.* Polish Academy of Sciences. Committee of Mining. Archives of Mining Sciences. Volume 52, Issue 2, Krakow, (2007), 247–255.

[7] R. Bellman, *Introduction to matrix analysis.* McGraw–Hill Book Company, Inc. New York Toronto London (1960).

[8] S.M. Ermakov, K.K. Shakenov, *On the applications of the Monte Carlo method to Navier–Stokes equations.* Bulletin of Leningrad State University. Series Mathematics, Mechanics, Astronomy. No. 6267-B86. Leningrad, (1986). 1–14.

[9] S.M. Ermakov, *Monte Carlo method and adjacent questions.* Nauka, Moscow (1975).

[10] S.M. Ermakov, V.V. Nekrutkin, A.S. Sipin, *Random process for solution of classical equations of mathematical physics.* Nauka, Moscow (1984).

[11] Harold J. Kushner, *Probability Methods of Approximations in Stochastic Control and for Elliptic Equations.* Academic Press. New York – San Francisco – London. (1977).

Kanat Shakenov
al-Farabi avenue, 71
al-Farabi Kazakh National University
050040, Almaty, Kazakhstan
e-mail: shakenov2000@mail.ru

Fourier Analysis
Trends in Mathematics, 357–374

Weak Monotonicity Concept and Its Applications

S. Tikhonov and M. Zeltser

Abstract. We introduce the concept of weak monotone sequences with respect to a given nondecreasing sequence $\{m_n\}$ as well as similar concept for functions. For such sequences/functions we prove several convergence tests for number series such as Cauchy, Maclaurin, Schlömilch, the Abel–Oliver nth term test. We also study Hardy's inequalities for weak monotone sequences/functions.

Mathematics Subject Classification (2010). Primary 40A05, 26A48; Secondary 26D15.

Keywords. Weak monotone sequences/functions, equiconvergence of sums and integrals, Hardy's inequality.

1. Introduction

Monotonicity of functions plays a crucial role in many questions in analysis. As examples, we mention convergence/divergence tests for number series $\sum_{n=1}^{\infty} F(n)$ by Cauchy, Maclaurin, Schlömilch, de la Vallée Poussin, etc. (see [2, 5]), or Hardy's inequalities (see, e.g., [4, 3]).

In [9], weak monotone sequences and functions were considered and several results on equiconvergence of number series were proved. By definition, a sequence $\mathbf{a} = \{a_n\}$ is weak monotone, written $\mathbf{a} \in WM$, if

$$a_k \leq C a_n \quad \text{for any} \quad n \leq k \leq 2n, \tag{1.1}$$

where $C > 0$ does not depend on n. Note that the class WM contains several known classes of monotone type sequences such as the class of quasi-monotone

The research of the first author was partially supported by MTM 2011-27637, 2009 SGR 1303, RFFI 13-01-00043, and NSH-979.2012.1. The research of the second author was supported by ETF 8627, Estonian Ministry of Education and Research (project SF0130010s12) and European Regional Development Fund (Centre of Excellence "Mesosystems: Theory and Applications", TK114).

sequences ($a_n n^{-\tau} \geq a_{n+1}(n+1)^{-\tau}$ for some $\tau > 0$; see [10]) or the wider class of general monotone sequences ($\sum_{k=n}^{2n} |a_k - a_{k+1}| \leq Ca_n$; see [11], [8]).

In this paper, we study a more general monotonicity concept.

Definition 1.1. Let $\mathbf{m} = \{m_n\}$ be a nondecreasing positive sequence with $m_1 \geq 1$. We say that a positive sequence $\mathbf{a} = \{a_n\}$ is weak monotone with respect to $\mathbf{m}$, written $\mathbf{a} \in WM(\mathbf{m})$, if

$$a_k \leq Ca_n \quad \text{for any} \quad n \leq k \leq n + m_n, \tag{1.2}$$

where $C > 0$ does not depend on n.

In particular, $WM \equiv WM(\mathbf{m})$ for $\mathbf{m} = \{m_n = n\}$. The basic embedding properties of the $WM(\mathbf{m})$ classes are given as follows.

1. For any $\mathbf{m} = \{m_n\}$, we have $M \subset WM(\mathbf{m}) \subset WM(\{1\})$, where M is the class of nonincreasing sequences.
2. If $m_n^{(2)} \leq Cm_n^{(1)}$ with some $C > 0$, we have $WM(\mathbf{m^{(1)}}) \subset WM(\mathbf{m^{(2)}})$.
3. If $m_n^{(1)} \asymp m_n^{(2)}$, i.e., $C_1 m_n^{(1)} \leq m_n^{(2)} \leq C_2 m_n^{(1)}$, then $WM(\mathbf{m^{(1)}}) \equiv WM(\mathbf{m^{(2)}})$
4. For any sequences $\mathbf{m^{(1)}}$ and $\mathbf{m^{(2)}}$ satisfying $m_n^{(2)}/m_n^{(1)} \to \infty$, there exists $\mathbf{a} = \mathbf{a}(\mathbf{m^{(1)}}, \mathbf{m^{(2)}})$ such that $\mathbf{a} \in WM(\mathbf{m^{(1)}}) \setminus WM(\mathbf{m^{(2)}})$.

Properties 1–3 follow from the definition. We only need to verify Property 4. To construct $\mathbf{a} \in WM(\mathbf{m^{(1)}}) \setminus WM(\mathbf{m^{(2)}})$, we take integers $\{n_k\}$ such that

$$n_k \geq n_{k-1} + (k-1)m_{n_{k-1}}^{(1)} \qquad \text{and} \qquad \frac{m_{n_k}^{(2)}}{m_{n_k}^{(1)}} \geq k, \qquad k \in \mathbb{N}. \tag{1.3}$$

Note that in view of (1.3) we have $\mu_k := [(n_{k+1} - n_k)/m_{n_k}^{(1)}] \geq k$. We set $a_1 := \cdots := a_{n_1 - 1} := 1$ and for $i = 0, \ldots, \mu_k$ and $j \in \mathbb{N}$ with $n_k + i m_{n_k}^{(1)} \leq j < \min\{n_k + (i+1)m_{n_k}^{(1)}; n_{k+1}\}$ we define

$$a_j := \frac{2^i}{2^{\mu_k} m_{n_k}^{(1)}}.$$

The defined sequence $\{a_j\}$ is a $WM(\mathbf{m^{(1)}})$ sequence and not $WM(\mathbf{m^{(2)}})$ sequence. The latter follows from

$$\frac{a_{n_k + [k m_{n_k}^{(1)}]}}{a_{n_k}} = \frac{2^k}{2^{\mu_k} m_{n_k}^{(1)}} : \frac{1}{2^{\mu_k} m_{n_k}^{(1)}} = 2^k$$

and $n_k + [k m_{n_k}^{(1)}] \leq n_k + m_{n_k}^{(2)}$. □

Let us give simple examples of $WM(\mathbf{m})$ sequences.

Example. Let $\alpha > 0$.

(1) $\{\ln^\alpha n\} \in WM \setminus WM(\{2^n\})$.
(2) $\{n^\alpha\} \in WM \setminus WM(\{n \ln n\})$.
(3) $\{e^{\ln^\alpha n}\} \in WM(\{\ln n\}) \setminus WM$ for $\alpha > 1$.

(4) $\{a_n\} \in WM(\{\ln n\}) \setminus WM(\{\ln^2 n\})$, where $a_1 = 1$ and

$$a_j := \frac{2^i}{2^{2^k}}, \quad 2^k + ik \le j < \min\{2^k + (i+1)k, 2^{k+1}\},$$
$$j \in \mathbb{N}, \quad i = 0, \dots, [2^k/k], \quad k \in \mathbb{N}.$$

Indeed, this sequence is a $WM(\{\ln k\})$ sequence. On the other hand, it is not in $WM(\{\ln^2 k\})$, since $a_{2^k+k^2}/a_{2^k} = 2^k \not\le C$. □

Without loss of generality we will further suppose that $\mathbf{m}$ is a sequence of integers, since by Property 3, we have $WM(\{m_n\}) = WM(\{[m_n]\})$.

The main goal of this paper is to study different convergence tests for number series and Hardy's inequalities for weak monotone sequences with respect to $\{m_n\}$. We will see that the class $WM = WM(\{n\})$ is the limiting class in many cases. Precisely, the Cauchy–Schlömilch condensation test reads differently in the cases $m_n \ge Cn$ and $m_n = o(n)$ as $n \to \infty$ (Sections 3 and 4). In Section 5, we study the Abel–Oliver n-term test for $WM(\{m_n\})$-sequences. In particular, we show that if the series $\sum_k a_k$ converges and $\{a_n\} \in WM(\{m_n\})$, then $a_n \min\{n, m_n\} \to 0$ and this result is the best possible.

Finally, in Section 6, we will show that the Hardy inequalities

$$\sum_{n=1}^{\infty} n^{\beta} \Big(\frac{1}{n} \sum_{k=1}^{n} a_k \Big)^p \le C \sum_{n=1}^{\infty} a_n^p n^{\beta}, \qquad \beta < p - 1,$$

and

$$\sum_{n=1}^{\infty} n^{\beta} \Big(\frac{1}{n} \sum_{k=n}^{\infty} a_k \Big)^p \le C \sum_{n=1}^{\infty} a_n^p n^{\beta}, \qquad \beta > p - 1,$$

where $0 < p < 1$ do not hold for $\{a_n\} \in WM(\{m_n\})$, where $m_n = o(n)$, unlike the case $\{a_n\} \in WM(\{m_n\})$, where $m_n \ge Cn$.

2. Integral test

To prove the Maclaurin–Cauchy integral test for equiconvergence of series and integrals, we introduce the weak monotonicity property for functions.

Definition 2.1. Let m be a measurable nondecreasing positive function on $(0, \infty)$. We say that a nonnegative function f defined on $(0, \infty)$ is $WM(m(x))$ if there exists a constant C such that

$$f(t) \le C f(x) \quad \text{for any} \quad t \in [x, x + m(x)], \tag{2.1}$$

where C does not depend on x.

It turns out that the monotonicity condition in the integral Maclaurin–Cauchy test can be replaced by the $WM(m(x))$ condition for any $m(x)$.

Theorem 2.2. *Let f be a $WM(m(x))$ function. Then the series*

$$\sum_{k=1}^{\infty} f(k) \tag{2.2}$$

and the integral

$$\int_1^{\infty} f(t)\,dt \tag{2.3}$$

converge or diverge simultaneously.

Proof. In view of properties of WM classes, we can suppose that $m(t) \geq 1$ for $t \geq 1$. Then the proof follows immediately from

$$\frac{1}{C} f(k+1) \leq \int_k^{k+1} f(t)dt \leq Cf(k). \qquad \square$$

The relationship between the discrete and continuous WM classes is given by the following

Remark 2.3. Let $f(x) = a_n$ for $n-1 < x \leq n$. Then for any nondecreasing positive function $m(\cdot)$ on $(0, \infty)$, we have

$$f \in WM(m(x)) \qquad \text{if and only if} \qquad \{a_n\} \in WM\big(\{m(n)\}\big).$$

3. Cauchy–Schlömilch type theorems

One of the widely applied tests for monotone sequences is the Cauchy condensation test: for monotone $\{a_n\}$ the series $\sum_n a_n$ and $\sum_n 2^n a_{2^n}$ are equiconvergent. The same result holds for sequences from the WM class as well (see [9]). A generalization of the Cauchy test was proved by Schlömilch (see [5, Ch. III, §14, 77] or [2, p. 44]). Let us formulate the Schlömilch test for WM sequences (see [9]).

Theorem A. *Let $\{a_n\}$ be a WM sequence, and let $u_1 \leq u_2 \leq \cdots$ be a sequence of integers such that $u_k \to \infty$ and*

$$u_{k+1} = O(u_k). \tag{3.1}$$

Then the series

$$\sum_k a_k \tag{3.2}$$

converges if and only if the series

$$\sum_{k=1}^{\infty} (u_{k+1} - u_k) a_{u_k} \tag{3.3}$$

converges.

In this section we study the Cauchy–Schlömilch type theorems for sequences from the class $WM(\{m_n\})$.

Theorem 3.1. *Let a nonnegative sequence* $\{a_k\} \in WM(\{m_k\})$ *and let* $1 = u_1 \le u_2 \le \cdots$ *be a sequence of integers such that* $u_k \to \infty$ *and*

$$u_{k+1} - u_k \le s m_{u_k} \tag{3.4}$$

for some $s \ge 1$.

(A) *Then the convergence of series (3.3) implies the convergence of series (3.2).*

(B) *If, additionally,*

$$u_{k+2} - u_{k+1} \le M(u_{k+1} - u_k) \tag{3.5}$$

for some $M > 0$*, then series (3.3) and (3.2) converge or diverge simultaneously.*

Proof. (A) In order to estimate the series

$$\sum_k a_k = \sum_k \sum_{i=u_k}^{u_{k+1}-1} a_i$$

for each $k \in \mathbb{N}$, we find $s_k \in \mathbb{N}$, $s_k \le s$, such that

$$(s_k - 1)m_{u_k} < u_{k+1} - u_k \le s_k m_{u_k}.$$

Then, by (1.2), we have

$$\begin{aligned}
\sum_{i=u_k}^{u_{k+1}-1} a_i &= \sum_{j=1}^{s_k-1} \sum_{i=u_k+(j-1)m_{u_k}}^{u_k+jm_{u_k}-1} a_i + \sum_{i=u_k+(s_k-1)m_{u_k}}^{u_{k+1}-1} a_i \\
&\le C \sum_{j=1}^{s_k-1} m_{u_k} a_{u_k+(j-1)m_{u_k}} \\
&\quad + C(u_{k+1} - u_k - (s_k - 1)m_{u_k}) a_{u_k+(s_k-1)m_{u_k}} \\
&\le C m_{u_k} \sum_{j=1}^{s_k-1} C^{j-1} a_{u_k} + C C^{s_k-1}(u_{k+1} - u_k - (s_k - 1)m_{u_k}) a_{u_k} \\
&\le C^{s_k}(s_k - 1)m_{u_k} a_{u_k} + C^{s_k}(u_{k+1} - u_k - (s_k - 1)m_{u_k}) a_{u_k} \\
&\le C^s a_{u_k}(u_{k+1} - u_k).
\end{aligned}$$

Therefore,

$$\sum_k a_k \le C^s \sum_k (u_{k+1} - u_k) a_{u_k}.$$

(B) First, by (1.2), we have

$$\begin{aligned}\sum_{i=u_k}^{u_{k+1}-1} a_i &\geq C^{-1}\sum_{j=1}^{s_k-1} m_{u_k}a_{u_k+jm_{u_k}} + C^{-1}(u_{k+1}-u_k-(s_k-1)m_{u_k})a_{u_{k+1}}\\ &\geq C^{-1}m_{u_k}\sum_{j=1}^{s_k-1} C^{-(s_k-j)}a_{u_{k+1}} + C^{-1}(u_{k+1}-u_k-(s_k-1)m_{u_k})a_{u_{k+1}}\\ &\geq C^{-s_k}(s_k-1)m_{u_k}a_{u_{k+1}} + C^{-s_k}(u_{k+1}-u_k-(s_k-1)m_{u_k})a_{u_{k+1}}\\ &\geq C^{-s}a_{u_{k+1}}(u_{k+1}-u_k).\end{aligned}$$

Further, (3.5) implies that

$$\sum_k a_k \geq C^{-s}\sum_k (u_{k+1}-u_k)a_{u_{k+1}} \geq C^{-s}M^{-1}\sum_k (u_{k+2}-u_{k+1})a_{u_{k+1}},$$

which completes the proof. □

Example. **(1)** Suppose $m_k = k^2$ and $u_k = 2^{2^k}$, then

$$u_{k+1}-u_k = 2^{2^{k+1}} - 2^{2^k} \leq 2^{2^{k+1}} = m_{u_k}.$$

So, if series (3.3), that is,

$$\sum_k (2^{2^{k+1}} - 2^{2^k})a_{2^{2^k}} \tag{3.6}$$

converges, then the series $\sum_k a_k$ converges as well.

The inverse implication does not hold in general. Consider the sequence $\{a_k\}$ with $a_k := 1/k^2$. Then the series $\sum_k a_k$ converges, but the series

$$\sum_k (2^{2^{k+1}} - 2^{2^k})a_{2^{2^k}} = \sum_k \left(1 - \frac{1}{2^{2^k}}\right)$$

diverges.

Note that condition (3.5) in not satisfied in this case:

$$\frac{u_{k+2}-u_{k+1}}{u_{k+1}-u_k} = 2^{2^k}\frac{2^{2^{k+1}}-1}{2^{2^k}-1} \to \infty.$$

(2) Let $u_k = k^2$ and $m_k = \sqrt{k}$. Then

$$u_{k+1}-u_k = 2k+1 \asymp m_{u_k} = k \quad \text{and} \quad \frac{u_{k+2}-u_{k+1}}{u_{k+1}-u_k} = \frac{2k+3}{2k+1} \leq 2,$$

so, by Theorem 3.1, the series

$$\sum_k k a_{k^2} \qquad \text{and} \qquad \sum_k a_k$$

converge or diverge simultaneously.

Remark 3.2. 1) Note that in the case $m_k \geq Ck$ condition (3.4) is weaker than condition (3.1): if $u_{k+1}/u_k \leq M$ for some $M > 0$ then

$$u_{k+1} - u_k = u_k \left(\frac{u_{k+1}}{u_k} - 1 \right) \leq (M-1)u_k \leq C^{-1}(M-1)m_{u_k}.$$

Moreover, condition (3.4) is strictly weaker than condition (3.1); take Example 3 (1).

In the case $m_k \leq Ck$ condition (3.1) is weaker than condition (3.4):

$$\frac{u_{k+1}}{u_k} - 1 = \frac{1}{u_k}(u_{k+1} - u_k) \leq \frac{1}{u_k} s m_{u_k} \leq C \frac{1}{u_k} s u_k = Cs.$$

Again, taking $u_k = k^2$ and $m_k = \ln k$, we see that condition (3.1) is strictly weaker than condition (3.4) in the case when $m_k \leq Ck$.

If $C_1 k \leq m_k \leq C_2 k$, then conditions (3.1) and (3.4) are equivalent.

2) Condition (3.1) cannot be omitted in Theorem A (see Example 3.6 (1) in [9]). In view of 1), we cannot omit condition (3.4) in Theorem 3.1.

3) Note that condition (3.5) in Theorem 3.1 generally cannot be replaced with (3.1) as the following example shows.

Example. We will construct an index sequence $\{u_k\}$ and a number sequence $\{a_k\}$ satisfying $WM(\{m_k\})$ as well as conditions (3.1) and (3.4) such that the series $\sum_k a_k$ converges while the series $\sum_k (u_{k+1} - u_k) a_{u_k}$ diverges and condition (3.5) is not satisfied. To define $\{u_k\}$, we construct an auxiliary index sequence $\{k_n\}$. We set $k_1 := u_1 := 1$ and suppose that $k_2, \ldots, k_n$ and $u_2, \ldots, u_{k_n}$ are already defined. We choose $N(n) \in \mathbb{N}$ such that

$$\frac{((n+1)!)^2}{u_{k_n} + (n!)^2(N(n)+1)} \leq 1 \tag{3.7}$$

and set $u_{k_n+i+1} := u_{k_n+i} + (n!)^2$ for $i = 0, \ldots, N(n)$ and $k_{n+1} := k_n + N(n) + 1$. So, we can rewrite condition (3.7) as

$$\frac{((n+1)!)^2}{u_{k_{n+1}}} \leq 1. \tag{3.8}$$

Now we define

$$a_s := 2^{i-1-N(n)}((n+1)!)^{-2}$$

for $u_{k_n} + (i-1)(n!)^2 < s \leq u_{k_n} + i(n!)^2$ and $i = 1, \ldots, N(n)+1$. Then

$$\sum_s a_s = \sum_n \sum_{i=0}^{N(n)} (n!)^2 \frac{2^i}{2^{N(n)}((n+1)!)^2} \lesssim \sum_n \frac{1}{(n+1)^2} < \infty.$$

On the other hand,

$$\sum_k (u_{k+1} - u_k) a_{u_k} \geq \sum_n (u_{k_n+1} - u_{k_n}) a_{u_{k_n}} = \sum_n (n!)^2 \frac{1}{(n!)^2} = \infty.$$

Further, we take $m_s = (n!)^2$ for $u_{k_n} \leq s < u_{k_{n+1}}$, and then $\{a_n\} \in WM(\{m_n\})$.

Moreover, condition (3.4) is fulfilled: for $i \in \mathbb{N}$ let $n \in \mathbb{N}$ be such that $k_n \leq i < k_{n+1}$, then

$$\frac{u_{i+1} - u_i}{m_{u_i}} = \frac{(n!)^2}{(n!)^2} = 1.$$

To verify that condition (3.1) is also fulfilled for any given $k \in \mathbb{N}$, we choose $n \in \mathbb{N}$ such that $k_n \leq k < k_{n+1}$. Then, in view of (3.8), we have

$$\frac{u_{k+1}}{u_k} = \frac{u_k + (n!)^2}{u_k} \leq 1 + \frac{(n!)^2}{u_{k_n}} \leq 2.$$

On the other hand, condition (3.5) does not hold because

$$\frac{u_{k_n+1} - u_{k_n}}{u_{k_n} - u_{k_n-1}} = \frac{(n!)^2}{((n-1)!)^2} = n^2. \qquad \square$$

Taking $u_k = 2^k$ in Theorem 3.1, we obtain equiconvergence of the series $\sum_k a_k$ and $\sum_k 2^k a_{2^k}$ for $\{a_n\} \in WM(\{m_n\})$ where $k \leq Cm_k$, i.e., the Cauchy condensation test. The next section deals with the Cauchy test for the case $k \geq Cm_k$.

4. The Cauchy condensation test for $WM(\{m_n\})$ series, where $m_n = o(n)$

If $m_k \geq Ck$ and $\{u_k\}$ satisfies condition (3.1), that is, $u_{k+1} = O(u_k)$, then in view of embedding properties, we get $WM(\{m_n\}) \subset WM$ and, therefore, we can use Theorem A instead of Theorem 3.1. If $\{u_k\}$ increases faster than condition (3.1) indicates but it still satisfies (3.4) (cf. Example 3 (1)), we can use Theorem 3.1.

The mentioned results deal with $\{m_k\}$ increasing rather rapidly ($m_k \geq Ck$). In the following proposition we obtain an analogue of the Cauchy theorem for sufficiently slowly increasing $\{m_k\}$, i.e., $m_k \leq Ck$. Since $WM(\{m_n\}) = WM(\{m_n/C\})$, we can assume that $m_k \leq k$.

Theorem 4.1. *Suppose $\{m_n\}$ is a nondecreasing positive sequence satisfying*

$$m_n \leq n \tag{4.1}$$

and $\{a_n\} \in WM(\{m_n\})$.

(A) *Then the convergence of the series*

$$\sum_k \widetilde{m}_k \sum_{i=0}^{t_k} a_{2^k + i\widetilde{m}_k}, \tag{4.2}$$

where $\widetilde{m}_k := m_{2^k}$ and $t_k := [2^k/\widetilde{m}_k]$, implies the convergence of the series $\sum_k a_k$.

(B) *If, additionally,*

$$m_{2k} \leq M m_k \tag{4.3}$$

for some $M > 0$, then the series (4.2) and $\sum_k a_k$ converge or diverge simultaneously.

Proof. (A) Note that $t_k \geq 1$ for $k \geq 1$. In order to apply Theorem 3.1, we define an index sequence $\{u_k\}$ as follows: set formally $t_0 := 0$ and

$$u_{t_0+\cdots+t_{i-1}+j+1} := 2^i + j\widetilde{m}_i \quad \text{for} \quad i \in \mathbb{N},\ j = 0,\dots,t_i - 1.$$

Given $k \in \mathbb{N}$ let $i \in \mathbb{N}$ and $j = 0,\dots,t_i-1$ be such that $k = t_1+\cdots+t_{i-1}+j+1$. If $k+1 \leq t_1 + \cdots + t_i$, then

$$u_{k+1} - u_k = \widetilde{m}_i = m_{2^i} \leq m_{2^i+j\widetilde{m}_i} = m_{u_k}.$$

If $k+1 > t_1 + \cdots + t_i$, then

$$\widetilde{m}_i \leq u_{k+1} - u_k = 2^{i+1} - (2^i + (t_i - 1)\widetilde{m}_i) < 2^i - \left(\frac{2^i}{\widetilde{m}_i} - 2\right)\widetilde{m}_i = 2\widetilde{m}_i. \quad (4.4)$$

Hence condition (3.4) is fulfilled. Therefore, by Theorem 3.1 (A), the convergence of the series $\sum_{k=1}^{\infty}(u_{k+1}-u_k)a_{u_k}$ implies the convergence of the series $\sum_k a_k$. Since, in view of (4.4), we have

$$\begin{aligned}\sum_{k=1}^{\infty}(u_{k+1}-u_k)a_{u_k} &= \sum_i \sum_{j=0}^{t_i-1}(u_{t_0+\cdots+t_{i-1}+j+2} - u_{t_0+\cdots+t_{i-1}+j+1})a_{u_{t_0+\cdots+t_{i-1}+j+1}} \\ &= \sum_i \left[\widetilde{m}_i \sum_{j=0}^{t_i-2} a_{2^i+j\widetilde{m}_i} + (2^i - (t_k - 1)\widetilde{m}_i)u_{2^i+(t_i-1)\widetilde{m}_i}\right] \\ &\leq 2\sum_i \widetilde{m}_i \sum_{j=0}^{t_i-1} a_{2^i+j\widetilde{m}_i} \leq 2\sum_i \widetilde{m}_i \sum_{j=0}^{t_i} a_{2^i+j\widetilde{m}_i},\end{aligned}$$

the required result follows.

(B) We assume that the series $\sum_k a_k$ converges. Let us first verify that condition (3.5) holds. Given $k \subset \mathbb{N}$ let $i \in \mathbb{N}$ be such that $t_1+\cdots+t_{i-1}+1 \leq k \leq t_1+\cdots+t_i$. If $k+2 \leq t_1 + \cdots + t_i$, then

$$\frac{u_{k+2}-u_{k+1}}{u_{k+1}-u_k} = \frac{\widetilde{m}_i}{\widetilde{m}_i} = 1.$$

If $k+1 \leq t_1 + \cdots + t_i$ but $k+2 > t_1 + \cdots + t_i$, then

$$\frac{u_{k+2}-u_{k+1}}{u_{k+1}-u_k} \leq \frac{2\widetilde{m}_i}{\widetilde{m}_i} = 2.$$

If $k+1 > t_1 + \cdots + t_i$, then by (4.3) we have

$$\frac{u_{k+2}-u_{k+1}}{u_{k+1}-u_k} \leq \frac{\widetilde{m}_{i+1}}{\widetilde{m}_i} \leq M.$$

Hence, condition (3.5) is satisfied, and Theorem 3.1 (B) yields the following statement: the convergence of the series $\sum_k a_k$ implies the convergence of the series

$\sum_{k=1}^{\infty}(u_{k+1}-u_k)a_{u_k}$. Now, by (4.4) and $WM(\{m_n\})$, we have

$$\sum_{k=1}^{\infty}(u_{k+1}-u_k)a_{u_k} \geq \sum_i \widetilde{m}_i \sum_{j=0}^{t_i-1} a_{2^i+j\widetilde{m}_i} \geq C^{-1}\sum_i \widetilde{m}_i \sum_{j=0}^{t_i} a_{2^i+j\widetilde{m}_i},$$

where in the last inequality we used the fact that $\{a_k\} \in WM(\{m_k\})$. Thus, series (4.2) converges. □

Note that we cannot omit the doubling condition (4.3) as we will see in Remark 5.2. Observe also that if $m_n = n$, we are again within the Cauchy condensation test for WM sequences.

5. Behavior of the nth term of series with $WM(\{m\})$ coefficients

In the case $\{m_n = n\}$, that is, $WM(\{m_n\}) = WM$, the Cauchy condensation theorem can be easily applied to obtain $na_n \to 0$ for a convergent series $\sum_k a_k$ (see Theorem 4.1 in [9]). It turns out that the corresponding result holds for any $\{m_n\}$ satisfying conditions of Theorem 4.1. Set

$$m'_n := \min\{n, m_n\}.$$

Theorem 5.1.

(A) *Let $\{a_k\}$ be a $WM(\{m_k\})$-sequence such that $\{m'_n\}$ satisfies the doubling condition* (4.3). *Suppose the series $\sum_k a_k$ converges, then $m'_n a_n \to 0$.*

(B) *Let $\{m_n\}$ be a nondecreasing index sequence and $\{u_n\}$ be a sequence with* $\limsup u_n = \infty$. *Then there exists a positive convergent series $\sum_k a_k$ with $\{a_k\} \in WM(\{m_k\})$ such that $u_n m'_n a_n \not\to 0$.*

Proof. The proof is in three steps.

Step 1. Let us first prove the following particular result of (A):

Let $\{m_n\}$ be a sequence satisfying conditions (4.1) *and* (4.3). *Suppose that $a_k \geq 0$ be a $WM(\{m_k\})$ sequence. If the series $\sum_k a_k$ converges, then $m_n a_n \to 0$.*

Indeed, by Theorem 4.1, the series (4.2) converges. Hence

$$\max_{i=0,\dots,t_k} \widetilde{m}_k a_{2^k+i\widetilde{m}_k}$$

is a null sequence. Given $j \in \mathbb{N}$, let i,k be integers such that $0 \leq i \leq t_k$ and $2^k + i\widetilde{m}_k \leq j < \min\{2^k+(i+1)\widetilde{m}_k; 2^{k+1}\}$.

Then

$$0 < m_j a_j \leq \widetilde{m}_{k+1} C a_{2^k+i\widetilde{m}_k} \leq CM\widetilde{m}_k a_{2^k+i\widetilde{m}_k}.$$

Hence $m_j a_j$ tends to zero, as required.

Step 2. To show (A), we first note that if $\{m_n\}$ satisfies the doubling condition (4.3), then $\{m'_n\}$ does as well. On the other hand, taking $m_n = 2^n$ we see that the inverse statement does not hold.

In view of embedding properties of the classes $WM(\mathbf{m})$, $\{a_k\} \in WM(\{m'_k\})$. Now we can use the statement from Step 1 for $\{m'_n\}$ getting $m'_n a_n \to 0$.

Step 3. Let us prove (B) by constructing specific sequences $\{a_k\}$ and $\{\nu_k\}$. Let ν_k be an integer such that

$$\nu_{k+1} > \nu_k \qquad \text{and} \qquad u_{\nu_{k+1}} > u_{\nu_k} > 4^k \qquad (k \in \mathbb{N}).$$

We set $a_1 := \cdots := a_{\nu_1} := 1$ and for $i = 0, \ldots, \mu_k := [(\nu_{k+1} - \nu_k)/m_{\nu_{k+1}}]$ and $j \in \mathbb{N}$ with $\nu_k + i m_{\nu_{k+1}} < j \le \min\{\nu_k + (i+1)m_{\nu_{k+1}}, \nu_{k+1}\}$ we define

$$a_j := \frac{2^i}{\sqrt{u_{\nu_{k+1}}} 2^{\mu_k} m_{\nu_{k+1}}}.$$

Note that if $m_{\nu_{k+1}} > \nu_{k+1}$, then $\mu_k = 0$ and

$$a_j = \frac{1}{\sqrt{u_{\nu_{k+1}}} \nu_{k+1}}, \quad \nu_k < j \le \nu_{k+1}.$$

Note also that

$$\max_{\nu_k < j \le \nu_{k+1}} a_j = \frac{1}{\sqrt{u_{\nu_{k+1}}} m_{\nu_{k+1}}} < \frac{1}{\sqrt{u_{\nu_k}} m_{\nu_k}} = \max_{\nu_{k-1} < j \le \nu_k} a_j.$$

Thus, the sequence $\{a_k\}$ is in $WM(\{m_k\})$. Moreover,

$$\sum_k a_k \le \nu_1 + \sum_k \frac{1}{2^{\mu_k} \sqrt{u_{\nu_{k+1}}}} \sum_{i=0}^{\mu_k} 2^i \le \nu_1 + \sum_k \frac{2}{\sqrt{u_{\nu_{k+1}}}} \le \nu_1 + \sum_k \frac{2}{2^k} < \infty.$$

On the other hand,

$$\max_{\nu_k < j \le \nu_{k+1}} a_j = \frac{1}{\sqrt{u_{\nu_{k+1}}} m_{\nu_{k+1}}} < \frac{1}{\sqrt{u_{\nu_k}} m_{\nu_k}} = \max_{\nu_{k-1} < j \le \nu_k} a_j,$$

which completes the proof. □

Example. 1) For $m_n = n$, Theorem 4.1 is the Cauchy condensation test and Theorem 5.1 is Theorem 4.1 in [9].

2) Let $\{m_n = \ln n\}$ and $\{a_k\} \in WM(\{m_k\})$. Then, by Theorems 4.1 and 5.1, the series

$$\sum_k a_k \qquad \text{and} \qquad \sum_k k \sum_{i=0}^{[2^k/k]} a_{2^k + ik} \tag{5.1}$$

are equiconvergent and their convergence implies

$$a_n \ln n \to 0.$$

Let us now show that the doubling condition (4.3) can not be omitted.

Remark 5.2. Let $\{m_n\}$ be a sequence with $m_n \le Cn$ but failing to satisfy the doubling condition (4.3). Then we can find a nonnegative sequence $\{a_k\} \in WM(\{m_k\})$ such that the series $\sum_k a_k$ converges, but $m_n a_n \not\to 0$, and series (4.2) diverges.

Proof. Since condition (4.3) is not fulfilled we can find an increasing index sequence $\{k_i\}$ such that

$$m_{2k_i} > 4^i m_{k_i}$$

and $k_{i+1} > 2k_i$. Now we find another index sequence $\{\nu_i\}$ such that

$$2^{\nu_i} \le k_i < 2^{\nu_i+1} \le 2k_i < 2^{\nu_i+2}.$$

Without loss of generality we can suppose that $2^{\nu_i+2} < 2^{\nu_{i+1}}$.

Since $\{m_k\}$ is nondecreasing, then $m_{2^{\nu_i+2}} > 4^i m_{2^{\nu_i}}$. We set $\hat{m}_k := m_{2^{\nu_1}}$ for $k \le 2^{\nu_1}$. Now by induction for $2^{\nu_i} < k \le 2^{\nu_{i+1}}$ we set $\hat{m}_k := 2\hat{m}_{2^{\nu_i}}$. Then the sequence $\{\hat{m}_k\}$ satisfies condition (4.3). Setting $u_n := m_n/\hat{m}_n$, we have

$$u_{2^{\nu_i+2}} = \frac{m_{2^{\nu_i+2}}}{\hat{m}_{2^{\nu_i+2}}} > \frac{4^i m_{2^{\nu_i}}}{2^{i+1} m_{2^{\nu_1}}} \ge 2^{i-1}.$$

Analyzing the proof of Theorem 5.1 (B), we can choose a positive sequence $\{a_k\} \in WM(\{m_k\})$ such that the series $\sum_k a_k$ converges and

$$m_{2^{\nu_i}} a_{2^{\nu_i}} = u_{2^{\nu_i}} \hat{m}_{2^{\nu_i}} a_{2^{\nu_i}} \to \infty.$$

So, $m_n a_n \not\to 0$ and, moreover, series (4.2) diverges. □

6. Hardy's inequalities for weak monotone sequences and functions

6.1. L_p-inequalities for Hardy's operators

Hardy proved the following inequalities for $p \ge 1$ and a nonnegative measurable function f on $(0, \infty)$ (see, e.g., [4]):

$$\int_0^\infty x^\beta \Big(\frac{1}{x}\int_0^x f(t)dt\Big)^p dx \le C(p,\beta) \int_0^\infty x^\beta f^p(x)dx, \qquad \beta < p-1, \tag{6.1}$$

and

$$\int_0^\infty x^\beta \Big(\frac{1}{x}\int_x^\infty f(t)dt\Big)^p dx \le C(p,\beta) \int_0^\infty x^\beta f^p(x)dx, \qquad \beta > p-1. \tag{6.2}$$

Their counterparts for series read as follows: for $p \ge 1$ and a nonnegative sequence $\{a_n\}$,

$$\sum_{n=1}^\infty n^\beta \Big(\frac{1}{n}\sum_{k=1}^n a_k\Big)^p \le C(p,\beta) \sum_{n=1}^\infty a_n^p n^\beta, \qquad \beta < p-1, \tag{6.3}$$

and

$$\sum_{n=1}^\infty n^\beta \Big(\frac{1}{n}\sum_{k=n}^\infty a_k\Big)^p \le C(p,\beta) \sum_{n=1}^\infty a_n^p n^\beta, \qquad \beta > p-1. \tag{6.4}$$

It is also known that any of inequalities (6.1)–(6.4) does not hold for $0 < p < 1$ but they do hold in the reverse direction.

Moreover, concerning the case $0 < p < 1$, an important observation was made by Konushkov [6] and Leindler [7]. They showed that inequalities (6.3) and (6.4) hold for $0 < p < 1$ for quasi-monotone sequences and, more generally, for the WM sequences. Also, Burenkov [3] obtained sharp constants in (6.1) and (6.2) with $0 < p < 1$ for monotone functions (see also [1]). Leindler [7] proved similar inequalities for WM functions but he considered operators $\frac{1}{x}\int_2^x f(t)dt$ and $\frac{1}{x}\int_x^\infty f(t)dt$ and integration over $(1,\infty)$. Using methods from [7] it is easy to show that *for WM functions inequalities* (6.1) *and* (6.2) *hold for* $0 < p < 1$. Similarly, *for WM sequences inequalities* (6.3) *and* (6.4) *hold for* $0 < p < 1$.

In this section, our main goal is to show that these results for $0 < p < 1$ are not true any longer for $f \in WM(m(\cdot))$, where $m(x)/x = o(1)$, or $\{a_n\} \in WM(\{m_n\})$, where $m_n = o(n)$. This is a fundamental difference between these cases and the cases $m(x) \geq Cx$ or $m_n \geq Cn$.

Set

$$Hf(x,y) = \int_x^y f(t)dt.$$

To construct counterexamples for inequalities (6.1) and (6.2), let us consider $f \in WM(m(x))$ such that $f(t) = 1$ on $0 < t < 1$ and $\beta > -1$. Then we can reduce this problem to the same one with integration over $(1,\infty)$, that is, to the inequalities

$$\int_1^\infty x^{\beta-p}\Big(Hf(1,x)\Big)^p dx \leq C(p,\beta)\int_1^\infty x^\beta f^p(x)dx, \qquad \beta < p-1, \tag{6.5}$$

and

$$\int_1^\infty x^{\beta-p}\Big(Hf(x,\infty)\Big)^p dx \leq C(p,\beta)\int_1^\infty x^\beta f^p(x)dx, \qquad \beta > p-1, \tag{6.6}$$

where $0 < p < 1$, do not hold in general for $f \in WM(m(x))$ where $m(x)/x \to 0$.

We will prove the following stronger result for the Hardy–Steklov operator $Hf(x,2x)$.

Theorem 6.1. *Let a nondecreasing function $m(x)$ on $(1,\infty)$ be such that*

$$\frac{m(x)}{x} \to 0 \qquad as \qquad x \to \infty. \tag{6.7}$$

Let $p \in (0,1)$. Then there is no such constant $C > 0$ satisfying the inequality

$$\int_1^\infty x^{\beta-p}\Big(Hf(x,2x)\Big)^p dx \leq C(p,\beta)\int_1^\infty x^\beta f^p(x)dx \tag{6.8}$$

for any nonnegative f from the class $WM(m(x))$.

Remark 6.2. Since we construct a counter-example as a step-function, the same counter-example can be applied for discrete Hardy inequalities (6.3), (6.4), and

$$\sum_{n=1}^\infty n^\beta \Big(\frac{1}{n}\sum_{k=n}^{2n} a_k\Big)^p \leq C(p,\beta)\sum_{n=1}^\infty a_n^p n^\beta, \tag{6.9}$$

with $\{a_n\} \in WM(\{m_n\})$, $m_n/n \to 0$.

Below we will write $a_n \lesssim b_n$ or $a_n \gtrsim b_n$ for the positive sequences $\{a_n\}$ and $\{b_n\}$ if $a_n \leq Cb_n$ or $a_n \geq Cb_n$, respectively. Moreover, $a_n \asymp b_n$ means that $a_n \lesssim b_n \lesssim a_n$.

Proof of Theorem 6.1. Let $m_n := m(n)$, $n \in \mathbb{N}$. Set

$$\alpha_n := \left(\frac{m_{2^n}}{2^n}\right)^{1-p} \qquad (n \in \mathbb{N}).$$

Let $\{\gamma_n\}$ be a decreasing positive sequence such that

$$\sum_n \gamma_n = \infty \qquad \text{and} \qquad \sum_n \alpha_n \gamma_n < \infty.$$

If $\sum_n \alpha_n < \infty$ we take $\gamma_n = n^{-1}$. If $\sum_n \alpha_n = \infty$, then, in view of (6.7), we can choose an index sequence $\{k_i\}$ such that

$$\sum_{k=k_i+1}^{k_{i+1}} \alpha_k < \frac{k_{i+1} - k_i}{i} \qquad \text{and} \qquad \sum_{k=k_{i+1}+1}^{k_{i+2}} \alpha_k > \sum_{k=k_i+1}^{k_{i+1}} \alpha_k.$$

Now we set

$$\gamma_k := \frac{1}{i^2 \sum_{k=k_i+1}^{k_{i+1}} \alpha_k} \qquad \text{for} \quad k_i < k \le k_{i+1}.$$

Then the sequence $\{\gamma_k\}$ is decreasing and

$$\sum_k \gamma_k = \sum_i \sum_{r=k_i+1}^{k_{i+1}} \frac{1}{i^2 \sum_{k=k_i+1}^{k_{i+1}} \alpha_k} \ge \sum_i (k_{i+1} - k_i) \frac{1}{i^2} \frac{i}{k_{i+1} - k_i} = \infty.$$

Also,

$$\sum_k \alpha_k \gamma_k = \sum_i \sum_{r=k_i+1}^{k_{i+1}} \alpha_r \frac{1}{i^2 \sum_{k=k_i+1}^{k_{i+1}} \alpha_k} = \sum_i \frac{1}{i^2} < \infty.$$

Let t_k be the smallest integer such that $2^k < (t_k+1)m_{2^k}$. Now, for $x \in [1, \infty)$ we consider

$$2^k + i m_{2^k} \le x < \min\{2^k + (i+1)m_{2^k}, 2^{k+1}\}, \qquad k \in \mathbb{N} \cup \{0\}, \quad i = 0, \dots, t_k$$

and define

$$f(x) := \frac{2^i 2^{\frac{k(p-\beta-1)}{p}} \gamma_k^{1/p}}{2^{t_k} m_{2^k}}.$$

Then $\{f(n)\} \in WM(\{m_n\})$ and, by Remark 2.3, $f(x) \in WM(m(x))$.

Let us first estimate $Hf(x/2, 2x)$. Given $x > 0$ let $i_0 = i_0(x)$ and $n = n(x)$ be integers such that $0 \le i_0 \le t_n$ and $2^n + i_0 m_{2^n} \le x < 2^n + (i_0+1)m_{2^n} \le 2^{n+1}$. Then

$$Hf(x/2, 2x) \ge Hf(2^n, 2^{n+1}) \ge \sum_{j=0}^{t_n - 1} m_{2^n} f(2^n + j m_{2^n})$$

$$= \sum_{j=0}^{t_n-1} m_{2^n} \frac{2^j 2^{\frac{n(p-\beta-1)}{p}} \gamma_n^{1/p}}{2^{t_n} m_{2^n}} \asymp 2^{\frac{n(p-\beta-1)}{p}} \gamma_n^{1/p}.$$

Further, using this estimate, we get

$$\begin{aligned}
&\int_1^\infty x^{\beta-p}\Big(Hf(x,2x)\Big)^p dx \\
&\quad \gtrsim \int_1^\infty x^{\beta-p}\Big(Hf(x,2x)\Big)^p dx + \int_2^\infty x^{\beta-p}\Big(Hf(x/2,x)\Big)^p dx \\
&\quad \gtrsim \int_2^\infty x^{\beta-p}\Big(Hf(x/2,2x)\Big)^p dx \ge \sum_{k=1}^\infty \int_{2^k}^{2^{k+1}} x^{\beta-p}\Big(Hf(x/2,2x)\Big)^p dx \\
&\quad \ge \sum_{k=1}^\infty \Big(Hf(2^k,2^{k+1})\Big)^p \int_{2^k}^{2^{k+1}} x^{\beta-p} dx \gtrsim \sum_{k=1}^\infty 2^{k(p-\beta-1)}\gamma_k 2^{k(\beta-p+1)} = \sum_{k=1}^\infty \gamma_k = \infty.
\end{aligned}$$

On the other hand, to estimate $\int_1^\infty f^p(t)t^\beta dt$, we have

$$\begin{aligned}
\int_1^\infty f^p(t)t^\beta dt &\le \sum_{k=0}^\infty \left[\sum_{i=0}^{t_k-1} \int_{2^k+im_{2^k}}^{2^k+(i+1)m_{2^k}} f^p(t)t^\beta dt + \int_{2^k+t_k m_{2^k}}^{2^{k+1}} f^p(t)t^\beta dt \right] \\
&\lesssim \sum_{k=0}^\infty \left[\sum_{i=0}^{t_k-1} \left\{(2^k+(i+1)m_{2^k})^{\beta+1} - (2^k+im_{2^k})^{\beta+1}\right\} f^p(2^k+im_{2^k}) \right. \\
&\quad \left. + \Big((2^k+(t_k+1)m_{2^k})^{\beta+1} - (2^k+t_k m_{2^k})^{\beta+1}\Big) f^p(2^k+t_k m_{2^k}) \right].
\end{aligned}$$

In view of (6.7), we get the following:

$$\begin{aligned}
(2^k+(i+1)m_{2^k})^{\beta+1} - (2^k+im_{2^k})^{\beta+1} &\lesssim (2^k+im_{2^k})^{\beta+1}\frac{m_{2^k}}{2^k+im_{2^k}} \\
&= (2^k+im_{2^k})^\beta m_{2^k} \asymp 2^{k\beta} m_{2^k}.
\end{aligned}$$

Hence,

$$\begin{aligned}
\int_1^\infty f^p(t)t^\beta dt &\lesssim \sum_{k=0}^\infty \sum_{i=0}^{t_k} 2^{k\beta} m_{2^k} \left(\frac{2^i 2^{\frac{k(p-\beta-1)}{p}} \gamma_k^{1/p}}{2^{t_k} m_{2^k}} \right)^p \\
&\asymp \sum_{k=0}^\infty 2^{k\beta} m_{2^k}^{1-p} 2^{k(p-\beta-1)}\gamma_k = \sum_{k=1}^\infty \left(\frac{m_{2^k}}{2^k}\right)^{1-p}\gamma_k = \sum_{k=1}^\infty \alpha_k\gamma_k < \infty,
\end{aligned}$$

which completes the proof. □

Remark 6.3. Note that the monotonicity assumption on m can be omitted in Theorem 6.1. Indeed, we use monotonicity of $m(\cdot)$ to verify that the function f is in $WM(m(x))$. Let us show that this can be proved without monotonicity assumption.

For $2^k \le x \le y < 2^{k+1}$ it is clear that $f(y) \le Cf(x)$ for $x \le y \le x+m(x)$. For $x \in [2^k + t_k m_{2^k}, 2^{k+1})$ and $y \in [2^{k+1}, 2^{k+1} + m_{2^{k+1}})$, taking into account the definition of t_k and (6.7), we have

$$\frac{f(y)}{f(x)} = \frac{2^{\frac{(k+1)(p-\beta-1)}{p}}\gamma_{k+1}^{1/p}}{2^{t_{k+1}}m_{2^{k+1}}} : \frac{2^{\frac{k(p-\beta-1)}{p}}\gamma_k^{1/p}}{m_{2^k}} \le \frac{2^{\frac{p-\beta-1}{p}}m_{2^k}}{2^{t_{k+1}}m_{2^{k+1}}}$$

$$\le \frac{2^{\frac{p-\beta-1}{p}}m_{2^k}}{t_{k+1}m_{2^{k+1}}} \lesssim \frac{2^{\frac{p-\beta-1}{p}}m_{2^k}}{2^{k+1}} \lesssim 1. \qquad \square$$

6.2. Pointwise inequalities for Hardy's operators

The key step to prove estimates (6.1)–(6.2) for monotone functions ([3, 7]) is to use the following auxiliary inequalities (see also [4]): for $0 < p < 1$,

$$\Big(Hf(x,\infty)\Big)^p \le p\int_x^\infty f^p(t)(t-x)^{p-1}dt, \tag{6.10}$$

$$\Big(Hf(a,x)\Big)^p \le p\int_a^x f^p(t)(t-x)^{p-1}dt. \tag{6.11}$$

For functions from the class $WM(m(x))$ where $m(x) \ge Cx$, their analogues are written as follows:

$$\Big(Hf(x,\infty)\Big)^p \le C\int_{x/2}^\infty f^p(t)t^{p-1}dt, \tag{6.12}$$

$$\Big(Hf(a,x)\Big)^p \le C\int_{a/2}^x f^p(t)t^{p-1}dt. \tag{6.13}$$

Let us show that these inequalities do not hold for $WM(m(x))$ with m satisfying (6.7).

Theorem 6.4. *Let a nondecreasing function $m(x)$ on $(1,\infty)$ satisfy (6.7). Let $p \in (0,1)$. Then there is no such constant $M > 0$ satisfying the inequality*

$$\Big(Hf(2,x)\Big)^p \le M\int_1^x f^p(x)x^{p-1}dx, \qquad 2 < x, \tag{6.14}$$

for any nonnegative f from the class $WM(m(x))$.

Proof. In view of (6.7), we can choose a positive sequence $\{\delta_k\}$ and an index sequence $\{r_k\}$ such that the series $\sum_k \delta_k$ converges, $\delta_{k+1} \le 2\delta_k$ and

$$\delta_{r_k}\left(\frac{2^{r_k}}{m_{2^{r_k}}}\right)^{1-p} \to \infty. \tag{6.15}$$

We set

$$\mu_k := \left(\delta_k\left(\frac{2^k}{m_{2^k}}\right)^{1-p}\right)^{1/p}.$$

Then, taking into account (6.15), we get

$$\sum_k \mu_k \ge \sum_k \mu_{r_k} = \sum_k \left(\delta_{r_k}\left(\frac{2^{r_k}}{m_{2^{r_k}}}\right)^{1-p}\right)^{1/p} = \infty.$$

Let t_k be the smallest integer such that $2^k < (t_k+1)m_{2^k}$ for any $k \geq 1$. Now, for $x \in [1,\infty)$ we consider $2^k + im_{2^k} \leq x < \min\{2^k+(i+1)m_{2^k}, 2^{k+1}\}$, $k \in \mathbb{N}\cup\{0\}$ and $i = 0, \dots, t_k$ and define

$$f(x) := \frac{2^i \mu_k}{2^{t_k} m_{2^k}}.$$

Then $\{f(n)\} \in WM(\{m_n\})$ and $f(x) \in WM(m(x))$.

First, we estimate $Hf(2,x)$ from below. Given $x > 0$ let i_0, n, $i_0 = i_0(x)$, $n = n(x)$ be integers such that $0 \leq i_0 \leq t_n$, $2^n + i_0 m_{2^n} \leq x < 2^n + (i_0+1)m_{2^n} \leq 2^{n+1}$. Then

$$Hf(2,x) \geq \sum_{k=2}^{n-1} \sum_{i=0}^{t_k-1} m_{2^k} \frac{2^i \mu_k}{2^{t_k} m_{2^k}} \geq \sum_{k=2}^{n-1} \mu_k.$$

To estimate $\int_1^x f^p(t)t^{p-1}dt$ from above, we have

$$\begin{aligned}
\int_1^x f^p(t)t^{p-1}dt &\leq \sum_{k=1}^{n} \left[\sum_{i=0}^{t_k-1} \int_{2^k+im_{2^k}}^{2^k+(i+1)m_{2^k}} f^p(t)t^{p-1}dt + \int_{2^k+t_k m_{2^k}}^{2^{k+1}} f^p(t)t^{p-1}dt \right] \\
&\lesssim \sum_{k=1}^{n} \left[\sum_{i=0}^{t_k-1} 2^{k(p-1)} m_{2^k} \left(\frac{2^i \mu_k}{2^{t_k} m_{2^k}} \right)^p + 2^{k(p-1)} m_{2^k} \left(\frac{2^{t_k} \mu_k}{2^{t_k} m_{2^k}} \right)^p \right] \\
&\asymp \sum_{k=1}^{n} 2^{k(p-1)} m_{2^k}^{1-p} \mu_k^p = \sum_{k=1}^{n} \delta_k.
\end{aligned}$$

Since $n = [\log_2 x]$, we get

$$\frac{Hf(2,x)^p}{\int_1^x f^p(x)x^{p-1}dx} \geq \frac{\sum_{k=2}^{[\log_2 x]-1} \mu_k}{\sum_{k=1}^{\infty} \delta_k} \to C\sum_{k=2}^{\infty} \mu_k = \infty,$$

so the result follows. □

Theorem 6.5. *Let a nondecreasing function $m(x)$ on $(1,\infty)$ satisfy (6.7). Let $p \in (0,1)$. Then there is no such constant $M > 0$ satisfying the inequality*

$$\Big(Hf(x,\infty)\Big)^p \leq M \int_{x/2}^{\infty} f^p(t)t^{p-1}dt, \qquad 2 < x, \tag{6.16}$$

for any nonnegative f from the class $WM(m(x))$.

Proof. Let $f(x)$ be defined as in Theorem 6.4.

To estimate $Hf(x,\infty)$, for given $x > 0$, let $i_0 = i_0(x)$, $n = n(x)$ be integers such that $0 \leq i_0 \leq t_n$, $2^n + i_0 m_{2^n} \leq x < 2^n + (i_0+1)m_{2^n} \leq 2^{n+1}$. Then

$$Hf(x,\infty) \geq \sum_{k=n+1}^{\infty} \sum_{i=0}^{t_k-1} m_{2^k} \frac{2^i \mu_k}{2^{t_k} m_{2^k}} \geq \sum_{k=n+1}^{\infty} \mu_k = \infty.$$

On the other hand,

$$\begin{aligned}\int_{x/2}^{\infty} f^p(t)t^{p-1}dt &\le \int_{2^{n-1}}^{\infty} f^p(t)t^{p-1}dt \\ &\lesssim \sum_{k=n-1}^{\infty}\left[\sum_{i=0}^{t_k-1} 2^{k(p-1)}m_{2^k}\left(\frac{2^i\mu_k}{2^{t_k}m_{2^k}}\right)^p + 2^{k(p-1)}m_{2^k}\left(\frac{2^{t_k}\mu_k}{2^{t_k}m_{2^k}}\right)^p\right] \\ &\lesssim \sum_{k=n-1}^{\infty} 2^{k(p-1)}m_{2^k}^{1-p}\mu_k^p = \sum_{k=n-1}^{\infty}\delta_k < \infty,\end{aligned}$$

which completes the proof. □

References

[1] J. Bergh, V. Burenkov, L.E. Persson, *Best constants in reversed Hardy's inequalities for quasimonotone functions*, Acta Sci. Math. **59** (1-2) (1994), 221–239.

[2] D.D. Bonar and M.J. Khoury, *Real Infinite Series*, MAA, Washington, DC, 2006.

[3] V.I. Burenkov, *On the best constant in Hardy's inequality with* $0 < p < 1$ *for monotone functions*, Proc. Steklov Inst. Math. 194, 59–63 (1993); translated from Tr. Mat. Steklova 194, 58–62 (1992).

[4] G.H. Hardy, J.E. Littlewood, G. Polya, *Inequalities*, Cambridge University Press, Cambridge, 1988.

[5] K. Knopp, *Theory and Application of Infinite Series*, Hafner Publ. Co., New York, 1971.

[6] A.A. Konyushkov, *Best approximations by trigonometric polynomials and Fourier coefficients*, Mat. Sb. (N.S.), 44(86):1 (1958), 53–84.

[7] L. Leindler, *Inequalities of Hardy–Littlewood type*, Anal. Math. **2** (1976), no. 2, 117–123.

[8] E. Liflyand, S. Tikhonov, *A concept of general monotonicity and applications*, Math. Nachr., Vol. 284, No. 8-9 (2011), 1083–1098.

[9] E. Liflyand, S. Tikhonov, M. Zeltser, *Extending tests for convergence of number series*, J. Math. Anal. Appl. **377**(1) (2011), 194–206.

[10] O. Szasz, *Quasi-Monotone Series*, Am. Jour. Math. **70**(1948), 203–206.

[11] S. Tikhonov, *Trigonometric series with general monotone coefficients*, J. Math. Anal. Appl. **326**(2007), 721–735.

S. Tikhonov
ICREA and Centre de Recerca Matemática (CRM)
Bellaterra (Barcelona), E-08193, Spain
e-mail: stikhonov@crm.es

M. Zeltser
Department of Mathematics, Tallinn University
Narva mnt. 25, 10120, Tallinn, Estonia
e-mail: mariaz@tlu.ee

Fourier Analysis
Trends in Mathematics, 375–390

Critical Exponent for the Semilinear Wave Equation with Scale Invariant Damping

Yuta Wakasugi

Abstract. In this paper we consider the critical exponent problem for the semilinear damped wave equation with time-dependent coefficients. We treat the scale invariant cases. In this case the asymptotic behavior of the solution is very delicate and the size of coefficient plays an essential role. We shall prove that if the power of the nonlinearity is greater than the Fujita exponent, then there exists a unique global solution with small data, provided that the size of the coefficient is sufficiently large. We shall also prove some blow-up results even in the case that the coefficient is sufficiently small.

Mathematics Subject Classification (2010). 35L71.

Keywords. Damped wave equation; time dependent coefficient; scale invariant damping; critical exponent.

1. Introduction

We consider the Cauchy problem for the semilinear damped wave equation

$$\begin{cases} u_{tt} - \Delta u + \dfrac{\mu}{1+t} u_t = |u|^p, & (t,x) \in (0,\infty) \times \mathbf{R}^n, \\ (u, u_t)(0,x) = (u_0, u_1)(x), & x \in \mathbf{R}^n, \end{cases} \tag{1.1}$$

where $\mu > 0$, $(u_0, u_1) \in H^1(\mathbf{R}^n) \times L^2(\mathbf{R}^n)$ and $1 < p \le \frac{n}{n-2}$ $(n \ge 3), 1 < p < \infty$ $(n = 1, 2)$. Our aim is to determine the critical exponent p_c, which is the number defined by the following property:

If $p_c < p$, all small data solutions of (1.1) are global; if $1 < p \le p_c$, the time-local solution cannot be extended time-globally for some data regardless of smallness.

We note that the linear part of (1.1) is invariant with respect to the hyperbolic scaling

$$\tilde{u}(t,x) = u(\lambda(1+t) - 1, \lambda x).$$

In this case the asymptotic behavior of solutions is very delicate. It is known that the size of the damping term μ plays an essential role. The damping term $\mu/(1+t)$ is known as the borderline between the *effective* and *non-effective* dissipation, here *effective* means that the solution behaves like that of the corresponding parabolic equation, and *non-effective* means that the solution behaves like that of the free wave equation. Concretely, for the linear damped wave equation

$$u_{tt} - \Delta u + (1+t)^{-\beta} u_t = 0, \tag{1.2}$$

if $-1 < \beta < 1$, then the solution u has the same L^p-L^q decay rates as those of the solution of the corresponding heat equation

$$-\Delta v + (1+t)^{-\beta} v_t = 0. \tag{1.3}$$

Moreover, if $-1/3 < \beta < 1$, then the lower frequency part of the solution u of (1.2) is asymptotically equivalent to that of a solution v of (1.3) in the L^2-sense (see [17]). This is called the *local diffusion phenomenon.* Wirth [15] also proved that when $\beta \le -1/3$, u is for each frequency asymptotically equivalent to that of v. This is called the *global diffusion phenomenon.* In particular, the constant coefficient case $\beta = 0$ has been investigated for a long time. We refer the reader to [9, 10, 11]. On the other hand, if $\beta > 1$, then the asymptotic profile of the solution of (1.2) is given by that of the free wave equation $\Box w = 0$ (see [16]). We mention that Wirth treated more general time-dependent damping terms and we refer the reader to [15, 16, 17] for detail.

Wirth [14] considered the linear problem

$$\begin{cases} u_{tt} - \Delta u + \dfrac{\mu}{1+t} u_t = 0, \\ (u, u_t)(0,x) = (u_0, u_1)(x). \end{cases} \tag{1.4}$$

He proved several L^p-L^q estimates for the solutions to (1.4). For example, if $\mu > 1$ it follows that

$$\|u(t,\cdot)\|_{L^q} \lesssim (1+t)^{\max\{-\frac{n-1}{2}(\frac{1}{p}-\frac{1}{q})-\frac{\mu}{2}, -n(\frac{1}{p}-\frac{1}{q})\}} (\|u_0\|_{H^s_p} + \|u_1\|_{H^{s-1}_p}),$$

$$\|(u_t, \nabla u)(t,\cdot)\|_{L^q} \lesssim (1+t)^{\max\{-\frac{n-1}{2}(\frac{1}{p}-\frac{1}{q})-\frac{\mu}{2}, -n(\frac{1}{p}-\frac{1}{q})-1\}} (\|u_0\|_{H^{s+1}_p} + \|u_1\|_{H^s_p}),$$

where $1 < p \le 2$, $1/p + 1/q = 1$ and $s = n(1/p - 1/q)$. This shows that if μ is sufficiently large, then the solution behaves like that of the corresponding heat equation

$$\frac{\mu}{1+t} v_t - \Delta v = 0 \tag{1.5}$$

as $t \to \infty$, and if μ is sufficiently small, then the solution behaves like that of the free wave equation in the above sense. We mention that for the wave equation with space-dependent damping $\Box u + V_0 \langle x \rangle^{-1} u_t = 0$, a similar asymptotic behavior is obtained by Ikehata, Todorova and Yordanov [7].

The Gauss kernel of (1.5) is given by

$$G_\mu(t,x) = \left(\frac{\mu}{2\pi((1+t)^2 - 1)} \right)^{\frac{n}{2}} e^{-\frac{\mu |x|^2}{2((1+t)^2-1)}}.$$

We can obtain the L^p-L^q estimates of the solution of (1.5). In fact, it follows that

$$\|v(t,\cdot)\|_{L^q} \lesssim (1+t)^{-n(\frac{1}{p}-\frac{1}{q})}\|v(0,\cdot)\|_{L^p}$$

for $1 \le p \le q \le \infty$. In particular, taking $q = 2$ and $p = 1$, we have

$$\|v(t,\cdot)\|_{L^2} \lesssim (1+t)^{-\frac{n}{2}}\|v(0,\cdot)\|_{L^1}.$$

From the point of view of the diffusion phenomenon, we expect that the same type estimate holds for the solution of (1.4) when μ is large. To state our results, we introduce an auxiliary function

$$\psi(t,x) := \frac{a|x|^2}{(1+t)^2}, \quad a = \frac{\mu}{2(2+\delta)}$$

with a positive parameter δ. We have the following linear estimate:

Theorem 1.1. *For any $\varepsilon > 0$, there exist constants $\delta > 0$ and $\mu_0 > 1$ having the following property: If $\mu \ge \mu_0$ and (u_0, u_1) satisfy*

$$I_0^2 := \int_{\mathbf{R}^n} e^{2\psi(0,x)}(u_0^2 + |\nabla u_0|^2 + u_1^2)dx < +\infty,$$

then the solution of (1.4) satisfies

$$\int_{\mathbf{R}^n} e^{2\psi}u^2 dx \le CI_0^2(1+t)^{-n+\varepsilon}, \tag{1.6}$$

$$\int_{\mathbf{R}^n} e^{2\psi}(u_t^2 + |\nabla u|^2)dx \le CI_0^2(1+t)^{-n-2+\varepsilon} \tag{1.7}$$

for $t \ge 0$, where $C = C_{\mu,\varepsilon}$ is a positive constant depending on μ, ε.

Remark 1.1. The constant μ_0 depends on ε. The relation is

$$\mu_0 \sim \varepsilon^{-2}.$$

Therefore, as ε is smaller, μ_0 must be larger. We can expect that ε can be removed and the same result holds for much smaller μ. However, we have no idea for the proof.

We also consider the critical exponent problem for (1.1). For the corresponding heat equation (1.5) with nonlinear term $|u|^p$, the critical exponent is given by

$$p_F := 1 + \frac{2}{n},$$

which is well known as the Fujita critical exponent (see [4]). Thus, we can expect that the critical exponent of (1.1) is also given by p_F if μ is sufficiently large.

For the damped wave equation with constant coefficient

$$u_{tt} - \Delta u + u_t = |u|^p,$$

Todorova and Yordanov [13] proved that the critical exponent is given by p_F. Lin, Nishihara and Zhai [8] (see also Nishihara [12]) extended this result to time-dependent coefficient cases

$$u_{tt} - \Delta u + (1+t)^{-\beta}u_t = |u|^p$$

with $-1 < \beta < 1$. They proved that p_F is still critical. Recently, D'Abbicco, Lucente and Reissig [3] extended this result to more general effective damping by using the linear decay estimates, which have been established by Wirth [17]. For the scale-invariant case (1.1), very recently, D'Abbicco [1] proved the existence of the global solution with small data for (1.1) in the case $n = 1, 2, \mu \geq n+2$ and $p_F < p$. He also obtained the decay rates of the solution without any loss ε.

Our main result is the following:

Theorem 1.2. *Let $p_F < p \leq n/(n-2)$ $(n \geq 3)$, $p_F < p < \infty$ $(n = 1, 2)$ and $0 < \varepsilon < 2n(p - p_F)/(p-1)$. Then there exist constants $\delta > 0$ and $\mu_0 > 1$ having the following property: if $\mu \geq \mu_0$ and*

$$I_0^2 := \int_{\mathbf{R}^n} e^{2\psi(0,x)}(u_0^2 + |\nabla u_0|^2 + u_1^2)dx$$

is sufficiently small, then there exists a unique solution $u \in C([0,\infty); H^1(\mathbf{R}^n)) \cap C^1([0,\infty); L^2(\mathbf{R}^n))$ of (1.1) satisfying

$$\int_{\mathbf{R}^n} e^{2\psi} u^2 dx \leq C_{\mu,\varepsilon} I_0^2 (1+t)^{-n+\varepsilon}, \tag{1.8}$$

$$\int_{\mathbf{R}^n} e^{2\psi}(u_t^2 + |\nabla u|^2)dx \leq C_{\mu,\varepsilon} I_0^2 (1+t)^{-n-2+\varepsilon} \tag{1.9}$$

for $t \geq 0$, where $C_{\mu,\varepsilon}$ is a positive constant depending on μ and ε.

Remark 1.2. Similarly as before, we note that μ_0 depends on ε. The relation is

$$\mu_0 \sim \varepsilon^{-2} \sim (p - p_F)^{-2}.$$

Therefore, as p is closer to p_F, μ_0 must be larger. Thus, we can expect that ε can be removed and the same result holds for much smaller μ. As mentioned above, D'Abbicco [1] obtained an affirmative result for this expectation when $n = 1, 2$. However, we have no idea for the higher-dimensional cases.

We prove Theorem 1.2 by a multiplier method which was essentially developed in [13]. Lin, Nishihara and Zhai [8] refined this method to fit the damping term $b(t) = (1+t)^{-\beta}$ with $-1 < \beta < 1$. They used the property $\beta < 1$ and so we cannot apply their method directly to our problem (1.1). Therefore, we need a further modification. Instead of the property $\beta < 1$, we assume that μ is sufficiently large and modify the parameters used in the calculation.

Remark 1.3. We can also treat other nonlinear terms, for example $-|u|^p, |u|^{p-1}u$.

We also have a blow-up result when $\mu > 1$ and $1 < p \leq p_F$. Following [5], we first introduce the definition of a weak solution.

Definition 1.3. Let $u_0, u_1 \in L^1_{\mathrm{loc}}(\mathbf{R}^n)$ and $T > 0$. We say that u is a local weak solution of the Cauchy problem (1.1) on $[0,T)$ if $u \in L^p_{\mathrm{loc}}([0,T) \times \mathbf{R}^n)$ and

$$\begin{aligned}\int_{[0,T)\times\mathbf{R}^n} & u(t,x)\left(\partial_t^2\psi(t,x) - \Delta\psi(t,x) - \partial_t\left(\frac{\mu}{1+t}\psi(t,x)\right)\right)dxdt \\ &= \int_{\mathbf{R}^n}(\mu\psi(0,x) - \partial_t\psi(0,x))u_0(x) + \psi(0,x)u_1(x)dx \\ &+ \int_{[0,T)\times\mathbf{R}^n}|u(t,x)|^p\psi(t,x)dxdt\end{aligned} \tag{1.10}$$

is valid for any $\psi \in C_0^2([0,T) \times \mathbf{R}^n)$. If T can be chosen as any positive number, we call u a global weak solution of (1.1).

We note that any local solution $u \in C([0,T); H^1(\mathbf{R}^n)) \cap C^1([0,T); L^2(\mathbf{R}^n))$ of (1.1) with initial data $(u_0, u_1) \in H^1(\mathbf{R}^n) \times L^2(\mathbf{R}^n)$ is also a weak solution of (1.1). This can be easily proved from the construction of the local solution (see, e.g., [6]).

Theorem 1.4. *Let $u_0, u_1 \in L^1_{\mathrm{loc}}(\mathbf{R}^n)$, $1 < p \le p_F$ and $\mu > 1$. Moreover, we assume that*

$$(\mu - 1)u_0 + u_1 \in L^1(\mathbf{R}^n) \quad \text{and} \quad \int_{\mathbf{R}^n}(\mu - 1)u_0 + u_1 dx > 0.$$

Then there is no global weak solution for (1.1).

Remark 1.4. Theorem 1.3 is essentially included in a recent work by D'Abbicco and Lucente [2]. In this paper we shall give a much simpler proof.

One of our novelty is blow-up results for non-effective damping cases. We also obtain blow-up results in the case $0 < \mu \le 1$.

Theorem 1.5. *Let $0 < \mu \le 1$ and*

$$1 < p \le 1 + \frac{2}{n + (\mu - 1)}.$$

We also assume $u_0 \in L^1_{\mathrm{loc}}(\mathbf{R}^n)$ and

$$u_1 \in L^1(\mathbf{R}^n), \quad \int_{\mathbf{R}^n} u_1(x)dx > 0.$$

Then there is no global weak solution for (1.1).

Remark 1.5. In Theorem 1.5, we do not put any assumption on the data u_0, and the blow-up result even holds for some $p \ge p_F$. We can interpret this phenomena as that the equation (1.1) loses the parabolic structure and recover the hyperbolic structure if μ is sufficiently small.

We prove this theorem by a test-function method developed by Qi S. Zhang [18]. In the same way of the proof of Theorem 1.5, we can treat the damping terms $(1+t)^{-\beta}$ with $\beta > 1$ (see Remark 3.1).

In the next section, we give a proof of Theorem 1.2. We can prove Theorem 1.1 by the almost same way, and so we omit the proof. In Section 3, we shall prove Theorems 1.4 and 1.5.

2. Proof of Theorem 1.2

In this section we prove our main result. First, we prepare some notation and terminology. We put

$$\|f\|_{L^p(\mathbf{R}^n)} := \left(\int_{\mathbf{R}^n} |f(x)|^p dx\right)^{1/p}.$$

By $H^1(\mathbf{R}^n)$ we denote the usual Sobolev space. For an interval I and a Banach space X, we define $C^r(I;X)$ as the Banach space whose element is an r-times continuously differentiable mapping from I to X with respect to the topology in X. The letter C indicates the generic constant, which may change from a line to the next line. We also use the symbols $\lesssim$ and $\sim$. The relation $f \lesssim g$ means $f \le Cg$ with some constant $C>0$ and $f \sim g$ means $f \lesssim g$ and $g \lesssim f$.

We first describe the local existence:

Proposition 2.1. *Let $1 < p \le n/(n-2)$ $(n \ge 3), 1 < p < \infty$ $(n = 1,2), \mu > 0$ and $\delta > 0$. Then there exists $T_m \in (0,+\infty]$ depending on I_0^2 such that the Cauchy problem* (1.1) *has a unique solution $u \in C([0,T_m);H^1(\mathbf{R}^n)) \cap C^1([0,T_m);L^2(\mathbf{R}^n))$ satisfying*

$$\sup_{0\le t\le T} \int_{\mathbf{R}^n} e^{2\psi(t,x)}(u^2 + |\nabla u|^2 + u_t^2)dx < +\infty$$

for any $T \in (0,T_m)$. Moreover, if $T_m < +\infty$, then we have

$$\liminf_{t\to T_m} \int_{\mathbf{R}^n} e^{2\psi(t,x)}(u^2 + |\nabla u|^2 + u_t^2)dx = +\infty.$$

We can prove this proposition by standard arguments (see [6]). We prove the a priori estimate for the following functional:

$$M(t) := \sup_{0\le\tau\le t} \left\{(1+\tau)^{n+2-\varepsilon} \int_{\mathbf{R}^n} e^{2\psi}(u_t^2 + |\nabla u|^2)dx + (1+\tau)^{n-\varepsilon} \int_{\mathbf{R}^n} e^{2\psi}u^2 dx\right\}.$$

We shall prove the following a priori estimate for $M(t)$:

Proposition 2.2. *Let $p_F < p \le (n+2)/(n-2)$ $(n \ge 3)$, $p_F < p < \infty$ $(n = 1,2)$ and $0 < \varepsilon < 2n(p-p_F)/(p-1)$. Then there exist constants $\delta > 0$ and $\mu_0 > 1$ having the following property: if $\mu \ge \mu_0$, (u_0,u_1) satisfies $I_0^2 < +\infty$ and $u \in C([0,T];H^1(\mathbf{R}^n)) \cap C^1([0,T];L^2(\mathbf{R}^n))$ is a solution of the Cauchy problem* (1.1) *for some $T > 0$ such that*

$$\sup_{0\le t\le T} \int_{\mathbf{R}^n} e^{2\psi(t,x)}(u^2 + |\nabla u|^2 + u_t^2)dx < +\infty,$$

then it follows that

$$M(t) \le CI_0^2 + CM(t)^{p+1},$$

where $C = C(p,\varepsilon,\mu)$ *is a positive constant depending on* p,ε,μ.

Proof. We put $b(t) = \frac{\mu}{1+t}$ and $f(u) = |u|^p$. By a simple calculation, we have

$$-\psi_t = \frac{2}{1+t}\psi, \quad \nabla\psi = \frac{2ax}{(1+t)^2}, \quad \frac{|\nabla\psi|^2}{-\psi_t} = \frac{b(t)}{2+\delta},$$

and

$$\Delta\psi = \frac{n}{2+\delta}\frac{b(t)}{1+t} =: \left(\frac{n}{2}-\delta_1\right)\frac{b(t)}{1+t}.$$

Here and after, δ_i $(i = 1,2,\dots)$ denote a positive constant depending only on δ such that

$$\delta_i \to 0^+ \quad \text{as} \quad \delta \to 0^+.$$

Multiplying (1.1) by $e^{2\psi}u_t$, we obtain

$$\begin{aligned}
&\frac{\partial}{\partial t}\left[\frac{e^{2\psi}}{2}(u_t^2+|\nabla u|^2)\right] - \nabla\cdot(e^{2\psi}u_t\nabla u)\\
&\quad + e^{2\psi}\left(b(t) - \frac{|\nabla\psi|^2}{-\psi_t} - \psi_t\right)u_t^2 + \underbrace{\frac{e^{2\psi}}{-\psi_t}|\psi_t\nabla u - u_t\nabla\psi|^2}_{T_1} \qquad (2.1)\\
&= \frac{\partial}{\partial t}\left[e^{2\psi}F(u)\right] + 2e^{2\psi}(-\psi_t)F(u),
\end{aligned}$$

where F is the primitive of f satisfying $F(0) = 0$. Using the Schwarz inequality, we can calculate

$$T_1 \ge e^{2\psi}\left(\frac{1}{5}(-\psi_t)|\nabla u|^2 - \frac{b(t)}{4(2+\delta)}u_t^2\right).$$

From this and integrating (2.1), we have

$$\begin{aligned}
&\frac{d}{dt}\int_{\mathbf{R}^n}\frac{e^{2\psi}}{2}(u_t^2+|\nabla u|^2)dx + \int_{\mathbf{R}^n}e^{2\psi}\left\{\left(\frac{b(t)}{4}-\psi_t\right)u_t^2 + \frac{-\psi_t}{5}|\nabla u|^2\right\}dx\\
&\qquad \le \frac{d}{dt}\int_{\mathbf{R}^n}e^{2\psi}F(u)dx + 2\int_{\mathbf{R}^n}e^{2\psi}(-\psi_t)F(u)dx. \qquad (2.2)
\end{aligned}$$

Noting that $e^{2\psi(t,x)}u_t(t,x)\nabla u(t,x) \in L^1(\mathbf{R}^n)$, we can use the divergence theorem and (2.2) is valid. On the other hand, by multiplying (1.1) by $e^{2\psi}u$, it follows that

$$\begin{aligned}
&\frac{\partial}{\partial t}\left[e^{2\psi}\left(uu_t + \frac{b(t)}{2}u^2\right)\right] - \nabla\cdot(e^{2\psi}u\nabla u)\\
&\quad + e^{2\psi}\left\{|\nabla u|^2 + \left(-\psi_t + \frac{1}{2(1+t)}\right)b(t)u^2 + \underbrace{2u\nabla\psi\cdot\nabla u}_{T_2} - 2\psi_t uu_t - u_t^2\right\}\\
&= e^{2\psi}uf(u).
\end{aligned}$$

We calculate

$$e^{2\psi}T_2 = 4e^{2\psi}u\nabla\psi\cdot\nabla u - 2e^{2\psi}u\nabla\psi\cdot\nabla u$$
$$= 4e^{2\psi}u\nabla\psi\cdot\nabla u - \nabla\cdot(e^{2\psi}u^2\nabla\psi) + 2e^{2\psi}u^2|\nabla\psi|^2 + e^{2\psi}(\Delta\psi)u^2$$

and have

$$\begin{aligned}
&\frac{\partial}{\partial t}\left[e^{2\psi}\left(uu_t + \frac{b(t)}{2}u^2\right)\right] - \nabla\cdot(e^{2\psi}(u\nabla u + u^2\nabla\psi))\\
&\quad + e^{2\psi}\Big\{\underbrace{|\nabla u|^2 + 4u\nabla u\cdot\nabla\psi + ((-\psi_t)b(t) + 2|\nabla\psi|^2)u^2}_{T_3}\\
&\quad + (n+1-2\delta_1)\frac{b(t)}{2(1+t)}u^2 - 2\psi_t uu_t - u_t^2\Big\} = e^{2\psi}uf(u).
\end{aligned} \tag{2.3}$$

The term T_3 is estimated as follows:

$$\begin{aligned}
T_3 &= \left(1 - \frac{4}{4+\delta/2}\right)|\nabla u|^2 + \frac{\delta}{2}|\nabla\psi|^2u^2 + \left|\frac{2}{\sqrt{4+\delta/2}}\nabla u + \sqrt{4+\delta/2}\nabla\psi\right|^2\\
&\geq \delta_2(|\nabla u|^2 + b(t)(-\psi_t)u^2).
\end{aligned}$$

Thus, we can rewrite (2.3) as

$$\begin{aligned}
&\frac{\partial}{\partial t}\left[e^{2\psi}\left(uu_t + \frac{b(t)}{2}u^2\right)\right] - \nabla\cdot(e^{2\psi}(u\nabla u + u^2\nabla\psi))\\
&\quad + e^{2\psi}\left\{\delta_2(|\nabla u|^2 + b(t)(-\psi_t)u^2) + (n+1-2\delta_1)\frac{b(t)}{2(1+t)}u^2 - 2\psi_t uu_t - u_t^2\right\}\\
&\leq e^{2\psi}uf(u).
\end{aligned}$$

Integrating the above inequality and then multiplying by a large parameter ν and adding $(1+t)\times$ (2.2), we obtain

$$\begin{aligned}
&\frac{d}{dt}\left[\int_{\mathbf{R}^n} e^{2\psi}\left\{\frac{1+t}{2}(u_t^2 + |\nabla u|^2) + \nu uu_t + \frac{\nu b(t)}{2}u^2\right\}dx\right]\\
&\quad + \int_{\mathbf{R}^n} e^{2\psi}\Big\{\Big(\underbrace{\frac{\mu}{4} - \nu - \frac{1}{2}}_{T_4} + (-\psi_t)(1+t)\Big)u_t^2 + \Big(\underbrace{\nu\delta_2 - \frac{1}{2}}_{T_5} + \frac{(-\psi_t)(1+t)}{5}\Big)|\nabla u|^2\\
&\quad + \nu\delta_2 b(t)(-\psi_t)u^2 + (n+1-2\delta_1)\frac{\nu b(t)}{2(1+t)}u^2 + \underbrace{2\nu(-\psi_t)uu_t}_{T_6}\Big\}dx\\
&\leq \frac{d}{dt}\left[(1+t)\int_{\mathbf{R}^n} e^{2\psi}F(u)dx\right] + C\int_{\mathbf{R}^n} e^{2\psi}(1 + (1+t)(-\psi_t))|u|^{p+1}dx.
\end{aligned}$$

We put the condition for μ and ν as

$$\frac{\mu}{4} - \nu - \frac{1}{2} > 0, \tag{2.4}$$

$$\nu\delta_2 - \frac{1}{2} > 0. \tag{2.5}$$

Then the terms T_4 and T_5 are positive. Using the Schwarz inequality, we can estimate T_6 as

$$|T_6| \le \frac{1}{2}(-\psi_t)(1+t)u_t^2 + \frac{2\nu^2}{1+t}(-\psi_t)u^2.$$

Now we put another condition

$$\mu \ge \frac{2\nu}{\delta_2}. \tag{2.6}$$

Then we obtain the following estimate:

$$\begin{aligned} &\frac{d}{dt}\hat{E}(t) + H(t) + (n+1-2\delta_1)\frac{\nu b(t)}{2(1+t)}J(t;u^2) \\ &\le \frac{d}{dt}[(1+t)J(t;F(u))] + C(J(t;|u|^{p+1}) + (1+t)J_\psi(t;|u|^{p+1})), \end{aligned} \tag{2.7}$$

where

$$\begin{aligned} \hat{E}(t) &:= \int_{\mathbf{R}^n} e^{2\psi}\left\{\frac{1+t}{2}(u_t^2 + |\nabla u|^2) + \nu u u_t + \frac{\nu b(t)}{2}u^2\right\}dx, \\ H(t) &= \int_{\mathbf{R}^n} e^{2\psi}\left\{\left(\frac{\mu}{4} - \nu - \frac{1}{2}\right)u_t^2 + \left(\nu\delta_2 - \frac{1}{2}\right)|\nabla u|^2\right\}dx, \\ J(t;u) &= \int_{\mathbf{R}^n} e^{2\psi}u\,dx, \qquad J_\psi(t;u) = \int_{\mathbf{R}^n} e^{2\psi}(-\psi_t)u\,dx. \end{aligned}$$

Multiplying (2.7) by $(1+t)^{n+1-\varepsilon}$, we have

$$\begin{aligned} &\frac{d}{dt}[(1+t)^{n+1-\varepsilon}\hat{E}(t)] - \underbrace{(n+1-\varepsilon)(1+t)^{n-\varepsilon}\hat{E}(t)}_{T_7} \\ &\quad + (1+t)^{n+1-\varepsilon}H(t) + (n+1-2\delta_1)(1+t)^{n+1-\varepsilon}\frac{\nu b(t)}{2(1+t)}J(t;u^2) \\ &\le \frac{d}{dt}[(1+t)^{n+2-\varepsilon}J(t;F(u))] \\ &\quad + C((1+t)^{n+1-\varepsilon}(J(t;|u|^{p+1}) + (1+t)^{n+2-\varepsilon}J_\psi(t;|u|^{p+1})). \end{aligned}$$

Now we estimate the bad term T_7. First, by the Schwarz inequality, one can obtain

$$|\nu u u_t| \le \frac{\nu}{4\delta_3 b(t)}u_t^2 + \delta_3\nu b(t)u^2,$$

where δ_3 is determined later. From this, T_7 is estimated as

$$\begin{aligned} T_7 &\le (n+1-\varepsilon)(1+t)^{n-\varepsilon} \\ &\times \int_{\mathbf{R}^n} e^{2\psi}\left\{\left(\frac{1+t}{2} + \frac{\nu(1+t)}{4\delta_3\mu}\right)u_t^2 + \frac{1+t}{2}|\nabla u|^2 + \frac{\nu b(t)}{2}(1+2\delta_3)u^2\right\}dx. \end{aligned}$$

We strengthen the conditions (2.4) and (2.5) as

$$\frac{\mu}{4} - \nu - \frac{1}{2} - (n+1-\varepsilon)\left(\frac{1}{2} + \frac{\nu}{4\delta_3\mu}\right) > 0, \tag{2.8}$$

$$\nu\delta_2 - \frac{1}{2}(n+2-\varepsilon) > 0. \tag{2.9}$$

Moreover, we take $\varepsilon = 3\delta_1$ and then choose δ_3 such that

$$(n+1-2\delta_1) - (n+1-3\delta_1)(1+2\delta_3) > 0.$$

Under these conditions, we can estimate T_7 and obtain

$$\begin{aligned}\frac{d}{dt}[(1+t)^{n+1-\varepsilon}\hat{E}(t)] \le \frac{d}{dt}[(1+t)^{n+2-\varepsilon}J(t;F(u))] \\ + C(1+t)^{n+1-\varepsilon}(J(t;|u|^{p+1}) + (1+t)^{n+2-\varepsilon}J_\psi(t;|u|^{p+1})).\end{aligned}$$

By integrating the above inequality, it follows that

$$\begin{aligned}(1+t)^{n+1-\varepsilon}\hat{E}(t) \le CI_0^2 + (1+t)^{n+2-\varepsilon}J(t;|u|^{p+1}) \\ + C\int_0^t (1+\tau)^{n+1-\varepsilon}(J(\tau;|u|^{p+1}) + (1+\tau)^{n+2-\varepsilon}J_\psi(\tau;|u|^{p+1})d\tau.\end{aligned}$$

By a simple calculation, we have

$$(1+t)E(t) + \frac{1}{1+t}J(t;u^2) \le C\hat{E}(t),$$

where

$$E(t) := \int_{\mathbf{R}^n} e^{2\psi}(u_t^2 + |\nabla u|^2)dx.$$

Thus, we obtain

$$\begin{aligned}&(1+t)^{n+2-\varepsilon}E(t) + (1+t)^{n-\varepsilon}J(t;u^2) \\ &\le CI_0^2 + (1+t)^{n+2-\varepsilon}J(t;|u|^{p+1}) \\ &\quad + C\int_0^t (1+\tau)^{n+1-\varepsilon}J(\tau;|u|^{p+1}) + (1+\tau)^{n+2-\varepsilon}J_\psi(\tau;|u|^{p+1})d\tau.\end{aligned} \tag{2.10}$$

Now we turn to estimate the nonlinear terms. We need the following lemma:

Lemma 2.3 (Gagliardo–Nirenberg). *Let* $1 < p \le (n+2)/(n-2)$ $(n \ge 3)$, $1 < p < \infty$ $(n = 1, 2)$ *and*

$$\sigma = \frac{n(p-1)}{2(p+1)} \in [0, 1].$$

Then for some constant $C = C(p, n) > 0$, *the inequality*

$$\|h\|_{L^{p+1}} \le C\|h\|_{L^2}^{1-\sigma}\|\nabla h\|_{L^2}^{\sigma} \quad \textit{for any} \quad h \in H^1(\mathbf{R}^n)$$

holds.

Noting that

$$J(t;|u|^{p+1}) = \int_{\mathbf{R}^n} \left|e^{\frac{2}{p+1}\psi}u\right|^{p+1} dx$$

and $\nabla(e^{\frac{2}{p+1}\psi}u) = \frac{2}{p+1}e^{\frac{2}{p+1}\psi}(\nabla\psi)u + e^{\frac{2}{p+1}\psi}\nabla u$, we apply the above lemma to $J(t;|u|^{p+1})$ and have

$$J(t;|u|^{p+1}) \le C\left(\int_{\mathbf{R}^n} e^{\frac{4}{p+1}\psi}u^2dx\right)^{\frac{1-\sigma}{2}(p+1)} \times\left(\int_{\mathbf{R}^n} e^{\frac{4}{p+1}\psi}|\nabla\psi|^2u^2dx + \int_{\mathbf{R}^n} e^{\frac{4}{p+1}\psi}|\nabla u|^2dx\right)^{\frac{\sigma}{2}(p+1)}.$$

We note that

$$e^{\frac{4}{p+1}\psi}|\nabla\psi|^2 = \frac{4a^2|x|^2}{(1+t)^4}e^{\frac{4}{p+1}\psi} \le C\frac{1}{(1+t)^2}e^{2\psi}$$

and obtain

$$J(t;|u|^{p+1}) \le C\left(\int_{\mathbf{R}^n} e^{2\psi}u^2dx\right)^{\frac{1-\sigma}{2}(p+1)} \times\left(\frac{1}{(1+t)^2}\int_{\mathbf{R}^n} e^{2\psi}u^2dx + \int_{\mathbf{R}^n} e^{2\psi}|\nabla u|^2dx\right)^{\frac{\sigma}{2}(p+1)}.$$

Therefore, we can estimate

$$(1+t)^{n+2-\varepsilon}J(t;|u|^{p+1}) \le (1+t)^{n+2-\varepsilon}\{(1+t)^{-(n-\varepsilon)}M(t)\}^{\frac{1-\sigma}{2}(p+1)} \times \{(1+t)^{-(n+2-\varepsilon)}M(t)\}^{\frac{\sigma}{2}(p+1)}.$$

By a simple calculation, if

$$\varepsilon < \frac{2n\left(p-\left(1+\frac{2}{n}\right)\right)}{p-1} \tag{2.11}$$

then we have

$$(1+t)^{n+2-\varepsilon}J(t;|u|^{p+1}) \le CM(t)^{p+1}.$$

In a similar way, we can estimate the other nonlinear terms. We note that the conditions (2.6), (2.8), (2.9), (2.11) are fulfilled by the determination of the parameters in the order

$$p \to \varepsilon \to \delta \to \nu \to \mu.$$

Consequently, we obtain the a priori estimate

$$M(t) \le CI_0^2 + CM(t)^{p+1}. \tag{2.12}$$

This proves Proposition 2.2. □

Proof of Theorem 1.2. In (2.12), we may assume that $C \ge 1$. We take I_0 sufficiently small so that it holds that

$$2CI_0^2 > CI_0^2 + C(2CI_0^2)^{p+1},$$

that is,

$$1 > 2C(2CI_0^2)^p.$$

Let M_1, M_2 $(M_1 < M_2)$ be the positive roots of the identity

$$M = CI_0^2 + CM^{p+1}.$$

By the continuity of $M(t)$ and

$$M(0) = I_0^2 \leq CI_0^2 < M_1,$$

we obtain

$$M(t) \leq M_1$$

for all $t \geq 0$. Thus, $M(t)$ is bounded and we have $T_m = +\infty$, where T_m is the maximal existence time of the local solution as in Proposition 2.1. The decay estimates (1.8) and (1.9) can be obtained immediately from the boundedness of $M(t)$. This completes the proof. □

3. Proof of Theorems 1.4 and 1.5

In this section we first give a proof of Theorem 1.4. We use a method by Lin, Nishihara and Zhai [8] to transform (1.1) into divergence form and then a test-function method by Qi S. Zhang [18].

Let $\mu > 1$. We multiply (1.1) by a positive function $g(t) \in C^2([0,\infty))$ and obtain

$$(gu)_{tt} - \Delta(gu) - (g'u)_t + (-g' + gb)u_t = g|u|^p.$$

We now choose $g(t)$ as the solution of the initial value problem for the ordinary differential equation

$$\begin{cases} -g'(t) + g(t)b(t) = 1, & t > 0, \\ g(0) = \frac{1}{\mu - 1}. \end{cases} \tag{3.1}$$

The solution $g(t)$ is explicitly given by

$$g(t) = \frac{1}{\mu - 1}(1+t).$$

Thus, we obtain the equation in divergence form

$$(gu)_{tt} - \Delta(gu) - (g'u)_t + u_t = g|u|^p. \tag{3.2}$$

Next, we apply a test function method. We first introduce test functions having the following properties:

$$\eta(t) \in C_0^\infty([0,\infty)), \quad 0 \leq \eta(t) \leq 1, \quad \eta(t) = \begin{cases} 1, & 0 \leq t \leq \frac{1}{2}, \\ 0, & t \geq 1, \end{cases} \tag{t1}$$

$$\phi(x) \in C_0^\infty(\mathbf{R}^n), \quad 0 \leq \phi(x) \leq 1, \quad \phi(x) = \begin{cases} 1, & |x| \leq \frac{1}{2}, \\ 0, & |x| \geq 1, \end{cases} \tag{t2}$$

$$\frac{\eta'(t)^2}{\eta(t)} \leq C \quad \left(\frac{1}{2} \leq t \leq 1\right), \quad \frac{|\nabla\phi(x)|^2}{\phi(x)} \leq C \quad \left(\frac{1}{2} \leq |x| \leq 1\right). \tag{t3}$$

Let R be a large parameter in $(0,\infty)$. We define the test function

$$\psi_R(t,x) := \eta_R(t)\phi_R(x) := \eta\left(\frac{t}{R}\right)\phi\left(\frac{x}{R}\right).$$

Let q be the dual of p, that is $q = \frac{p}{p-1}$. Suppose that u is a global weak solution of (1.1) with initial data (u_0, u_1) satisfying

$$(\mu - 1)u_0 + u_1 \in L^1(\mathbf{R}^n), \quad \int_{\mathbf{R}^n} ((\mu - 1)u_0 + u_1)dx > 0.$$

We define

$$I_R := \int_{Q_R} g(t)|u(t,x)|^p \psi_R(t,x)^q dxdt,$$

where $Q_R := [0, R] \times B_R$ and $B_R := \{x \in \mathbf{R}^n; |x| \le R\}$. According to the idea of the transformation the equation into divergence form (3.2), in the definition of the weak solution (1.10), we substitute $g(t)\psi_R(t,x)^q$ into $\psi(t,x)$ and obtain

$$\begin{aligned} I_R = & -\frac{1}{\mu - 1}\int_{B_R} ((\mu - 1)u_0 + u_1)\phi_R^q dx \\ & + \int_{Q_R} gu\partial_t^2(\psi_R^q)dxdt + \int_{Q_R} (g'u - u)\partial_t(\psi_R^q)dxdt - \int_{Q_R} gu\Delta(\psi_R^q)dxdt \\ =: & -\frac{1}{\mu - 1}\int_{B_R} ((\mu - 1)u_0 + u_1)\phi_R^q dx + J_1 + J_2 + J_3. \end{aligned}$$

By the assumption on the data (u_0, u_1) it follows that

$$I_R < J_1 + J_2 + J_3$$

for large R. We shall estimate J_1, J_2 and J_3, respectively. We use the notation

$$\hat{Q}_R := [R/2, R] \times B_R, \quad \tilde{Q}_R := [0, R] \times (B_R \setminus B_{R/2}).$$

We first estimate J_3. By the conditions (t1)–(t3) and the Hölder inequality we have the following estimate:

$$\begin{aligned} |J_3| & \lesssim R^{-2}\int_{\tilde{Q}_R} g(t)|u|\psi_R^{q-1}dxdt \\ & \lesssim R^{-2}\left(\int_{\tilde{Q}_R} g(t)|u|^p\psi_R^q(t,x)dxdt\right)^{1/p}\left(\int_{\tilde{Q}_R} g(t)dxdt\right)^{1/q} \\ & \lesssim \tilde{I}_R^{1/p} R^{\frac{n+2}{q}-2}, \end{aligned}$$

where

$$\tilde{I}_R := \int_{\tilde{Q}_R} g(t)|u|^p\psi_R^q(t,x)dxdt.$$

In a similar way, we can estimate J_1 and J_2 as

$$|J_1| + |J_2| \lesssim \hat{I}_R^{1/p} R^{\frac{n+2}{q}-2}, \quad \hat{I}_R := \int_{\hat{Q}_R} g(t)|u|^p\psi_R^q(t,x)dxdt.$$

Hence, we obtain

$$I_R \lesssim (\tilde{I}_R^{1/p} + \hat{I}_R^{1/p})R^{\frac{n+2}{q}-2}, \tag{3.3}$$

in particular $I_R^{1-1/p} \lesssim R^{\frac{n+2}{q}-2}$. If $1 < p < p_F$, by letting $R \to \infty$ we have $I_R \to 0$ and, hence $u \equiv 0$. Therefore, by the definition of the weak solution (1.10), we have

$$\int_{\mathbf{R}^n} ((\mu - 1)u_0 + u_1)\phi_R(x)dx = 0$$

for any $R > 0$, which contradicts the assumption on the data. If $p = p_F$, we have only $I_R \le C$ with some constant C independent of R. This implies that $g(t)|u|^p$ is integrable on $(0, \infty) \times \mathbf{R}^n$ and, hence

$$\lim_{R\to\infty} (\tilde{I}_R + \hat{I}_R) = 0.$$

By (3.3), we obtain $\lim_{R\to\infty} I_R = 0$. Therefore, u must be 0. This also leads a contradiction.

Proof of Theorem 1.5. The proof is almost the same as above. Let $0 < \mu \le 1$. Instead of (3.1), we consider the ordinary differential equation

$$-g'(t) + g(t)b(t) = 0 \tag{3.4}$$

with $g(0) > 0$. We can easily solve this and have

$$g(t) = g(0)(1 + t)^{\mu}.$$

Then we have

$$(gu)_{tt} - \Delta(gu) - (g'u)_t = g|u|^p. \tag{3.5}$$

Using the same test function $\psi_R(t, x)$ as above, we can calculate

$$I_R := \int_{Q_R} g(t)|u|^p\psi_R^q dxdt = -\int_{B_R} g(0)u_1\phi_R^q dx + \sum_{k=1}^{3} J_k,$$

where

$$J_1 = \int_{Q_R} gu\partial_t^2(\psi_R^q)dxdt, \quad J_2 = \int_{Q_R} g'u\partial_t(\psi_R^q)dxdt, \quad J_3 = -\int_{Q_R} gu\Delta(\psi_R^q)dxdt.$$

We note that the term of u_0 vanishes and so we put the assumption only for u_1. We first estimate J_2. Noting $g'(t) = \mu g(0)(1 + t)^{\mu-1}$, we have

$$|J_2| \lesssim \frac{1}{R}\int_{\hat{Q}_R} (1 + t)^{\mu-1}|u|\psi_R^{q-1}dxdt.$$

By noting that $(1+t)^{\mu-1} \sim g(t)^{1/p}(1+t)^{\mu/q-1}$ and the Hölder inequality, it follows that

$$\begin{aligned}
|J_2| &\lesssim \frac{1}{R}\left(\int_{\hat{Q}_R} g|u|^p\psi_R^q dxdt\right)^{1/p}\left(\int_{R/2}^{R}\int_{B_R} (1 + t)^{\mu-q}dxdt\right)^{1/q} \\
&\lesssim \frac{1}{R}\hat{I}_R^{1/p}(1 + R)^{(\mu-q)/q}\left(\int_{R/2}^{R}\int_{B_R} dxdt\right)^{1/q} \lesssim \hat{I}_R^{1/p}R^{-1+(n+1+\mu-q))/q},
\end{aligned}$$

where $\hat{I}_R$ is defined as before. A simple calculation shows

$$-1 + (n + 1 + \mu - q)/q \leq 0 \Leftrightarrow p \leq 1 + \frac{2}{n + (\mu - 1)}.$$

In the same way, we can estimate J_1 and J_3 as

$$|J_1| + |J_3| \lesssim (\hat{I}_R^{1/p} + \tilde{I}_R^{1/p}) R^{-2+(n+\mu+1)/q},$$

where $\tilde{I}_R$ is same as before. It is also easy to see that

$$-2 + (n + \mu + 1)/q \leq 0 \Leftrightarrow p \leq 1 + \frac{2}{n + (\mu - 1)}.$$

Finally, we have

$$I_R \lesssim \tilde{I}_R^{1/p} + \hat{I}_R^{1/p}$$

if $p \leq 1 + 2/(n + (\mu - 1))$. The rest of the proof is same as before. □

Remark 3.1. We can apply the proof of Theorem 1.5 to the wave equation with non-effective damping terms

$$\begin{cases} u_{tt} - \Delta u + b(t)u_t = |u|^p, \\ (u, u_t)(0, x) = (u_0, u_1)(x), \end{cases}$$

where

$$b(t) = (1 + t)^{-\beta}$$

with $\beta > 1$. We can easily solve (3.1) and have

$$g(t) = g(0) \exp\left(\frac{1}{-\beta + 1}((1 + t)^{-\beta+1} - 1)\right).$$

We note that $g(t) \sim 1$. The same argument implies that if

$$1 < p \leq 1 + \frac{2}{n - 1}, \quad \int_{\mathbf{R}^n} u_1 dx > 0,$$

then there is no global solution. We note that the exponent $1 + 2/(n - 1)$ is greater than the Fujita exponent. This shows that when $\beta > 1$, the equation loses the parabolic structure even in the nonlinear cases. One can expect that the critical exponent is given by the well-known Strauss critical exponent. However, this problem is completely open due to the author's knowledge.

Acknowledgement

The author has generous support from Professors Tatsuo Nishitani and Kenji Nishihara. In particular, Prof. Nishihara gave him an essential idea for the proof of Theorem 1.4. He is also grateful to an anonymous referee for many useful suggestions and comments.

References

[1] M. D'Abbicco, *The threshold between effective and noneffective damping for semilinear wave equations*, arXiv:1211.0731v2.

[2] M. D'Abbicco, S. Lucente, *A modified test function method for damped wave equations*, arXiv:1211.0453v1.

[3] M. D'Abbicco, S. Lucente, M. Reissig, *Semi-Linear wave equations with effective damping*, Chin. Ann. Math., Ser. B **34** (2013), 345–380.

[4] H. Fujita, *On the blowing up of solutions of the Cauchy problem for* $u_t = \Delta u + u^{1+\alpha}$, J. Fac. Sci. Univ. Tokyo Sec. I, **13** (1966), 109–124.

[5] M. Ikeda, Y. Wakasugi, *Small data blow-up of* L^2*-solution for the nonlinear Schrödinger equation without gauge invariance*, Differential Integral Equations, **26** (2013), 1275–1285.

[6] R. Ikehata, K. Tanizawa, *Global existence of solutions for semilinear damped wave equations in* $\mathbf{R}^N$ *with noncompactly supported initial data*, Nonlinear Anal., **61** (2005), 1189–1208.

[7] R. Ikehata, G. Todorova, B. Yordanov, *Optimal decay rate of the energy for wave equations with critical potential*, J. Math. Soc. Japan **65** (2013), 183–236.

[8] J. Lin, K. Nishihara, J. Zhai, *Critical exponent for the semilinear wave equation with time-dependent damping*, Discrete Contin. Dyn. Syst., **32** (2012), 4307–4320.

[9] A. Matsumura, *On the asymptotic behavior of solutions of semi-linear wave equations*, Publ. Res. Inst. Math. Sci. Kyoto Univ., **12** (1976), 169–189.

[10] T. Narazaki, L^p-L^q *estimates for damped wave equations and their applications to semi-linear problem*, J. Math. Soc. Japan, **56** (2004), 585–626.

[11] K. Nishihara, L^p-L^q *estimates of solutions to the damped wave equation in 3-dimensional space and their application*, Math. Z., **244** (2003), 631–649.

[12] K. Nishihara, *Asymptotic behavior of solutions to the semilinear wave equation with time-dependent damping*, Tokyo J. Math., **34** (2011), 327–343.

[13] G. Todorova, B. Yordanov, *Critical exponent for a nonlinear wave equation with damping*, J. Differential Equations, **174** (2001), 464–489.

[14] J. Wirth, *Solution representations for a wave equation with weak dissipation*, Math. Meth. Appl. Sci., **27** (2004), 101–124.

[15] J. Wirth, *Asymptotic properties of solutions to wave equations with time-dependent dissipation*, PhD thesis, TU Bergakademie Freiberg, 2004.

[16] J. Wirth, *Wave equations with time-dependent dissipation I. Non-effective dissipation*, J. Differential Equations, **222** (2006), 487–514.

[17] J. Wirth, *Wave equations with time-dependent dissipation II. Effective dissipation*, J. Differential Equations **232** (2007), 74–103.

[18] Qi S. Zhang, *A blow-up result for a nonlinear wave equation with damping: the critical case*, C. R. Acad. Sci. Paris Sér. I Math., **333** (2001), 109–114.

Yuta Wakasugi
Department of Mathematics, Graduate School of Science
Osaka University, Toyonaka, Osaka, 560-0043, Japan
e-mail: y-wakasugi@cr.math.sci.osaka-u.ac.jp

Fourier Analysis
Trends in Mathematics, 391–415

Semilinear Hyperbolic Equations in Curved Spacetime

Karen Yagdjian

Abstract. This is a survey of the author's recent work rather than a broad survey of the literature. The survey is concerned with the global in time solutions of the Cauchy problem for matter waves propagating in the curved spacetimes, which can be, in particular, modeled by cosmological models. We examine the global in time solutions of some class of semililear hyperbolic equations, such as the Klein–Gordon equation, which includes the Higgs boson equation in the Minkowski spacetime, de Sitter spacetime, and Einstein & de Sitter spacetime. The crucial tool for the obtaining those results is a new approach suggested by the author based on the integral transform with the kernel containing the hypergeometric function.

Mathematics Subject Classification (2010). Primary 35L71,35L53; Secondary 81T20, 35C15.

Keywords. de Sitter spacetime; Klein–Gordon equation; global solutions; Huygens' principle; Higuchi bound.

1. Introduction

This survey is concerned with the global in time solutions of the Cauchy problem for matter waves propagating in the curved spacetimes, which can be, in particular, modeled by cosmological models. We are motivated by the significant importance of the qualitative description of the global solutions of the partial differential equations arising in the cosmological problems for understanding of the structure of the universe and fundamental particles physics. On the other hand, the physical implications of the mathematical results given here are out of the scope of this paper. More precisely, in this survey we examine the global in time solutions of some class of semililear hyperbolic equations, and, in particular, the Klein–Gordon equation, which includes the Higgs boson equation in the Minkowski spacetime, de Sitter spacetime, and Einstein & de Sitter spacetime. The Higgs boson plays a fundamental role in unified theories of weak, strong, and electromagnetic interactions [45].

The Klein–Gordon equation arising in relativistic physics and, in particular, general relativity and cosmology, as well as, in more recent quantum field theories, is a covariant equation that is considered in the curved pseudo-Riemannian manifolds. (See, e.g., Birrell and Davies [7], Parker and Toms [34], Weinberg [45].) The latest astronomical observational discovery that the expansion of the universe is speeding supports the model of the expanding universe that is mathematically described by the manifold with a metric tensor depending on time and spatial variables.

The homogeneous and isotropic cosmological models possess highest symmetry, which makes them more amenable to rigorous study. Among them, FLRW (Friedmann–Lemaître–Robertson–Walker) models are mentioned, which have the flat metric of the slices of constant time. The FLRW spacetime metric can be written in the form $ds^2 = -dt^2 + a^2(t)(dx^2 + dy^2 + dz^2)$, with an appropriate scale factor $a(t)$. (See, e.g., [26, 36, 43].) In particular, the metric in de Sitter spacetime in the Lemaître–Robertson coordinates [33], [43] has this form with the cosmic scale factor $a(t) = e^t$. The time dependence of the function $a(t)$ is determined by the Einstein field equations for gravity with the *cosmological constant* Λ,

$$R_{\mu\nu} - \frac{1}{2}g_{\mu\nu}R = 8\pi G T_{\mu\nu} - \Lambda g_{\mu\nu} .$$

The unknown of principal importance in the Einstein equations is a metric g. It comprises the basic geometrical feature of the gravitational field, and consequently explains the phenomenon of the mutual gravitational attraction of substance.

The metric of the Einstein & de Sitter universe (EdeS universe) is a particular member of the FLRW metrics

$$ds^2 = -dt^2 + a^2(t)\left[\frac{dr^2}{1-Kr^2} + r^2 d\Omega^2\right] ,$$

where $K = -1, 0$, or $+1$, for a hyperbolic, flat or spherical spatial geometry, respectively. The Einstein & de Sitter model of the universe is the simplest non-empty expanding model with the line-element

$$ds^2 = -dt^2 + a_0^2 t^{4/3}\left(dx^2 + dy^2 + dz^2\right)$$

in comoving coordinates [16]. It was first proposed jointly by Einstein & de Sitter (the EdeS model) [15]. The observations of the microwave radiation fit in with this model [14]. The result of this case also correctly describes the early epoch, even in a universe with curvature different from zero [9, Sec. 8.2]. Even though the EdeS spacetime is conformally flat, its causal structure is quite different from asymptotically flat geometries. In particular, and unlike Minkowski or Schwarzschild spacetimes, the past particle horizons exist. The EdeS spacetime is a good approximation to the large scale structure of the universe during the matter dominated phase, when the averaged (over space and time) energy density evolves adiabatically and pressures are vanishingly small, as, e.g., immediately after inflation. This justifies why such a metric is adopted to model the collapse of overdensity perturbations in the early matter dominated phase that followed inflation.

The matter waves in the spacetime are described by the function ϕ, which satisfies equations of motion. In the model of universe with curved spacetime the equation for the scalar field with potential function V is the covariant wave equation

$$\square_g \phi = V'(\phi) \quad \text{or} \quad \frac{1}{\sqrt{|g|}} \frac{\partial}{\partial x^i} \left(\sqrt{|g|} g^{ik} \frac{\partial \phi}{\partial x^k} \right) = V'(\phi) \,,$$

with the usual summation convention. Written explicitly in the Lemaître–Robertson coordinates in the de Sitter spacetime it, in particular, for

$$V'(\phi) = -\mu^2 \phi + \lambda |\phi|^{p-1} \phi, \quad p > 1,$$

has the form

$$\phi_{tt} + n\phi_t - e^{-2t} \Delta \phi = \mu^2 \phi - \lambda |\phi|^{p-1} \phi \,, \tag{1.1}$$

where $\mu > 0$ and $\lambda > 0$. Here Δ is the Laplace operator on the flat metric, $\Delta := \sum_{i=1}^{n} \frac{\partial^2}{\partial x_i^2}$. The equation for the Higgs real-valued scalar field in the de Sitter spacetime is a special case of (1.1) when $p = 3$, $n = 3$:

$$\phi_{tt} + 3\phi_t - e^{-2t} \Delta \phi = \mu^2 \phi - \lambda \phi^3 \,.$$

Scalar fields play a fundamental role in the standard model of particle physics, as well as its possible extensions. In particular, scalar fields generate spontaneous symmetry breaking and provide masses to gauge bosons and chiral fermions by the Brout–Englert–Higgs mechanism [18] using a Higgs-type potential [27].

In the spacetime with the constant metric tensor g the differential operator in the equation contains only the second-order derivatives. For the equation

$$\phi_{tt} - \Delta \phi = \mu^2 \phi - \lambda |\phi|^{p-1} \phi$$

the existence of a weak global solution in the energy space is known (see, e.g., Proposition 3.2 [23]) under certain conditions. The equation

$$\phi_{tt} - \Delta \phi = \mu^2 \phi - \lambda \phi^3 \tag{1.2}$$

for the Higgs scalar field in the Minkowski spacetime has the time-independent flat solution $\phi_N(x) = \frac{\mu}{\sqrt{\lambda}} \tanh\left(\frac{\mu}{\sqrt{2}} N \cdot (x - x_0) \right)$, $N, x_0, x \in \mathbb{R}^3$. The unit vector N defines the direction of the propagation of the wave front. This solution, after Lorentz transformation, gives rise to a traveling solitary wave of the form

$$\phi_{N,v}(x,t) = \frac{\mu}{\sqrt{\lambda}} \tanh\left(\frac{\mu}{\sqrt{2}} [N \cdot (x - x_0) \pm v(t - t_0)] \frac{1}{\sqrt{1 - v^2}} \right), \; N, x_0, x \in \mathbb{R}^3,$$

$t \geq t_0$, if $0 < v < 1$, where v is the initial velocity. The set of zeros of the solitary wave $\phi = \phi_{N,v}(x,t)$, that is, the set given by $N \cdot (x - x_0) \pm v(t - t_0) = 0$, is the moving boundary of the *wall*. The existence of standing waves $\phi = \exp(i\omega t) v(x)$, which are exponentially small at infinity $|x| = \infty$, and of corresponding solitary waves for the equation (1.2) with $\mu^2 < 0$ and $\lambda < 0$ is known (see, e.g., [41]).

The covariant linear wave equation in the Einstein & de Sitter spacetime written in the coordinates is

$$\left(\frac{\partial}{\partial t}\right)^2 \psi - t^{-4/3} \sum_{i=1,2,3} \left(\frac{\partial}{\partial x^i}\right)^2 \psi + \frac{2}{t}\frac{\partial}{\partial t}\psi = f\,. \tag{1.3}$$

In this survey we investigate the initial value problem for this equation and give the representation formulas for the fundamental solutions in the case of arbitrary dimension $n \in \mathbb{N}$ of the spatial variable $x \in \mathbb{R}^n$. The equation (1.3) is strictly hyperbolic in the domain with $t > 0$. On the surface $t = 0$ its coefficients have singularities that make the study of the initial value problem difficult. Then, the speed of propagation is $t^{-\frac{2}{3}}$ for every $t \in \mathbb{R} \setminus \{0\}$. The classical works on the Tricomi and Gellerstedt equations (see, e.g., [8, 10, 13, 46]) appeal to the singular Cauchy problem for the Euler–Poisson–Darboux equation, and to the Asgeirsson mean value theorem when handling a high-dimensional case.

Open Problem 1.1. Microlocal analysis (the representation of fundamental solutions via Fourier integral operators; the wave front sets of solutions) for the wave equation in the Einstein & de Sitter spacetime.

2. Method of investigation. Integral transform

We suggested in [48] a novel approach to study second-order hyperbolic equations with variable coefficients. That approach avoids explicit appeal to the Fourier integral operators, and it seems to be more immediate than the one that uses the Euler–Poisson–Darboux equation. It is used in a series of papers [48]–[56], [21] to investigate in a unified way several equations such as the linear and semilinear Tricomi and Tricomi-type equations, Gellerstedt equation, the wave equation in EdeS spacetime, the wave and the Klein–Gordon equations in the de Sitter and anti-de Sitter spacetimes. The listed equations play an important role in the gas dynamics, elementary particle physics, quantum field theory in curved spaces, and cosmology. For all above-mentioned equations, we have obtained among other things, fundamental solutions, representation formulas for the initial-value problem, L_p-L_q-estimates, local and global solutions for the semilinear equations, blow up phenomena, sign-changing phenomena, self-similar solutions and number of other results.

More precisely, in that method the solution $v = v(x,t;b)$ to the Cauchy problem

$$v_{tt} - \triangle v = 0, \quad (t,x) \in \mathbb{R}^{1+n}, \; v(x,0;b) = \varphi(x,b), \; v_t(x,0) = 0, \; x \in \mathbb{R}^n, \tag{2.1}$$

with the parameter $b \in B \subseteq \mathbb{R}$ is utilized. Denote that solution by $v_\varphi = v_\varphi(x,t;b)$. There are well known explicit representation formulas for the solution of the last problem. (See, e.g., [39].) In particular, if φ is independent of the second variable b, then $v_\varphi(x,t;b)$ does not depend on b and we write $v_\varphi(x,t)$.

The starting point of that approach [48] is the Duhamel principle, which we revise in order to prepare the ground for generalization. Our *first observation* is that we obtain the following representation

$$u(x,t) = \int_{t_0}^{t} d\tau \int_{0}^{t-\tau} w_f(x,z;\tau)\, dz\,, \tag{2.2}$$

of the solution of the Cauchy problem $u_{tt} - \Delta u = f(x,t)$ in $\mathbb{R}^{n+1}$, and $u(x,t_0) = 0$, $u_t(x,t_0) = 0$ in $\mathbb{R}^n$, where the function $w_f = w_f(x;t;\tau)$ is the solution of the problem (2.1). This formula allows us to solve problem with the source term if we solve the problem for the *same* equation without source term but with the first initial datum.

The *second observation* is that in (2.2) the upper limit $t-\tau$ of the inner integral is generated by the propagation phenomena with the speed which equals to one. In fact, that is a distance function between the points at time t and τ.

Our *third observation* is that the solution operator $G : f \longmapsto u$ can be regarded as a composition of two operators. The first one

$$\mathcal{WE} : f \longmapsto w$$

is a Fourier Integral Operator (FIO), which is a solution operator of the Cauchy problem with the first initial datum for wave equation in the Minkowski spacetime. The second operator

$$\mathcal{K} : w \longmapsto u$$

is the integral operator given by (2.2). We regard the variable z in (2.2) as a "subsidiary time". Thus, $G = \mathcal{K} \circ \mathcal{WE}$ and we arrive at the diagram:

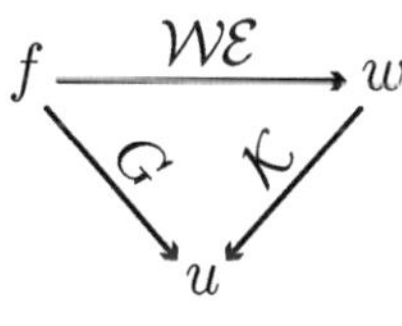

Based on this diagram, we generated a class of operators for which we obtained explicit representation formulas for the solutions. That means also that we have representations for the *fundamental solutions of the partial differential operator*. In fact, this diagram brings into a single hierarchy several different partial differential operators. Indeed, if we take into account the propagation cone by introducing the distance function $\phi(t)$, and if we provide the integral operator with the kernel $K(t;r,b)$ as follows:

$$\mathcal{K}[w](x,t) = 2\int_{t_0}^{t} db \int_{0}^{|\phi(t)-\phi(b)|} K(t;r,b)w(x,r;b)dr, \quad x\in\mathbb{R}^n,\ t>t_0, \tag{2.3}$$

then we actually can generate new representations for the solutions of different well-known equations. Below we give some examples of the operators with the variable coefficients. (See also [54].)

1^0 Tricomi-type equations. This operator is generated by the kernel $K(t;r,b) = 2E(0,t;r,b)$, where the function $E(x,t;r,b)$ [48] is defined by

$$E(x,t;r,b) := c_k \left((\phi(t)+\phi(b))^2-(x-r)^2\right)^{-\gamma} \times F\left(\gamma,\gamma;1;\frac{(\phi(t)-\phi(b))^2-(x-r)^2}{(\phi(t)+\phi(b))^2-(x-r)^2}\right), \tag{2.4}$$

with $\gamma := k/(2k+2)$, $c_k = (k+1)^{-k/(k+1)}2^{-1/(k+1)}$, $k \neq -1$, $k \in \mathbb{R}$, and the distance function is $\phi(t) = t^{k+1}/(k+1)$, while $F\big(a,b;c;\zeta\big)$ is the Gauss hypergeometric function. It is proved in [48] that for the smooth function $f = f(x,t)$, the function

$$u(x,t) = 2c_k \int_0^t db \int_0^{\phi(t)-\phi(b)} \left((\phi(t)+\phi(b))^2-r^2\right)^{-\gamma} \times F\left(\gamma,\gamma;1;\frac{(\phi(t)-\phi(b))^2-r^2}{(\phi(t)+\phi(b))^2-r^2}\right) w(x,r;b)dr, \quad t>0,$$

solves the Tricomi-type equation ($l = 2k \in \mathbb{N}$) (for the Tricomi equation $l = 1$)

$$u_{tt} - t^l\Delta u = f(x,t) \quad \text{in} \quad \mathbb{R}^{n+1}_+ := \{(x,t)\,|\,x\in\mathbb{R}^n,\, t>0\}, \tag{2.5}$$

and takes vanishing initial values

$$u(x,0) = 0, \quad u_t(x,0) = 0 \quad \text{in } \mathbb{R}^n. \tag{2.6}$$

2^0 The wave equation in the FLRW-models: de Sitter spacetime. In this example $K(t;r,b) = 2E(0,t;r,b)$, where the function $E(x,t;r,b)$ [52] is defined by

$$E(x,t;r,b) := \left((e^{-b}+e^{-t})^2-(x-r)^2\right)^{-\frac{1}{2}} \times F\left(\frac{1}{2},\frac{1}{2};1;\frac{(e^{-t}-e^{-b})^2-(x-r)^2}{(e^{-t}+e^{-b})^2-(x-r)^2}\right), \tag{2.7}$$

and $\phi(t) := 1-e^{-t}$. For simplicity, in (2.7) we use the notation $x^2 = x\cdot x = |x|^2$ for $x \in \mathbb{R}^n$. It is proved in [52] that, defined by the integral transform (2.3) with the kernel (2.7) the function

$$u(x,t) = 2\int_0^t db \int_0^{e^{-b}-e^{-t}} \left((e^{-b}+e^{-t})^2-r^2\right)^{-\frac{1}{2}} \times F\left(\frac{1}{2},\frac{1}{2};1;\frac{(e^{-t}-e^{-b})^2-r^2}{(e^{-t}+e^{-b})^2-r^2}\right) w(x,r;\tau)\,dr$$

solves the wave equation in the FLRW spaces arising in the de Sitter model of the universe (see, e.g., [33]), $u_{tt} - e^{-2t}\Delta u = f(x,t)$ in $\mathbb{R}^{n+1}_+$, and takes vanishing initial data (2.6).

3^0 The wave equation in the FLRW-models: anti-de Sitter spacetime. The third example we obtain if we set $K(t;r,b) = 2E(0,t;r,b)$, where the function $E(x,t;r,b)$ is defined by (see [51])

$$E(x,t;r,b) := \left((e^b+e^t)^2-(x-r)^2\right)^{-\frac{1}{2}} \times F\left(\frac{1}{2},\frac{1}{2};1;\frac{(e^t-e^b)^2-(x-r)^2}{(e^t+e^b)^2-(x-r)^2}\right), \tag{2.8}$$

while the distance function is $\phi(t) := e^t - 1$. In that case the function $u = u(x,t)$ produced by the integral transform (2.3) with $t_0 = 0$ and the kernel (2.8), solves the wave equation in the FLRW space arising in the anti-de Sitter model of the universe (see, e.g., [33]), $u_{tt} - e^{2t}\Delta u = f(x,t)$ in $\mathbb{R}^{n+1}_+$. Moreover, it takes vanishing initial values (2.6).

4^0 The wave equation in the Einstein & de Sitter spacetime. If we allow negative $k \in \mathbb{R}$ in (2.4), then we obtain another way to get new operators of the above described hierarchy. In fact, in the hierarchy of the hypergeometric functions $F(a,b;c;\zeta)$ the simplest non-constant function is $F(-1,-1;1;\zeta) = 1+\zeta$. The exponent l leading to $F(-1,-1;1;\zeta)$ is exactly the exponent $l = -4/3$ of the *wave equation* (and of the *metric tensor*) *in the Einstein & de Sitter spacetime*. In that case the kernel is $K(t;r,b) = \frac{1}{18}\left(9t^{2/3}+9b^{2/3}-r^2\right)$. Consequently, the function

$$u(x,t) = \int_0^t db \int_0^{3t^{1/3}-3b^{1/3}} \frac{1}{18}\left((3t^{1/3})^2+(3b^{1/3})^2-r^2\right) w(x,r;b)dr, \tag{2.9}$$

$x \in \mathbb{R}^n$, $t > 0$, solves (see [21]) the equation

$$u_{tt} - t^{-4/3}\Delta u = f \quad \text{in} \quad \mathbb{R}^{n+1}_+, \tag{2.10}$$

and takes vanishing initial data (2.6) provided that $w = \mathcal{WE}(f)$. Because of the singularity in the coefficient of equation (2.10), the Cauchy problem is not well posed. In order to obtain a well-posed problem the initial conditions must be modified to the weighted initial value conditions

$$\begin{cases} \lim\limits_{t\to 0} u(x,t) = \varphi_0(x), & x \in \mathbb{R}^n, \\ \lim\limits_{t\to 0} \left(u_t(x,t) + 3t^{-1/3} \triangle \varphi_0(x)\right) = \varphi_1(x), & x \in \mathbb{R}^n. \end{cases}$$

The operator of equation (2.10) coincides with the principal part of (1.3). We remind that the EdeS model of the universe is the simplest non-empty expanding model (see, e.g., Section 4.3[16]). The last equation belongs to the family of the non-Fuchsian partial differential equations. There is very advanced theory of such equations (see, e.g., [32]), but according to our knowledge the weighted initial value problem suggested in [21] is the new one.

More examples on the integral transforms and representation formulas are given below.

3. Huygens' principle for the Klein–Gordon equation in the de Sitter spacetime

In this section we show that the Klein–Gordon equation in the de Sitter spacetime, which is the curved manifold due to the cosmological constant, obeys the Huygens principle only if the physical mass m of the scalar field and the dimension $n \geq 2$ of the spatial variable are tied by the equation $m^2 = (n^2-1)/4$. Recall (see, e.g., [25]) that a hyperbolic equation is said to satisfy Huygens' principle if the solution vanishes at all points which cannot be reached from the support of initial data by a null geodesic, that is, there is no tail. The tails are important within cosmological context. (See, e.g., [17, 19, 24] and references therein.)

Moreover, we define the *incomplete Huygens principle*, which is the Huygens principle restricted to the vanishing second initial datum, and then prove that the massless scalar field in the de Sitter spacetime obeys the incomplete Huygens principle and does not obey the Huygens principle, for the dimensions $n = 1, 3$, only.

In quantum field theory for the massive scalar field, the equation of motion is the Klein–Gordon equation generated by the metric g:

$$\frac{1}{\sqrt{|g|}}\frac{\partial}{\partial x^i}\left(\sqrt{|g|}g^{ik}\frac{\partial\phi}{\partial x^k}\right) = m^2\phi + V'(\phi)\,.$$

In physical terms this equation describes a local self-interaction for a scalar particle. In the de Sitter universe the equation for the scalar field with mass m and potential V written out explicitly in coordinates is

$$\phi_{tt} + n\phi_t - e^{-2t}\bigtriangleup\phi + m^2\phi = -V'(\phi)\,. \tag{3.1}$$

For the solution Φ of the Cauchy problem for the linear Klein–Gordon equation

$$\Phi_{tt} + n\Phi_t - e^{-2t}\bigtriangleup\Phi + m^2\Phi = 0\,, \quad \Phi(x,0) = \varphi_0(x)\,, \quad \Phi_t(x,0) = \varphi_1(x)\,, \tag{3.2}$$

the following formula is obtained in [52]:

$$\begin{aligned}\Phi(x,t) &= e^{-\frac{n-1}{2}t}v_{\varphi_0}(x,\phi(t))\\ &\quad + e^{-\frac{n}{2}t}\int_0^1 v_{\varphi_0}(x,\phi(t)s)\big(2K_0(\phi(t)s,t) + nK_1(\phi(t)s,t)\big)\phi(t)\,ds\\ &\quad + 2e^{-\frac{n}{2}t}\int_0^1 v_{\varphi_1}(x,\phi(t)s)K_1(\phi(t)s,t)\phi(t)\,ds, \quad x\in\mathbb{R}^n,\ t>0\,,\end{aligned} \tag{3.3}$$

provided that the mass m is large, that is, $m^2 \geq n^2/4$. Here, $\phi(t) := 1 - e^{-t}$ and the function $v(x,t)$ is defined by (2.1). Next we proceed to the definition of $K_0(z,t)$ and $K_1(z,t)$. (See also Section 3 [52].)

We introduce the following notations. First, we define a *chronological future* $D_+(x_0,t_0)$ and a *chronological past* $D_-(x_0,t_0)$ of the point (x_0,t_0), $x_0\in\mathbb{R}^n$, $t_0\in\mathbb{R}$, as follows: $D_\pm(x_0,t_0) := \{(x,t)\in\mathbb{R}^{n+1}\,;\,|x-x_0| \leq \pm(e^{-t_0}-e^{-t})\}$. We define also the *characteristic conoid* (ray cone) by $C_{\pm}(x_0,t_0) := \{(x,t) \in \mathbb{R}^{n+1}\,;\, |x-x_0| = \pm(e^{-t_0}-e^{-t})\}$. Then, we define for $(x_0,t_0) \in \mathbb{R}^n \times \mathbb{R}$ the function

$$E(x,t;x_0,t_0;M) := 4^{-M} e^{M(t_0+t)} \Big((e^{-t}+e^{-t_0})^2 - (x-x_0)^2 \Big)^{-\frac{1}{2}+M} \times F\Big(\frac{1}{2}-M, \frac{1}{2}-M; 1; \frac{(e^{-t_0}-e^{-t})^2-(x-x_0)^2}{(e^{-t_0}+e^{-t})^2-(x-x_0)^2}\Big), \tag{3.4}$$

$$E(x,t;x_0,t_0) := E(x,t;x_0,t_0;-iM), \tag{3.5}$$

in $D_+(x_0,t_0) \cup D_-(x_0,t_0)$, where $F\big(a,b;c;\zeta\big)$ is the hypergeometric function. The kernels $K_0(z,t)$, $K_1(z,t)$, $K_0(z,t;M)$, and $K_0(z,t;M)$ are defined by

$$\begin{aligned} K_0(z,t;M) & \\ := 4^{-M} e^{tM} & \big((1+e^{-t})^2 - z^2\big)^M \frac{1}{[(1-e^{-t})^2-z^2]\sqrt{(1+e^{-t})^2-z^2}} \\ & \times \Bigg[\big(e^{-t}-1+M(e^{-2t}-1-z^2)\big) F\Big(\frac{1}{2}-M, \frac{1}{2}-M; 1; \frac{(1-e^{-t})^2-z^2}{(1+e^{-t})^2-z^2}\Big) \\ & + (1-e^{-2t}+z^2)\Big(\frac{1}{2}+M\Big) F\Big(-\frac{1}{2}-M, \frac{1}{2}-M; 1; \frac{(1-e^{-t})^2-z^2}{(1+e^{-t})^2-z^2}\Big)\Bigg], \end{aligned} \tag{3.6}$$

$$K_1(z,t;M) := 4^{-M} e^{Mt} \big((1+e^{-t})^2 - z^2\big)^{-\frac{1}{2}+M} \times F\left(\frac{1}{2}-M, \frac{1}{2}-M; 1; \frac{(1-e^{-t})^2-z^2}{(1+e^{-t})^2-z^2}\right). \tag{3.7}$$

For (3.3) we set $M = \sqrt{m^2 - \frac{n^2}{4}}$.

For the case of small mass, $m^2 \le n^2/4$, a similar formula is obtained in [53]. More precisely, if we denote $M = \sqrt{\frac{n^2}{4} - m^2}$, and define

$$K_0(z,t) := K_0(z,t;-iM), \tag{3.8}$$

$$K_1(z,t) := K_1(z,t;-iM). \tag{3.9}$$

Then for the solution Φ of the Cauchy problem (3.2), there is a representation

$$\begin{aligned} \Phi(x,t) &= e^{-\frac{n-1}{2}t} v_{\varphi_0}(x,\phi(t)) \\ &+ e^{-\frac{n}{2}t} \int_0^1 v_{\varphi_0}(x,\phi(t)s)\big(2K_0(\phi(t)s,t;M) + nK_1(\phi(t)s,t;M)\big)\phi(t)\,ds \\ &+ 2e^{-\frac{n}{2}t} \int_0^1 v_{\varphi_1}(x,\phi(t)s) K_1(\phi(t)s,t;M)\phi(t)\,ds, \quad x \in \mathbb{R}^n,\ t>0\,. \end{aligned} \tag{3.10}$$

According to [54], the fundamental solutions (the retarded and advanced Green functions) of the operator have similar representations.

Suppose now that we are looking for the simplest possible kernels $K_0(z,t;M)$ and $K_1(z,t;M)$ of the integral transforms. Surprisingly that perspective shades a light on the quantum field theory in the de Sitter universe and reveals a new unexpected link between the Higuchi bound [28] and the Huygens principle.

Indeed, in the hierarchy of the hypergeometric functions the simplest one is the constant, $F(0,0;1;\zeta) = 1$. The parameter M leading to such function $F(0,0;1;\zeta) = 1$ is $M = \frac{1}{2}$, and, consequently, $m^2 = \frac{n^2-1}{4}$.

The next simple non-constant function of that hierarchy is $F(-1,-1;1;\zeta) = 1+\zeta$. The parameter M leading to such function is $M = \frac{3}{2}$, and, consequently, $m^2 = \frac{n^2-9}{4}$.

In the case of $n = 3$ the only real masses, which simplify the kernels, that is, make F polynomial, are $m = \sqrt{2}$ and $m = 0$. These are exactly the endpoints of the interval $(0, \sqrt{2})$ that, in the case of $n = 3$, is known in the quantum field theory as the so-called Higuchi bound [28]. In fact, the interval $(0, \sqrt{2})$ plays a significant role in the linear quantum field theory [28], in a completely different context than the explicit representation of the solutions of the Cauchy problem. More precisely, the Higuchi bound [28, 1, 5, 11, 12], arises in the quantization of free massive fields with the spin-2 in the de Sitter spacetime with $n = 3$. It is the forbidden mass range for spin-2 field theory in de Sitter spacetime because of the appearance of negative norm states. Thus, the point $m = \sqrt{2}$ is exceptional for the quantum fields theory in the de Sitter spacetime. In particular, for massive spin-2 fields, it is known [11, 28] that the norm of the helicity zero mode changes sign across the line $m^2 = 2$. The region $m^2 < 2$ is therefore unitarily forbidden. It is noted in [1] that all canonically normalized helicity $-0, \pm 1, \pm 2$ modes of massive graviton on the de Sitter universe satisfy the Klein–Gordon equation for a massive scalar field with the same effective mass.

In the case of $n \in \mathbb{N}$ we obtain for the physical mass several points, $m^2 = \frac{n^2}{4} - \left(\frac{1}{2}+k\right)^2$, $k = 0, 1, \dots, \left[\frac{n-1}{2}\right]$, which make F polynomial. We will call these points *the knot points*. For $n = 1$ only the massless field $m = 0$ has a knot point.

We state below that the largest knot point, and, in particular, *the right end-point of the Higuchi bound if $n = 3$, is the only value of the mass of the particle which produces a scalar field that obeys the Huygens principle.*

An exemplar equation satisfying Huygens' principle is the wave equation in $n+1$-dimensional Minkowski spacetime for odd $n \geq 3$. According to Hadamard's conjecture (see, e.g., [25, 4, 30]) this is the only (modulo transformations of coordinates and unknown function) Huygensian linear second-order hyperbolic equation. There exists an extensive literature on the Huygens principle in the four-dimensional spacetime of constant curvature (see, e.g., [19, 40] and references therein). In [58] is suggested a new proof of the following statement.

Theorem 3.1 ([58]). *The value $m = \sqrt{n^2-1}/2$ is the only value of the physical mass m, such that the solutions of the equation*

$$\Phi_{tt} + n\Phi_t - e^{-2t} \bigtriangleup \Phi + m^2\Phi = 0, \tag{3.11}$$

obey the Huygens principle, whenever the wave equation in the Minkowski spacetime does, that is, $n \geq 3$ is an odd number.

In fact, for $m = \sqrt{n^2-1}/2$ the equation by means of the transformations of coordinates and the unknown function can be reduced to the wave equation on the Minkowski spacetime. It is easily seen that the value $m = \sqrt{n^2-1}/2$ is also the conformal value, i.e., the value under which a conformal change of the metric turns the problem into one on a compact in time cylinder.

Even if the equation is not Huygensian (not tail-free for some admissible data), one might nevertheless be interested in data that produce tail-free solution.

Definition 3.2 ([58]). We say that the equation obeys the incomplete Huygens principle with respect to the first initial datum, if the solution with the second datum $\varphi_1 = 0$ vanishes at all points which cannot be reached from the support of initial data by a null geodesic.

Theorem 3.3 ([58]). *Suppose that equation* (3.11) *does not obey the Huygens principle. Then, it obeys the incomplete Huygens principle with respect to the first initial datum, if and only if the equation is massless, $m = 0$, and either $n = 1$ or $n = 3$.*

By combining Theorem 3.1 and Theorem 3.3 we arrive at the following interesting conclusion.

Corollary 3.4. [58] *Assume that the equations $\Phi_{tt} + n\Phi_t - c_1^2 e^{-2t} \triangle \Phi + m_1^2 \Phi = 0$ and $\Phi_{tt} + n\Phi_t - c_2^2 e^{-2t} \triangle \Phi + m_2^2 \Phi = 0$, where c_1, c_2 are positive numbers, obey the incomplete Huygens principle. Then they describe the fields with different mass, $m_1 \neq m_2$, (in fact, $\frac{\sqrt{n^2-1}}{2}$ and 0) if and only if the dimension n is 3.*

Thus, in the de Sitter spacetime the existence of two different scalar fields (in fact, with $m = 0$ and $m^2 = (n^2-1)/4$), which obey the incomplete Huygens principle, is equivalent to the condition $n = 3$. The dimension $n = 3$ of the last corollary agrees with the experimental data. In fact, Paul Ehrenfest in 1917 addressed the question: "Why has our space just three dimensions?"

4. Global solutions of the semilinear system of Klein–Gordon equations in de Sitter spacetime

In this section we discuss global existence of small data solutions of the Cauchy problem for the semilinear system of Klein–Gordon equations in the de Sitter spacetime. Unlike the same problem in the Minkowski spacetime, we have no restriction on the order of the nonlinearity and the structure of the nonlinear term, provided that the spectrum of the mass matrix of the fields, which describes the linear interactions of the fields, is in the positive half-line and has no intersection with some open bounded interval.

A large amount of work has been devoted to the Cauchy problem for the scalar semilinear Klein–Gordon equation in the Minkowski spacetime. The existence of global weak solutions has been obtained by Jörgens [31], Segal [37, 38], Pecher [35], Brenner [6], Strauss [42], Ginibre and Velo [22, 23] for the equation

$$u_{tt} - \Delta u + m^2 u = |u|^\alpha u.$$

For global solvability, the exact relation between n and $\alpha > 0$ was finally established. More precisely, consider the Cauchy problem for the nonlinear Klein–Gordon equation

$$u_{tt} - \Delta u = -V'(u)\,,$$

where $V' = V'(u)$ is a nonlinear function, a typical form of which is the sum of two powers

$$V'(u) = \lambda_0 u + \lambda |u|^\alpha u$$

with $\alpha \geq 0$ and $\lambda \geq 0$. For this equation, a conservation of energy is valid. For finite energy solutions scaling arguments suggest the assumption $\alpha < 4/(n-1)$. In [23] the existence and uniqueness of strong global solutions in the energy space $H_{(1)} \oplus L^2$ are proved for arbitrary space dimension n under assumptions on V' that cover the case of a sum of powers $\lambda|u|^\alpha u$ with $0 \leq \alpha < 4/(n-1)$, $n \geq 2$, and $\lambda > 0$ for the highest α. Some of the results can be extended to the case $\alpha = 4/(n-1)$ (see, e.g., [22], [23, Sec. 4]).

In this section we consider the model of interacting fields, which can be described by the system of Klein–Gordon equations with different masses, containing interaction via mass matrix and the semilinear term. The model obeys the following system

$$\Phi_{tt} + n\Phi_t - e^{-2t} \bigtriangleup \Phi + \mathbf{M}\Phi = F(\Phi)\,. \tag{4.1}$$

Here F is a vector-valued function of the vector-valued function Φ. We assume that the mass matrix $\mathbf{M}$ is real valued, diagonalizable, and it has eigenvalues $m_1^2, \ldots, m_l^2$, $i = 1, 2, \ldots, l$. By the similarity transformation with the real-valued matrix $\mathbf{O}$ the mass matrix $\mathbf{M}$ can be diagonalized. Therefore, we use the change of the unknown function as follows:

$$\Psi = e^{\frac{n}{2}t}\mathbf{O}\Phi, \qquad \Phi = e^{-\frac{n}{2}t}\mathbf{O}^{-1}\Psi\,.$$

Then the system (4.1) takes the form

$$\Psi_{tt} - e^{-2t} \bigtriangleup \Psi + \mathcal{M}^2\Psi = e^{\frac{n}{2}t}\mathbf{O}F(e^{-\frac{n}{2}t}\mathbf{O}^{-1}\Psi)\,, \tag{4.2}$$

where the diagonal matrix $\mathcal{M}$, with nonnegative real part $\Re\mathcal{M} \geq 0$, is

$$\mathcal{M}^2 := \mathbf{O}\mathbf{M}\mathbf{O}^{-1} - \frac{n^2}{4}\mathbf{I}\,, \qquad \mathbf{I} \ \text{ is the identity matrix.}$$

The matrix $\mathcal{M}^2$ will be called the *curved mass matrix* of the particles, which is also sometimes referred to as the *effective mass matrix*. It is convenient to use the diagonal matrix $M = \mathrm{diag}(|m_i^2 - \frac{n^2}{4}|^{1/2})$. We distinguish the following three cases: the case of *large mass matrix* $\mathbf{M}$ that is $\mathcal{M}^2 \geq 0$ ($m_i^2 \geq \frac{n^2}{4}$, $i = 1, 2, \ldots, l$); the case of *dimensional mass matrix* $\mathbf{M}$ that is $\mathcal{M}^2 = 0$ ($m_i^2 = \frac{n^2}{4}$, $i = 1, 2, \ldots, l$); and the case of *small mass matrix* $\mathbf{M}$ that is $\mathcal{M}^2 < 0$ ($m_i^2 < \frac{n^2}{4}$, $i = 1, 2, \ldots, l$). We also call the mass matrix $\mathbf{M}$ *critical* if $\mathcal{M}^2 = -\frac{1}{4}I$. They lead to three different equations: the Klein–Gordon equation with the *real curved mass matrix* $\mathcal{M}$,

$$\Psi_{tt} - e^{-2t} \bigtriangleup \Psi + M^2\Psi = e^{\frac{n}{2}t}\mathbf{O}F(e^{-\frac{n}{2}t}\mathbf{O}^{-1}\Psi)\,;$$

the wave equation with the *zero curved mass matrix*

$$\Psi_{tt} - e^{-2t} \bigtriangleup \Psi = e^{\frac{n}{2}t}\mathbf{O}F(e^{-\frac{n}{2}t}\mathbf{O}^{-1}\Psi);$$

and the Klein–Gordon equation with the *imaginary curved mass matrix* $\mathcal{M}$,

$$\Psi_{tt} - e^{-2t} \bigtriangleup \Psi - M^2\Psi = e^{\frac{n}{2}t}\mathbf{O}F(e^{-\frac{n}{2}t}\mathbf{O}^{-1}\Psi).$$

Let $H_{(s)}(\mathbb{R}^n)$ be the Sobolev space. We use the notation $\| \cdot \|_{H_{(s)}(\mathbb{R}^n)}$ for both the norm of a vector-valued function and for the norm of its components. To estimate the nonlinear term $F(\Phi)$ we use the following Lipschitz condition:

Condition ($\mathcal{L}$). *The function F is said to be Lipschitz continuous in the space $H_{(s)}(\mathbb{R}^n)$ if there are constants $\alpha \geq 0$ and C such that*

$$\|F(\Phi_1) - F(\Phi_2)\|_{H_{(s)}(\mathbb{R}^n)} \leq C\|\Phi_1 - \Phi_2\|_{H_{(s)}(\mathbb{R}^n)}\Big(\|\Phi_1\|^{\alpha}_{H_{(s)}(\mathbb{R}^n)} + \|\Phi_2\|^{\alpha}_{H_{(s)}(\mathbb{R}^n)}\Big)$$

for all $\Phi_1, \Phi_2 \in H_{(s)}(\mathbb{R}^n)$.

If $s > n/2$, then any polynomial is Lipschitz continuous in the space $H_{(s)}(\mathbb{R}^n)$. For more examples of the Lipschitz continuous in the space $H_{(s)}(\mathbb{R}^n)$ functions with low s see, for example, [35], [39]. Define the complete metric space

$$X(R,s,\gamma) := \Big\{\Phi \in C([0,\infty); H_{(s)}(\mathbb{R}^n)) \;\big|\; \|\Phi\|_X := \sup_{t\in[0,\infty)} e^{\gamma t} \|\Phi(x,t)\|_{H_{(s)}(\mathbb{R}^n)} \leq R\Big\}$$

with the metric

$$d(\Phi_1,\Phi_2) := \sup_{t\in[0,\infty)} e^{\gamma t} \| \Phi_1(x,t) - \Phi_2(x,t) \|_{H_{(s)}(\mathbb{R}^n)}.$$

The first result of this section is the following theorem.

Theorem 4.1 ([57]). *Assume that the nonlinear term $F(\Phi)$ is Lipschitz continuous in the space $H_{(s)}(\mathbb{R}^n)$, $s > n/2 \geq 1$, $\alpha > 0$, and $F(0) = 0$. Assume also that the system has a large mass matrix. Then, there exists $\varepsilon_0 > 0$ such that, for every given vector-valued functions $\varphi_0, \varphi_1 \in H_{(s)}(\mathbb{R}^n)$, such that*

$$\|\varphi_0\|_{H_{(s)}(\mathbb{R}^n)} + \|\varphi_1\|_{H_{(s)}(\mathbb{R}^n)} \leq \varepsilon, \qquad \varepsilon < \varepsilon_0,$$

there exists a global solution $\Phi \in C^1([0,\infty); H_{(s)}(\mathbb{R}^n))$ of the Cauchy problem

$$\Phi_{tt} + n\Phi_t - e^{-2t}\Delta\Phi + \mathbf{M}\Phi = F(\Phi), \tag{4.3}$$

$$\Phi(x,0) = \varphi_0(x), \quad \Phi_t(x,0) = \varphi_1(x). \tag{4.4}$$

The solution $\Phi(x,t)$ belongs to the space $X(2\varepsilon, s, 0)$, that is,

$$\sup_{t\in[0,\infty)} \|\Phi(x,t)\|_{H_{(s)}(\mathbb{R}^n)} < 2\varepsilon.$$

For the scalar equation this theorem implies Theorem 0.1 [55]. In fact, for the scalar equation if $F(\Phi) = \pm|\Phi|^{\alpha}\Phi$ or $F(\Phi) = \pm|\Phi|^{\alpha+1}$, then, according to Theorem 0.1 [55], the small data Cauchy problem is globally solvable for every $\alpha \in (0,\infty)$ if $m \in (0, \sqrt{n^2-1}/2) \cup [n/2,\infty)$ and the condition ($\mathcal{L}$) is fulfilled.

Conjecture 4.2 ([57]). The open interval $(\sqrt{n^2-1}/2, n/2)$ is a forbidden physical mass interval for the small data global solvability of the Cauchy problem for all $\alpha \in (0,\infty)$.

Consider the particular case of the scalar equation with the spatial dimension $n=3$. In this case the interval $(\sqrt{n^2-1}/2, n/2)$ for the physical mass is reduced to $(\sqrt{2}, 3/2)$. The interval $(0,\sqrt{2})$ is the Higuchi bound. This is why we pay special attention to the system of equations with the mass matrix $\mathbf{M}$ which is orthogonally similar to the matrix $\frac{n^2-1}{4}\mathbf{I}$. We call such mass matrix $\mathbf{M}$ critical. We also call the mass matrix $\mathbf{M}$ *semi-critical mass matrix* if the spectrum $\sigma(\mathbf{M})$ of the mass matrix $\mathbf{M}$ is a subset of $(0,(n^2-1)/4]$. For the system with the semi-critical mass matrix $\mathbf{M}$ we prove the global existence, which is not known in the critical case even for the scalar equation.

Theorem 4.3 ([57]). *Assume that the nonlinear term $F(\Phi)$ is Lipschitz continuous in the space $H_{(s)}(\mathbb{R}^n)$, $s>n/2\geq 1$, $\alpha>0$, and $F(0)=0$. Assume also that the mass matrix $\mathbf{M}$ is semi-critical, that is $\sigma(\mathbf{M})\subset(0,(n^2-1)/4]$. Then, there exists $\varepsilon_0>0$ such that, for every given vector-valued functions $\varphi_0,\varphi_1\in H_{(s)}(\mathbb{R}^n)$, such that*

$$\|\varphi_0\|_{H_{(s)}(\mathbb{R}^n)}+\|\varphi_1\|_{H_{(s)}(\mathbb{R}^n)}\leq\varepsilon, \quad \varepsilon<\varepsilon_0\,,$$

there exists a global solution $\Phi\in C^1([0,\infty);H_{(s)}(\mathbb{R}^n))$ of the Cauchy problem (4.3), (4.4). The solution $\Phi(x,t)$ belongs to the space $X(2\varepsilon,s,\gamma)$, where

$$\gamma<\frac{1}{\alpha+1}\left(\frac{n}{2}-\max\left\{\sqrt{\frac{n^2}{4}-\lambda}\,;\lambda\in\sigma(\mathbf{M})\right\}\right),$$

that is,

$$\sup_{t\in[0,\infty)} e^{\gamma t}\|\Phi(x,t)\|_{H_{(s)}(\mathbb{R}^n)}<2\varepsilon.$$

Baskin [3] discussed small data global solutions for the scalar Klein–Gordon equation on asymptotically de Sitter spaces, which is a compact manifold with boundary. More precisely, in [3] the Cauchy problem is considered for the semilinear equation $\Box_g u+m^2u=f(u)$, $u(x,t_0)=\varphi_0(x)\in H_{(1)}(\mathbb{R}^n)$, $u_t(x,t_0)=\varphi_1(x)\in L^2(\mathbb{R}^n)$, where mass is large, $m^2>n^2/4$. In Theorem 1.3 [3] the existence of the global solution for small energy data is stated. (For references on the asymptotically de Sitter spaces, see [2], [44].)

5. The scalar equation. Case of large mass

We extract a linear part of the system (4.2) as an initial model that must be treated first. That linear system is diagonal, which allows us to restrict ourselves to one scalar equation

$$u_{tt}-e^{-2t}\,\triangle u+\mathcal{M}^2u=f\,, \tag{5.1}$$

where $\mathcal{M}$ is a non-negative number throughout this section. The equation (5.1) is strictly hyperbolic. That implies the well-posedness of the Cauchy problem for

(5.1) in several function spaces. The coefficients of the equation are analytic functions and, consequently, Holmgren's theorem implies local uniqueness in the space of distributions. Moreover, the speed of propagation is equal to e^{-t} for every $t \in \mathbb{R}$. The second-order strictly hyperbolic equation (5.1) possesses two fundamental solutions resolving the Cauchy problem. They can be written in terms of Fourier integral operators [29], which give a complete description of the wave front sets of the solutions. The distance $e^{-t}|\xi|$ between two characteristic roots of the equation (5.1) tends to zero as $t \to +\infty$. Thus, the operator is not uniformly strictly hyperbolic. The finite integrability of the characteristic roots leads to the existence of a so-called *horizon* for that equation. The equation (5.1) is neither Lorentz invariant nor invariant with respect to scaling and that brings additional difficulties.

L^p-L^q estimates for equation with source. We consider the equation with $n \geq 2$. The solution $u = u(x,t)$ to the Cauchy problem

$$u_{tt} - e^{-2t}\Delta u + M^2 u = f, \quad u(x,0) = 0, \quad u_t(x,0) = 0, \tag{5.2}$$

with $f \in C^\infty(\mathbb{R}^{n+1})$ and with vanishing initial data is given by the next expression

$$u(x,t) = 2\int_0^t db \int_0^{e^{-b}-e^{-t}} dr\ v_f(x,r;b)E(r,t;0,b),$$

where the function $v_f(x,t;b)$ is a solution to the Cauchy problem

$$v_{tt} - \Delta v = 0\,, \quad v(x,0;b) = f(x,b)\,, \quad v_t(x,0;b) = 0\,. \tag{5.3}$$

Thus, for the solution Φ of the equation

$$\Phi_{tt} + n\Phi_t - e^{-2t} \bigtriangleup \Phi + m^2\Phi = f\,, \tag{5.4}$$

due to the relation $u = e^{\frac{n}{2}t}\Phi$, we obtain

$$\Phi(x,t) = 2e^{-\frac{n}{2}t}\int_0^t db \int_0^{e^{-b}-e^{-t}} dr\, e^{\frac{n}{2}b} v_f(x,r;b)E(r,t;0,b). \tag{5.5}$$

For the solution $u = u(x,t)$ of the Cauchy problem (5.2) according to Corollary 9.3 [52] one has estimate

$$\begin{aligned}&\|(-\bigtriangleup)^{-s}u(x,t)\|_{L^q(\mathbb{R}^n)}\\ &\quad\leq C_M\int_0^t \|f(x,b)\|_{L^p(\mathbb{R}^n)}e^b\left(e^{-b}-e^{-t}\right)^{1+2s-n(\frac{1}{p}-\frac{1}{q})}(1+t-b)^{1-\operatorname{sgn}M}\,db\,,\end{aligned}$$

provided that $1 < p \leq 2$, $\frac{1}{p}+\frac{1}{q}=1$, $\frac{1}{2}(n+1)\left(\frac{1}{p}-\frac{1}{q}\right) \leq 2s \leq n\left(\frac{1}{p}-\frac{1}{q}\right) < 2s+1$. Thus, for the solution Φ (5.5) of the equation (5.4), we obtain

$$\begin{aligned}&\|(-\bigtriangleup)^{-s}\Phi(x,t)\|_{L^q(\mathbb{R}^n)}\\ &\quad\leq C_M e^{-\frac{n}{2}t}\int_0^t e^{\frac{n}{2}b}\|f(x,b)\|_{L^p(\mathbb{R}^n)}e^b\left(e^{-b}-e^{-t}\right)^{1+2s-n(\frac{1}{p}-\frac{1}{q})}(1+t-b)^{1-\operatorname{sgn}M}\,db.\end{aligned}$$

In particular, for $s=0$ and $p=q=2$, we have

$$\|\Phi(x,t)\|_{L^2(\mathbb{R}^n)} \le C_M e^{-\frac{n}{2}t} \int_0^t e^{\frac{n}{2}b} \|f(x,b)\|_{L^2(\mathbb{R}^n)} (1+t-b)^{1-\operatorname{sgn}M}\, db.$$

Here the rates of exponential factors are independent of $\mathcal{M}$ and, consequently, of the mass m.

***L^p-L^q* estimates for equations without source.** According to Theorem 10.1 [52] the solution $u=u(x,t)$ of the Cauchy problem

$$u_{tt} - e^{-2t} \bigtriangleup u + M^2 u = 0\,, \quad u(x,0) = \varphi_0(x)\,, \quad u_t(x,0) = \varphi_1(x)\,,$$

satisfies the following L^p-L^q estimate

$$\|(-\bigtriangleup)^{-s} u(x,t)\|_{L^q(\mathbb{R}^n)} \le C_M (1+t)^{1-\operatorname{sgn}M} (1-e^{-t})^{2s-n(\frac{1}{p}-\frac{1}{q})} \\ \times \left\{ e^{\frac{t}{2}} \|\varphi_0\|_{L^p(\mathbb{R}^n)} + (1-e^{-t}) \|\varphi_1\|_{L^p(\mathbb{R}^n)} \right\}$$

for all $t \in (0,\infty)$, provided that $1<p\le 2$, $\frac{1}{p}+\frac{1}{q}=1$, $\frac{1}{2}(n+1)\big(\frac{1}{p}-\frac{1}{q}\big) \le 2s \le n\big(\frac{1}{p}-\frac{1}{q}\big) < 2s+1$. In particular, for large t we obtain the following *no decay* estimate

$$\|(-\bigtriangleup)^{-s} u(x,t)\|_{L^q(\mathbb{R}^n)} \le C_M (1+t)^{1-\operatorname{sgn}M} \left\{ e^{\frac{t}{2}} \|\varphi_0\|_{L^p(\mathbb{R}^n)} + \|\varphi_1\|_{L^p(\mathbb{R}^n)} \right\}.$$

Thus, for the solution Φ of the Cauchy problem (3.2), due to the relation $u = e^{\frac{n}{2}t}\Phi$, we obtain the decay estimate

$$\|(-\bigtriangleup)^{-s} \Phi(x,t)\|_{L^q(\mathbb{R}^n)} \le C_M e^{-\frac{n}{2}t} (1+t)^{1-\operatorname{sgn}M} (1-e^{-t})^{2s-n(\frac{1}{p}-\frac{1}{q})} \\ \times \left\{ e^{\frac{t}{2}} \|\varphi_0\|_{L^p(\mathbb{R}^n)} + (1-e^{-t}) \|\varphi_1\|_{L^p(\mathbb{R}^n)} \right\}$$

for all $t>0$.

6. The scalar equation. Imaginary curved mass

In this section we consider the linear part of the scalar equation

$$u_{tt} - e^{-2t} \bigtriangleup u - M^2 u = -e^{\frac{n}{2}t} V'(e^{-\frac{n}{2}t} u), \tag{6.1}$$

with $M \ge 0$. The equation (6.1) covers two important cases. The first one is the Higgs boson equation, which has $V'(\phi) = \lambda\phi^3$ and $M^2 = \mu m^2 + n^2/4$ with $\lambda>0$ and $\mu>0$, while $n=3$. The second case is the case of the small physical mass, that is $0 \le m \le \frac{n}{2}$. For the last case $M^2 = \frac{n^2}{4} - m^2$. The solution $u=u(x,t)$ to the Cauchy problem

$$u_{tt} - e^{-2t} \Delta u - M^2 u = f, \quad u(x,0) = 0, \quad u_t(x,0) = 0, \tag{6.2}$$

with $f \in C^\infty(\mathbb{R}^{n+1})$ and with vanishing initial data is given in [53] by the next expression

$$u(x,t) = 2 \int_0^t db \int_0^{e^{-b}-e^{-t}} dr\; v(x,r;b) E(r,t;0,b;M),$$

where the function $v(x,t;b)$ is a solution to the Cauchy problem for the wave equation (5.3).

The solution $u = u(x,t)$ to the Cauchy problem

$$u_{tt} - e^{-2t} \bigtriangleup u - M^2 u = 0\,, \quad u(x,0) = \varphi_0(x)\,, \quad u_t(x,0) = \varphi_1(x)\,,$$

with $\varphi_0, \varphi_1 \in C_0^\infty(\mathbb{R}^n)$, $n \geq 2$, can be represented (see [53]) as follows:

$$u(x,t) = e^{\frac{t}{2}} v_{\varphi_0}(x, \phi(t)) + 2 \int_0^1 v_{\varphi_0}(x, \phi(t)s) K_0(\phi(t)s, t; M) \phi(t)\, ds$$
$$+ 2 \int_0^1 v_{\varphi_1}(x, \phi(t)s) K_1(\phi(t)s, t; M) \phi(t)\, ds, \quad x \in \mathbb{R}^n,\ t > 0\,,$$

where $\phi(t) := 1 - e^{-t}$. Here, for $\varphi \in C_0^\infty(\mathbb{R}^n)$ and for $x \in \mathbb{R}^n$, the function $v_\varphi(x,t)$ is the solution of the Cauchy problem (2.1). Thus, for the solution Φ of the Cauchy problem

$$\Phi_{tt} + n\Phi_t - e^{-2t} \bigtriangleup \Phi + m^2 \Phi = f, \quad \Phi(x,0) = 0, \quad \Phi_t(x,0) = 0,$$

due to the relation $u = e^{\frac{n}{2}t}\Phi$, we obtain with $f \in C^\infty(\mathbb{R}^{n+1})$ and with vanishing initial data the next expression

$$\Phi(x,t) = 2e^{-\frac{n}{2}t} \int_0^t db \int_0^{e^{-b}-e^{-t}} dr\, e^{\frac{n}{2}b} v_f(x,r;b) E(r,t;0,b;M), \qquad (6.3)$$

where the function $v_f(x,t;b)$ is a solution to the Cauchy problem for the wave equation (5.3). Thus, for the solution Φ of the Cauchy problem (3.2), due to the relation $u = e^{\frac{n}{2}t}\Phi$, we obtain (3.10). In fact, the representation formulas of this section have been used in [56] to establish sign-changing properties of the solutions of the Higgs boson equation.

The critical case of $m^2 = (n^2 - 1)/4$. In this case we have

$$E\left(x,t;x_0,t_0;\frac{1}{2}\right) = \frac{1}{2} e^{\frac{1}{2}(t_0+t)}, \quad E\left(z,t;0,b;\frac{1}{2}\right) = \frac{1}{2} e^{\frac{1}{2}(b+t)}\,,$$

while

$$K_0\left(z,t;\frac{1}{2}\right) = -\frac{1}{4} e^{\frac{1}{2}t}, \qquad K_1\left(z,t;\frac{1}{2}\right) = \frac{1}{2} e^{\frac{1}{2}t}\,.$$

For the solution (5.5) of the equation (5.4) with the source term it follows

$$\Phi(x,t) = e^{-\frac{n-1}{2}t} \int_0^t e^{\frac{n+1}{2}b} V_f(x, e^{-b} - e^{-t}; b)\, db.$$

where $v(x,r;b)$ is defined by (5.3). We denote by $V_f(x,t;b)$ the solution of the problem $V_{tt} - \Delta V = 0$, $V(x,0) = 0$, $V_t(x,0) = f(x,b)$. Further, for the solution Φ (3.10) of the equation without source term we have

$$\Phi(x,t) = e^{-\frac{n-1}{2}t} \left(\frac{\partial V_{\varphi_0}}{\partial t}\right)(x, 1 - e^{-t})$$
$$+ \frac{n-1}{2} e^{-\frac{n-1}{2}t} V_{\varphi_0}(x, 1 - e^{-t}) + e^{-\frac{n-1}{2}t} V_{\varphi_1}(x, 1 - e^{-t}),$$

$x \in \mathbb{R}^n$, $t>0$, where we denote by V_φ the solution of the problem $V_{tt} - \triangle V = 0$, $V(x,0) = 0$, $V_t(x,0) = \varphi(x)$. Thus, in particular, we arrive at the next theorem.

Theorem 6.1 ([57]). *The solutions of the equation $\Phi_{tt} + n\Phi_t - e^{-2t}\Delta\Phi + \mathbf{M}\Phi = 0$, obey the strong Huygens Principle, if and only if $n \geq 3$ is an odd number and the mass matrix $\mathbf{M}$ is the diagonal matrix $\frac{n^2-1}{4}\mathbf{I}$.*

Theorem 6.2 ([53]). *Suppose that $m^2 = (n^2-1)/4$. If $\varphi_0 = \varphi_1 = 0$ and $\frac{1}{2}(n+1)(\frac{1}{p}-\frac{1}{q}) - 1 \leq 2s \leq n(\frac{1}{p}-\frac{1}{q})$, then for the solution $\Phi = \Phi(x,t)$ of the equation (5.4) the following estimate holds*

$$\|(-\triangle)^{-s}\Phi(x,t)\|_{L^q(\mathbb{R}^n)} \leq Ce^{-\frac{n-1}{2}t}\int_0^t e^{\frac{n+1}{2}b}(e^{-b}-e^{-t})^{1+2s-n(\frac{1}{p}-\frac{1}{q})}\|f(x,b)\|_{L^p(\mathbb{R}^n)}db, \quad t>0\,.$$

For the solution $\Phi = \Phi(x,t)$ of the Cauchy problem (3.2): if $f \equiv 0$, $\varphi_0 = 0$, and $\frac{1}{2}(n+1)(\frac{1}{p}-\frac{1}{q}) - 1 \leq 2s \leq n(\frac{1}{p}-\frac{1}{q})$, then

$$\|(-\triangle)^{-s}\Phi(x,t)\|_{L^q(\mathbb{R}^n)} \leq Ce^{-\frac{n-1}{2}t}(1-e^{-t})^{1+2s-n(\frac{1}{p}-\frac{1}{q})}\|\varphi_1\|_{L^p(\mathbb{R}^n)}, \quad t>0\,,$$

while if $f \equiv 0$, $\varphi_1 = 0$, and $\frac{1}{2}(n+1)(\frac{1}{p}-\frac{1}{q}) \leq 2s \leq n(\frac{1}{p}-\frac{1}{q})$, then

$$\|(-\triangle)^{-s}\Phi(x,t)\|_{L^q(\mathbb{R}^n)} \leq Ce^{-\frac{n-1}{2}t}(1-e^{-t})^{2s-n(\frac{1}{p}-\frac{1}{q})}\|\varphi_0\|_{L^p(\mathbb{R}^n)}, \quad t>0\,.$$

To complete the list of the L^p-L^q estimates we quote below results (Theorems 2.2, 2.6 from [55]), which are applicable to the scalar equation with noncritical mass. The solution $\Phi = \Phi(x,t)$ of the Cauchy problem

$$\Phi_{tt} + n\Phi_t - e^{-2t}\triangle\Phi \pm m^2\Phi = 0\,, \quad \Phi(x,0) = \varphi_0(x)\,, \quad \Phi_t(x,0) = \varphi_1(x)\,,$$

with either $M = \sqrt{\frac{n^2}{4}-m^2}$ and $m < \sqrt{n^2-1}/2$ for the case of "plus", or $M = \sqrt{\frac{n^2}{4}+m^2}$ for the case of "minus", satisfies the following L^p-L^q estimate

$$\|(-\triangle)^{-s}\Phi(x,t)\|_{L^q(\mathbb{R}^n)} \leq C_{M,n,p,q,s}(1-e^{-t})^{2s-n(\frac{1}{p}-\frac{1}{q})}e^{(M-\frac{n}{2})t}\Big\{\|\varphi_0\|_{L^p(\mathbb{R}^n)} + (1-e^{-t})\|\varphi_1\|_{L^p(\mathbb{R}^n)}\Big\}$$

for all $t \in (0,\infty)$, provided that $1 < p \leq 2$, $\frac{1}{p}+\frac{1}{q} = 1$, $\frac{1}{2}(n+1)\left(\frac{1}{p}-\frac{1}{q}\right) \leq 2s \leq n\left(\frac{1}{p}-\frac{1}{q}\right) < 2s+1$.

Further, let $\Phi = \Phi(x,t)$ be the solution of the Cauchy problem

$$\Phi_{tt} + n\Phi_t - e^{-2t}\triangle\Phi \pm m^2\Phi = f\,, \quad \Phi(x,0) = 0\,, \quad \Phi_t(x,0) = 0\,, \tag{6.4}$$

with either $M = \sqrt{\frac{n^2}{4}-m^2}$ and $m < \sqrt{n^2-1}/2$ for the case of "plus", or $M = \sqrt{\frac{n^2}{4}+m^2}$ for the case of "minus". Then $\Phi = \Phi(x,t)$ satisfies the following L^p-L^q

estimate:

$$\begin{aligned}
&\|(-\triangle)^{-s}\Phi(x,t)\|_{L^q(\mathbb{R}^n)}\\
&\quad\le C_M e^{-Mt}e^{-\frac{n}{2}t}e^{-t[2s-n(\frac{1}{p}-\frac{1}{q})]}\\
&\quad\quad\times\int_0^t e^{\frac{n}{2}b}e^{Mb}(e^{t-b}-1)^{1+2s-n(\frac{1}{p}-\frac{1}{q})}(e^{t-b}+1)^{2M-1}\|f(x,b)\|_{L^p(\mathbb{R}^n)}\,db\,,
\end{aligned}$$

as well as

$$\begin{aligned}
&\|(-\triangle)^{-s}\Phi(x,t)\|_{L^q(\mathbb{R}^n)}\\
&\quad\le C_M e^{-(\frac{n}{2}-M)t}\int_0^t e^{(\frac{n}{2}-M)b}e^{-b(2s-n(\frac{1}{p}-\frac{1}{q}))}\|f(x,b)\|_{L^p(\mathbb{R}^n)}\,db
\end{aligned}$$

for all $t \in (0,\infty)$, and for the above written range of the parameters p, q, s.

7. Global existence. Small data solutions

The Cauchy problem (6.2) for the scalar equation was studied in [53]. For $F(\Phi) = c|\Phi|^{\alpha+1}$, $c \neq 0$, Theorem 1.1 [53], implies nonexistence of a global solution even for arbitrary small initial data $\varphi_0(x)$ and $\varphi_1(x)$ under some conditions on n, α, and M. Theorem 3.1 [55] gives the blow up result for the equation with imaginary physical mass. That theorem shows that instability of the trivial solution occurs in a very strong sense, that is, an arbitrarily small perturbation of the initial data can make the perturbed solution blowing up in finite time. If we allow large initial data, then, according to Theorem 1.2 [53], the concentration of the mass, due to the non-dispersion property of the de Sitter spacetime, leads to the nonexistence of the global solution, which cannot be recovered even by adding an exponentially decaying factor in the nonlinear term.

In this section we are going to study the global existence of solutions for the system of semilinear Klein–Gordon equations. The first step toward such result is to establish the L_p-L_q-estimates for the equation with source term. For the scalar equation this estimate is proved in [55]. Although we want to prove the global existence for two different cases, for the system with the semi-critical mass matrix and for the system of equations with the large mass matrix, the consideration can be done in the single framework.

We reduce the Cauchy problem to the integral equation. The main tool for such reduction is the fundamental solution for the interacting fields, which can be described by the system of Klein–Gordon equations containing interaction via mass matrix and the semilinear term. The model obeys the system (4.1). By the similarity transformation $\mathbf{O}$ the mass matrix $\mathbf{M}$ can be diagonalized, therefore we use a change of unknown function

$$\Psi = \mathbf{O}\Phi\,, \quad \Phi = \mathbf{O}^{-1}\Psi,$$

and arrive at

$$\Psi_{tt} + nH\Psi_t - e^{-2Ht}\triangle\Psi + \widetilde{\mathbf{M}}\Psi = \widetilde{F}(\Psi)\,,$$

where

$$\widetilde{\mathbf{M}} := \mathbf{O}M\mathbf{O}^{-1} = \mathrm{diag}\{m_1^2, \dots, m_l^2\}, \quad \widetilde{F}(\Psi) := \mathbf{O}F(\mathbf{O}^{-1}\Psi).$$

Let us consider the linear diagonal system

$$\Psi_{tt} + nH\Psi_t - e^{-2Ht} \bigtriangleup \Psi + \widetilde{\mathbf{M}}\Psi = \widetilde{f}\,.$$

Here $\widetilde{f}$ is a vector-valued function with the components f_i, $i = 1, \dots, l$. Then, the solution of the Cauchy problem for the last system with the initial conditions

$$\Psi(x,0) = 0, \qquad \Psi_t(x,0) = 0,$$

is

$$\Psi(x,t) = 2e^{-\frac{n}{2}t} \int_0^t db \int_0^{e^{-b}-e^{-t}} dr\, e^{\frac{n}{2}b} \widetilde{E}(r,t;0,b)\widetilde{v}(x,r;b),$$

where the components v_i, $i = 1, \dots, l$, of the vector-valued function $\widetilde{v}(x,t;b)$ are solutions to the Cauchy problem for the wave equation

$$v_{tt} - \Delta v = 0\,, \quad v(x,0;b) = f_i(x,b)\,, \quad v_t(x,0;b) = 0, \quad i = 1, \dots, l\,. \tag{7.1}$$

The kernel $\widetilde{E}(r,t;0,b)$ is a diagonal matrix with the elements $E_i(r,t;0,b)$, $i = 1, \dots, l$, which are defined either by (2.4) with corresponding mass terms m_i, $i = 1, \dots, l$, or by (3.4), in accordance with the value of mass $m_i^2 \ge n^2/4$ or $m_i^2 < n^2/4$, respectively. Then, the solution Ψ of the Cauchy problem for the equation

$$\Psi_{tt} + nH\Psi_t - e^{-2Ht} \bigtriangleup \Psi + \widetilde{\mathbf{M}}\Psi = 0$$

with the initial conditions

$$\Psi(x,0) = \widetilde{\psi}_0(x), \quad \Psi_t(x,0) = \widetilde{\psi}_1(x),$$

with the vector-valued functions $\widetilde{\psi}_0, \widetilde{\psi}_1 \in C_0^\infty(\mathbb{R}^n)$, $n \ge 2$, can be represented as follows:

$$\begin{aligned}
\Psi(x,t) = {} & e^{-\frac{n-1}{2}t}\widetilde{v}_{\widetilde{\psi}_0}(x,\phi(t)) \\
& + e^{-\frac{n}{2}t} \int_0^1 \big(2\widetilde{K}_0(\phi(t)s,t) + n\widetilde{K}_1(\phi(t)s,t)\big)\widetilde{v}_{\widetilde{\psi}_0}(x,\phi(t)s)\phi(t)\, ds \\
& + 2e^{-\frac{n}{2}t} \int_0^1 \widetilde{K}_1(\phi(t)s,t)\widetilde{v}_{\widetilde{\psi}_1}(x,\phi(t)s)\phi(t)\, ds, \quad x \in \mathbb{R}^n,\ t > 0\,,
\end{aligned}$$

where $\phi(t) := 1 - e^{-t}$ and the kernels $\widetilde{K}_0$, $\widetilde{K}_1$, are the diagonal matrices with the elements $\widetilde{K}_{0i}(z,t)$, $i = 1, \dots, l$, and $\widetilde{K}_{1i}(z,t)$, which are defined either by (3.8) and (3.9) with the corresponding mass terms m_i, $i = 1, \dots, l$, or by the diagonal matrices with the elements $\widetilde{K}_{0i}(z,t;M)$, $i = 1, \dots, l$, and $\widetilde{K}_{1i}(z,t;M)$, which are defined by (3.6) and (3.7), in accordance with the value of mass $m_i^2 \ge n^2/4$ or $m_i^2 < n^2/4$, respectively. Here, for the vector-valued function $\widetilde{\psi} \in C_0^\infty(\mathbb{R}^n)$ and for $x \in \mathbb{R}^n$, the vector-valued function $\widetilde{v}_{\widetilde{\psi}}(x,t)$ is a solution of the Cauchy problem $\widetilde{v}_{tt} - \Delta\widetilde{v} = 0$, $\widetilde{v}(x,0) = \widetilde{\psi}(x)$, $\widetilde{v}_t(x,0) = 0$. We study the Cauchy problem through the integral equation. To determine that integral equation we appeal to the operator

$$\widetilde{G} := \widetilde{\mathcal{K}} \circ \widetilde{\mathcal{W}E}\,,$$

where the operator $\widetilde{\mathcal{W}E}$ is defined by (7.1), that is,

$$\widetilde{\mathcal{W}E}[f](x,t;b) = \widetilde{v}(x,t;b),$$

and the vector-valued function $\widetilde{v}(x,t;b)$ is a solution to the Cauchy problem for the wave equation, while $\widetilde{\mathcal{K}}$ is introduced either by (5.5), for the large mass matrix, or by (6.3), for the small mass matrix. Hence,

$$\widetilde{G}[f](x,t) = 2e^{-\frac{n}{2}t}\int_0^t db \int_0^{e^{-b}-e^{-t}} dr\, e^{\frac{n}{2}b}\,\widetilde{E}(r,t;0,b;M)\widetilde{\mathcal{W}E}[f](x,r;b).$$

Thus, the Cauchy problem (4.3), (4.4) leads to the following integral equation

$$\Psi(x,t) = \Psi_0(x,t) + \widetilde{G}[\widetilde{F}(\Psi)](x,t)\,. \tag{7.2}$$

Every solution $\Phi = \Phi(x,t)$ to the equation (4.3) generates the function $\Psi = \Psi(x,t)$, which solves the last integral equation with some function $\Psi_0(x,t)$, that is generated by the solution of the Cauchy problem (3.2).

Solvability of the integral equation associated with the Klein–Gordon equation. We are going to apply Banach's fixed-point theorem. In order to estimate the nonlinear term we use the Lipschitz condition $(\mathcal{L})$, which imposes some restrictions on n, α, s. Then we consider the equation (7.2), where the vector-valued function $\Psi_0 \in C([0,\infty); L^q(\mathbb{R}^n))$ is given. The solvability of the integral equation (7.2) depends on the operator $\widetilde{G}$. We start with the case of the Sobolev space $H_{(s)}(\mathbb{R}^n)$ with $s > n/2$, which is an algebra. In the next theorem the operator $\widetilde{\mathcal{K}}$ is generated by the linear part of the equation (4.3).

Theorem 7.1 ([57]). *Assume that $F(\Psi)$ is Lipschitz continuous in the space $H_{(s)}(\mathbb{R}^n)$, $s > n/2$, and also that $\alpha > 0$.*

(i) *Let the spectrum of the mass matrix $\mathbf{M}$ be $\{m_1^2,\dots,m_l^2\} \subset (0,(n^2-1)/4]$, and $m = \min\{m_1,m_2,\dots,m_l\}$. Then for every given function $\Psi_0(x,t) \in X(R,s,\gamma_0)$ such that*

$$\sup_{t\in[0,\infty)} e^{\gamma_0 t}\|\Psi_0(x,t)\|_{H_{(s)}(\mathbb{R}^n)} < \varepsilon\,, \quad \textit{where } \gamma_0 \le (n/2) - \sqrt{(n^2/4)-m^2}\,,$$

and for sufficiently small ε the integral equation (7.2) has a unique solution $\Psi(x,t) \in X(R,s,\gamma)$ with $0<\gamma<\gamma_0/(\alpha+1)$. For the solution one has

$$\sup_{t\in[0,\infty)} e^{\gamma t}\|\Psi(x,t)\|_{H_{(s)}(\mathbb{R}^n)} < 2\varepsilon\,.$$

(ii) *If the eigenvalues of the mass matrix are large, $\frac{n}{2} \le m_i$, $i=1,\dots,l$, then for every given function $\Psi_0(x,t) \in X(R,s,0)$ such that*

$$\sup_{t\in[0,\infty)} \|\Psi_0(x,t)\|_{H_{(s)}(\mathbb{R}^n)} < \varepsilon,$$

and for sufficiently small ε the integral equation (7.2) has a unique solution $\Psi(x,t) \in X(R,s,0)$, and

$$\sup_{t\in[0,\infty)} \|\Psi(x,t)\|_{H_{(s)}(\mathbb{R}^n)} < 2\varepsilon.$$

8. Asymptotic at infinity

For $\varphi \in C_0^\infty(\mathbb{R}^n)$ let $V_\varphi(x,t)$ be a solution of the Cauchy problem

$$V_{tt} - \Delta V = 0, \qquad V(x,0) = 0, \quad V_t(x,0) = \varphi(x).$$

Denote,

$$V_\varphi^{(k)}(x) = \frac{(-1)^k}{k!} \left[\left(\frac{\partial}{\partial t} \right)^k V_\varphi(x,t) \right]_{t=1} \in C_0^\infty(\mathbb{R}^n), \quad k = 1, 2, \dots .$$

Then, for every integer $N \geq 1$ we have

$$V_\varphi(x, 1 - e^{-t}) = \sum_{k=0}^{N-1} V_\varphi^{(k)}(x) e^{-kt} + R_{V_\varphi,N}(x,t), \quad R_{V_\varphi,N} \in C^\infty,$$

where with the constant $C(\varphi)$ the remainder $R_{V_\varphi,N}$ satisfies the inequality

$$|R_{V_\varphi,N}(x,t)| \leq C(\varphi) e^{-Nt} \quad \text{for all} \quad x \in \mathbb{R}^n \quad \text{and all} \quad t \in [0,\infty).$$

Further, we introduce the polynomial in z with the smooth in $x \in \mathbb{R}^n$ coefficients as follows:

$$\Phi_{\text{asypt}}^{(N)}(x,z) = z^{\frac{n-1}{2}} \sum_{k=0}^{N-1} \left(\frac{n-1}{2} V_{\varphi_0}^{(k)}(x) - (k+1) V_{\varphi_0}^{(k+1)}(x) + V_{\varphi_1}^{(k)}(x) \right) z^k.$$

where $x \in \mathbb{R}^n$, $z \in \mathbb{C}$. Thus, we arrive at the next theorem.

Theorem 8.1 ([58]). *Suppose that $m = \sqrt{n^2-1}/2$. Then, for every integer positive N the solution of the equation (3.11) with the initial values $\varphi_0, \varphi_1 \in C_0^\infty(\mathbb{R}^n)$ has the following asymptotic expansion at infinity:*

$$\Phi(x,t) = \Phi_{\text{asypt}}^{(N)}(x, e^{-t}) + O(e^{-Nt - \frac{n-1}{2}t})$$

for large t uniformly for $x \in \mathbb{R}^n$, in the sense that for every integer positive N the following estimate is valid:

$$\|\Phi(x,t) - \Phi_{\text{asypt}}^{(N)}(x, e^{-t})\|_{L^\infty(\mathbb{R}^n)} \leq C(\varphi_0, \varphi_1) e^{-Nt - \frac{n-1}{2}t} \qquad \textit{for large} \ \ t.$$

Unlike to the result by Vasy [44] the last inequality does not have the logarithmic term.

Open Problem 8.2. Write the complete asymptotic expansion in the non-critical case.

Acknowledgment

I am grateful to the editor of this volume, Prof. Ville Turunen, for the invitation to contribute to this volume. I am indebted to the anonymous referee whose numerous remarks, comments and suggestions improved the readability of the text.

References

[1] L. Alberte, *Massive Gravity on Curved Background,* International Journal of Modern Physics D. Vol. **21**, No. 6 (2012), 1250058.

[2] D. Baskin, *A parametrix for the fundamental solution of the Klein–Gordon equation on asymptotically de Sitter spaces.* Journal of Functional Analysis **259** (2010), 1673–1719.

[3] D. Baskin, *Strichartz estimates on asymptotically de Sitter spaces.* Available via arXiv: http://arxiv.org/pdf/1011.4696v4.pdf.

[4] Y. Berest, *Hierarchies of Huygens' operators and Hadamard's conjecture.* Acta Appl. Math. **53**, no. 2 (1998), 125–185.

[5] F. Berkhahn, D.D. Dietrichb and S. Hofmann, *Consistency of relevant cosmological deformations on all scales.* Physics Journal of Cosmology and Astroparticle Physics, 2011, September, doi:10.1088/1475-7516/2011/09/024.

[6] P. Brenner, *On L^p-L^q estimates for the wave-equation.* Math. Zeitschrift **145** (1975), 251–254.

[7] N.D. Birrell, P.C.W. Davies, *Quantum fields in curved space.* Cambridge, Cambridge University Press, New York, 1984.

[8] R.W. Carroll, R.E. Showalter, *Singular and degenerate Cauchy problems.* Mathematics in Science and Engineering. Vol. 127. Academic Press, Harcourt Brace Jovanovich, Publishers, New York-London, 1976.

[9] T.-P. Cheng, *Relativity, Gravitation And Cosmology: A Basic Introduction.* Oxford University Press, Oxford, New York, 2005.

[10] S. Delache, J. Leray, *Calcul de la solution élémentaire de l'opérateur d'Euler–Poisson–Darboux et de l'opérateur de Tricomi–Clairaut, hyperbolique, d'ordre 2.* Bull. Soc. Math. France **99** (1971), 313–336.

[11] S. Deser, A. Waldron, *Stability of Massive Cosmological Gravitons.* Phys. Lett. B **508** (2001), 347–353.

[12] S. Dengiz, B. Tekin, *Higgs mechanism for new massive gravity and Weyl-invariant extensions of higher-derivative theories.* Physical Review D **84** (2011), 024033.

[13] J.B. Diaz, H.F. Weinberger, *A solution of the singular initial value problem for the Euler–Poisson–Darboux equation.* Proc. Amer. Math. Soc. **4** (1953), 703–715.

[14] P.A.M. Dirac, *The large numbers hypothesis and the Einstein theory of gravitation.* Proc. Roy. Soc. London Ser. A **365**, no. 1720 (1979), 19–30.

[15] A. Einstein, W. de Sitter, *On the relation between the expansion and the mean density of the universe.* Proc. Natn. Acad. Sci. U.S.A. **18** (1932), 213–214.

[16] G. Ellis, H. van Elst, *Cargèse Lectures 1998: Cosmological Models.* Available via arXiv: http://arxiv.org/pdf/gr-qc/9812046v5.pdf

[17] G.F.R. Ellis, D.W. Sciama, *Global and non-global problems in cosmology. General relativity.* (Papers in honor of J.L. Synge), pp. 35–59. Clarendon Press, Oxford, 1972.

[18] F. Englert, R. Brout, *Broken Symmetry and the Mass of Gauge Vector Mesons.* Phys. Rev. Lett. **13**, no. 9 (1964), 321–323.

[19] V. Faraoni, E. Gunzig, *Tales of tails in cosmology.* International Journal of Modern Physics D, April, Vol. **08**, No. 02 (1999), 177–188.

[20] F.G. Friedlander, *The wave equation on a curved space-time.* Cambridge Monographs on Mathematical Physics, No. 2. Cambridge University Press, Cambridge-New York-Melbourne, 1975.

[21] A. Galstian, T. Kinoshita, K. Yagdjian, *A note on wave equation in Einstein and de Sitter space-time.* J. Math. Phys. **51**, no. 5 (2010), 052501.

[22] J. Ginibre, G. Velo, *The global Cauchy problem for the nonlinear Klein–Gordon equation.* Math. Z. **189**, no. 4 (1985), 487–505.

[23] J. Ginibre, G. Velo, *The global Cauchy problem for the nonlinear Klein–Gordon equation. II.* Ann. Inst. H. Poincaré Anal. Non Linéaire. **6**, no. 1 (1989), 15–35.

[24] R.J. Gleiser, R.H. Price, J. Pullin, *Late-time tails in the Kerr spacetime.* Classical Quantum Gravity **25**, no. 7, (2008), 072001.

[25] P. Günther, *Huygens' principle and Hadamard's conjecture.* Math. Intelligencer **13**, no. 2 (1991), 56–63.

[26] S.W. Hawking, G.F.R. Ellis, *The large scale structure of space-time.* Cambridge Monographs on Mathematical Physics, No. 1. London-New York: Cambridge University Press, 1973.

[27] P.W. Higgs, *Broken symmetries and the masses of gauge bosons.* Phys. Rev. Lett. **13**, no. 16 (1964), 508–509.

[28] A. Higuchi, *Forbidden mass range for spin-2 field theory in de Sitter spacetime.* Nuclear Phys. B **282**, no. 2 (1987), 397–436.

[29] L. Hörmander, *The analysis of linear partial differential operators. IV. Fourier integral operators.* Grundlehren der Mathematischen Wissenschaften, **275**. Springer-Verlag, Berlin, 1994.

[30] N.Kh. Ibragimov, A.O. Oganesyan, *Hierarchy of Huygens equations in spaces with a nontrivial conformal group.* Uspekhi Mat. Nauk **46**, no. 3 (279), (1991), 111–146.

[31] K. Jörgens, *Das Anfangswertproblem im Grossen für eine Klasse nichtlinearer Wellengleichungen.* Math. Z. **77** (1961), 295–308.

[32] T. Mandai, *Characteristic Cauchy problems for some non-Fuchsian partial differential operators.* J. Math. Soc. Japan **45**, no. 3 (1993), 511–545.

[33] C. Møller, *The theory of relativity.* Oxford: Clarendon Press, 1952.

[34] L.E. Parker, D.J. Toms, *Quantum field theory in curved spacetime. Quantized fields and gravity.* Cambridge Monographs on Mathematical Physics. Cambridge University Press, Cambridge, 2009.

[35] H. Pecher, *L^p-Abschätzungen und klassische Lösungen für nichtlineare Wellengleichungen. I,* Math. Zeitschrift **150** (1976), 159–183.

[36] P.J.E. Peebles, *Principles of physical cosmology.* Princeton University Press, Princeton, N.J., 1993.

[37] I.E. Segal, *Non-linear semi-groups.* Ann. of Math. (2) **78** (1963), 339–364.

[38] I.E. Segal, *The global Cauchy problem for a relativistic scalar field with power interaction.* Bulletin de la Société Mathématique de France. **91** (1963), 129–135.

[39] J. Shatah, M. Struwe, *Geometric wave equations.* Courant Lecture Notes in Mathematics, 2. New York University, Courant Institute of Mathematical Sciences, New York; American Mathematical Society, Providence, RI, 1998.

[40] S. Sonego, V. Faraoni, *Huygens' principle and characteristic propagation property for waves in curved space-times.* J. Math. Phys. **33**, no. 2 (1992), 625–632.

[41] W.A. Strauss, *Existence of solitary waves in higher dimensions.* Comm. Math. Phys. **55**, no. 2 (1977), 149–162.

[42] W.A. Strauss, *Nonlinear scattering theory at low energy: sequel.* J. Funct. Anal. **43**, no. 3 (1981), 281–293.

[43] R.C. Tolman, *Relativity, thermodynamics, and cosmology.* Clarendon Press, Oxford, 1934.

[44] A. Vasy, *The wave equation on asymptotically de Sitter-like spaces.* Adv. Math. **223**, no. 1 (2010), 49–97.

[45] S. Weinberg, *The quantum theory of fields.* vol. 2. New York: Cambridge University Press, 1996.

[46] A. Weinstein, *The singular solutions and the Cauchy problem for generalized Tricomi equations.* Comm. Pure Appl. Math. **7** (1954), 105–116.

[47] K. Yagdjian, *The Cauchy Problem for Hyperbolic Operators. Multiple Characteristics. Micro-Local Approach.* Akademie Verlag, Berlin, 1997.

[48] K. Yagdjian, *A note on the fundamental solution for the Tricomi-type equation in the hyperbolic domain.* J. Differential Equations **206** (2004), 227–252.

[49] K. Yagdjian, *Global existence for the n-dimensional semilinear Tricomi-type equations.* Comm. Partial Diff. Equations **31** (2006), 907–944.

[50] K. Yagdjian, *Self similar solutions of semilinear wave equation with variable speed of propagation.* J. Math. Anal. Appl. **336** (2007), 1259–1286.

[51] K. Yagdjian, A. Galstian, *Fundamental solutions of the wave equation in Robertson–Walker spaces.* J. Math. Anal. Appl. **346**, no. 2 (2008), 501–520.

[52] K. Yagdjian, A. Galstian, *Fundamental Solutions for the Klein–Gordon Equation in de Sitter Spacetime.* Comm. Math. Phys. **285** (2009), 293–344.

[53] K. Yagdjian, *The semilinear Klein–Gordon equation in de Sitter spacetime.* Discrete Contin. Dyn. Syst. Ser. S **2**, no. 3 (2009), 679–696.

[54] K. Yagdjian, *Fundamental Solutions for Hyperbolic Operators with Variable Coefficients.* Rend. Istit. Mat. Univ. Trieste **42** Suppl. (2010), 221–243.

[55] K. Yagdjian, *Global existence of the scalar field in de Sitter spacetime.* J. Math. Anal. Appl. **396**, no. 1 (2012), 323–344.

[56] K. Yagdjian, *On the global solutions of the Higgs boson equation.* Comm. Partial Differential Equations **37**, no. 3 (2012), 447–478.

[57] K. Yagdjian, *Global Solutions of Semilinear System of Klein–Gordon Equations in de Sitter Spacetime.* Progress in Partial Differential Equations. Springer Proceedings in Mathematics & Statistics, Volume 44, 2013, pp. 409–444.

[58] K. Yagdjian, *Huygens' Principle for the Klein–Gordon equation in the de Sitter spacetime.* J. Math. Phys. 54, no. 9 (2013), 091503.

Karen Yagdjian
Department of Mathematics, University of Texas-Pan American
1201 W. University Drive, Edinburg, TX 78539, USA
e-mail: `yagdjian@utpa.edu`

MIX
Papier aus verantwortungsvollen Quellen
Paper from responsible sources
FSC® C105338

If you have any concerns about our products,
you can contact us on
ProductSafety@springernature.com

In case Publisher is established outside the EU,
the EU authorized representative is:
Springer Nature Customer Service Center GmbH
Europaplatz 3, 69115 Heidelberg, Germany

Printed by Libri Plureos GmbH
in Hamburg, Germany